Student's Solutions Manual Part II

Calculus and Analytic Geometry

8th Edition

Student's Solutions Manual
Part II

Calculus and Analytic Geometry
8th Edition

Thomas/Finney

Thomas L. Cochran
Michael Schneider
Belleville Area College

ADDISON-WESLEY PUBLISHING COMPANY
Reading, Massachusetts Menlo Park, California New York
Don Mills, Ontario Wokingham, England Amsterdam Bonn
Sydney Singapore Tokyo Madrid San Juan Milan Paris

Reproduced by Addison-Wesley from camera-ready copy supplied by the author.

ISBN 0-201-53306-5
1 2 3 4 5 6 7 8 9 10-BA-9695949392

The authors would like to dedicate this book to their children, Jane, J.T., and Connie Cochran and Michael and John Schneider

Tom Cochran
Michael Schneider

The authors would like to thank Bruce Sisko, Belleville Area College, for the fine job he did in proofreading and checking our work.

The authors have attempted to make this manual as error free as possible but nobody's "perfeck". If you find errors, we would appreciate knowing them. You are more than welcome to write us at:

Belleville Area College
2500 Carlyle Road
Belleville, Illinois 62221

TABLE OF CONTENTS

Student's Solutions Manual
Part II

Calculus and
Analytic Geometry
8th Edition

CHAPTER 8

INFINITE SERIES

8.1 INFINITE SERIES

1. $a_1 = 0,\ a_2 = -\frac{1}{4},\ a_3 = -\frac{2}{9},\ a_4 = -\frac{3}{16}$

3. $a_1 = 1,\ a_2 = -\frac{1}{3},\ a_3 = \frac{1}{5},\ a_4 = -\frac{1}{7}$

5. $1, \frac{3}{2}, \frac{7}{4}, \frac{15}{8}, \frac{31}{16}, \frac{63}{32}, \frac{127}{64}, \frac{255}{128}, \frac{511}{256}, \frac{1023}{512}$

7. $2, 1, -\frac{1}{2}, -\frac{1}{4}, \frac{1}{8}, \frac{1}{16}, -\frac{1}{32}, -\frac{1}{64}, \frac{1}{128}, \frac{1}{256}$

9. 1,1,2,3,5,8,13,21,34,55

11. $\lim_{n\to\infty} a_n = \lim_{n\to\infty} \left(2 + (0.1)^n\right) = 2 \Rightarrow$ converges

13. $\lim_{n\to\infty} a_n = \lim_{n\to\infty} \frac{1-2n}{1+2n} = -1 \Rightarrow$ converges

15. $\lim_{n\to\infty} a_n = \lim_{n\to\infty} \frac{1-5n^4}{n^4+8n^3} = -5 \Rightarrow$ converges

17. $\lim_{n\to\infty} a_n = \lim_{n\to\infty} \frac{n^2-2n+1}{n-1} = \lim_{n\to\infty} n-1 = \infty \Rightarrow$ diverges

19. $\lim_{n\to\infty} a_n = \lim_{n\to\infty} \left(1 + (-1)^n\right)$ does not exist $\Rightarrow$ diverges

21. $\lim_{n\to\infty} a_n = \lim_{n\to\infty} \left(\frac{n+1}{2n}\right)\left(1 - \frac{1}{n}\right) = \frac{1}{2} \Rightarrow$ converges

23. $\lim_{n\to\infty} a_n = \lim_{n\to\infty} \frac{(-1)^{(n+1)}}{2n-1} = 0$

25. $\lim_{n\to\infty} a_n = \lim_{n\to\infty} \frac{\sin n}{n} = 0$, by the Sandwich Theorem for Sequences

27. $\lim_{n\to\infty} a_n = \lim_{n\to\infty} \sqrt{\frac{2n}{n+1}} = \sqrt{\lim_{n\to\infty} \frac{2n}{n+1}} = \sqrt{2}$

29. $\lim_{n\to\infty} a_n = \lim_{n\to\infty} \tan^{-1} n = \frac{\pi}{2}$

31. $\lim_{n\to\infty} a_n = \lim_{n\to\infty} \frac{n}{2^n} = \lim_{n\to\infty} \frac{1}{2^n \ln 2} = 0$

33. $\lim_{n\to\infty} a_n = \lim_{n\to\infty} \frac{\ln(n+1)}{n} = 0$, due to the growth rates in section 7.6

35. $\lim_{n\to\infty} a_n = \lim_{n\to\infty} 8^{1/n} = \lim_{n\to\infty} \exp\left(\frac{\ln 8}{n}\right) = e^0 = 1$

37. $\lim_{n\to\infty} a_n = \lim_{n\to\infty} \left(1 + \frac{7}{n}\right)^n = e^7$, due to table 9.1 part 5

39. $\lim_{n\to\infty} a_n = \lim_{n\to\infty} \frac{1}{(0.9)^n} = \lim_{n\to\infty} \left(\frac{10}{9}\right)^n = \infty \Rightarrow$ diverges

41. $\lim_{n\to\infty} a_n = \lim_{n\to\infty} \sqrt[n]{10\,n} = \lim_{n\to\infty} \exp\left(\frac{\ln(10\,n)}{n}\right) = \exp\left(\lim_{n\to\infty} \frac{\ln 10\,n}{n}\right) = \exp\left(\lim_{n\to\infty} \frac{1}{n}\right) = e^0 = 1$

43. $\lim_{n\to\infty} a_n = \lim_{n\to\infty} \left(\frac{3}{n}\right)^{1/n} = \frac{\lim_{n\to\infty} \sqrt[n]{3}}{\lim_{n\to\infty} \sqrt[n]{n}} = \frac{1}{1} = 1$, due to table 8.1, parts 2 and 3

45. $\lim_{n\to\infty} a_n = \lim_{n\to\infty} \frac{\ln n}{n^{1/n}} = \frac{\lim_{n\to\infty} \ln n}{\lim_{n\to\infty} \sqrt[n]{n}} = \frac{\infty}{1} = \infty$, due to table 8.1, part 2

47. $\lim_{n\to\infty} b_n = \lim_{n\to\infty} \left(\frac{1}{3}\right)^n = 0$, due to table 8.1, part 4 and $\lim_{n\to\infty} c_n = \lim_{n\to\infty} \frac{1}{\sqrt{2^n}} = 0$

$\therefore \lim_{n\to\infty} a_n = \lim_{n\to\infty} \left(\left(\frac{1}{3}\right)^n + \frac{1}{\sqrt{s^n}}\right) = \lim_{n\to\infty} \left(\frac{1}{3}\right)^n + \lim_{n\to\infty} \frac{1}{\sqrt{2^n}} = 0 + 0 = 0 \Rightarrow$ converges

49. $\lim_{n\to\infty} a_n = \lim_{n\to\infty} \frac{n!}{n^n} = \lim_{n\to\infty} \frac{1 \cdot 2 \cdot 3 \cdots (n-1)(n)}{n \cdot n \cdot n \cdots n \cdot n} < \lim_{n\to\infty} \frac{1}{n} = 0$ and $\frac{n!}{n^n} \ge 0 \Rightarrow$

$\lim_{n\to\infty} \frac{n!}{n^n} = 0 \Rightarrow$ converges

51. $\lim_{n\to\infty} a_n = \lim_{n\to\infty} \left(\frac{1}{n}\right)^{1/\ln(n)} = \lim_{n\to\infty} \exp\left(\frac{\ln(1/n)}{\ln(n)}\right) = \exp\left(\lim_{n\to\infty} \frac{\ln(1/n)}{\ln(n)}\right) = \exp\left(\lim_{n\to\infty} -1\right) = e^{-1}$

53. $\lim_{n\to\infty} a_n = \lim_{n\to\infty} \frac{n!}{10^{6n}} = \frac{1}{\lim_{n\to\infty} \frac{\left(10^6\right)^n}{n!}} = \infty \Rightarrow$ diverges

55. $\lim_{n\to\infty} a_n = \lim_{n\to\infty} \tanh n = 1 \Rightarrow$ converges

57. $\lim_{n\to\infty} a_n = \ln\left(\lim_{n\to\infty} \left(1 + \frac{1}{n}\right)^n\right) = \ln e = 1 \Rightarrow$ converges

59. $\lim_{n\to\infty} a_n = \lim_{n\to\infty} \exp\left(\frac{\ln((3n+1)/(3n-1))}{1/n}\right) = \lim_{n\to\infty} \exp\left(\frac{6n^2}{(3n+1)(3n-1)}\right) = e^{2/3} \Rightarrow$ converges

61. $\lim_{n\to\infty} a_n = x \lim_{n\to\infty} \sqrt[n]{\frac{1}{2n+1}} = x \lim_{n\to\infty} \exp\left(\frac{\ln(2n+1)}{n}\right) = x \lim_{n\to\infty} \exp\left(\frac{-2}{2n+1}\right) = x\, e^0 = x \Rightarrow$ converges where $x > 0$

63. $\lim_{n\to\infty} a_n = \lim_{n\to\infty} \frac{\int_1^n \frac{1}{x}\,dx}{n} = \lim_{n\to\infty} \frac{\frac{1}{n}}{1} = 0 \Rightarrow$ converges

65. $\lim_{n\to\infty} a_n = \lim_{n\to\infty} \exp\left(\frac{\ln\left(n^2+n\right)}{n}\right) = \lim_{n\to\infty} \exp\left(\frac{2n+1}{n^2+n}\right) = e^o = 1 \Rightarrow$ converges

66. $\lim_{n\to\infty} a_n = \lim_{n\to\infty} \frac{(200)\,(\ln n)^{199}}{n} = \lim_{n\to\infty} \frac{(200)(199)(\ln n)^{198}}{n} \cdots \lim_{n\to\infty} \frac{200!}{n} = 0 \Rightarrow$ converges

67. $\lim_{n\to\infty} a_n = \lim_{n\to\infty} \frac{n^2 \sin(1/n)}{2n-1} = \lim_{n\to\infty} \frac{\sin\left(\frac{1}{n}\right)}{\frac{2}{n} - \frac{1}{n^2}} = \lim_{n\to\infty} \frac{-\left(\cos\left(\frac{1}{n}\right)\right)\left(\frac{1}{n^2}\right)}{-\frac{2}{n^2} + \frac{2}{n^3}} = \lim_{n\to\infty} \frac{-\cos(1/n)}{-2+2/n} = \frac{1}{2} \Rightarrow$

converges

69. $\lim_{n\to\infty} a_n = \lim_{n\to\infty} \left(n - \sqrt{n^2-n}\right) = \lim_{n\to\infty} \left(\left(n - \sqrt{n^2-n}\right)\left(\frac{n+\sqrt{n^2-n}}{n+\sqrt{n^2-n}}\right)\right) = \lim_{n\to\infty} \frac{1}{1+\sqrt{1-\frac{1}{n}}} =$

$\frac{1}{2} \Rightarrow$ converges

71. $\left|\sqrt[n]{0.5} - 1\right| < 10^{-3} \Rightarrow -\frac{1}{1000} < \left(\frac{1}{2}\right)^{1/n} - 1 < \frac{1}{1000} \Rightarrow n > \frac{\ln(1/2)}{\ln(999/1000)} \Rightarrow n > 692.8 \rightarrow N = 693$

73. $(0.9)^n < 10^{-3} \Rightarrow n \ln(0.9) < -3 \ln 10 \Rightarrow n > 65.56 \Rightarrow N = 66$

75. $x_1 = 1$, $x_2 = 1 + \cos(1) \approx 1.540302305$, $x_3 = 1 + \cos(1) + \cos(1 + \cos(1)) \approx 1.570791601$,
$x_4 = 1 + \cos(1) + \cos(1 + \cos(1)) + \cos(1 + \cos(1) + \cos(1))) \approx 1.570796327$

77. a) $(1)^2 - 2(1) = -1$; $(3)^2 - 2(2)^2 = 1$; If $a^2 - 2b^2 = 1$, then $(a+2b)^2 - 2(a+b)^2 = a^2 + 4ab + 4b^2 - 2a^2 - 4ab - 2b^2 = 2b^2 - a^2 - 1$. If $a^2 - 2b^2 + -1$, then $(a+2b)^2 - 2(a+b)^2 = 2b^2 - a^2 = -(a^2 - 2b^2) = -(-1) = 1$.

b) $r_n - 2 = \left(\frac{a+2b}{a+b}\right)^2 - 2 = \frac{a^2 + 4ab + 4b^2 - 2(a^2 + 2ab + b^2)}{a^2 + 2ab + b^2} = \frac{-\left(a^2 - 2b^2\right)}{(a+b)^2} = \frac{\pm 1}{(y_n)^2} \Rightarrow$

$r_n = \sqrt{2 \pm \left(\frac{1}{y_n}\right)^2}$ $\therefore$ $\lim_{n\to\infty} r_n = \lim_{n\to\infty} \sqrt{2 \pm \left(\frac{1}{y_n}\right)^2} = \sqrt{2}$. In the first and second fractions $y_n \geq n$.

Let $\frac{a}{b}$ represent the $(n-1)^{th}$ fraction where $\frac{a}{b} \geq 1$ and $b \geq n-1$ for n a positive integer greater than or equal to three. Now the nth fraction is $\frac{a+2b}{a+b}$ and $a + b \geq 2b \geq 2n - 2 \geq n \Rightarrow y_n \geq n$.

$\therefore$ by mathematical induction $y_n \geq n$.

79. a) $f(x) = x^2 - 2$; the sequence converges to $1.414213562 \approx \sqrt{2}$

b) $f(x) = \tan(x) - 1$; the sequence converges to $0.7853981635 \approx \frac{\pi}{4}$

c) $f(x) = e^x$; the sequence $1, 0, -1, -2, -3, -4, -5 \cdots$ diverges

81. a) If $a = 2n + 1$, then $b = \left\lfloor \frac{a^2}{2} \right\rfloor = \left\lfloor \frac{4n^2 + 4n + 1}{2} \right\rfloor = \left\lfloor 2n^2 + 2n + \frac{1}{2} \right\rfloor = 2n^2 + 2n$, $c = \left\lceil \frac{a^2}{2} \right\rceil = \left\lceil 2n^2 + 2n + \frac{1}{2} \right\rceil = 2n^2 + 2n + 1$ and $a^2 + b^2 = (2n+1)^2 + (2n^2 + 2n)^2 = 4n^2 + 4n + 1 + 4n^4 + 8n^3 + 4n^2 = 4n^4 + 8n^3 + 8n^2 + 4n + 1 = (2n^2 + 2n + 1)^2 = c^2$.

b) $\lim\limits_{a \to \infty} \frac{\left\lfloor \frac{a^2}{2} \right\rfloor}{\left\lceil \frac{a^2}{2} \right\rceil} = \lim\limits_{a \to \infty} \frac{2n^2 + 2n}{2n^2 + 2n + 1} = 1$ or $\lim\limits_{a \to \infty} \frac{\left\lfloor \frac{a^2}{2} \right\rfloor}{\left\lceil \frac{a^2}{2} \right\rceil} = \lim\limits_{a \to \infty} \sin\theta = \lim\limits_{\theta \to \pi/2} \sin\theta = 1$

83. $\lim\limits_{n \to \infty} a_n = \lim\limits_{n \to \infty} nf\left(\frac{1}{n}\right) = \lim\limits_{x \to 0^+} \frac{f(x)}{x} = \lim\limits_{x \to 0^+} f'(x) = f'(0)$

85. If $f(x) = e^x - 1$, then $\lim\limits_{n \to \infty} a_n = e^0 = 1$

87. a) $\lim\limits_{n \to \infty} \frac{\ln n}{n^c} = \lim\limits_{n \to \infty} \frac{\frac{1}{n}}{cn^{c-1}} = \frac{1}{c}\left(\lim\limits_{n \to \infty} \frac{1}{n^c}\right) = 0$

b) For all $\varepsilon > 0$ there exists a $N = \varepsilon^{-(\ln \varepsilon)/c}$ such that whenever $n > \varepsilon^{-(\ln \varepsilon)/c} \Rightarrow \ln n > -\frac{\ln \varepsilon}{c} \Rightarrow \ln n^c > \ln \frac{1}{\varepsilon} \Rightarrow n^c > \frac{1}{\varepsilon} \Rightarrow \frac{1}{n^c} < \varepsilon \Rightarrow \left| \frac{1}{n^c} - 0 \right| < \varepsilon$. $\therefore \lim\limits_{n \to \infty} \left(\frac{1}{n^c}\right) = 0$

89. The conclusion is $|f(a_n) - f(L)| < \varepsilon \Rightarrow f(a_n) \to f(L)$.

91. $G(x) = \sqrt{x}$; $2 \to 1.00000132$ in 20 iterations, $0.1 \to 0.999998902$ in 22 iterations $\Rightarrow$ a root is 1

93. $G(x) = \cos x$; $0.1 \to -0.73908456$ in 35 iterations

95. $G(x) = 0.1 + \sin x$; $-2 \to 0.853748068$ in 44 iterations

97. If x_1 = initial guess > 0, then $x_n = (x_1)^{(1/2)^{n-1}}$, where $n \geq 2$. $\therefore$ as $n \to \infty \Rightarrow x_n \to 1$.

99. a) $g(x) = 2x + 3 \Rightarrow g^{-1}(x) = \frac{x-3}{2}$ and when the iterative method is applied to $g^{-1}(x)$ we have $2 \to -2.99999881$ in 23 iterations $\Rightarrow -3$ is the fixed point

b) $g(x) = 1 - 4x \Rightarrow g^{-1}(x) = \frac{1-x}{4}$ and when the iterative method is applied to $g^{-1}(x)$ we have $2 \to 0.199999571$ in 12 iterations $\Rightarrow 0.2$ is the fixed point

8.2 INFINITE SERIES

1. $s_n = \frac{a(1-r^n)}{(1-r)} = \frac{2(1-(1/3)^n)}{1-1/3} \Rightarrow \lim_{n\to\infty} s_n = \frac{2}{1-1/3} = 3$

3. $s_n = \frac{a(1-r^n)}{(1-r)} = \frac{1-(-1/2)^n}{1-(-1/2)} \Rightarrow \lim_{n\to\infty} s_n = \frac{1}{3/2} = \frac{2}{3}$

5. $\frac{1}{(n+1)(n+2)} = \frac{1}{n+1} - \frac{1}{n+2} \Rightarrow s_n = \left(\frac{1}{2}-\frac{1}{3}\right) + \left(\frac{1}{3}-\frac{1}{4}\right) + \ldots + \left(\frac{1}{n+1}-\frac{1}{n+2}\right) = \frac{1}{2} - \frac{1}{n+2} \Rightarrow$
$\lim_{n\to\infty} s_n = \frac{1}{2}$

7. $1 - \frac{1}{4} + \frac{1}{16} - \frac{1}{64} + \cdots$ The sum of this geometric series is $\frac{1}{1-\frac{1}{4}} = \frac{4}{5}$

9. $\frac{7}{4} + \frac{7}{16} + \frac{7}{64} + \ldots$ The sum of this geometric series is $\frac{7/4}{1-1/4} = \frac{7}{3}$.

11. $(5+1) + \left(\frac{5}{2}+\frac{1}{3}\right) + \left(\frac{5}{4}+\frac{1}{9}\right) + \left(\frac{5}{8}+\frac{1}{27}\right) + \ldots$ is the sum of two geometric series; the sum $\frac{5}{1-1/2} + \frac{1}{1-1/3} = 10 + \frac{3}{2} = \frac{23}{2}$.

13. $(1+1) + \left(\frac{1}{2}-\frac{1}{5}\right) + \left(\frac{1}{4}+\frac{1}{25}\right) + \left(\frac{1}{8}-\frac{1}{125}\right) + \ldots$ is the sum of two geometric series; the sum $\frac{1}{1-1/2} + \frac{1}{1+1/5} = 2 + \frac{5}{6} = \frac{17}{6}$.

15. $\frac{4}{(4n-3)(4n+1)} = \frac{1}{4n-3} - \frac{1}{4n+1} \Rightarrow s_n = \left(1-\frac{1}{5}\right) + \left(\frac{1}{5}-\frac{1}{9}\right) + \left(\frac{1}{9}-\frac{1}{13}\right) + \ldots + \left(\frac{1}{4n-7}-\frac{1}{4n-3}\right)$
$+ \left(\frac{1}{4n-3}-\frac{1}{4n+1}\right) = 1 - \frac{1}{4n+1} \Rightarrow \lim_{n\to\infty} s_n = \lim_{n\to\infty} 1 - \frac{1}{4n+1} = 1$

17. $\frac{4}{(4n-3)(4n+1)} = \frac{1}{4n-3} - \frac{1}{4n+1} \Rightarrow s_n = \left(\frac{1}{9}-\frac{1}{13}\right) + \left(\frac{1}{13}-\frac{1}{17}\right) + \left(\frac{1}{17}-\frac{1}{21}\right) + \ldots +$
$\left(\frac{1}{4n-7}-\frac{1}{4n-3}\right) + \left(\frac{1}{4n-3}-\frac{1}{4n+1}\right) \Rightarrow \lim_{n\to\infty} s_n = \lim_{n\to\infty} \frac{1}{9} - \frac{1}{4n+1} = \frac{1}{9}$

19. A convergent geometric series with a sum of $\frac{1}{1-1/\sqrt{2}} = 2 + \sqrt{2}$.

21. A convergent geometric series with a sum of $\frac{3/2}{1-(-1/2)} = 1$.

23. $\lim_{n\to\infty} \cos(n\pi) \neq 0 \Rightarrow$ divergence.

25. A convergent geometric series with a sum of $\frac{1}{1-1/e^2} = \frac{e^2}{e^2-1}$.

27. $\lim_{n\to\infty} a_n = \lim_{n\to\infty} (-1)^{n+1} n \neq 0 \Rightarrow$ divergence.

29. The difference of two convergent geometric series with a sum of $\frac{1}{1-2/3} - \frac{1}{1-1/3} = 3 - \frac{3}{2} = \frac{3}{2}$.

31. $\lim_{n\to\infty} a_n = \lim_{n\to\infty} \frac{n!}{1000^n} = \frac{1}{\lim_{n\to\infty} \frac{1000^n}{n!}} = \infty \neq 0 \Rightarrow$ divergence.

33. A convergent geometric series with a sum of $\frac{1}{1-\frac{e}{\pi}} = \frac{\pi}{\pi - e}$.

35. $\sum_{n=1}^{\infty} \ln\left(\frac{n}{n+1}\right) = \sum_{n=1}^{\infty} \ln(n) - \ln(n+1) \Rightarrow s_n = \big(\ln(1) - \ln(2)\big) + \big(\ln(2) - \ln(3)\big) +$
$\big(\ln(3) - \ln(4)\big) + \ldots + \big(\ln(n-1) - \ln(n)\big) + \big(\ln(n) - \ln(n+1)\big) = \ln(1) - \ln(n+1) = -\ln(n+1) \Rightarrow$
$\lim_{n\to\infty} s_n = -\infty$, $\Rightarrow$ the series diverges

37. $S_n = \left(1 - \frac{1}{\sqrt{2}}\right) + \left(\frac{1}{\sqrt{2}} - \frac{1}{\sqrt{3}}\right) + \left(\frac{1}{\sqrt{3}} - \frac{1}{2}\right) + \cdots + \left(\frac{1}{\sqrt{n-1}} - \frac{1}{\sqrt{n}}\right) + \left(\frac{1}{\sqrt{n}} - \frac{1}{\sqrt{n+1}}\right) = 1 - \frac{1}{\sqrt{n+1}} \Rightarrow$
$\lim_{n\to\infty} S_n = 1$ $\therefore$ the series converges to 1.

39. $\frac{1}{1+x} = \frac{1}{1-(-x)} \Rightarrow a = 1, r = -x$ where $|x| < 1$

41. $\frac{6}{3-x} = \frac{2}{1-\frac{x}{3}} \Rightarrow a = 2, r = \frac{x}{3}$ where $|x| < 3$

43. If $|2x| < 1$, then $|x| < \frac{1}{2}$ and the sum is $\frac{1}{1-2x}$.

45. If $|x+1| < 1$, then $-2 < x < 0$ and the sum is $\frac{1}{2+x}$.

47. If $|\sin x| < 1$, then x is a real except when $x = k(\pi)$ where k is an integer. The sum is $\frac{1}{1-\sin x}$.

49. $0.\overline{23} = \frac{23}{100} + \frac{23}{100}\left(\frac{1}{10^2}\right) + \frac{23}{100}\left(\frac{1}{10^2}\right)^2 + \cdots = \frac{23}{99}$

51. $0.\overline{7} = \frac{7}{10} + \frac{7}{10}\left(\frac{1}{10}\right) + \frac{7}{10}\left(\frac{1}{10}\right)^2 + \cdots = \frac{7}{9}$

53. $0.0\overline{6} = \frac{6}{100} + \frac{6}{100}\left(\frac{1}{10}\right) + \frac{6}{100}\left(\frac{1}{10}\right) + \frac{6}{100}\left(\frac{1}{10}\right)^2 + \cdots = \frac{1}{15}$

55. $1.24\overline{123} = \frac{31}{25} + \frac{123}{10^5} + \frac{123}{10^5}\left(\frac{1}{10^3}\right) + \frac{123}{10^5}\left(\frac{1}{10^5}\right)^2 + \cdots = \frac{31}{25} + \frac{123}{99900} = \frac{41333}{33300}$

57. distance $= 4 + 2\left[(4)\left(\frac{3}{4}\right) + (4)\left(\frac{3}{4}\right)^2 + \ldots\right] = 4 + 2\left[\frac{3}{1-3/4}\right] = 28$ m

59. a) $\sum_{n=-2}^{\infty} \frac{1}{(n+4)(n+5)}$ b) $\sum_{n=0}^{\infty} \frac{1}{(n+2)(n+3)}$ c) $\sum_{n=5}^{\infty} \frac{1}{(n-3)(n-2)}$

61. If n is odd, then $S_n = 1$ while $S_n = 0$ when n is even. $\therefore$ $S_n = \frac{1+(-1)^{n+1}}{2}$.

63. area $= 2^2 + \left(\sqrt{2}\right)^2 + (1)^2 + \left(\frac{1}{\sqrt{2}}\right)^2 + \ldots = 4 + 2 + 1 + \frac{1}{2} + \ldots = \frac{4}{1-1/2} = 8$ m^2

65. a) $L_1 = 3, L_2 = 3\left(\frac{4}{3}\right), L_3 = 3\left(\frac{4}{3}\right)^2, \ldots, L_n = 3\left(\frac{4}{3}\right)^n \therefore \lim_{n \to \infty} L_n = \lim_{n \to \infty} 3\left(\frac{4}{3}\right)^n = \infty$

b) $A_1 = \frac{1}{2}(1)\left(\frac{\sqrt{3}}{2}\right) = \frac{\sqrt{3}}{4}, A_2 = A_1 + 3\left(\frac{1}{2}\right)\left(\frac{1}{3}\right)\left(\frac{\sqrt{3}}{6}\right) = \frac{\sqrt{3}}{4} + \frac{\sqrt{3}}{12}, A_3 = A_1 + A_2 + 12\left(\frac{1}{2}\right)\left(\frac{1}{9}\right)\left(\frac{\sqrt{3}}{18}\right) = \frac{\sqrt{3}}{4} + \frac{\sqrt{3}}{12} + \frac{\sqrt{3}}{27}, A_4 = A_1 + A_2 + 48\left(\frac{1}{2}\right)\left(\frac{1}{27}\right)\left(\frac{\sqrt{3}}{54}\right), \ldots, A_n = \frac{\sqrt{3}}{4} + \frac{27\sqrt{3}}{64}\left(\frac{4}{9}\right)^2 + \frac{27\sqrt{3}}{64}\left(\frac{4}{9}\right)^3 + \cdots = \frac{\sqrt{3}}{4} + \sum_{n=1}^{\infty} \frac{27\sqrt{3}}{64}\left(\frac{4}{9}\right)^n = \frac{2\sqrt{3}}{5}$

67. Both $\sum_{n=1}^{\infty} \left(\frac{1}{4}\right)^n$ and $\sum_{n=1}^{\infty} \left(\frac{1}{8}\right)^n$ converge, but $\sum_{n=1}^{\infty} \frac{(1/4)^n}{(1/8)^n} = \sum_{n=1}^{\infty} 2^n$ diverges.

69. Let $\sum a_n = \sum_{n=0}^{\infty} \left(\frac{1}{2}\right)^n$ and $\sum b_n = \sum_{n=0}^{\infty} \left(\frac{1}{3}\right)^n$. $\therefore A = \sum a_n = 2$ and $B = \sum b_n = \frac{3}{2}$ while $\sum_{n=0}^{\infty} \left(\frac{1}{2}\right)^n \left(\frac{1}{3}\right)^n = = \sum_{n=0}^{\infty} \left(\frac{1}{6}\right)^n = \frac{6}{5} \neq 3 = AB.$

8.3 SERIES WITHOUT NEGATIVE TERMS: THE COMPARISON AND INTEGRAL TESTS

1. converges, a geometric series with r = 1/10
3. converges, by the Comparison Test for Convergence, since $\frac{\sin^2 n}{2^n} \le \frac{1}{2^n}$
5. converges, by the Comparison Test for Convergence, since $\frac{1 + \cos n}{n^2} \le \frac{2}{n^2}$
7. diverges, by the Comparison Test for Convergence, since $\frac{1}{n} < \frac{\ln n}{n}$ for $n \ge 3$
9. converges, a geometric series with r = 2/3

11. diverges, by the Comparison Test for Convergence, since $\frac{1}{n+1} < \frac{1}{1 + \ln n}$ and $\sum_{n=1}^{\infty} \frac{1}{n+1}$

diverges by the Limit Comparison Test when compared with $\sum_{n=1}^{\infty} \frac{1}{n}$

13. diverges, $\lim_{n\to\infty} a_n = \lim_{n\to\infty} \frac{2^n}{n+1} = \lim_{n\to\infty} \frac{2^n \ln 2}{1} = \infty \neq 0$

15. converges, by the Limit Comparison Test when compared with $\sum_{n=1}^{\infty} \frac{1}{n^{3/2}}$

17. diverges, by the Limit Comparison Test when compared with $\sum_{n=1}^{\infty} \frac{1}{n}$

19. diverges, $\lim_{n\to\infty} a_n = \lim_{n\to\infty} \left(1 + \frac{1}{n}\right)^n = e \neq 0$

21. converges: $\sum_{n=1}^{\infty} \frac{1-n}{n\,2^n} = \sum_{n=1}^{\infty} \frac{1}{n\,2^n} + \sum_{n=1}^{\infty} \frac{-1}{2^n}$, the sum of two convergent series.

$\sum_{n=1}^{\infty} \frac{1}{n\,2^n}$ converges, by the Comparison Test for Convergence, since $\frac{1}{n\,2^n} < \frac{1}{2^n}$.

23. converges, by the Comparison Test for Convergence, since $\frac{1}{3^{n-1}+1} < \frac{1}{3^{n-1}}$

25. converges, by the Comparison Test for Convergence, since $\frac{\tan^{-1} n}{n^{1.1}} < \frac{\pi/2}{n^{1.1}}$

27. converges, by the Integral Test, since $\int_1^{\infty} \frac{dx}{x\left(1 + \ln^2 x\right)} = \lim_{b\to\infty} \int_1^b \frac{1/x}{1 + \ln^2 x}\,dx =$

$\lim_{b\to\infty} \left[\tan^{-1}(\ln x)\right]_1^b = \lim_{b\to\infty} \tan^{-1}(\ln b) = \frac{\pi}{2}$

29. converges, by the Limit Comparison test when compared with $\sum_{n=1}^{\infty} \frac{1}{n^2}$ for

$$\frac{1}{1 + 2 + 3 + \cdots + n} = \frac{1}{\frac{n(n-1)}{2}} = \frac{2}{n(n-1)}$$

31. converges, by the Integral Test, since $\int_1^{\infty} \operatorname{sech} x\, dx = 2 \lim_{b \to \infty} \int_1^b \frac{e^x}{1 + (e^x)^2}\, dx =$

$2 \lim_{b \to \infty} \left[\tan^{-1} e^x\right]_1^b = 2 \lim_{b \to \infty} \left[\tan^{-1} e^b - \tan^{-1} e\right] = \pi - 2 \tan^{-1} e$

33. diverges, by the Comparison Test for Convergence since, $1 > \frac{2}{3} \cdot \frac{4}{5} \cdot \frac{6}{7} \cdots \frac{2n-2}{2n-1} \Rightarrow$

$\frac{1 \cdot 3 \cdot 5 \cdots (2n-1)}{2 \cdot 4 \cdot 6 \cdots (2n)} > \frac{2 \cdot 4 \cdot 6 \cdots (2n-2)}{2 \cdot 4 \cdot 6 \cdots 2n} = \frac{1}{2n}$ and $\frac{1}{2} \sum_{n=1}^{\infty} \frac{1}{n}$ diverges

35. There are $(13)(365)(24)(60)(60)(10^9)$ seconds in 13 billion years. $s_n \le 1 + \ln n$ where $n = (13)(365)(24)(60)(60)(10^9) \Rightarrow s_n \le 1 + \ln\left((13)(365)(24)(60)(60)(10^9)\right) = 1 + \ln(13) + \ln(365) + \ln(24) + 2 \ln(60) + 9 \ln(10) \approx 41.55$

37. If $\sum_{n=1}^{\infty} a_n$ converges and $\frac{a_n}{n} \le a_n$, then by the Comparison Test for Convergence,

$\sum_{n=1}^{\infty} \frac{a_n}{n}$ converges.

39. If $\{S_n\}$ is nonincreasing with lower bound M, then $\{-S_n\}$ is a nondecreasing sequence with upper bound $-M$. By Theorem 6, $\{-S_n\}$ converges, and hence, $\{S_n\}$ converges. If $\{S_n\}$ has no lower bound, then $\{-S_n\}$ has no upper bound and diverges. Hence, $\{S_n\}$ also diverges.

41. a) $a_{2^n} = \frac{1}{2^n \ln(2^n)} = \frac{1}{2^n\, n \ln(2)}$. $\sum_{n=2}^{\infty} 2^n a_{(2^n)} = \sum_{n=2}^{\infty} 2^n \frac{1}{2^n\, n \ln(2)} = \frac{1}{\ln(2)} \sum_{n=2}^{\infty} \frac{1}{n}$ which

diverges. $\therefore \sum_{n=2}^{\infty} \frac{1}{n \ln n}$ is divergent.

b) $a_{2^n} = \frac{1}{2^{np}}$. $\sum_{n=1}^{\infty} 2^n a_{2^n} = \sum_{n=1}^{\infty} 2^n \frac{1}{2^{np}} = \sum_{n=1}^{\infty} \frac{1}{(2^n)^{p-1}}$ which converges when $p > 1$, but

diverges for $p \le 1$.

43. a) If $f(x) = \frac{1}{x}$ and $\int_1^{n+1} f(x)\,dx \le a_1 + a_2 + a_3 + \cdots + a_n \le a_1 + \int_1^n f(x)\,dx$, then

$[\ln x]_1^{n+1} \le 1 + \frac{1}{2} + \frac{1}{3} + \cdots + \frac{1}{n} \le 1 + [\ln x]_1^n \Rightarrow \ln(n+1) \le 1 + \frac{1}{2} + \frac{1}{3} + \cdots + \frac{1}{n} \le 1 + \ln n \Rightarrow$

$\ln(n+1) - \ln n \le a_n \le 1$, where $a_n = 1 + \frac{1}{2} + \frac{1}{3} + \cdots + \frac{1}{n} - \ln n$

b) From the graph of $f(x) = \frac{1}{x}$, it is clear that $\frac{1}{n+1} < \int_n^{n+1} \frac{1}{x}\,dx = \ln(n+1) - \ln n \Rightarrow$

$0 > \frac{1}{n+1} - (\ln(n+1) - \ln n) = \left(1 + \frac{1}{2} + \frac{1}{3} + \cdots + \frac{1}{n+1} - \ln(n+1)\right) - \left(1 + \frac{1}{2} + \frac{1}{3} + \cdots + \frac{1}{n} - \ln n\right) =$

$a_{n+1} - a_n \Rightarrow a_n > a_{n+1}$ $\therefore$ $\{a_n\}$ is decreasing

8.4 SERIES WITHOUT NEGATIVE TERMS: THE RATIO AND ROOT TESTS

1. converges, by the Ratio Test for $\lim_{n\to\infty} \left| \frac{(n+1)^2}{2^{n+1}} \frac{2^n}{n^2} \right| = \frac{1}{2} < 1$

3. converges, by the Ratio Test for $\lim_{n\to\infty} \left| \frac{(n+1)^{10}}{10^{n+1}} \frac{10^n}{n^{10}} \right| = \frac{1}{10} < 1$

5. diverges, for $\lim_{n\to\infty} a_n = \lim_{n\to\infty} \left(\frac{n-2}{n}\right)^n = \lim_{n\to\infty} \left(1 + \frac{-2}{n}\right)^n = e^{-2} \ne 0$

7. diverges, by the Ratio Test for $\lim_{n\to\infty} \left| \frac{(n+1)!\, e^{-(n+1)}}{n!\, e^{-n}} \right| = \infty$

9. diverges, $\lim_{n\to\infty} a_n = \lim_{n\to\infty} \left(1 - \frac{3}{n}\right)^n = e^{-3} \ne 0$

11. converges, by the Comparison Test for Convergence, since $\frac{\ln n}{n^3} < \frac{n}{n^2} = \frac{1}{n^2}$ for $n \ge 2$

13. diverges, by the Comparison Test for Convergence, since $\frac{\ln n}{n} > \frac{1}{n}$ for $n \ge 3$

15. converges, by the Ratio Test for $\lim_{n\to\infty} \left| \frac{(n+2)(n+3)}{(n+1)!} \frac{n!}{(n+1)(n+2)} \right| = 0 < 1$

17. converges, by the Ratio Test for $\lim_{n\to\infty} \left| \frac{(n+4)!}{3!\,(n+1)!\,3^{n+1}} \frac{3!\,n!\,3^n}{(n+3)!} \right| = \frac{1}{3} < 1$

19. converges, by the Ratio Test for $\lim_{n\to\infty} \left| \frac{1}{(2n+3)!} \frac{(2n+1)!}{1} \right| = 0 < 1$

$$\exp\left[\lim_{n\to\infty} \frac{\ln\left(\frac{n}{n+1}\right)}{\frac{1}{n}}\right] = \exp\left[\lim_{n\to\infty} \frac{-n^2}{n(n+1)}\right] = e^{-1} = \frac{1}{e} < 1$$

21. converges, by the Root Test for $\lim_{n\to\infty} \sqrt[n]{\frac{n}{(\ln n)^n}} = \lim_{n\to\infty} \frac{\sqrt[n]{n}}{\ln n} = \lim_{n\to\infty} \frac{1}{\ln n} = 0 < 1$

23. converges, by the Comparison Test for Convergence, since $\frac{n! \ln n}{n(n+2)!} = \frac{\ln n}{n(n+1)(n+2)} <$

$$\frac{1}{(n+1)(n+2)} < \frac{1}{n^2} \text{ and } \sum_{n=1}^{\infty} \frac{1}{n^2} \text{ is a convergence p-series}$$

25. diverges, by the Root Test for $\lim_{n\to\infty} \sqrt[n]{\frac{(n!)^n}{\left(n^n\right)^2}} = \lim_{n\to\infty} \frac{n!}{n^2} = \infty$

27. converges, by the Root Test for $\lim_{n\to\infty} \sqrt[n]{\frac{n^n}{2^{n^2}}} = \lim_{n\to\infty} \frac{n}{2^n} = \lim_{n\to\infty} \frac{1}{2^n \ln 2} = 0$

29. converges, by the ratio Test for $\lim_{n\to\infty} \left| \frac{1\cdot 3 \cdots (2n-1)(2n+1)}{4^{n+1}\, 2^{n+1}\, (n+1)!} \cdot \frac{4^n\, 2^n\, n!}{1\cdot 3 \cdots (2n-1)} \right| =$

$$\frac{1}{8} \lim_{n\to\infty} \left(\frac{2n+1}{n+1}\right) = \frac{1}{4} < 1$$

31. converges, by the ratio Test for $\lim_{n\to\infty} \frac{\left(\frac{1+\sin n}{n}\right) a_n}{a_n} = 0 < 1$

33. diverges, the given series is $\sum_{n=1}^{\infty} \frac{3}{n}$

35. converges, by the Ratio Test for $\lim_{n\to\infty} \left| \frac{a_{n+1}}{a_n} \right| = \lim_{n\to\infty} \left| \frac{(1+\ln n)\, a_n}{n\, a_n} \right| = \lim_{n\to\infty} \frac{1+\ln n}{n} = 0 < 1$

37. converges, by the Ratio Test for $\lim_{n\to\infty} \left| \frac{2^{n+1} (n+1)!\,(n+1)!}{(2n+2)!} \frac{(2n)!}{2^n\,(n!)(n!)} \right| =$

$$\lim_{n\to\infty} \frac{2(n+1)^2}{(2n+2)(2n+1)} = \frac{1}{2} < 1$$

39. coverges, by the Comparison Test for Convergence, since $a_1 = 1 = \frac{12}{(1)(3)(2)^2}$, $a_2 = \frac{1 \cdot 2}{3 \cdot 4} = \frac{12}{(2)(4)(3)^2}$, $a_3 = \frac{2 \cdot 3}{4 \cdot 5}\,\frac{1 \cdot 2}{3 \cdot 4} = \frac{12}{(3)(5)(4)^2}$, $a_4 = \frac{3 \cdot 4}{5 \cdot 6}\,\frac{2 \cdot 3}{4 \cdot 5}\,\frac{1 \cdot 2}{3 \cdot 4} = \frac{12}{(4)(6)(5)^2}, \ldots \Rightarrow$

$1 + \sum_{n=1}^{\infty} \frac{12}{(n+1)(n+3)(n+2)^2}$ represents the given series and $\frac{12}{(n+1)\,(n+3)\,(n+2)^2} < \frac{12}{n^4}$

41. diverges, since if $a_n \to L$ as $n \to \infty$, then $L = 1 + \frac{1}{L} \Rightarrow L = \frac{-1 \pm \sqrt{5}}{2}$ but $L > 0 \Rightarrow L = -1 + \sqrt{5} \neq 0$

43. Ratio: $\lim_{n \to \infty} \left| \frac{1}{(n+1)^p} \div \frac{1}{n^p} \right| = \left(\lim_{n \to \infty} \frac{n}{n+1} \right)^p = 1^p = 1 \Rightarrow$ no conclusion

Root: $\lim_{n \to \infty} \sqrt[n]{\frac{1}{n^p}} = \lim_{n \to \infty} \frac{1}{\left(\sqrt[n]{n}\right)^p} = \frac{1}{(1)^p} = 1 \Rightarrow$ no conclusion

8.5 ALTERNATING SERIES AND ABSOLULTE CONVERGENCE

1. converges absolutely, by the Absolute Convergence Theorem, since $\sum_{n=1}^{\infty} |a_n|$ is a convergent p–series

3. diverges, $\lim_{n \to \infty} a_n \neq 0$

5. converges, by the Alternating Series Theorem, since $f(x) = \frac{\sqrt{x}+1}{x+1} \Rightarrow f'(x) = \frac{1 - 2x - 2\sqrt{x}}{2\sqrt{x}\,(x+1)^2} < 0 \Rightarrow$

f(x) is decreasing, $\lim_{n \to \infty} a_n = \lim_{n \to \infty} \frac{\sqrt{n}+1}{n+1} = 0$

7. $\sum_{n=2}^{\infty} (-1)^n \log_n 2 = \sum_{n=2}^{\infty} (-1)^n \frac{\ln 2}{\ln n}$ converges, by the Alternating Series Theorem, since $f(x) = \frac{1}{\ln x} \Rightarrow f'(x) = \frac{-1}{x\,(\ln x)^2} < 0 \Rightarrow$ f(x) is decreasing, $\lim_{n \to \infty} a_n = \lim_{n \to \infty} \frac{\ln 2}{\ln n} = 0$

9. diverges, $\lim_{n \to \infty} a_n = \lim_{n \to \infty} \frac{3\sqrt{n+1}}{\sqrt{n}+1} = \lim_{n \to \infty} \frac{3\sqrt{1+1/n}}{\sqrt{1}+1/\sqrt{n}} = 3 \neq 0$

11. diverges, $\lim_{n \to \infty} a_n = \lim_{n \to \infty} \left(\sqrt{n + \sqrt{n}} - \sqrt{n}\right) = \lim_{n \to \infty} \left[\left(\sqrt{n + \sqrt{n}} - \sqrt{n}\right)\left(\frac{\sqrt{n + \sqrt{n}} + \sqrt{n}}{\sqrt{n + \sqrt{n}} + \sqrt{n}}\right)\right] =$

$\lim_{n \to \infty} \frac{1}{\sqrt{1 + \frac{1}{\sqrt{n}}} + 1} = \frac{1}{2} \neq 0$

13. converges absolutely, by the Absolute Convergence Theorem, since $\sum_{n=1}^{\infty} |a_n|$ is a convergent geometric series

15. converges absolutely, by the Absolute Convergence Theorem, since $\sum_{n=1}^{\infty} |a_n|$ converges, by the Limit Comparison Test when compared with $\sum_{n=1}^{\infty} \frac{1}{n^2}$

17. converges conditionally, since $f(x) = \frac{1}{x + 3} \Rightarrow f'(x) = \frac{-1}{(x + 3)^2} < 0 \Rightarrow f(x)$ is decreasing and $\lim_{n \to \infty} \frac{1}{n + 3} = 0 \Rightarrow$ the given series converges, by the Alternating Series Test, but $\sum_{n=1}^{\infty} \frac{1}{n + 3}$ diverges, by the Limit Comparison Test when compared with $\sum_{n=1}^{\infty} \frac{1}{n}$

19. diverges, $\lim_{n \to \infty} a_n = \lim_{n \to \infty} \frac{3 + n}{5 + n} = 1 \neq 0$

21. converges conditionally, since $f(x) = \frac{1}{x^2} + \frac{1}{x} \Rightarrow f'(x) = -\left(\frac{1}{x^3} + \frac{1}{x}\right) < 0 \Rightarrow f(x)$ is decreasing and $\lim_{n \to \infty} a_n = \lim_{n \to \infty} \frac{1}{n^2} + \frac{1}{n} = 0 \Rightarrow$ the given series converges, by the Alternating Series Test, but $\sum_{n=1}^{\infty} \frac{1 + n}{n^2} = \sum_{n=1}^{\infty} \frac{1}{n^2} + \sum_{n=1}^{\infty} \frac{1}{n}$, the sum of a convergent and divergent series, diverges

23. converges absolutely, by the Ratio Test for $\lim_{n \to \infty} \left| \frac{(n + 1)^2 \left(\frac{2}{3}\right)^{n+1}}{n^2 \left(\frac{2}{3}\right)^n} \right| = \frac{2}{3} < 1$

25. converges absolutely, by the Integral Test, since $\int_1^{\infty} \arctan x \left(\frac{1}{1+x^2}\right) dx =$

$$\lim_{t \to \infty} \int_1^t \arctan x \left(\frac{1}{1+x^2}\right) dx = \lim_{t \to \infty} \left[\frac{(\arctan t)^2}{2}\right]_1^t = \frac{3\pi^2}{32}$$

27. diverges, the given series is $\frac{1}{2}\left(\sum_{n=1}^{\infty} \frac{1}{n}\right)$

29. diverges, $\lim_{n \to \infty} a_n = \lim_{n \to \infty} \frac{(-1)^n n}{n+1} = \lim_{n \to \infty} (-1)^n \neq 0$

31. converges absolutely, by the Absolute Convergence Theorem, since $\left|\frac{-1}{n^2 + 2n + 1}\right| < \frac{1}{n^2}$

33. converges absolutely, by the Ratio Test for $\lim_{n \to \infty} \left|\frac{(100)^{n+1}}{(n+1)!} \frac{n!}{100^n}\right| = \lim_{n \to \infty} \frac{100}{n+1} = 0 < 1$

35. converges absolutely, by the Absolute Convergence Theorem, since $\left|\frac{\cos n\pi}{n\sqrt{n}}\right| = \left|\frac{(-1)^{n+1}}{n^{3/2}}\right| = \frac{1}{n^{3/2}}$, a convergent p–series

37. converges conditionally, since $f(x) = \sqrt{x+1} - \sqrt{x} \Rightarrow f'(x) = \frac{1}{2}\left(\frac{1}{\sqrt{x+1}} - \frac{1}{\sqrt{x}}\right) < 0 \Rightarrow f(x)$ is decreasing and $\lim_{n \to \infty} a_n = \lim_{n \to \infty} \frac{1}{\sqrt{n} + \sqrt{n+1}} = 0 \Rightarrow$ the given series converges, by the Alternating Series Test, but $\sum_{n=1}^{\infty} \frac{1}{\sqrt{n} + \sqrt{n+1}} = \sum_{n=1}^{\infty} \sqrt{n+1} - \sum_{n=1}^{\infty} \sqrt{n}$, the difference of two diverging series

39. diverges, since $\lim_{n \to \infty} a_n = \lim_{n \to \infty} \frac{(2n)!}{2^n n!\, n} = \lim_{n \to \infty} \frac{(n+1)(n+2)\cdots 2n}{2^n n} =$

$$\lim_{n \to \infty} \frac{(n+1)(n+2)\cdots(n+(n-1))}{2^{n-1}} > \lim_{n \to \infty} \left(\frac{2n-1}{2}\right)^{n-1} = \lim_{n \to \infty} \left(n - \frac{1}{2}\right)^{n-1} = \infty \neq 0$$

41. converges conditionally, since $\frac{\sqrt{n+1} - \sqrt{n}}{1} \cdot \frac{\sqrt{n+1} + \sqrt{n}}{\sqrt{n+1} + \sqrt{n}} = \frac{1}{\sqrt{n+1} + \sqrt{n}}$ and $\left\{\frac{1}{\sqrt{n+1} + \sqrt{n}}\right\}$ is a decreasing sequence which converges to 0 so $\sum_{i=1}^{n} \frac{(-1)^n}{\sqrt{n+1} + \sqrt{n}}$ converges, but

$$\frac{1}{\sqrt{n+1} + \sqrt{n}} \geq \frac{1}{3\sqrt{n}} \Rightarrow \sum_{i=1}^{n} \frac{1}{\sqrt{n+1} + \sqrt{n}} \text{ diverges}$$

43. converges absolutely, by the Absolute Convergence Test, since $\text{sech}(n) = \dfrac{2}{e^n + e^{-n}} =$

$\dfrac{2e^n}{e^{2n}+1} < \dfrac{2\,e^n}{e^{2n}} = \dfrac{2}{e^n}$, a term from a convergent geometric series

45. $|\text{error}| < \left| (-1)^6 \frac{1}{5} \right| = 0.2$

47. $|\text{error}| < \left| (-1)^6 \frac{(0.01)^5}{5} \right| = 2.\ \text{x}\ 10^{-11}$

49. $\dfrac{1}{(2n)!} < \dfrac{5}{10^6} \Rightarrow (2n)! > \dfrac{10^6}{5} = 200000 \Rightarrow 2n = 10 \Rightarrow n = 5 \Rightarrow 1 - \dfrac{1}{2!} + \dfrac{1}{4!} - \dfrac{1}{6!} + \dfrac{1}{8!} \approx$

$0.540302579 \approx 0.54030$

51. a) $a_n \geq a_{n+1}$ fails, since $\frac{1}{3} < \frac{1}{2}$

b) $\left(\frac{1}{3} + \frac{1}{9} + \frac{1}{27} + \ldots\right) - \left(\frac{1}{2} + \frac{1}{4} + \frac{1}{8} + \ldots\right) = \dfrac{1/3}{1 - 1/3} - \dfrac{1/2}{1 - 1/2} = \frac{1}{2} - 1 = -\frac{1}{2}$

53. The unused terms are $\displaystyle\sum_{j=n+1}^{\infty} (-1)^{j+1} a_j = (-1)^{n+1}\left(a_{n+1} - a_{n+2}\right) + (-1)^{n+3}\left(a_{n+3} - a_{n+4}\right) + \ldots =$

$(-1)^{n+1}\left[\left(a_{n+1} - a_{n+2}\right) + \left(a_{n+3} - a_{n+4}\right) + \ldots\right]$. Each grouped term is positive, hence the remainder has the same sign as $(-1)^{n+1}$, which is the sign of the first unused term.

55. If $a_n = b_n = (-1)^n \frac{1}{\sqrt{n}}$, then $\displaystyle\sum_{n=1}^{\infty} (-1)^n \frac{1}{\sqrt{n}}$ converges but $\displaystyle\sum_{n=1}^{\infty} a_n b_n = \sum_{n=1}^{\infty} \frac{1}{n}$ diverges.

57. a) If $\displaystyle\sum_{n=1}^{\infty} |a_n|$ converges, then $\displaystyle\sum_{n=1}^{\infty} a_n$ converges and $\displaystyle\frac{1}{2}\sum_{n=1}^{\infty} a_n + \frac{1}{2}\sum_{n=1}^{\infty} |a_n| = \sum_{n=1}^{\infty} \frac{a_n + |a_n|}{2}$ converges where $b_n = \dfrac{a_n + |a_n|}{2}$.

b) If $\displaystyle\sum_{n=1}^{\infty} |a_n|$ converges, then $\displaystyle\sum_{n=1}^{\infty} a_n$ converges and $\displaystyle\frac{1}{2}\sum_{n=1}^{\infty} a_n - \frac{1}{2}\sum_{n=1}^{\infty} |a_n| = \sum_{n=1}^{\infty} \frac{a_n - |a_n|}{2}$ converges where $c_n = \dfrac{a_n - |a_n|}{2}$.

59. $S_1 = -\frac{1}{2}$, $S_2 = -\frac{1}{2} + 1$, $S_3 = -\frac{1}{2} + 1 - \frac{1}{4} - \frac{1}{4} - \frac{1}{6} - \frac{1}{8} - \frac{1}{10} - \frac{1}{12} - \frac{1}{14} - \frac{1}{16} - \frac{1}{18} - \frac{1}{20} - \frac{1}{22} \approx -0.5099$, $S_4 = -\frac{1}{2} + 1 - \frac{1}{4} - \frac{1}{4} - \frac{1}{6} - \frac{1}{8} - \frac{1}{10} - \frac{1}{12} - \frac{1}{14} - \frac{1}{16} - \frac{1}{18} - \frac{1}{20} - \frac{1}{22} + \frac{1}{3} \approx -0.1766$, $S_5 = -\frac{1}{2} + 1 - \frac{1}{4} - \frac{1}{4} - \frac{1}{6} - \frac{1}{8} - \frac{1}{10} - \frac{1}{12} - \frac{1}{14} - \frac{1}{16} - \frac{1}{18} - \frac{1}{20} - \frac{1}{22} + \frac{1}{3} - \frac{1}{24} - \frac{1}{26} - \frac{1}{28} - \frac{1}{30} - \frac{1}{32} - \frac{1}{34} - \frac{1}{36} - \frac{1}{38} - \frac{1}{40} - \frac{1}{42} - \frac{1}{44} \approx -0.512$, $S_6 = -\frac{1}{2} + 1 - \frac{1}{4} - \frac{1}{4} - \frac{1}{6} - \frac{1}{8} - \frac{1}{10} - \frac{1}{12} - \frac{1}{14} - \frac{1}{16} - \frac{1}{18} - \frac{1}{20} - \frac{1}{22} + \frac{1}{3} - \frac{1}{24} - \frac{1}{26} - \frac{1}{28} - \frac{1}{30} - \frac{1}{32} - \frac{1}{34} - \frac{1}{36} - \frac{1}{38} - \frac{1}{40} - \frac{1}{42} - \frac{1}{44} - \frac{1}{46} - \frac{1}{48} - \frac{1}{50} - \frac{1}{52} - \frac{1}{54} - \frac{1}{56} - \frac{1}{58} - \frac{1}{60} - \frac{1}{62} - \frac{1}{64} - \frac{1}{66} \approx -0.511065$

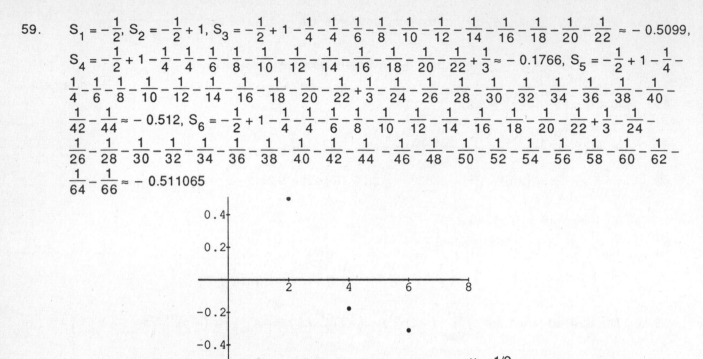

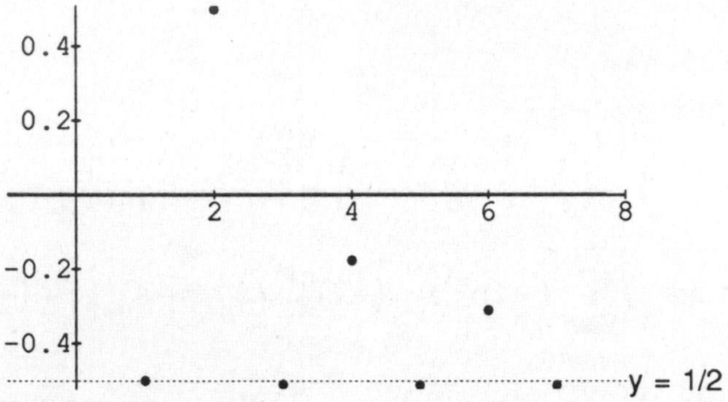

Graph 8.5.59

8.6 POWER SERIES

1. $\lim_{n \to \infty} \left| \frac{x^{n+1}}{x^n} \right| < 1 \Rightarrow -1 < x < 1$; when $x = -1$ we have $\sum_{n=1}^{\infty} (-1)^n$, a divergent series; when $x = 1$ we have $\sum_{n=1}^{\infty} 1$, a divergent series $\therefore$

a) $1, -1 < x < 1$ b) $-1 < x < 1$ c) { }

3. $\lim_{n \to \infty} \left| \frac{(x+1)^{n+1}}{(x+1)^n} \right| < 1 \Rightarrow -2 < x < 0$; when $x = -2$ we have $\sum_{n=1}^{\infty} 1$, a divergent series; when $x = 0$ we have $\sum_{n=1}^{\infty} (-1)^n$, a divergent series $\therefore$

a) $1, -2 < x < 0$ b) $-2 < x < 0$ c) { }

5. $\lim_{n \to \infty} \left| \frac{(x-2)^{n+1}}{10^{n+1}} \frac{10^n}{(x-2)^n} \right| < 1 \Rightarrow -8 < x < 12$; when $x = -8$ we have $\sum_{n=1}^{\infty} (-1)^n$, a divergent series; when $x = 12$ we have $\sum_{n=1}^{\infty} 1$, a divergent series $\therefore$

a) $10, -8 < x < 12$ b) $-8 < x < 12$ c) { }

7. $\lim_{n \to \infty} \left| \frac{(n+1)\,x^{n+1}}{n+3} \frac{n+2}{n\,x^n} \right| < 1 \Rightarrow -1 < x < 1$; when $x = -1$ we have $\sum_{n=1}^{\infty} (-1)^n \frac{n}{n+2}$, a divergent series; when $x = 1$ we have $\sum_{n=1}^{\infty} \frac{n}{n+2}$, a divergent series $\therefore$

a) $1, -1 < x < 1$ b) $-1 < x < 1$ c) { }

9. $\lim_{n \to \infty} \left| \frac{x^{n+1}}{(n+1)\sqrt{n+1}\,3^{n+1}} \cdot \frac{n\sqrt{n}\,3^n}{x^n} \right| < 1 \Rightarrow \frac{|x|}{3} \left(\lim_{n \to \infty} \frac{n}{n+1} \right) \left(\sqrt{\lim_{n \to \infty} \frac{n}{n+1}} \right) < 1 \Rightarrow$

$-3 < x < 3$; when $x = -3$ we have $\sum_{i=1}^{n} \frac{(-1)^n}{n^{3/2}}$ an absolutely convergent series; when $x = 3$ we have

$\sum_{i=1}^{n} \frac{1}{n^{3/2}}$ a convergent p–series $\therefore$

a) 3, [– 3,3] b) [– 3,3] c) { }

11. $\lim_{n \to \infty} \left| \frac{x^{n+1}}{(n+1)!} \frac{n!}{x^n} \right| < 1 \Rightarrow |x| \lim_{n \to \infty} \frac{1}{n+1} < 1$ for all $x \Rightarrow$

a) ∞, For all x b) For all x c) { }

13. $\lim_{n \to \infty} \left| \frac{x^{2n+3}}{(n+1)!} \frac{n!}{x^{2n+1}} \right| < 1 \Rightarrow x^2 \lim_{n \to \infty} \left| \frac{1}{n+1} \right| < 1$ for all x

a) ∞, For all x b) For all x c) { }

15. $\lim_{n\to\infty}\left|\frac{x^{n+1}}{\sqrt{n^2+2n+4}}\frac{\sqrt{n^2+3}}{x^n}\right| < 1 \Rightarrow |x|\sqrt{\lim_{n\to\infty}\frac{n^2+3}{n^2+2n+4}} < 1 \Rightarrow$

$-1 < x < 1$; when $x = -1$ we have $\sum_{n=1}^{\infty}\frac{(-1)^n}{\sqrt{n^2+3}}$, a convergent series;

when $x = 1$ we have $\sum_{n=1}^{\infty}\frac{1}{\sqrt{n^2+3}}$, a divergent series $\therefore$

a) $1, -1 \le x < 1$ b) $-1 < x < 1$ c) $\{-1\}$

17. $\lim_{n\to\infty}\left|\frac{(n+1)\,x^{n+1}}{4^{n+1}\left(n^2+2n+2\right)}\cdot\frac{4^n\left(n^2+1\right)}{n\,x^n}\right| < 1 \Rightarrow \frac{|x|}{4}\lim_{n\to\infty}\left|\frac{(n+1)(n^2+1)}{n(n^2+2n+2)}\right| < 1 \Rightarrow$

$-4 < x < 4$; when $x = -4$ we have $\sum_{n=1}^{\infty}\frac{n\,(-1)^n}{n^2+1}$ a conditionally convergent series; when

$x = 4$ we have $\sum_{n=1}^{\infty}\frac{n}{n^2+1}$ a divergent series $\therefore$

a) $4, [-4,4)$ b) $(-4,4)$ c) $\{-4\}$

19. $\lim_{n\to\infty}\left|\frac{\sqrt{n+1}\,x^{n+1}}{3^{n+1}}\frac{3^n}{\sqrt{n}\,x^n}\right| < 1 \Rightarrow \frac{|x|}{3}\sqrt{\lim_{n\to\infty}\frac{n+1}{n}} < 1 \Rightarrow -3 < x < 3$; both series:

$\sum_{n=1}^{\infty}\sqrt{n}\,(-1)^n$, when $x = -3$ and $\sum_{n=1}^{\infty}\sqrt{n}$, when $x = 3$ diverge $\therefore$

a) $3, -3 < x < 3$ b) $-3 < x < 3$ c) $\{\}$

21. $\lim_{n\to\infty}\left|\frac{\left(1+\frac{1}{n+1}\right)^{n+1}x^{n+1}}{\left(1+\frac{1}{n}\right)^n x^n}\right| < 1 \Rightarrow |x|\lim_{n\to\infty}\left(\frac{n(n+2)}{(n+1)^2}\right)^n\left(1+\frac{1}{n+1}\right) < 1 \Rightarrow -1 < x < 1;$

both series: $\sum_{n=1}^{\infty}(-1)^n\left(1+\frac{1}{n}\right)^n$, when $x = -1$ and $\sum_{n=1}^{\infty}\left(1+\frac{1}{n}\right)^n$, when $x = 1$ diverge $\therefore$

a) $1, -1 < x < 1$ b) $-1 < x < 1$ c) $\{\}$

23. $\lim_{n\to\infty}\left|\frac{(n+1)^{n+1}x^{n+1}}{n^n x^n}\right| < 1 \Rightarrow |x|\left(\lim_{n\to\infty}\left(1+\frac{1}{n}\right)^n\right)\left(\lim_{n\to\infty}(n+1)\right) < 1 \Rightarrow$

$e|x| \lim_{n\to\infty}(n+1) < 1 \Rightarrow$

a) 0, x = 0 b) x = 0 c) { }

25. $\lim_{n\to\infty}\left|\frac{(x-2)^{n+1}}{(n+1)2^{n+1}}\cdot\frac{n2^n}{(x-2)^n}\right| < 1 \Rightarrow \frac{|x-2|}{2}\lim_{n\to\infty}\left|\frac{n}{n+1}\right| < 1 \Rightarrow 0 < x < 4$; when x = 0 we have $\sum_{n=1}^{\infty}\frac{-1}{n}$ a divergent series; when x = 4 we have $\sum_{n=1}^{\infty}\frac{(-1)^n}{n}$ a convergent alternating series ∴

a) 2, (0,4] b) 0 < x < 4 c) {4}

27. $\lim_{n\to\infty}\left|\frac{x^{n+1}}{(n+1)\ln(n+1)}\cdot\frac{n\ln(n)}{x^n}\right| < 1 \Rightarrow |x|\left(\lim_{n\to\infty}\frac{n}{n+1}\right)\left(\lim_{n\to\infty}\frac{\ln(n)}{\ln(n+1)}\right) < 1 \Rightarrow -1 < x < 1$; when x = − 1 we have $\sum_{n=2}^{\infty}\frac{(-1)^n}{n\ln n}$ a convergent alternating series for if $f(x) = \frac{1}{x\ln x}$, $f'(x) = -\frac{\ln(x)+1}{x^2\ln^2 x} \le 0$ and $\lim_{n\to\infty}\frac{1}{x\ln x} = 0$; when x = 1 we have $\sum_{n=2}^{\infty}\frac{1}{n\ln n}$ a divergent logarithmic p-series ∴

a) 1, [− 1,1) b) (− 1,1) c) {− 1}

29. $\lim_{n\to\infty}\left|\frac{1}{|x|^{n+1}}\cdot\frac{|x|^n}{1}\right| < 1 \Rightarrow \frac{1}{|x|}\lim_{n\to\infty} 1 < 1 \Rightarrow |x| > 1 \Rightarrow x > 1$ or $x < -1$; when $x = \pm 1$ we have $\sum_{n=1}^{\infty} 1$ a divergent series ∴

a) ∞, (− ∞,− 1) ∪(1,∞) b) x < − 1 or x > 1 c) { }

31. $\lim_{n\to\infty}\left|\frac{\cosh(n+1)\,x^{n+1}}{\cosh(n)\,x^n}\right| < 1 \Rightarrow |x|\lim_{n\to\infty}\left|\frac{1+e^{-2(n+1)}}{e^{-1}+e^{-(2n+1)}}\right| < 1 \Rightarrow -\frac{1}{e} < x < \frac{1}{e}$; when $x = \pm\frac{1}{e}$ both series: $\sum_{n=1}^{\infty}\frac{\cosh n}{(-e)^n}$ and $\sum_{n=1}^{\infty}\frac{\cosh n}{e^n}$ diverge, since $\lim_{n\to\infty} a_n \ne 0$ ∴

a) $\frac{1}{e}, \left(-\frac{1}{e}, \frac{1}{e}\right)$ b) $-\frac{1}{e} < x < \frac{1}{e}$ c) { }

33. $\lim_{n \to \infty} \left| \dfrac{\tanh(n+1)\, x^{n+1}}{\tanh(n)\, x^n} \right| < 1 \Rightarrow |x| \lim_{n \to \infty} \left(\dfrac{1 - e^{-2(n+1)}}{1 + e^{-2(n+1)}} \cdot \dfrac{1 - e^{-2n}}{1 - e^{-2n}} \right) < 1 \Rightarrow |x| < 1 \Rightarrow -1 < x < 1;$

when $x = \pm 1$ both series: $\sum_{n=1}^{\infty} (-1)^n \tanh(n)$ and $\sum_{n=1}^{\infty} \tanh(n)$ diverge $\therefore$

a) $1, (-1,1)$ b) $-1 < x < 1$ c) $\{\}$

35. $\lim_{n \to \infty} \left| \dfrac{(x-1)^{2n+2}}{2^{n+1}} \cdot \dfrac{2^n}{(x-1)^{2n}} \right| < 1 \Rightarrow |x - 1| < \sqrt{2} \Rightarrow 1 - \sqrt{2} < x,\ 1 + \sqrt{2}$; when $x = 1 \pm \sqrt{2}$ we have

$\sum_{n=1}^{\infty} \dfrac{1}{f(i)}$ a divergent series $\Rightarrow$ the interval of convergence is $(1 - \sqrt{2}, 1 + \sqrt{2})$ and $\sum_{n=1}^{\infty} \left(\dfrac{(x-1)^2}{2} \right)^n =$

$\dfrac{1}{1 - \dfrac{(x-1)^2}{2}} = \dfrac{2}{1 + 2x - x^2}$

37. $\lim_{n \to \infty} \left| \dfrac{\left(x^2 - 1\right)^{n+1}}{2^{n+1}} \cdot \dfrac{2^n}{\left(x^2 - 1\right)^n} \right| < 1 \Rightarrow \left| x^2 - 1 \right| < 2 \Rightarrow -\sqrt{3} < x < \sqrt{3}$; when $x = \pm\sqrt{3}$ we have

$\sum_{n=0}^{\infty} 1^n$ a divergent series $\Rightarrow$ the interval of convegence is $(-\sqrt{3}, \sqrt{3})$ and $\sum_{n=0}^{\infty} \left(\dfrac{x^2 - 1}{2} \right)^n =$

$\dfrac{1}{1 - \left(\dfrac{x^2 - 1}{2} \right)} = \dfrac{2}{3 - x^2}$

39. $\lim_{n \to \infty} \left| \dfrac{3^{n+1}}{\left(2 - x^2\right)^{n+1}} \cdot \dfrac{\left(2 - x^2\right)^n}{3^n} \right| < 1 \Rightarrow 3 < \left| 2 - x^2 \right| \Rightarrow x > \sqrt{5}$ or $x < -\sqrt{5}$; when $x = \pm\sqrt{5}$

we have $\sum_{n=1}^{\infty} (-1)^n$ a divergent series; $\therefore$ the intervals of convergence are $(-\infty, -\sqrt{5}) \cup (\sqrt{5}, \infty)$ and

$\sum_{n=1}^{\infty} \left(\dfrac{3}{2 - x^2} \right)^n = \dfrac{1}{1 - \dfrac{3}{2 - x^2}} = \dfrac{x^2 - 2}{x^2 + 1}$

$\sum_{n=1}^{\infty} \dfrac{1}{f(i)} = \dfrac{1}{1 - \dfrac{1}{x - 1}} = \dfrac{x - 1}{x - 2}$

41. $\lim_{n\to\infty}\left|\frac{(x-3)^{n+1}}{2^{n+1}}\,\frac{2^n}{(x-3)^n}\right| < 1 \Rightarrow |x-3| < 2 \Rightarrow 1 < x < 5$; both series: $\sum_{n=1}^{\infty}(1)$,

when $x = 1$ and $\sum_{n=1}^{\infty}(-1)^n$, when $x = 5$ diverge $\therefore$ the interval of convergence is $1 < x < 5$;

the sum of this convergent geometric series is $\frac{1}{1+\frac{x-3}{2}} = \frac{2}{x-1}$;

$f(x) = 1 - \frac{1}{2}(x-3) + \frac{1}{4}(x-3)^2 + \ldots + \left(-\frac{1}{2}\right)^n (x-3)^n + \ldots = \frac{2}{x-1} \Rightarrow$

$f'(x) = -\frac{1}{2} + \frac{1}{2}(x-3) + \ldots + \left(-\frac{1}{2}\right)^n n(x-3)^{n-1} + \ldots$ is convergent when $1 < x < 5$, and diverges when $x = 1$ or 5; its sum is $\frac{-2}{(x-1)^2}$, the derivative of $\frac{2}{x-1}$

43. a) $\ln|\sec x| + C = \int \tan x\,dx = \int x + \frac{x^3}{3} + \frac{2x^5}{15} + \frac{17x^7}{315} + \frac{62x^9}{2835} + \ldots\,dx = \frac{x^2}{2} + \frac{x^4}{12} + \frac{x^6}{45} + \frac{17x^8}{2520} + \frac{31x^{10}}{14175} + \ldots + C$, but $x = 0 \Rightarrow C = 0 \Rightarrow \ln|\sec x| = \frac{x^2}{2} + \frac{x^4}{12} + \frac{x^6}{45} + \frac{17x^8}{2520} + \frac{31x^{10}}{14175} + \ldots$, when $-\frac{\pi}{2} < x < \frac{\pi}{2}$

b) $\sec^2 x = \frac{d(\tan x)}{dx} = \frac{d\left(x + \frac{x^3}{3} + \frac{2x^5}{15} + \frac{17x^7}{315} + \frac{62x^9}{2835} + \ldots\right)}{dx} = 1 + x^2 + \frac{2x^4}{3} + \frac{17x^6}{45} + \frac{62x^8}{105} + \ldots$, when $-\frac{\pi}{2} < x < \frac{\pi}{2}$

c) $\sec^2 x = (\sec x)(\sec x) = \left(1 + \frac{x^2}{2} + \frac{5x^4}{24} + \frac{61x^6}{720} + \ldots\right)\left(1 + \frac{x^2}{2} + \frac{5x^4}{24} + \frac{61x^6}{720} + \ldots\right) = 1 + \left(\frac{1}{2} + \frac{1}{2}\right)x^2 + \left(\frac{5}{24} + \frac{1}{4} + \frac{5}{24}\right)x^4 + \left(\frac{61}{720} + \frac{5}{48} + \frac{5}{48} + \frac{61}{720}\right)x^6 + \ldots = 1 + x^2 + \frac{2x^4}{3} + \frac{17x^6}{45} + \ldots$

45. a) If $f(x) = \sum_{n=0}^{\infty} a_n x^n$, then $f^{(k)}(x) = \sum_{n=k}^{\infty} n(n-1)(n-2)\cdots(n-(k-1))\,a_n x^{n-k}$ and

$f^{(k)}(0) = k!\,a_k \Rightarrow a_k = \frac{f^{(k)}(0)}{k!}$ and likewise if $f(x) = \sum_{n=0}^{\infty} b_n x^n$, then $b_k = \frac{f^{(k)}(0)}{k!} \Rightarrow$

$a_k = b_k$ for every nonnegative integer k.

b) If $f(x) = \sum_{n=0}^{\infty} a_n x^n = 0$ for all x, then $f^{(k)}(x) = 0$ for all x and from part a) $a_k = 0$ for every nonnegative integer k.

47. The series $\sum_{n=1}^{\infty} \frac{x^n}{n}$ has an interval of convergence os [– 1,1) where the left endpoint is conditionally convergent; the series $\sum_{n=1}^{\infty} \frac{x^n}{n^2}$ has an interval of convergence of [–1,1] where the left endpoint is absolutely convergent.

8.7 TAYLOR SERIES AND MACLAURIN SERIES

1. $f(x) = \ln x$ $\quad f'(x) = \frac{1}{x}$ $\quad f''(x) = -\frac{1}{x^2}$ $\quad f'''(x) = \frac{2}{x^3}$

$f(1) = \ln 1 = 0$ $\quad f'(1) = 1$ $\quad f''(1) = -1$ $\quad f'''(1) = 2$

$P_0(x) = 0,\ P_1(x) = x - 1,\ P_2(x) = (x-1) - \frac{1}{2}(x-1)^2,\ P_3(x) = (x-1) - \frac{1}{2}(x-1)^2 + \frac{1}{3}(x-1)^3$

3. $f(x) = \frac{1}{x} + x^{-1}$ $\quad f'(x) = -x^{-2}$ $\quad f''(x) = 2x^{-3}$ $\quad f'''(x) = -6x^{-4}$

$f(2) = \frac{1}{2}$ $\quad f'(2) = -\frac{1}{4}$ $\quad f''(2) = \frac{1}{4}$ $\quad f'''(2) = -\frac{3}{8}$

$P_0(x) = \frac{1}{2},\ P_1(x) = \frac{1}{2} - \frac{1}{4}(x-2),\ P_2(x) = \frac{1}{2} - \frac{1}{4}(x-2) + \frac{1}{8}(x-2)^2,$

$P_3(x) = \frac{1}{2} - \frac{1}{4}(x-2) + \frac{1}{8}(x-2)^2 - \frac{1}{16}(x-2)^3$

5. $f(x) = \sin x \Rightarrow f'(x) = \cos x \Rightarrow f''(x) = -\sin x \Rightarrow f'''(x) = -\cos x \Rightarrow f(\pi/4) = \sin \pi/4 = \frac{\sqrt{2}}{2},$

$f'(\pi/4) = \cos \pi/4 = \frac{\sqrt{2}}{2},\ f''(\pi/4) = -\sin \pi/4 = -\frac{\sqrt{2}}{2},\ f'''(\pi/4) = -\cos \pi/4 = -\frac{\sqrt{2}}{2} \Rightarrow P_0 = \frac{\sqrt{2}}{2},$

$P_1(x) = \frac{\sqrt{2}}{2} + \frac{\sqrt{2}}{2}\left(x - \frac{\pi}{4}\right),\ P_2(x) = \frac{\sqrt{2}}{2} + \frac{\sqrt{2}}{2}\left(x - \frac{\pi}{4}\right) - \frac{\sqrt{2}}{4}\left(x - \frac{\pi}{4}\right)^2,$

$P_3(x) = \frac{\sqrt{2}}{2} + \frac{\sqrt{2}}{2}\left(x - \frac{\pi}{4}\right) - \frac{\sqrt{2}}{4}\left(x - \frac{\pi}{4}\right)^2 - \frac{\sqrt{2}}{12}\left(x - \frac{\pi}{4}\right)^3$

7. $f(x) = \sqrt{x} = x^{1/2} \Rightarrow f'(x) = (1/2)x^{-1/2} \Rightarrow f''(x) = (-1/4)x^{-3/2} \Rightarrow f'''(x) = (3/8)x^{-5/2} \Rightarrow$

$f(4) = \sqrt{4} = 4^{1/2} = 2,\ f'(4) = (1/2)4^{-1/2} = \frac{1}{4},\ f''(4) = (-1/4)4^{-3/2} = -\frac{1}{32},\ f'''(4) = (3/8)4^{-5/2} = \frac{3}{256} \Rightarrow$

$P_0(x) = 2, P_1(x) = 2 + \frac{1}{4}(x-4),\ P_2(x) = 2 + \frac{1}{4}(x-4) - \frac{1}{64}(x-4)^2,$

$P_3(x) = 2 + \frac{1}{4}(x-4) - \frac{1}{64}(x-4)^2 + \frac{1}{512}(x-4)^3$

9. $e^x = \sum_{n=0}^{\infty} \frac{x^n}{n!} \Rightarrow e^{-x} = \sum_{n=0}^{\infty} \frac{(-x)^n}{n!} = 1 - x + \frac{x^2}{2!} - \frac{x^3}{3!} + \frac{x^4}{4!} - \ldots$

11. $\sin x = \sum_{n=0}^{\infty} \frac{(-1)^n x^{2n+1}}{(2n+1)!} \Rightarrow \sin 3x = \sum_{n=0}^{\infty} \frac{(-1)^n (3x)^{2n+1}}{(2n+1)!} = 3x - \frac{(3x)^3}{3!} + \frac{(3x)^5}{5!} - \ldots$

13. $\cos(-x) = \cos(x) = \sum_{n=0}^{\infty} \frac{(-1)^n x^{2n}}{(2n)!} = 1 - \frac{x^2}{2!} + \frac{x^4}{4!} - \frac{x^6}{6!} + \ldots$, since cosine is an even function

15. $\cosh x = \frac{e^x + e^{-x}}{2} = \frac{1}{2}\left[\left(1 + x + \frac{x^2}{2!} + \frac{x^3}{3!} + \frac{x^4}{4!} + \ldots\right) + \left(1 - x + \frac{x^2}{2!} - \frac{x^3}{3!} + \frac{x^4}{4!} - \ldots\right)\right] =$
$1 + \frac{x^2}{2!} + \frac{x^4}{4!} + \frac{x^6}{6!} + \ldots$

17. $\frac{x^2}{2} - 1 + \cos x = \frac{x^2}{2} - 1 + \sum_{n=0}^{\infty} \frac{(-1)^n x^{2n}}{(2n)!} = \frac{x^2}{2} - 1 + 1 - \frac{x^2}{2!} + \frac{x^4}{4!} - \frac{x^6}{6!} + \ldots = \frac{x^4}{4!} - \frac{x^6}{6!} + \frac{x^8}{8!} - \frac{x^{10}}{10!} + \ldots$

19. $f(x) = e^{\tan x} \Rightarrow f'(x) = e^{\tan x}\sec^2 x \Rightarrow f''(x) = e^{\tan x}\sec^4 x + 2\,e^{\tan x}\sec^2 x \tan x =$
$e^{\tan x}\sec^2 x\left(\sec^2 x + 2\tan x\right) \Rightarrow f'''(x) = e^{\tan x}\sec^2 x\left(\sec^2 x + 2\tan x\right)^2 +$
$e^{\tan x}\sec^2 x\left(2\sec^2 x \tan x + 2\sec^2 x\right) = e^{\tan x}\sec^2 x\left[\left(\sec^2 x + 2\tan x\right)^2 + \left(2\sec^2 x \tan x + 2\sec^2 x\right)\right]$
$\therefore\ e^{\tan x} = 1 + x + \frac{x^2}{2} + \frac{f'''(c)}{3!}x^3$ where c is between 0 and x

21. $f(x) = \ln(\cos x) \Rightarrow f'(x) = -\frac{\sin x}{\cos x} = -\tan x \Rightarrow f''(x) = -\sec^2 x \Rightarrow f'''(x) = -2\sec^2 x \tan x$
$\therefore\ \ln(\cos x) = -x^2 + \frac{f'''(c)}{3!}x^3$ where c is between 0 and x

23. $f(x) = \sinh x \Rightarrow f'(x) = \cosh x \Rightarrow f''(x) = \sinh x \Rightarrow f'''(x) = \cosh x\ \therefore\ \cosh x = x + \frac{f'''(c)}{3!}x^3$
where c is between 0 and x

25. If $e^x = \sum_{n=0}^{\infty} \frac{f^{(n)}(a)}{n!}(x-a)^n$ and $f(x) = e^x$, we have $f^{(n)}(a) = e^a$ for all n = 0, 1, 2, 3,...;
$e^x = e^a\left[\frac{(x-a)^0}{0!} + \frac{(x-a)^1}{1!} + \frac{(x-a)^2}{2!} + \ldots\right] = e^a\left[1 + (x-a) + \frac{(x-a)^2}{2!} + \ldots\right]$, at x = a

27. $\left|R_3(x)\right| = \left|\frac{-\cos c}{5!}(x-0)^5\right| < \left|\frac{x^5}{5!}\right| < 5 \times 10^{-4} \Rightarrow -5 \times 10^{-4} < \frac{x^5}{5!} < 5 \times 10^{-4} \Rightarrow$
$-\sqrt[5]{5!\left(5 \times 10^{-4}\right)} < x < \sqrt[5]{5!\left(5 \times 10^{-4}\right)} \Rightarrow -0.569679052 < x < 0.569679052$

29. $\sin x = x + R_1(x)$, when $|x| < 10^{-3} \Rightarrow \left|R_1(x)\right| = \left|\frac{-\cos c}{3!}x^3\right| < \left|\frac{(1)\,x^3}{3!}\right| < \frac{\left(10^{-3}\right)^3}{3!} = 1.67 \times 10^{-10}$
From exercise 45 in section 9.5, $R_1(x)$ has the same sign as $-\frac{x^3}{3!}$. $x < \sin x \Rightarrow 0 < \sin x - x =$
$R_1(x)$, which has the same sign as $-\frac{x^3}{3!} \Rightarrow x < 0 \Rightarrow -10^{-3} < x < 0$.

31. $\left|R_2(x)\right| = \left|\frac{e^c x^3}{3!}\right| < \frac{3^{(0.1)}(0.1)^3}{3!} = 0.00018602$, where c is between 0 and x.

33. If we approximate sinh x with $x + \frac{x^3}{3!}$ and $|x| < 0.5$, then the $|\text{error}| < \left|\frac{(\cosh c)\, x^5}{5!}\right| < \frac{(0.5)^5}{5!} = \frac{1}{(32)(120)} = 0.000260416$, where c is between 0 and x.

35. sin x, when x = 0.1; the sum is $\sin(0.1) \approx 0.099833416$

37. $\sin x = x - \frac{x^3}{3!} + \frac{x^5}{5!} - \frac{x^7}{7!} + \ldots$; $\frac{d(\sin x)}{dx} = \frac{d\left(x - \frac{x^3}{3!} + \frac{x^5}{5!} - \frac{x^7}{7!} + \ldots\right)}{dx} = 1 - \frac{x^2}{2!} + \frac{x^4}{4!} - \frac{x^6}{6!} + \ldots = \cos x$;

$\frac{d(\cos x)}{dx} = \frac{d\left(1 - \frac{x^2}{2!} + \frac{x^4}{4!} - \frac{x^6}{6!} + \ldots\right)}{dx} = -x + \frac{x^3}{3!} - \frac{x^5}{5!} + \ldots = -\sin x$; $\frac{d\left(e^x\right)}{dx} =$

$\frac{d\left(1 + x + \frac{x^2}{2!} + \frac{x^3}{3!} + \frac{x^4}{4!} + \ldots\right)}{dx} = 1 + x + \frac{x^2}{2!} + \frac{x^3}{3!} + \frac{x^4}{4!} + \ldots$

39. $2[\cos x][\sin x] = 2\left[1 - \frac{x^2}{2!} + \frac{x^4}{4!} - \frac{x^6}{6!} + \ldots\right]\left[x - \frac{x^3}{3!} + \frac{x^5}{5!} - \frac{x^7}{7!} + \ldots\right] =$

$2\left[x - \frac{4x^3}{3!} + \frac{16x^5}{5!} - \frac{64x^7}{7!} + \frac{256x^9}{9!} - \ldots\right] = 2x - \frac{8x^3}{3!} + \frac{32x^5}{5!} - \frac{128\,x^7}{7!} + \frac{512\,x^9}{9!} - \ldots$

41. $\cos x = 1 - \frac{x^2}{2!} + \frac{x^4}{4!} - \frac{x^6}{6!} + \ldots \Rightarrow 1 - \cos x = \frac{x^2}{2!} - \frac{x^4}{4!} + \frac{x^6}{6!} - \frac{x^8}{8!} \ldots = \frac{1 - \cos x}{x^2} = \frac{1}{2} - \frac{x^2}{4!} + \frac{x^6}{6!} - \frac{x^8}{8!} + \ldots$

If L is the sum of the series representing $\frac{1 - \cos x}{x^2}$, $L - s_1 = \frac{1 - \cos x}{x^2} - \frac{1}{2} < 0$ and $\frac{1 - \cos x}{x^2} - \left(\frac{1}{2} - \frac{x^2}{4!}\right) > 0$, by the Alternating Series Estimation Theorem. $\therefore \frac{1}{2} - \frac{x^2}{4!} < \frac{1 - \cos x}{x^2} < \frac{1}{2}$

43. $\sin^2 x = \left(\frac{1 - \cos 2x}{2}\right) = \frac{1}{2} - \frac{1}{2}\cos 2x = \frac{1}{2} - \frac{1}{2}\left(1 - \frac{(2x)^2}{2!} + \frac{(2x)^4}{4!} - \frac{(2x)^6}{6!} + \cdots\right) =$

$\frac{2\,x^2}{2!} - \frac{2^3\,x^4}{4!} + \frac{2^5\,x^6}{6!} - \cdots$, $\frac{d}{dx}\left(\sin^2 x\right) = \frac{d}{dx}\left(\frac{2\,x^2}{2!} - \frac{2^3\,x^4}{4!} + \frac{2^5\,x^6}{6!} - \cdots\right) = 2x - \frac{(2x)^3}{3!} + \frac{(2x)^5}{5!} - \frac{(2x)^7}{7!} + \cdots = \sin 2x$

45. a) $e^{-i\pi} = \cos(-\pi) + i\sin(-\pi) = -1 + i(0) = -1$

b) $e^{i\pi/4} = \cos\left(\frac{\pi}{4}\right) + i\sin\left(\frac{\pi}{4}\right) = \frac{1}{\sqrt{2}} + \frac{i}{\sqrt{2}} = \left(\frac{1}{\sqrt{2}}\right)(1 + i)$

c) $e^{-i\pi/2} = \cos\left(-\frac{\pi}{2}\right) + i\sin\left(-\frac{\pi}{2}\right) = 0 + i(-1) = -i$

47. $e^x = 1+x+\frac{x^2}{2!}+\frac{x^3}{3!}+\frac{x^4}{4!}+\ldots \Rightarrow e^{i\theta} = 1+i\theta+\frac{(i\theta)^2}{2!}+\frac{(i\theta)^3}{3!}+\frac{(i\theta)^4}{4!}+\ldots$ and $e^{-i\theta} =$
$1-i\theta+\frac{(-i\theta)^2}{2!}+\frac{(-i\theta)^3}{3!}+\frac{(-i\theta)^4}{4!}+\ldots = 1-i\theta+\frac{(i\theta)^2}{2!}-\frac{(i\theta)^3}{3!}+\frac{(i\theta)^4}{4!}-\ldots,$

$$\frac{e^{i\theta}+e^{-i\theta}}{2} = \frac{\left(1+i\theta+\frac{(i\theta)^2}{2!}+\frac{(i\theta)^3}{3!}+\frac{(i\theta)^4}{4!}+\ldots\right)+\left(1-i\theta+\frac{(i\theta)^2}{2!}-\frac{(i\theta)^3}{3!}+\frac{(i\theta)^4}{4!}-\ldots\right)}{2} =$$

$1-\frac{\theta^2}{2!}+\frac{\theta^4}{4!}-\frac{\theta^6}{6!}+\ldots = \cos\theta;$

$$\frac{e^{i\theta}-e^{-i\theta}}{2} = \frac{\left(1+i\theta+\frac{(i\theta)^2}{2!}+\frac{(i\theta)^3}{3!}+\frac{(i\theta)^4}{4!}+\ldots\right)-\left(1-i\theta+\frac{(i\theta)^2}{2!}-\frac{(i\theta)^3}{3!}+\frac{(i\theta)^4}{4!}-\ldots\right)}{2i} =$$

$\theta-\frac{\theta^3}{3!}+\frac{\theta^5}{5!}-\frac{\theta^7}{7!}+\ldots = \sin\theta$

49. $e^x \sin x = \left(1+x+\frac{x^2}{2!}+\frac{x^3}{3!}+\frac{x^4}{4!}+\ldots\right)\left(x-\frac{x^3}{3!}+\frac{x^5}{5!}-\frac{x^7}{7!}+\ldots\right) = (1)\,x+(1)\,x^2+\left(-\frac{1}{6}+\frac{1}{2}\right)x^3+$
$\left(-\frac{1}{6}+\frac{1}{6}\right)x^4+\left(\frac{1}{120}-\frac{1}{12}+\frac{1}{24}\right)x^5+\ldots = x+x^2+\frac{1}{3}x^3-\frac{1}{30}x^5\ \ldots;\ e^x\cdot e^{ix} = e^{(1+ix)} =$
$e^x(\cos x+i\sin x) = e^x\cos x+i\left(e^x\sin x\right) \Rightarrow e^x\sin x$ is the series of the imaginary part of $e^{(1+ix)}$;
$e^x\sin x$ will converge for all x

51. a) $e^{i\theta_1}e^{i\theta_2} = \left(\cos\theta_1+i\sin\theta_1\right)\left(\cos\theta_2+i\sin\theta_2\right) = \cos\theta_1\cos\theta_2-\sin\theta_1\sin\theta_2+$
$i\left(\sin\theta_1\cos\theta_2+\sin\theta_2\cos\theta_1\right) = \cos\left(\theta_1+\theta_2\right)+i\sin\left(\theta_1+\theta_2\right) = e^{i(\theta_1+\theta_2)}$
b) $e^{-i\theta} = \cos(-\theta)+i\sin(-\theta) = \cos\theta-i\sin\theta = (\cos\theta-i\sin\theta)\left(\frac{\cos\theta+i\sin\theta}{\cos\theta+i\sin\theta}\right) =$
$\frac{1}{\cos\theta+i\sin\theta} = \frac{1}{e^{i\theta}}$

53. If $f(x) = \sum_{n=0}^{\infty} a_n x^n$, then $f^{(k)}(x) = \sum_{n=k}^{\infty} n(n-)(n-2)\cdots(n)(k-1)\,a_k\,x^{n-k}$ and $f^{(k)}(0) = k!\,a_k \Rightarrow$
$a_k = \frac{f^{(k)}(0)}{k!}$ for k a nonnegative integer. ∴ the coefficients of f(x) are identical with the corresponding coefficients in the Maclaurin series of f(x) and the statement follow.

55. a) Suppose f(x) is a continuous periodic function, with period p, which is not bounded. Then for all $x\ \varepsilon\ [x_o, x_o+p]$ we have a maximum, M, for f(x). Since f(x) is not bounded there exists a $x_1\ \varepsilon\ [x_o+p, x_o+2p]$ such that $f\left(x_1\right) > M$, a contradiction to the periodicity of f(x). ∴ every continuous periodic function is bounded.
b) The doninate term in a nth order Taylor polynomial generated by cos x is $\frac{\sin(a)}{n!}(x-a)^n$ or $\frac{\cos(a)}{n!}(x-a)^n$. In both cases as |x| increases the absolute value of these dominate terms → ∞ causing the graph of $P_n(x)$ to move away from cos x.

8.8 CALCULATIONS WITH TAYLOR SERIES

1. $f(x) = \cos x \Rightarrow f(\pi/3) = 0.5,\ f'(\pi/3) = -\frac{\sqrt{3}}{2},\ f''(\pi/3) = -0.5,\ f'''(\pi/3) = \frac{\sqrt{3}}{2},$
$f^{(4)}(\pi/3) = 0.5;\ \cos x = \frac{1}{2} - \frac{\sqrt{3}}{2}(x - \pi/3) - \frac{1}{4}(x - \pi/3)^2 + \frac{\sqrt{3}}{12}(x - \pi/3)^3 + \ldots$

3. $e^x = 1 + x + \frac{x^2}{2!} + \frac{x^3}{3!} + \ldots$

5. $f(x) = \cos x \Rightarrow f(22\pi) = 1,\ f'(22\pi) = 0,\ f''(22\pi) = -1,\ f'''(22\pi) = 0, f^{(4)}(22\pi) = 1,\ f^{(5)}(22\pi) = 0,$
$f^{(6)}(22\pi) = -1;\ \cos x = 1 - \frac{1}{2}(x - 22\pi)^2 + \frac{1}{4!}(x - 22\pi)^4 - \frac{1}{6!}(x - 22\pi)^6 + \ldots$

7. $\int_0^{0.2} \sin x^2\, dx = \int_0^{0.2} \left(x^2 - \frac{x^6}{3!} + \frac{x^{10}}{5!} - \cdots\right) dx = \left[\frac{x^3}{3} - \frac{x^7}{7 \cdot 3!} + \cdots\right]_0^{0.2} \approx 0.0027$ with
error $|E| \le \frac{(.2)^7}{7 \cdot 3!} \approx 0.0000003$

9. $\int_0^{0.1} x^2 e^{-x^2} dx = \int_0^{0.1} x^2\left(1 - x^2 + \frac{x^4}{2!} - \frac{x^6}{3!} + \cdots\right) dx = \int_0^{0.1} \left(x^2 - x^4 + \frac{x^6}{2!} - \cdots\right) dx =$
$\left[\frac{x^3}{3} - \frac{x^5}{5} + \frac{x^7}{7 \cdot 2!} - \cdots\right]_0^{0.1} \approx 0.00033$ with error $|E| \le \frac{(0.1)^5}{5} \approx 0.000002$

11. $\int_0^{0.4} \frac{1 - e^{-x}}{x} dx = \int_0^{0.4} \frac{1}{x}\left(1 - \left(1 - x + \frac{x^2}{2} - \frac{x^3}{3!} + \cdots\right)\right) dx = \int_0^{0.4} \left(1 - \frac{x}{2!} + \frac{x^2}{3!} - \frac{x^3}{4!} + \cdots\right) dx =$
$\left[1 - \frac{x^2}{2 \cdot 2!} + \frac{x^3}{3 \cdot 3!} - \frac{x^4}{4 \cdot 4!} + \cdots\right]_0^{0.4} \approx 0.96356$ with error $|E| \le \frac{(0.4)^4}{4 \cdot 4!} \approx 0.000266$

13. $\int_0^{0.1} \frac{1}{\sqrt{1 + x^4}} dx = \int_0^{0.1} \left(1 - \frac{x^4}{2} + \frac{3x^8}{8} - \cdots\right) dx = \left[x - \frac{x^5}{10} + \frac{x^9}{24} - \cdots\right]_0^{0.1} \approx 0.1$ with error
$|E| \le \frac{(0.1)^5}{10} = 0.000001$

15. $\int_0^{0.1} \frac{\sin x}{x} dx = \int_0^{0.1} 1 - \frac{x^2}{3!} + \frac{x^4}{5!} - \frac{x^6}{7!} + \ldots dx = \left[x - \frac{x^3}{3 \cdot 3!} + \frac{x^5}{5 \cdot 5!} - \frac{x^7}{7 \cdot 7!} + \ldots\right]_0^{0.1} \approx$
$\left[x - \frac{x^3}{3 \cdot 3!} + \frac{x^5}{5 \cdot 5!}\right]_0^{0.1} \approx 0.099944461$

17. $\left(1 + x^4\right)^{1/2} = (1)^{1/2} + \frac{1/2}{1}(1)^{-1/2}\left(x^4\right) + \frac{(1/2)(-1/2)}{2!}(1)^{-3/2}\left(x^4\right)^2 + \frac{(1/2)(-1/2)(-3/2)}{3!}(1)^{-5/2}\left(x^4\right)^3 +$
$\frac{(1/2)(-1/2)(-3/2)(-5/2)}{4!}(1)^{-7/2}\left(x^4\right)^4 + \ldots = 1 + \frac{x^4}{2} - \frac{x^8}{8} + \frac{x^{12}}{16} - \frac{5x^{16}}{128} + \ldots;$
$\int_0^{0.1} 1 + \frac{x^4}{2} - \frac{x^8}{8} + \frac{x^{12}}{16} - \frac{5x^{16}}{128} + \ldots dx = \left[x + \frac{x^5}{10} - \frac{x^9}{72} + \frac{x^{13}}{208} - \frac{5x^{17}}{2176} + \ldots\right]_0^{0.1} \approx 0.100001$

19. $\ln\left(\frac{1+x}{1-x}\right) = \ln(1+x) - \ln(1-x) = \left(x - \frac{x^2}{2} + \frac{x^3}{3} - \frac{x^4}{4} + \ldots\right) - \left(-x - \frac{x^2}{2} - \frac{x^3}{3} - \frac{x^4}{4} - \ldots\right) =$
$2\left(x + \frac{x^3}{3} + \frac{x^5}{5} + \ldots\right)$

21. $\tan^{-1}x = x - \frac{x^3}{3} + \frac{x^5}{5} - \frac{x^7}{7} + \frac{x^9}{9} - \ldots + \frac{(-1)^{n-1}x^{2n-1}}{2n-1} + \ldots$ and the $|\text{error}| =$
$\left|\frac{(-1)^{n-1}x^{2n-1}}{2n-1}\right| = \frac{1}{2n-1}$, when $x = 1$; $\frac{1}{2n-1} < \frac{1}{10^3} \Rightarrow n > \frac{1001}{2} = 500.5 \Rightarrow$ the first term not used is $501^{st} \Rightarrow$ we must use 500 terms

23. $\tan^{-1}x = x - \frac{x^3}{3} + \frac{x^5}{5} - \frac{x^7}{7} + \frac{x^9}{9} - \ldots + \frac{(-1)^{n-1}x^{2n-1}}{2n-1} + \ldots$; when the series representing $48\tan^{-1}\left(\frac{1}{18}\right)$ has an error of magnitude less than 10^{-6}, then the series representing $48\tan^{-1}\left(\frac{1}{18}\right) + 32\tan^{-1}\left(\frac{1}{57}\right) - 20\tan^{-1}\left(\frac{1}{239}\right)$ will also have an error of magnitude less than 10^{-6}; $\frac{\left(\frac{1}{18}\right)^{2n-1}}{2n-1} < \frac{1}{10^6} \Rightarrow n \geq 3 \Rightarrow$ 3 terms

25. a) $\left(1 - x^2\right)^{-1/2} \approx 1 + \frac{x^2}{2} + \frac{3x^4}{8} + \frac{5x^6}{16} \Rightarrow \sin^{-1}x \approx x + \frac{x^3}{6} + \frac{3x^5}{40} + \frac{5x^7}{112}$;
$\lim_{n \to \infty} \left|\frac{1\cdot3\cdot5\cdots(2n-1)(2n+1)\,x^{2n+3}}{2\cdot4\cdot6\cdots(2n)(2n+2)(2n+3)} \cdot \frac{2\cdot4\cdot6\cdots(2n)(2n+1)}{1\cdot3\cdot5\cdots(2n-1)\,x^{2n+1}}\right| < 1 \Rightarrow$
$x^2 \lim_{n \to \infty} \left|\frac{(2n+1)(2n+1)}{(2n+2)(2n+3)}\right| < 1 \Rightarrow |x| < 1 \Rightarrow$ the radius of convergence is 1

b) since $\frac{d}{dx}\left(\cos^{-1}x\right) = -\left(1 - x^2\right)^{-1/2} \Rightarrow \cos^{-1}x = \frac{\pi}{2} - \sin^{-1}x \approx \frac{\pi}{2} - \left(x + \frac{x^3}{6} + \frac{3x^5}{40} + \frac{5x^7}{112}\right) \approx$
$\frac{\pi}{2} - x - \frac{x^3}{6} - \frac{3x^5}{40} - \frac{5x^7}{112}$

27. $\frac{d}{dx}\left(\frac{-1}{1+x}\right) = \frac{d}{dx}\left(\frac{-1}{1-(-x)}\right) = \frac{d}{dx}\left((-1) + (-1)(-x) + (-1)(-x)^2 + (-1)(-x)^3 + (-1)(-x)^4 + \cdots\right) =$
$1 - 2x + 3x^2 - 4x^3 + \cdots = \frac{1}{(1+x)^2}$

29. $\ln(\sec x) = \int_0^1 \tan t\,dt = \int_0^x \left(t + \frac{t^3}{3} + \frac{2t^5}{15} + \cdots\right)dt \approx \frac{x^2}{2} + \frac{x^4}{12} + \frac{x^6}{45}$

8.M MISCELLANEOUS EXERCISES

1. $\lim_{n\to\infty} a_n = \lim_{n\to\infty}\left(1+\frac{(-1)^n}{n}\right) = 1$, converges to 1

3. $a_1 = \cos\left(\frac{\pi}{2}\right) = 0$, $a_2 = \cos(\pi) = -1$, $a_3 = \cos\left(\frac{3\pi}{2}\right) = 0$, $a_4 = \cos(2\pi) = 1, \ldots \Rightarrow$ the sequence diverges

5. $\lim_{n\to\infty} a_n = \lim_{n\to\infty} \frac{n}{\ln n^2} = \lim_{n\to\infty} \frac{1}{2/n} = \infty$, diverges

7. $\lim_{n\to\infty} a_n = \lim_{n\to\infty} \sqrt[2n]{\frac{3^n}{n}} = \sqrt{\lim_{n\to\infty}\left(\frac{3^n}{n}\right)^{1/n}} = \sqrt{\lim_{n\to\infty}\frac{3}{\sqrt[n]{n}}} = \sqrt{3}$, converges

9. $\lim_{n\to\infty} a_n = \lim_{n\to\infty} \frac{(-4)^n}{n!} = 0$, by table 9.1, converges to 0

11. $\lim_{n\to\infty} a_n = \lim_{n\to\infty} n\left(2^{1/n}-1\right) = \lim_{n\to\infty}\frac{2^{1/n}-1}{1/n} = \lim_{n\to\infty}\frac{-\frac{\ln 2}{n^2}\cdot 2^{1/n}}{-\frac{1}{n^2}} = \lim_{n\to\infty}(\ln 2)\,2^{1/n} = \ln 2$, converges

13. $\lim_{n\to\infty} a_n = \lim_{n\to\infty}\frac{1}{\left(1+\frac{1}{n}\right)^n} = \frac{1}{e}$, converges

15. $\sum_{n=1}^{\infty}\ln\left(\frac{n}{n+1}\right) = \sum_{n=1}^{\infty}\ln(n)-\ln(n+1) \Rightarrow s_n = \left(\ln(1)-\ln(2)\right)+\left(\ln(2)-\ln(3)\right)+\left(\ln(3)-\ln(4)\right)+\ldots+\left(\ln(n-1)-\ln(n)\right)+\left(\ln(n)-\ln(n+1)\right) = \ln(1)-\ln(n+1) = -\ln(n+1) \Rightarrow \lim_{n\to\infty} s_n = -\infty$, $\Rightarrow$ the series diverges

17. $\sum_{n=0}^{\infty} e^{-n} = \sum_{n=0}^{\infty}\left(\frac{1}{e}\right)^n = 1+\frac{1}{e}+\left(\frac{1}{e}\right)^2+\left(\frac{1}{e}\right)^3+\ldots = \frac{1}{1-\frac{1}{e}} = \frac{e}{e-1}$, a convergent geometric series

19. diverges, a p–series where $p = \frac{1}{2}$

21. $f(x) = \frac{1}{x^{1/2}} \Rightarrow f'(x) = -\frac{1}{2x^{1/2}} < 0 \Rightarrow$ f(x) is decreasing, $\lim_{n\to\infty} a_n = \lim_{n\to\infty}\frac{(-1)^n}{\sqrt{n}} = 0$ and $a_{n+1} < a_n \Rightarrow$

$\sum_{n=1}^{\infty}\frac{(-1)^n}{\sqrt{n}}$ converges, by the Alternating Series Theorem; the series $\sum_{n=1}^{\infty}\frac{1}{\sqrt{n}}$ diverges $\Rightarrow$

the given series converges conditionally

23. the given series does not converge absolutlely, since $\frac{1}{\ln(n+1)} > \frac{1}{n+1}$ and $\sum_{n=1}^{\infty} \frac{1}{n+1}$ diverges;

$f(x) = \frac{1}{\ln(x+1)} \Rightarrow f'(x) = -\frac{1}{\left(\ln(x+1)\right)^2 (x+1)} < 0 \Rightarrow$ f(x) is decreasing, $\lim_{n \to \infty} a_n =$

$\lim_{n \to \infty} \frac{1}{\ln(n+1)} = 0$ and $a_{n+1} < a_n \Rightarrow$ the given series converges conditionally, by

the Alternating Series Test

25. $\lim_{n \to \infty} \frac{\frac{1}{n\sqrt{n^2+1}}}{\frac{1}{n^2}} = \sqrt{\lim_{n \to \infty} \frac{n^2}{n^2+1}} = \sqrt{1} = 1 \Rightarrow$ converges absolutely, by the Limit

Comparison Test

27. converges, by the Ratio Test, since $\lim_{n \to \infty} \left| \frac{n+2}{(n+1)!} \frac{n!}{n+1} \right| = \lim_{n \to \infty} \frac{n+2}{(n+1)^2} = 0$

29. converges absolutely, by the Ratio Test, since $\lim_{n \to \infty} \left| \frac{3^{n+1}}{(n+1)!} \frac{n!}{3^n} \right| = \lim_{n \to \infty} \left| \frac{3}{n+1} \right| = 0$

31. converges, since $\lim_{n \to \infty} \frac{\frac{1}{n^{3/2}}}{\frac{1}{\sqrt{n(n+1)(n+2)}}} = \sqrt{\lim_{n \to \infty} \frac{n(n+1)(n+2)}{n^3}} = 1$, by the Limit

Comparison Test

33. converges, since $\frac{1}{(3n-2)^{(2n+1)/2}} < \frac{1}{(3n-2)^{3/2}}$ and $\sum_{n=1}^{\infty} \frac{1}{(3n-2)^{3/2}}$ converges by the Limit

Comparison Test for $\lim_{n \to \infty} \frac{\frac{1}{n^{3/2}}}{\frac{1}{(3n-2)^{3/2}}} = \lim_{n \to \infty} \left(\frac{3n-2}{n}\right)^{3/2} = 3^{3/2}$

35. diverges, since $\lim_{n \to \infty} a_n = \lim_{n \to \infty} (-1)^n \tanh n = \lim_{b \to \infty} (-1)^n \left(\frac{1-e^{-2n}}{1+e^{-2n}}\right) = \lim_{n \to \infty} (-1)^n \neq 0$

37. converges absolutely, by the Limit Comparison Test, since $\lim_{n \to \infty} \frac{\ln\left(1+e^{-n}\right)}{e^{-n}} =$

$\lim_{n \to \infty} \frac{-e^{-n}/(1+e^{-n})}{-e^{-n}} = 1$, where $\sum_{n=1}^{\infty} \frac{1}{e^n}$ is a convergent geometric series

39. converges, by the Comparison Test, since $\frac{\ln(n!)}{n^4} < \frac{1}{n^2}$, a term of a convergence p–series, for $n! < n^n \Rightarrow \ln(n!) < n\ln(n) \Rightarrow \frac{\ln(n!)}{n^4} < \frac{\ln(n)}{n^3} < \frac{n}{n^3} = \frac{1}{n^2}$

41. $\lim_{n\to\infty}\left|\frac{(x+2)^{n+1}}{(n+1)\,3^{n+1}}\cdot\frac{n\,3^n}{(x+2)^n}\right| < 1 \Rightarrow \frac{|x+2|}{3}\lim_{n\to\infty}\left|\frac{n}{n+1}\right| < 1 \Rightarrow -5 < x < 1$; when $x = -5$ we have $\sum_{n=1}^{\infty}\frac{(-1)^n}{n}$ the alternating harmonic series and when $x = 1$ we have $\sum_{n=1}^{\infty}\frac{1}{n}$ the harmonic series $\therefore$

a) 3, [– 5,1) b) (– 5,1) c) {– 5}

43. $\lim_{n\to\infty}\left|\frac{x^{n+1}}{(n+1)^{n+1}}\cdot\frac{n^n}{x^n}\right| < 1 \Rightarrow |x|\lim_{n\to\infty}\left|\left(\frac{n}{n+1}\right)^n\left(\frac{1}{n+1}\right)\right| < 1 \Rightarrow \frac{|x|}{e}\lim_{n\to\infty}\left(\frac{1}{n+1}\right) < 1 \Rightarrow$ $|x|\cdot 0 < 1 \;\therefore$

a) ∞, for all x b) for all x c) { }

45. $\lim_{n\to\infty}\left|\frac{(x-1)^{n+1}}{(n+1)^2}\cdot\frac{n^2}{(x-1)^n}\right| < 1 \Rightarrow |x-1|\lim_{n\to\infty}\frac{n^2}{(n+1)^2} < 1 \Rightarrow 0 < x < 2$; when $x = 0$ we have $\sum_{n=1}^{\infty}\frac{(-1)^{2n-1}}{n^2}$ an absolutely convergent series, when $x = 2$ we have $\sum_{n=1}^{\infty}\frac{(-1)^{n-1}}{n^2}$ an absolutely convergent series $\therefore$

a) 1, [0,2] b) [0,2] c) { }

47. $\lim_{n\to\infty}\left|\frac{\operatorname{csch}(n+1)\,x^{n+1}}{\operatorname{csch}(n)\,x^n}\right| < 1 \Rightarrow |x|\lim_{n\to\infty}\left|\frac{e^{-1}-e^{-2n-1}}{1-e^{-2(n+1)}}\right| < 1 \Rightarrow -e < x < e$; when $x = \pm e$ both series: $\sum_{n=1}^{\infty}(-e)^n\operatorname{csch}(n)$ and $\sum_{n=1}^{\infty}e^n\operatorname{csch}(n)$ diverge $\therefore$

a) e, (– e,e) b) (– e,e) c) { }

49. $\lim_{n\to\infty}\left|\frac{(n+2)\,x^{2n+1}}{3^{n+1}}\cdot\frac{3^n}{(n+1)\,x^{2n-1}}\right| < 1 \Rightarrow \frac{x^2}{3}\lim_{n\to\infty}\left|\frac{n+2}{n+1}\right| < 1 \Rightarrow -\sqrt{3} < x < \sqrt{3}$; when $x = \pm\sqrt{3}$ both series: $\sum_{n=1}^{\infty}-\frac{n+1}{\sqrt{3}}$ and $\sum_{n=1}^{\infty}\frac{n+1}{\sqrt{3}}$ diverge $\therefore$

a) $\sqrt{3}$, $(-\sqrt{3},\sqrt{3})$ b) $(-\sqrt{3},\sqrt{3})$ c) { }

51. $\lim_{n \to \infty} \left| \frac{2 \cdot 5 \cdot 8 \cdots (3n-1)(3n+2)\, x^{n+1}}{2 \cdot 4 \cdot 6 \cdots (2n)(2n+2)} \cdot \frac{2 \cdot 4 \cdot 6 \cdots (2n)}{2 \cdot 5 \cdot 8 \cdots (3n-1)\, x^n} \right| < 1 \Rightarrow |x| \lim_{n \to \infty} \left| \frac{3n+2}{2n+2} \right| < 1 \Rightarrow$

$|x| < \frac{2}{3} \Rightarrow$ the radius of convergence is $\frac{2}{3}$

53. The given series is in the form $1 - x + x^2 - x^3 + \ldots + (-x)^n + \ldots = \frac{1}{1+x}$, where $x = \frac{1}{4}$.

The sum is $\frac{1}{1 + 1/4} = \frac{4}{5}$.

55. The given series is in the form $x - \frac{x^3}{3!} + \frac{x^5}{5!} - \ldots + (-1)^n \frac{x^{2n+1}}{(2n+1)!} + \ldots = \sin x$, where $x = \pi$.

The sum is $\sin \pi = 0$.

57. The given series is in the form $1 + x + \frac{x^2}{2!} + \frac{x^3}{3!} + \ldots + \frac{x^n}{n!} + \ldots = e^x$, where $x = \ln 2$.

The sum is $e^{\ln(2)} = 2$.

59. $f(x) = \sqrt{3 + x^2} = \left(3 + x^2\right)^{1/2} \Rightarrow f(-1) = 2,\ f'(-1) = -\frac{1}{2},\ f''(-1) = \frac{3}{8},\ f'''(-1) = \frac{9}{32}$;

$\sqrt{3 + x^2} = 2 - \frac{(x+1)}{2 \cdot 1!} + \frac{3(x+1)^2}{2^3 \cdot 2!} + \frac{9(x+1)^3}{2^5 \cdot 3!} + \ldots$

61. Yes, 0.876726215; if the sequence of terms generated by $x_{n+1} = \sqrt{\sin(x_n)}$ convergent to L,

then $L = \sqrt{\sin(L)} \Rightarrow L^2 = \sin L$

63. $\sum_{k=2}^{n} \ln\left(1 - \frac{1}{k^2}\right) = \sum_{k=2}^{n} \left[\ln\left(1 + \frac{1}{k}\right) + \ln\left(1 - \frac{1}{k}\right)\right] = \sum_{k=2}^{n} \left[\ln(k+1) - \ln(k) + \ln(k-1) - \ln(k)\right] =$

$\ln\left(\frac{n+1}{2n}\right)$. $\therefore \lim_{n \to \infty} \ln\left(\frac{n+1}{2n}\right) = \ln\left(\frac{1}{2}\right)$

65. $\sum_{n=1}^{\infty} \left(x_{n+1} - x_n\right) = \lim_{n \to \infty} \sum_{k=1}^{n} \left(x_{k+1} - x_k\right) = \lim_{n \to \infty} \left(x_{n+1} - x_1\right) = \lim_{n \to \infty} \left(x_{n+1}\right) - x_1 \ \therefore$ both the series and sequence must either converge or divege.

67. a) Each A_{n+1} fits into the corresponding upper triangular region, whose vertices are:

(n,f(n) − f(n + 1)), (n + 1,f(n + 1)) and (n,f(n)) along the line whose slope is f(n + 2) − f(n + 1).
All the A_n's fit into the first upper triangular region whose area is $\frac{f(1) - f(2)}{2}$.

$\therefore \sum_{n=1}^{\infty} A_n < \frac{f(1) - f(2)}{2}$.

b) If $A_k = \frac{f(k+1) - f(k)}{2} - \int_k^{k+1} f(x)\,dx$, then $\sum_{k=1}^{n-1} A_k =$

$$\frac{f(1) + f(2) + f(2) + f(32) + f(3) + \cdots + f(n-1) + f(n)}{2} - \int_1^2 f(x)\,dx - \int_2^3 f(x)\,dx - \cdots -$$

$$\int_{n-1}^{n} f(x)\,dx = \frac{f(1) + f(n)}{2} + \sum_{k=2}^{n-1} f(k) - \int_1^n f(x)\,dx \Rightarrow \sum_{k=1}^{n-1} A_k = \sum_{k=1}^{n} f(k) - \frac{f(1) + f(n)}{2} -$$

$\int_1^n f(x)\,dx < \frac{f(1) - f(2)}{2}$, from part a). The sequence $\left\{\sum_{k=1}^{n-1} A_k\right\}$ is bounded and increasing and therefore it converges and the limit in question must exist.

c) From part b) we have $\sum_{k=1}^{n} f(k) - \int_1^n f(x)\,dx < f(1) - \frac{f(2)}{2} + \frac{f(n)}{2}$ and

$$\lim_{n \to \infty} \left[\sum_{k=1}^{n} f(k) - \int_1^n f(x)\,dx\right] < \lim_{n \to \infty} \left[f(1) - \frac{f(2)}{2} + \frac{f(n)}{2}\right] = f(1) - \frac{f(2)}{2}.$$ The sequence

$\left\{\sum_{k=1}^{n} f(k) - \int_1^n f(x)\,dx <\right\}$ is bounded and incresing and therefore it converges and the limit in question must exist.

69. $x_{n+1} = x_n - \frac{(x_n - 1)^{40}}{40(x_n - 1)^{39}} = \frac{39}{40} x_n + \frac{1}{40}$. If the sequence $\{x_n\}$ has a limit say L, then $L = \frac{39}{40} L + \frac{1}{40} \Rightarrow$

$L = 1$.

71. If $\frac{u_{n+1}}{u_n} = \frac{4n^2 - 4n + 1}{4n^2 + 2n} = 1 - \frac{3/2}{n} + \frac{2n/(2n+1)}{n^2}$, then $C = -\frac{3}{2} < 1$ and $\sum_{n=1}^{\infty} u_n$ diverges.

73. $\sum_{n=3}^{\infty} \frac{1}{n \ln n (\ln(\ln n))^p}$ converges if $p > 1$ and diverges otherwise by the Integral Test for,

when $p = 1$ we have $\lim_{b \to \infty} \int_3^b \frac{1/x}{\ln x}\,dx = \lim_{b \to \infty} \left[\ln(\ln x)\right]_3^b = \infty$, when $p \neq 1$ we have

$$\lim_{b \to \infty} \int_3^b (\ln(\ln x))^{-p} \frac{1/x}{\ln x}\,dx = \lim_{b \to \infty} \left[\frac{(\ln(\ln x))^{1-p}}{1-p}\right]_3^b = \begin{cases} \frac{\ln(\ln 3)^{1-p}}{p-1}, & p > 1 \\ \infty, & p < 1 \end{cases}$$

75. a) If $\sum_{n=1}^{\infty} a_n = L$, then $a_n^2 \le a_n \sum_{n=1}^{\infty} a_n = a_n L \Rightarrow \sum_{n=1}^{\infty} a_n^2 \le L \sum_{n=1}^{\infty} a_n \quad \therefore \sum_{n=1}^{\infty} a_n^2$ converges, by the Comparison Test.

b) converges, by the Limit Comparison Test, since $\lim_{n \to \infty} \dfrac{\frac{a_n}{1-a_n}}{a_n} = \lim_{n \to \infty} \dfrac{1}{1-a_n} = 1$ and $\sum_{n=1}^{\infty} a_n$ converges

77. a) $\lim_{n \to \infty} \left| \dfrac{1 \cdot 4 \cdot 7 \cdots (3n-2)(3n+1)\, x^{3n+3}}{(3n+3)!} \cdot \dfrac{(3n)!}{1 \cdot 4 \cdot 7 \cdots (3n-2)\, x^{3n}} \right| < 1 \Rightarrow \left|x^3\right| \cdot 0 < 1 \Rightarrow$ the radius of convergence is ∞

b) $y = 1 + \sum_{n=1}^{\infty} \dfrac{1 \cdot 4 \cdot 7 \cdots (3n-2)}{(3n)!} x^{3n} \Rightarrow \dfrac{dy}{dx} = \sum_{n=1}^{\infty} \dfrac{1 \cdot 4 \cdot 7 \cdots (3n-2)}{(3n-1)!} x^{3n-1} \Rightarrow$

$$\frac{d^2y}{dx^2} = \sum_{n=1}^{\infty} \frac{1 \cdot 4 \cdot 7 \cdots (3n-2)}{(3n-2)!} x^{3n-2} = x + \sum_{n=2}^{\infty} \frac{1 \cdot 4 \cdot 7 \cdots (3n-5)}{(3n-3)!} x^{3(n-1)} =$$

$$x\left(1 + \sum_{n=1}^{\infty} \frac{1 \cdot 4 \cdot 7 \cdots (3n-2)}{(3n)!} x^{3n}\right) = xy + 0 \Rightarrow a = 1 \text{ and } b = 0$$

79. a) $y = 1 - \left(1 - \dfrac{x^2}{2!} + \dfrac{x^4}{4!} - \dfrac{x^6}{6!} + \cdots\right) = \dfrac{x^2}{2!} - \dfrac{x^4}{4!} + \dfrac{x^6}{6!} - \cdots$ and $\ln(1-y) = -y - \dfrac{y^2}{2} - \dfrac{y^3}{3} - \dfrac{y^4}{4} - \cdots \Rightarrow$

$$\ln(1 - \cos x) \approx -\left(\frac{x^2}{2!} - \frac{x^4}{4!} + \frac{x^6}{6!}\right) - \frac{\left(\frac{x^2}{2!} - \frac{x^4}{4!}\right)^2}{2} - \frac{\left(\frac{x^2}{2!}\right)^3}{3} = -\frac{x^2}{2} - \frac{x^4}{12} - \frac{x^6}{45}$$

b) $f(x) = \ln(\cos x) \Rightarrow f'(x) = -\tan x \Rightarrow f''(x) = -\sec^2 x \Rightarrow f'''(x) = -2\sec^2 x \tan x \Rightarrow$ $f^{(4)}(x) = -2\sec^2 x - 6\sec^2 x \tan^2 x \Rightarrow f^{(5)}(x) = -16 \tan x \sec^2 x - 24 \tan^3 x \sec^2 x \Rightarrow$ $f^{(6)}(x) = -16\sec^2 x - 120 \tan^2 x \sec^2 x - 120 \tan^4 x \sec^2 x \Rightarrow \ln(\cos x) \approx 0 + 0x - \dfrac{1}{2!}x^2 - \dfrac{2}{4!}x^4 - \dfrac{16}{6!}x^6 = -\dfrac{x^2}{2} - \dfrac{x^4}{12} - \dfrac{x^6}{45}$

81. If $f(x) = (1-x)^{-1} \Rightarrow f'(x) = (1-x)^{-2} \Rightarrow f''(x) = 2(1-x)^{-3} \cdots f^{(n)}(x) = n!(1-x)^{-(n+1)}$, then $\dfrac{1}{1-x} = \sum_{n=0}^{\infty} (-1)^{n+1}(x-2)^n$. $\lim_{n \to \infty} \left| \dfrac{(x-2)^{n+1}}{(x-2)^n} \right| < 1 \Rightarrow 1 < x < 3 \Rightarrow 1$ is the radius of convergence and the interval of convergence is (1,3).

83. If $f(x) = \cos x \Rightarrow f'(x) = -\sin x \Rightarrow f''(x) = -\cos x \Rightarrow f^{(4)}(x) = \sin x$, then $\cos x =$

$$\sum_{n=0}^{\infty} \frac{(-1)^{n+1}}{(2n)!}(x-\pi)^{2n}.\ \lim_{n \to \infty} \left| \frac{(x-\pi)^{2n+2}}{(2n+2)!} \cdot \frac{(2n)!}{(x-\pi)^{2n}} \right| < 1 \Rightarrow (x-\pi)^2 \cdot 0 < 1 \Rightarrow \text{the radius of}$$

convergence is ∞ and the interval of convergence is all x.

85. The $\{x_n\} \to \frac{\pi}{2}$ from below $\Rightarrow \{\varepsilon_n\} = \left\{\frac{\pi}{2} - x_n\right\} > 0$. $\varepsilon_{n+1} - \frac{1}{6}\varepsilon_n^3 < 0$, by the Alternating Series Estimation Theorem. $\therefore 0 < \varepsilon_{n+1} < \frac{1}{6}\varepsilon_n^3$

87. $$\int_0^1 x \sin(x^3)\, dx \approx \int_0^1 x\left(x^3 - \frac{x^9}{3!} + \frac{x^{15}}{5!} - \frac{x^{21}}{7!} + \frac{x^{27}}{9!}\right) dx = \int_0^1 \left(x^4 - \frac{x^{10}}{3!} + \frac{x^{16}}{5!} - \frac{x^{22}}{7!} + \frac{x^{28}}{9!}\right) dx =$$

$$\left[\frac{x^5}{5} - \frac{x^{11}}{11 \cdot 3!} + \frac{x^{17}}{17 \cdot 5!} - \frac{x^{23}}{23 \cdot 7!} + \frac{x^{29}}{29 \cdot 9!}\right]_0^1 = 0.185330149$$

89. $$\left[\tan^{-1} t\right]_x^{\infty} = \int_x^{\infty} \frac{1/t^2}{1 - \left(1/t^2\right)} dt = \lim_{b \to \infty} \int_x^b \left(\frac{1}{t^2} - \frac{1}{t^4} + \frac{1}{t^6} - \frac{1}{t^8} + \cdots\right) dt =$$

$$\lim_{b \to \infty} \left[-\frac{1}{t} + \frac{1}{3t^3} - \frac{1}{5t^5} + \cdots\right]_x^b = \lim_{b \to \infty} \left[\left(-\frac{1}{b} + \frac{1}{3b^3} - \frac{1}{5b^5} + \cdots\right) - \left(-\frac{1}{x} + \frac{1}{3x^3} - \frac{1}{5x^5} + \cdots\right)\right] =$$

$$\frac{\pi}{2} - \tan^{-1} x = \frac{1}{x} - \frac{1}{3x^3} + \frac{1}{5x^5} - \cdots \Rightarrow \tan^{-1} x = \frac{\pi}{2} - \frac{1}{x} + \frac{1}{3x^3} - \frac{1}{5x^5} + \cdots;$$

$$\left[\tan^{-1} t\right]_{-\infty}^{x} = \int_{-\infty}^{x} \frac{1/t^2}{1 - \left(1/t^2\right)} dt = \lim_{b \to -\infty} \int_b^x \left(\frac{1}{t^2} - \frac{1}{t^4} + \frac{1}{t^6} - \frac{1}{t^8} + \cdots\right) dt =$$

$$\lim_{b \to -\infty} \left[-\frac{1}{t} + \frac{1}{3t^3} - \frac{1}{5t^5} + \cdots\right]_b^x = \lim_{b \to -\infty} \left[\left(-\frac{1}{x} + \frac{1}{3x^3} - \frac{1}{5x^5} + \cdots\right) - \left(-\frac{1}{b} + \frac{1}{3b^3} - \frac{1}{5b^5} + \cdots\right)\right] =$$

$$\tan^{-1} x + \frac{\pi}{2} = -\frac{1}{x} + \frac{1}{3x^3} - \frac{1}{5x^5} + \cdots \Rightarrow \tan^{-1} x = -\frac{\pi}{2} - \frac{1}{x} + \frac{1}{3x^3} - \frac{1}{5x^5} + \cdots$$

91. The convergence of $\sum_{n=1}^{\infty} |a_n|$ implies that $\lim_{n \to \infty} |a_n| = 0$. Let $N > 0$ be such that $|a_n| < \frac{1}{2} \Rightarrow$

$$1 - |a_n| > \frac{1}{2} \Rightarrow \frac{|a_n|}{1 - |a_n|} < 2\,|a_n| \text{ for all } n > N. \text{ Now } \left|\ln\left(1 + a_n\right)\right| = \left|a_n - \frac{a_n^2}{2} + \frac{a_n^3}{3} - \frac{a_n^4}{4} + \cdots\right| \le$$

$$|a_n| + \left|\frac{a_n^2}{2}\right| + \left|\frac{a_n^3}{3}\right| + \left|\frac{a_n^4}{4}\right| + \cdots < |a_n| + |a_n|^2 + |a_n|^3 + |a_n|^4 + \cdots = \frac{|a_n|}{1 - |a_n|} < 2\,|a_n|.$$

$$\therefore \sum_{n=N}^{\infty} \ln\left(1 + a_n\right) \le 2 \sum_{n=N}^{\infty} |a_n| \text{ a convergent series, hence the statement follows.}$$

93. a) $\frac{1}{(1-x)^2} = \frac{d}{dx}\left(\frac{1}{1-x}\right) = \frac{d}{dx}\left(1 + x + x^2 + x^3 + \cdots\right) = 1 + 2x + 3x^2 + 4x^3 + \cdots = \sum_{n=1}^{\infty} n\,x^{n-1}$

b) from part a) we have $\sum_{n=1}^{\infty} n\left(\frac{5}{6}\right)^{n-1}\left(\frac{1}{6}\right) = \left(\frac{1}{6}\right)\left(\frac{1}{(1-5/6)^2}\right) = 6$

c) from part a) we have $\sum_{n=1}^{\infty} n\,p^{n-1}q = \frac{q}{(1-p)^2} = \frac{q}{q^2} = \frac{1}{q}$

95. a) $S_{2n+1} = \frac{c_1}{1} + \frac{c_2}{2} + \frac{c_3}{3} + \cdots + \frac{c_{2n+1}}{2n+1} = \frac{t_1}{1} + \frac{t_2 - t_1}{2} + \frac{t_3 - t_2}{3} + \cdots + \frac{t_{2n+1} - t_{2n}}{2n+1} = t_1\left(1 - \frac{1}{2}\right) +$

$t_2\left(\frac{1}{2} - \frac{1}{3}\right) + \cdots + t_{2n}\left(\frac{1}{2n} - \frac{1}{2n+1}\right) + \frac{t_{2n+1}}{2n+1} = \sum_{k=1}^{2n} \frac{t_k}{k(k+1)} + \frac{t^{2n+1}}{2n+1}$.

b) If $\{c_n\} = \{(-1)^n\}$, then $\sum_{n=1}^{\infty} \frac{(-1)^n}{n}$ converges

c) If $\{c_n\} = \{1, -1, -1, 1, 1, -1, -1, 1, 1, \cdots\}$, then the series $1 - \frac{1}{2} - \frac{1}{3} + \frac{1}{4} + \frac{1}{5} - \frac{1}{6} - \frac{1}{7} + \cdots$ converges by the foregoing theorem.

97. $\lim_{x \to 0} \frac{e^x - (1+x)}{x^2} = \lim_{x \to 0} \frac{1}{x^2}\left[\left(1 + x + \frac{x^2}{2} + \frac{x^3}{3!} + \cdots\right) - 1 - x\right] = \frac{1}{2}$

99. $\lim_{\theta \to \infty} \theta \sin\left(\frac{1}{\theta}\right) = \lim_{\theta \to \infty} \theta\left[\frac{1}{\theta} - \frac{1}{3!}\left(\frac{1}{\theta^3}\right) + \cdots\right] = 1$

101. $\lim_{x \to 0} \frac{\sin x}{e^x - 1} = \lim_{x \to 0} \frac{x - \frac{x^3}{3} + \frac{x^5}{5} - \cdots}{x + \frac{x^2}{2} + \frac{x^3}{3!} + \cdots} = \lim_{x \to 0} \frac{1 - \frac{x^2}{3} + \frac{x^4}{5} - \cdots}{1 + \frac{x}{2} + \frac{x^2}{3!} + \cdots} = 1$

103. $\lim_{u \to 0} \frac{e^u - e^{-u} - 2u}{u - \sin u} = \lim_{u \to 0} \frac{\left(1 + u + \frac{u^2}{2} + \frac{u^3}{6} + \cdots\right) - \left(1 - u + \frac{u^2}{2} - \frac{u^3}{6} + \cdots\right) - 2u}{u - \left(u - \frac{u^3}{6} + \frac{u^5}{120} - \cdots\right)} = \lim_{u \to 0} \frac{x^3/3}{x^3/6} = 2$

105. $\lim_{\theta \to 0} \frac{\tan\theta - \sin\theta}{\theta^3 \cos\theta} = \lim_{\theta \to 0} \frac{\left(\theta + \frac{\theta^3}{3} + \frac{2\theta^5}{15} + \cdots\right) - \left(\theta - \frac{\theta^3}{6} + \frac{\theta^5}{120} - \cdots\right)}{\theta^3\left(1 - \frac{\theta^2}{2} + \frac{\theta^4}{24} - \cdots\right)} =$

$\lim_{\theta \to 0} \frac{\frac{\theta^3}{2} + \frac{\theta^5}{8} + \cdots}{\theta^3 - \frac{\theta^5}{2} + \cdots} = \frac{1}{2}$

107. $$\lim_{h \to 0} \frac{\ln\left(1 + h^2\right)}{1 - \cos h} = \lim_{h \to 0} \frac{h^2 - \frac{h^4}{2} + \frac{h^6}{3} - \cdots}{1 - \left(1 - \frac{h^2}{2} + \frac{h^4}{24} - \cdots\right)} = \lim_{h \to 0} \frac{1 - \frac{h^2}{2} + \frac{h^4}{3} - \cdots}{\frac{1}{2} - \frac{h^2}{24} + \cdots} = 2$$

109. First we must write $\ln x$ as a series in terms of $(x - 1)$ i.e. $\ln x = 0 + (x - 1) - \frac{1}{2}(x - 1)^2 + \cdots$.

$$\therefore \lim_{x \to 1} \frac{\ln x}{x - 1} = \lim_{x \to 1} \frac{(x - 1) - \frac{1}{2}(x - 1)^2 + \cdots}{x - 1} = \lim_{x \to 1} \left[1 - \frac{1}{2}(x - 1) + \cdots\right] = 1$$

111. $$\lim_{x \to 0} \left(\frac{1}{\sin x} - \frac{1}{x}\right) = \lim_{x \to 0} \frac{x - \sin x}{x \sin x} = \lim_{x \to 0} \frac{x - \left(x - \frac{x^3}{3!} + \frac{x^5}{5!} - \cdots\right)}{x \cdot \left(x - \frac{x^3}{3!} + \frac{x^5}{5!} - \cdots\right)} = \lim_{x \to 0} x \cdot \frac{\frac{1}{3!} - \frac{x^2}{5!} + \cdots}{1 - \frac{x^2}{3!} + \cdots} = 0$$

113. $$\lim_{x \to 0} \frac{\ln(1 - x) - \sin x}{1 - \cos^2 x} = \lim_{x \to 0} \frac{\left(-x - \frac{x^2}{2} - \frac{x^3}{3} - \cdots\right) - \left(x - \frac{x^3}{6} + \frac{x^5}{120} - \cdots\right)}{1 - \left(1 - x^2 + \frac{x^4}{3} - \cdots\right)} =$$

$$\lim_{x \to 0} \frac{-2x - \frac{x^2}{2} - \frac{x^3}{6} - \cdots}{x^2 - \frac{x^4}{3} + \frac{2x^6}{45} - \cdots} = \lim_{x \to 0} \frac{-2 - \frac{x}{2} - \frac{x^2}{6} - \cdots}{x - \frac{x^3}{3} + \frac{2x^5}{45} - \cdots} = \infty$$

CHAPTER 9

CONIC SECTIONS, PARAMETRIZED CURVES, AND POLAR COORDINATES

SECTION 9.1 EQUATIONS FOR CONIC SECTIONS

1. $x = \frac{y^2}{8} \Rightarrow 4p = 8 \Rightarrow p = 2.$ $\therefore$ Focus is (2,0), directrix is $x = -2$.

3. $y = -\frac{x^2}{6} \Rightarrow 4p = 6 \Rightarrow p = \frac{3}{2}.$ $\therefore$ Focus is $(0, -\frac{3}{2})$, directrix is $y = \frac{3}{2}$.

5. $\frac{x^2}{4} - \frac{y^2}{9} = 1 \Rightarrow c = \sqrt{4+9} = \sqrt{13} \Rightarrow$ Foci are $(\pm\sqrt{13}, 0)$. $e = \frac{c}{a} = \frac{\sqrt{13}}{2} \Rightarrow \frac{a}{e} = \frac{2}{\frac{\sqrt{13}}{2}} = \frac{4}{\sqrt{13}} \Rightarrow$

   Directrices are $x = \pm\frac{4}{\sqrt{13}}$. Asymptotes are $y = \pm\frac{3}{2}x$.

7. $\frac{x^2}{2} + y^2 = 1 \Rightarrow c = \sqrt{2-1} = 1 \Rightarrow$ Foci are $(\pm1, 0)$. $e = \frac{c}{a} = \frac{1}{\sqrt{2}} \Rightarrow$ Directrices are $x = \pm\frac{\sqrt{2}}{\frac{1}{\sqrt{2}}} = \pm 2$.

9. $16x^2 + 25y^2 = 400 \Rightarrow \frac{x^2}{25} + \frac{y^2}{16} = 1$

   $\Rightarrow c = \sqrt{a^2 - b^2} = \sqrt{25 - 16} = 3.$

   $e = \frac{c}{a} = \frac{3}{5}$

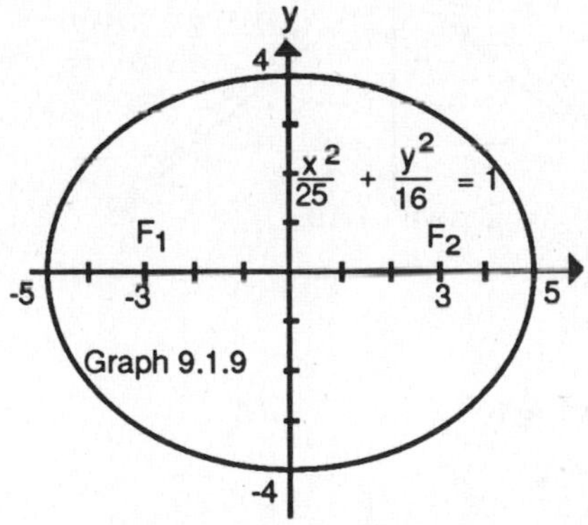

Graph 9.1.9

11. $2x^2 + y^2 = 2 \Rightarrow x^2 + \frac{y^2}{2} = 1$

   $\Rightarrow c = \sqrt{a^2 - b^2} = \sqrt{2-1} = 1$

   $e = \frac{c}{a} = \frac{1}{\sqrt{2}}$

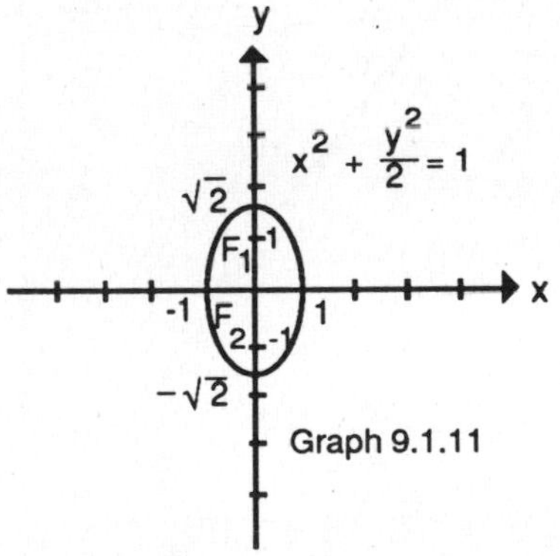

Graph 9.1.11

13. $3x^2 + 2y^2 = 6 \Rightarrow \frac{x^2}{2} + \frac{y^2}{3} = 1$

   $\Rightarrow c = \sqrt{a^2 - b^2} = \sqrt{3-2} = 1$

   $e = \frac{c}{a} = \frac{1}{\sqrt{3}}$

   (Graph on next page.)

15. $6x^2 + 9y^2 = 54 \Rightarrow \frac{x^2}{9} + \frac{y^2}{6} = 1$

   $c = \sqrt{a^2 - b^2} = \sqrt{9-6} = \sqrt{3}$

   $e = \frac{c}{a} = \frac{\sqrt{3}}{3}$

   (Graph on next page.)

13.

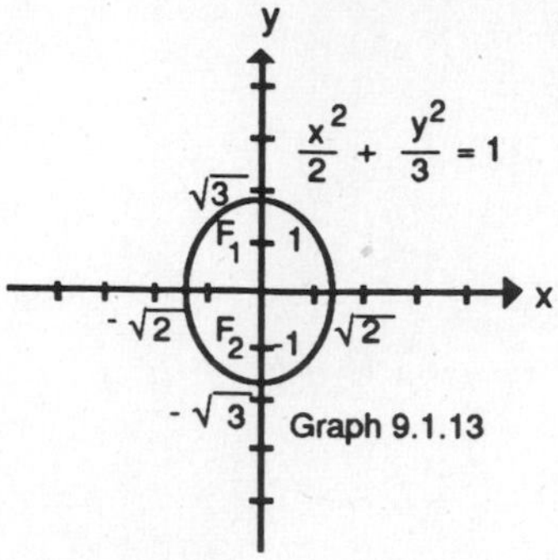

Graph 9.1.13

15.

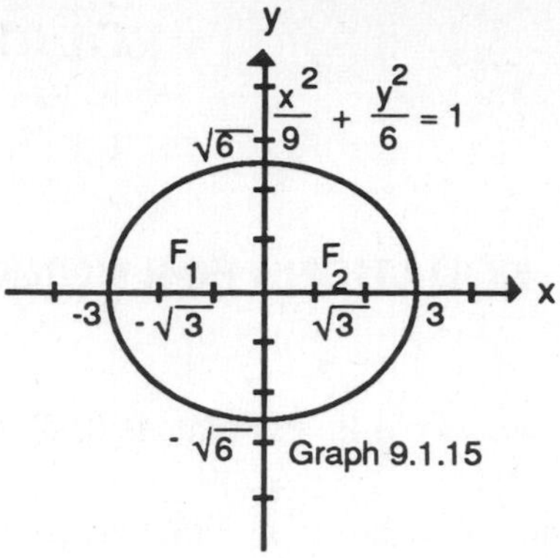

Graph 9.1.15

17. $x^2 - y^2 = 1 \Rightarrow c = \sqrt{a^2 + b^2} = \sqrt{1 + 1} = \sqrt{2}$

$e = \frac{c}{a} = \frac{\sqrt{2}}{1} = \sqrt{2}$

Asymptotes are $y = \pm x$

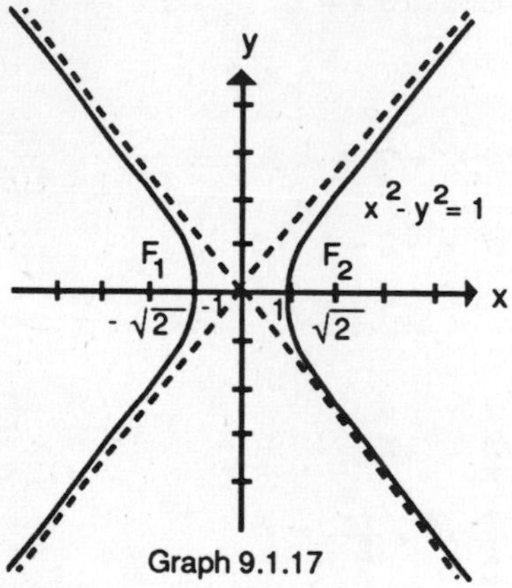

Graph 9.1.17

19. $y^2 - x^2 = 8 \Rightarrow \frac{y^2}{8} - \frac{x^2}{8} = 1$

$\Rightarrow c = \sqrt{a^2 + b^2} = \sqrt{8 + 8} = 4$

$e = \frac{c}{a} = \frac{4}{\sqrt{8}} = \sqrt{2}$ Asymptotes are $y = \pm x$.

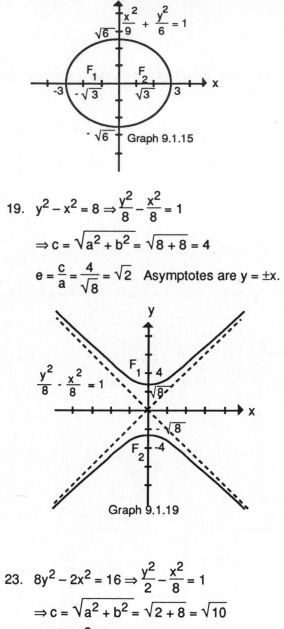

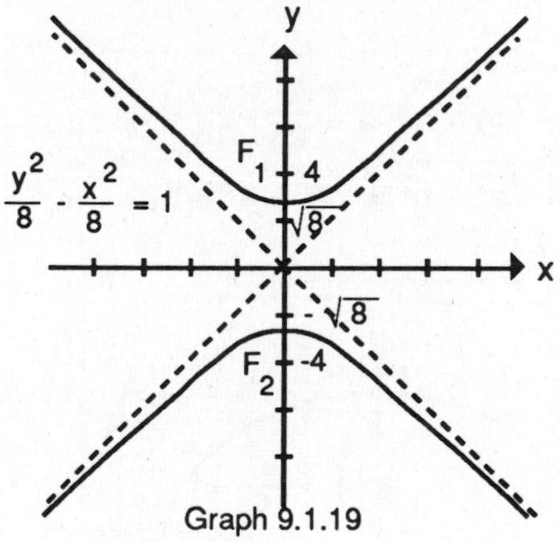

Graph 9.1.19

21. $8x^2 - y^2 = 16 \Rightarrow \frac{x^2}{2} - \frac{y^2}{8} = 1$

$\Rightarrow c = \sqrt{a^2 + b^2} = \sqrt{2 + 8} = \sqrt{10}$

$e = \frac{c}{a} = \frac{\sqrt{10}}{\sqrt{2}} = \sqrt{5}$

Asymptotes are $y = \pm 2x$.

(The graph is on the next page.)

23. $8y^2 - 2x^2 = 16 \Rightarrow \frac{y^2}{2} - \frac{x^2}{8} = 1$

$\Rightarrow c = \sqrt{a^2 + b^2} = \sqrt{2 + 8} = \sqrt{10}$

$e = \frac{c}{a} = \frac{2}{\sqrt{3}}$

$e = \frac{c}{a} = \frac{\sqrt{10}}{\sqrt{2}} = \sqrt{5}$ Asymptotes are $y = \pm\frac{x}{2}$.

(The graph is on the next page.)

21. (Graph)

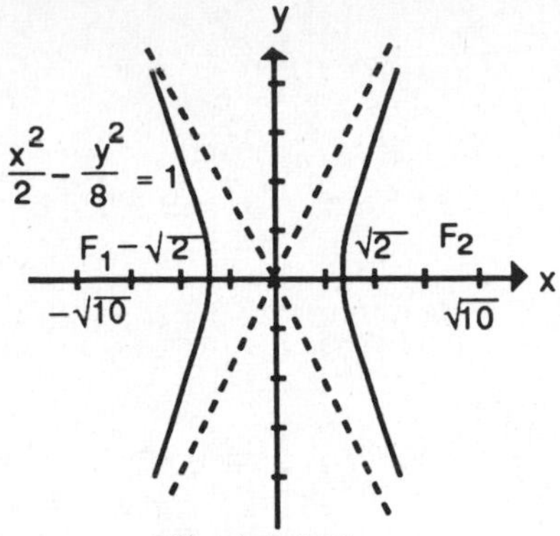

Graph 9.1.21

23. (Graph)

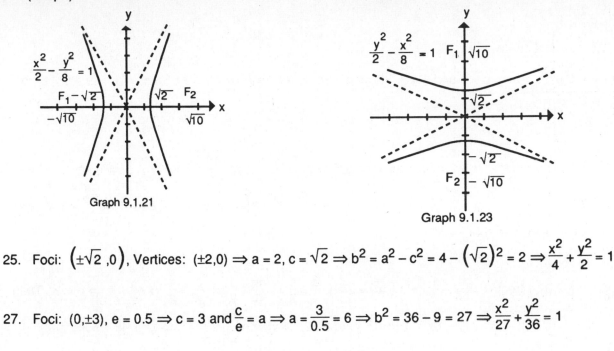

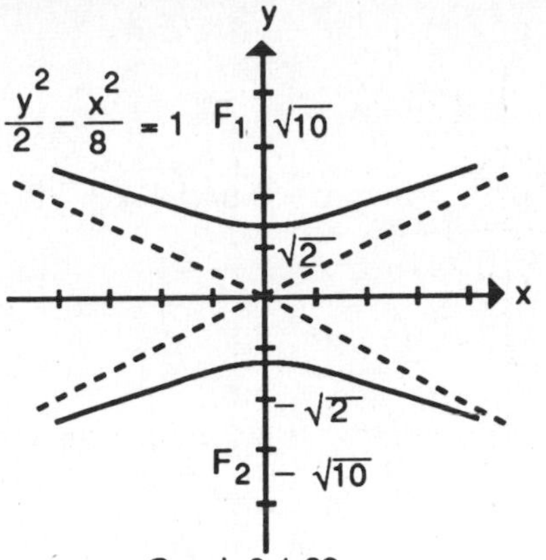

Graph 9.1.23

25. Foci: $(\pm\sqrt{2},0)$, Vertices: $(\pm 2,0) \Rightarrow a=2,\ c=\sqrt{2} \Rightarrow b^2=a^2-c^2=4-(\sqrt{2})^2=2 \Rightarrow \frac{x^2}{4}+\frac{y^2}{2}=1$

27. Foci: $(0,\pm 3)$, $e=0.5 \Rightarrow c=3$ and $\frac{c}{e}=a \Rightarrow a=\frac{3}{0.5}=6 \Rightarrow b^2=36-9=27 \Rightarrow \frac{x^2}{27}+\frac{y^2}{36}=1$

29. Vertices: $(0,\pm 70)$, $e=0.1 \Rightarrow a=70$ and $c=ae=70(0.1)=7 \Rightarrow b^2=4900-49=4851 \Rightarrow \frac{x^2}{4851}+\frac{y^2}{4900}=1$

31. Focus: $(\sqrt{5},0)$, Directrix: $x=\frac{9}{\sqrt{5}} \Rightarrow c=\sqrt{5}=ae$ and $\frac{a}{e}=\frac{9}{\sqrt{5}} \Rightarrow \frac{ae}{e^2}=\frac{9}{\sqrt{5}} \Rightarrow \frac{\sqrt{5}}{e^2}=\frac{9}{\sqrt{5}} \Rightarrow \frac{1}{e^2}=\frac{9}{5} \Rightarrow$
$e=\frac{\sqrt{5}}{3}$. Then $PF=\frac{\sqrt{5}}{3}PD \Rightarrow \sqrt{(x-\sqrt{5})^2+(y-0)^2}=\frac{\sqrt{5}}{3}\left|x-\frac{9}{\sqrt{5}}\right| \Rightarrow (x-\sqrt{5})^2+y^2=\frac{5}{9}\left(x-\frac{9}{\sqrt{5}}\right)^2$
$\Rightarrow x^2-2\sqrt{5}\,x+5+y^2=\frac{5}{9}\left(x^2-\frac{18}{\sqrt{5}}x+\frac{81}{5}\right) \Rightarrow \frac{4}{9}x^2+y^2=4 \Rightarrow \frac{x^2}{9}+\frac{y^2}{4}=1$

33. Focus: $(-4,0)$, Directrix: $x=-16 \Rightarrow c=ae=4$ and $\frac{a}{e}=16 \Rightarrow \frac{ae}{e^2}=16 \Rightarrow \frac{4}{e^2}=16 \Rightarrow \frac{1}{e^2}=4 \Rightarrow e=\frac{1}{2}$.
Then $PF=\frac{1}{2}PD \Rightarrow \sqrt{(x+4)^2+(y-0)^2}=\frac{1}{2}|x+16| \Rightarrow (x+4)^2+y^2=\frac{1}{4}(x+16)^2 \Rightarrow x^2+8x+16+y^2=$
$\frac{1}{4}(x^2+32x+256) \Rightarrow \frac{3}{4}x^2+y^2=48 \Rightarrow \frac{x^2}{64}+\frac{y^2}{48}=1$.

35. Foci: $(0,\pm\sqrt{2})$, Asymptotes: $y=\pm x \Rightarrow c=\sqrt{2}$ and $\frac{b}{a}=1 \Rightarrow a=b \Rightarrow c^2=a^2+b^2=2a^2 \Rightarrow 2=2a^2 \Rightarrow$
$a=1 \Rightarrow b=1 \Rightarrow y^2-x^2=1$.

37. Vertices: $(\pm 3,0)$, Asymptotes: $y=\pm\frac{4}{3}x \Rightarrow a=3$ and $\frac{b}{a}=\frac{4}{3} \Rightarrow b=\frac{4}{3}a=\frac{4}{3}(3)=4 \Rightarrow \frac{x^2}{9}-\frac{y^2}{16}=1$.

39. Vertices: $(0,\pm 1)$, $e = 3 \Rightarrow a = 1$ and $e = \frac{c}{a} = 3 \Rightarrow c = 3a = 3 \Rightarrow c^2 - a^2 = b^2 \Rightarrow b^2 = 9 - 1 = 8 \Rightarrow$ $y^2 - \frac{x^2}{8} = 1$.

41. Foci: $(\pm 3,0)$, $e = 3 \Rightarrow c = 3$ and $e = \frac{c}{a} = 3 \Rightarrow c = 3a \Rightarrow 3 = 3a \Rightarrow a = 1 \Rightarrow b^2 = c^2 - a^2 = 9 - 1 = 8 \Rightarrow$ $x^2 - \frac{y^2}{8} = 1$.

43. Focus: $(4,0)$, Directrix: $x = 2 \Rightarrow c = ae = 4$ and $\frac{a}{e} = 2 \Rightarrow \frac{ae}{e^2} = 2 \Rightarrow \frac{4}{e^2} = 2 \Rightarrow \frac{1}{e^2} = \frac{1}{2} \Rightarrow e = \sqrt{2}$. Then PF = $\sqrt{2}$ PD $\Rightarrow \sqrt{(x-4)^2 + (y-0)^2} = \sqrt{2}\,|x-2| \Rightarrow (x-4)^2 + y^2 = 2(x-2)^2 \Rightarrow x^2 - 8x + 16 + y^2 = 2(x^2 - 4x + 4)$ $\Rightarrow -x^2 + y^2 = -8 \Rightarrow \frac{x^2}{8} - \frac{y^2}{8} = 1$.

45. Focus: $(-2,0)$, Directrix: $x = -\frac{1}{2} \Rightarrow c = ae = 2$ and $\frac{a}{e} = \frac{1}{2} \Rightarrow \frac{ae}{e^2} = \frac{1}{2} \Rightarrow \frac{2}{e^2} = \frac{1}{2} \Rightarrow e^2 = 4 \Rightarrow e = 2$. Then PF = 2PD $\Rightarrow \sqrt{(x+2)^2 + (y-0)^2} = 2\left|x + \frac{1}{2}\right| \Rightarrow (x+2)^2 + y^2 = 4\left(x + \frac{1}{2}\right)^2 \Rightarrow x^2 + 4x + 4 + y^2 =$ $4\left(x^2 + x + \frac{1}{4}\right) \Rightarrow -3x^2 + y^2 = -3 \Rightarrow x^2 - \frac{y^2}{3} = 1$.

47.

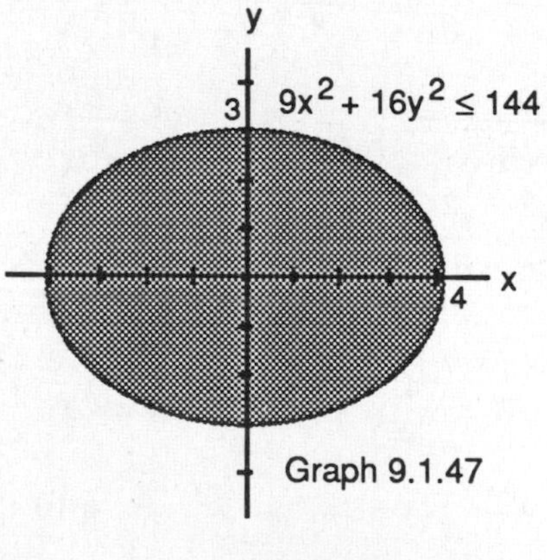

Graph 9.1.47

49.

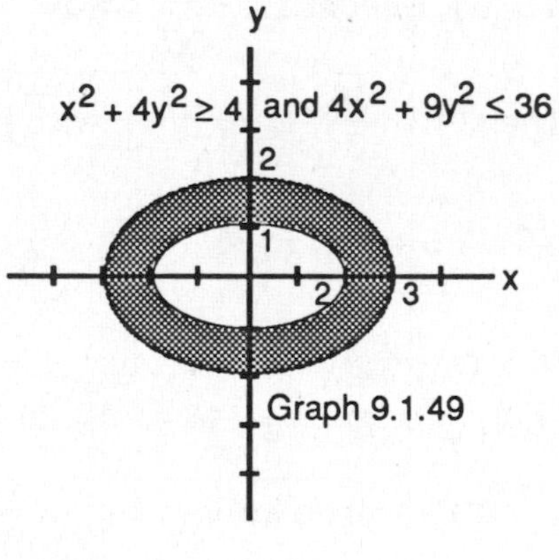

Graph 9.1.49

51.

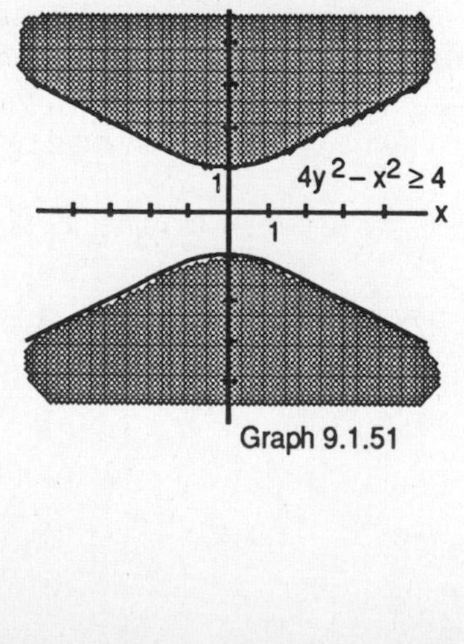

Graph 9.1.51

53. Volume of the Parabolic Solid: $V_1 = \int_0^{b/2} 2\pi x\left(h - \frac{4h}{b^2}x^2\right)dx = 2\pi h \int_0^{b/2}\left(x - \frac{4x^3}{b^2}\right)dx =$

$2\pi h\left[\frac{x^2}{2} - \frac{x^4}{b^2}\right]_0^{b/2} = \frac{\pi h b^2}{8}$ Volume of the Cone: $V_2 = \frac{1}{3}\pi\left(\frac{b}{2}\right)^2 h = \frac{1}{3}\pi\left(\frac{b^2}{4}\right)h = \frac{\pi h b^2}{12}$

$\therefore V_1 = \frac{3}{2}V_2$

55. $(x-2)^2 + (y-1)^2 = 5 \Rightarrow 2(x-2) + 2(y-1)\frac{dy}{dx} = 0 \Rightarrow \frac{dy}{dx} = -\frac{x-2}{y-1}$. $y = 0 \Rightarrow (x-2)^2 + (0-1)^2 = 5 \Rightarrow$ $(x-2)^2 = 4 \Rightarrow x = 4$ or $x = 0$. $\therefore$ the circle crosses the x–axis at (4,0) and (0,0). $x = 0 \Rightarrow (0-2)^2 + (y-1)^2 = 5 \Rightarrow (y-1)^2 = 1 \Rightarrow y = 2$ or $y = 0$. $\therefore$ the circle crosses the y–axis at (0,2) and (0,0). At (4,0), $\frac{dy}{dx} = -\frac{4-2}{0-1} = 2 \Rightarrow$ the tangent line is $y = 2(x-4)$ or $y = 2x - 8$. At (0,0), $\frac{dy}{dx} = -\frac{0-2}{0-1} = -2 \Rightarrow$ the tangent line is $y = -2x$. At (0,2), $\frac{dy}{dx} = -\frac{0-2}{2-1} = 2 \Rightarrow$ the tangent line is $y - 2 = 2x$ or $y = 2x + 2$.

57.

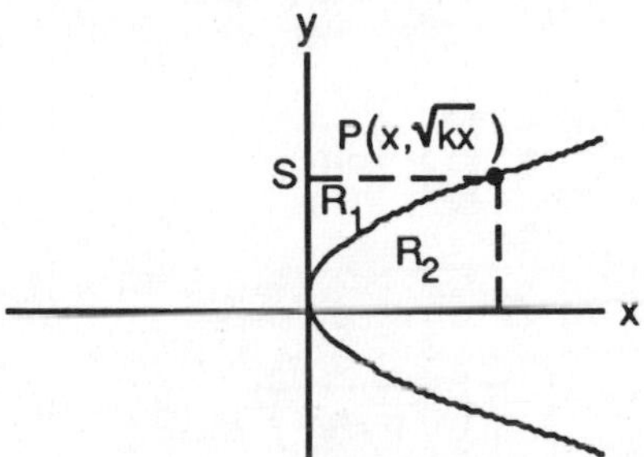

Graph 9.1.57

a) $y^2 = kx \Rightarrow x = \frac{y^2}{k}$. Then the volume of the solid formed by revolving R_1 about the y–axis is $V_1 = \int_0^{\sqrt{kx}} \pi\left(\frac{y^2}{k}\right)^2 dy = \frac{\pi}{k^2}\int_0^{\sqrt{kx}} y^4\,dy = \frac{\pi x^2\sqrt{kx}}{5}$. The volume of the right circular cylinder formed by revolving PQ about the y–axis is $V_2 = \pi x^2\sqrt{kx} \Rightarrow$ the volume of the solid formed by revolving R_2 about the y–axis is $V_3 = V_2 - V_1 = \frac{4\pi x^2\sqrt{kx}}{5}$. $\therefore$ we can see the ratio of V_3 to V_1 is 4:1.

b) The volume of the solid formed by revolving R_2 about the x–axis is $V_1 = \int_0^x \pi\left(\sqrt{kt}\right)^2 dt = \pi k\int_0^x t\,dt$ $= \frac{\pi k x^2}{2}$ where $y = \sqrt{kx}$ is the upper half of the parabola. The volume of the right circular cylinder formed by revolving PS about the x–axis is $V_2 = \pi\left(\sqrt{kx}\right)^2 x = \pi k x^2 \Rightarrow$ the volume of the solid formed by revolving R_1 about the x–axis is $V_3 = V_2 - V_1 = \pi k x^2 - \frac{\pi k x^2}{2} = \frac{\pi k x^2}{2}$. $\therefore$ the ratio of V_3 to V_1 is 1:1.

59. Let $y = \sqrt{1 - \frac{x^2}{4}}$ on the interval $0 \le x \le 2$. Then the area of the inscribed rectangle is given by

$A(x) = 2x\left(2\sqrt{1 - \frac{x^2}{4}}\right) = 4x\sqrt{1 - \frac{x^2}{4}}$ since the length is 2x and the height is 2y.

Then $A'(x) = 4\sqrt{1 - \frac{x^2}{4}} - \frac{x^2}{\sqrt{1 - \frac{x^2}{4}}}$. Let $A'(x) = 0 \Rightarrow 4\sqrt{1 - \frac{x^2}{4}} - \frac{x^2}{\sqrt{1 - \frac{x^2}{4}}} = 0 \Rightarrow 4\left(1 - \frac{x^2}{4}\right) - x^2 = 0$

$\Rightarrow x^2 = 2 \Rightarrow x = \sqrt{2}$ (Only the positive square root is in the interval.) Since $A(0) = 0$, $A(2) = 0$, $A(\sqrt{2}) = 4$ is the maximum area when the length is $2\sqrt{2}$ and the height is $\sqrt{2}$.

61. a) Around the x–axis: $9x^2 + 4y^2 = 36 \Rightarrow y^2 = 9 - \frac{9}{4}x^2 \Rightarrow y = \pm\sqrt{9 - \frac{9}{4}x^2}$, use the positive root.

$$V = 2\int_0^2 \pi\left(\sqrt{9 - \frac{9}{4}x^2}\right)^2 dx = 2\int_0^2 \pi\left(9 - \frac{9}{4}x^2\right) dx = 2\pi\left[9x - \frac{3}{4}x^3\right]_0^2 = 24\pi$$

b) Around the y–axis: $9x^2 + 4y^2 = 36 \Rightarrow x^2 = 4 - \frac{4}{9}y^2 \Rightarrow x = \pm\sqrt{4 - \frac{4}{9}y^2}$, use the positive root.

$$V = 2\int_0^3 \pi\left(\sqrt{4 - \frac{4}{9}y^2}\right)^2 dy = 2\int_0^3 \pi\left(4 - \frac{4}{9}y^2\right) dy = 2\pi\left[4y - \frac{4}{27}y^3\right]_0^3 = 16\pi$$

63. $x^2 - y^2 = 1 \Rightarrow x = \pm\sqrt{1 + y^2}$ on the interval $-3 \le y \le 3$. $\therefore$ Volume $= \int_{-3}^{3} \pi\left(\sqrt{1 + y^2}\right)^2 dy =$

$$2\int_0^3 \pi\left(\sqrt{1 + y^2}\right)^2 dy = 2\pi\int_0^3 (1 + y^2)dy = 2\pi\left[y + \frac{y^3}{3}\right]_0^3 = 24\pi$$

65. $\frac{dr_A}{dt} = \frac{dr_B}{dt} \Rightarrow \int \frac{dr_A}{dt} = \int \frac{dr_B}{dt} \Rightarrow r_A + C_1 = r_B + C_2 \Rightarrow r_A - r_B = C$, a constant $\Rightarrow$ The points, P(t), lie on a hyperbola with foci A and B.

67. $\lim_{x \to \infty}\left[\frac{b}{a}x - \frac{b}{a}\sqrt{x^2 - a^2}\right] = \frac{b}{a}\lim_{x \to \infty}\left(x - \sqrt{x^2 - a^2}\right) = \frac{b}{a}\lim_{x \to \infty}\left[\frac{\left(x - \sqrt{x^2 - a^2}\right)\left(x + \sqrt{x^2 - a^2}\right)}{x + \sqrt{x^2 - a^2}}\right] =$

$\frac{b}{a}\lim_{x \to \infty}\left[\frac{x^2 - (x^2 - a^2)}{x + \sqrt{x^2 - a^2}}\right] = \frac{b}{a}\lim_{x \to \infty}\left[\frac{a^2}{x + \sqrt{x^2 - a^2}}\right]$

SECTION 9.2 THE GRAPHS OF QUADRATIC EQUATIONS IN X AND Y

1. $x^2 - 3xy + y^2 - x = 0 \Rightarrow B^2 - 4AC = (-3)^2 - 4(1)(1) = 5 > 0 \Rightarrow$ Hyperbola

3. $3x^2 - 7xy + \sqrt{17}\,y^2 = 1 \Rightarrow B^2 - 4AC = (-7)^2 - 4(3)\sqrt{17} = -0.477 < 0 \Rightarrow$ Ellipse

5. $x^2 + 2xy + y^2 + 2x - y + 2 = 0 \Rightarrow B^2 - 4AC = (2)^2 - 4(1)(1) = 0 \Rightarrow$ Parabola

7. $x^2 + 4xy + 4y^2 - 3x = 6 \Rightarrow B^2 - 4AC = 4^2 - 4(1)(4) = 0 \Rightarrow$ Parabola

9. $xy + y^2 - 3x = 5 \Rightarrow B^2 - 4AC = 1^2 - 4(0)(1) = 1 > 0 \Rightarrow$ Hyperbola

11. $3x^2 - 5xy + 2y^2 - 7x - 14y = -1 \Rightarrow B^2 - 4AC = (-5)^2 - 4(3)(2) = 1 > 0 \Rightarrow$ Hyperbola

13. $x^2 - 3xy + 3y^2 + 6y = 7 \Rightarrow B^2 - 4AC = (-3)^2 - 4(1)(3) = -3 < 0 \Rightarrow$ Ellipse

15. $6x^2 + 3xy + 2y^2 + 17y + 2 = 0 \Rightarrow B^2 - 4AC = 3^2 - 4(6)(2) = -39 < 0 \Rightarrow$ Ellipse

17. $\cot 2\alpha = \frac{A-C}{B} = \frac{0}{1} = 0 \Rightarrow 2\alpha = \frac{\pi}{2} \Rightarrow \alpha = \frac{\pi}{4}$. $\therefore$ $x = x'\cos\alpha - y'\sin\alpha$, $y = x'\sin\alpha + y'\cos\alpha \Rightarrow$
$x = x'\frac{\sqrt{2}}{2} - y'\frac{\sqrt{2}}{2}$, $y = x'\frac{\sqrt{2}}{2} + y'\frac{\sqrt{2}}{2} \Rightarrow \left(\frac{\sqrt{2}}{2}x' - \frac{\sqrt{2}}{2}y'\right)\left(\frac{\sqrt{2}}{2}x' + \frac{\sqrt{2}}{2}y'\right) = 2 \Rightarrow \frac{1}{2}x'^2 - \frac{1}{2}y'^2 = 2 \Rightarrow x'^2 - y'^2 = 4$
$\Rightarrow$ Hyperbola

19. $\cot 2\alpha = \frac{A-C}{B} = \frac{3-1}{2\sqrt{3}} = \frac{1}{\sqrt{3}} \Rightarrow 2\alpha = \frac{\pi}{3} \Rightarrow \alpha = \frac{\pi}{6}$. $\therefore$ $x = x'\cos\alpha - y'\sin\alpha$, $y = x'\sin\alpha + y'\cos\alpha \Rightarrow$
$x = \frac{\sqrt{3}}{2}x' - \frac{1}{2}y'$, $y = \frac{1}{2}x' + \frac{\sqrt{3}}{2}y' \Rightarrow 3\left(\frac{\sqrt{3}}{2}x' - \frac{1}{2}y'\right)^2 + 2\sqrt{3}\left(\frac{\sqrt{3}}{2}x' - \frac{1}{2}y'\right)\left(\frac{1}{2}x' + \frac{\sqrt{3}}{2}y'\right) + \left(\frac{1}{2}x' + \frac{\sqrt{3}}{2}y'\right)^2 -$
$8\left(\frac{\sqrt{3}}{2}x' - \frac{1}{2}y'\right) + 8\sqrt{3}\left(\frac{1}{2}x' + \frac{\sqrt{3}}{2}y'\right) = 0 \Rightarrow 4x'^2 + 16y' = 0$, Parabola

21. $\cot 2\alpha = \frac{A-C}{B} = \frac{1-1}{-2} = 0 \Rightarrow 2\alpha = \frac{\pi}{2} \Rightarrow \alpha = \frac{\pi}{4}$. $\therefore$ $x = x'\cos\alpha - y'\sin\alpha$, $y = x'\sin\alpha + y'\cos\alpha \Rightarrow$
$x = x'\frac{\sqrt{2}}{2} - y'\frac{\sqrt{2}}{2}$, $y = x'\frac{\sqrt{2}}{2} + y'\frac{\sqrt{2}}{2} \Rightarrow \left(\frac{\sqrt{2}}{2}x' - \frac{\sqrt{2}}{2}y'\right)^2 - 2\left(\frac{\sqrt{2}}{2}x' - \frac{\sqrt{2}}{2}y'\right)\left(\frac{\sqrt{2}}{2}x' + \frac{\sqrt{2}}{2}y'\right) + \left(\frac{\sqrt{2}}{2}x' + \frac{\sqrt{2}}{2}y'\right)^2 = 2$
$\Rightarrow y'^2 = 1$, Parallel Horizontal Lines.

23. $\cot 2\alpha = \frac{A-C}{B} = \frac{\sqrt{2}-\sqrt{2}}{2\sqrt{2}} = 0 \Rightarrow 2\alpha = \frac{\pi}{2} \Rightarrow \alpha = \frac{\pi}{4}$. $\therefore$ $x = x'\cos\alpha - y'\sin\alpha$, $y = x'\sin\alpha + y'\cos\alpha \Rightarrow$
$x = x'\frac{\sqrt{2}}{2} - y'\frac{\sqrt{2}}{2}$, $y = x'\frac{\sqrt{2}}{2} + y'\frac{\sqrt{2}}{2} \Rightarrow \sqrt{2}\left(\frac{\sqrt{2}}{2}x' - \frac{\sqrt{2}}{2}y'\right)^2 + 2\sqrt{2}\left(\frac{\sqrt{2}}{2}x' - \frac{\sqrt{2}}{2}y'\right)\left(\frac{\sqrt{2}}{2}x' + \frac{\sqrt{2}}{2}y'\right) +$

23. (Continued)

$\sqrt{2}\left(\frac{\sqrt{2}}{2}x' + \frac{\sqrt{2}}{2}y'\right)^2 - 8\left(\frac{\sqrt{2}}{2}x' - \frac{\sqrt{2}}{2}y'\right) + 8\left(\frac{\sqrt{2}}{2}x' + \frac{\sqrt{2}}{2}y'\right) = 0 \Rightarrow 2\sqrt{2}\,x'^2 + 8\sqrt{2}\,y' = 0$, Parabola

25. $\cot 2\alpha = \frac{A - C}{B} = \frac{3 - 3}{2} = 0 \Rightarrow 2\alpha = \frac{\pi}{2} \Rightarrow \alpha = \frac{\pi}{4}$. $\therefore$ $x = x'\cos\alpha - y'\sin\alpha$, $y = x'\sin\alpha + y'\cos\alpha \Rightarrow$ $x = x'\frac{\sqrt{2}}{2} - y'\frac{\sqrt{2}}{2}$, $y = x'\frac{\sqrt{2}}{2} + y'\frac{\sqrt{2}}{2} \Rightarrow 3\left(\frac{\sqrt{2}}{2}x' - \frac{\sqrt{2}}{2}y'\right)^2 + 2\left(\frac{\sqrt{2}}{2}x' - \frac{\sqrt{2}}{2}y'\right)\left(\frac{\sqrt{2}}{2}x' + \frac{\sqrt{2}}{2}y'\right) + 3\left(\frac{\sqrt{2}}{2}x' + \frac{\sqrt{2}}{2}y'\right)^2$ $= 19 \Rightarrow 4x'^2 + 2y'^2 = 19$, Ellipse.

27. $\tan 2\alpha = \frac{-1}{1 - 3} = \frac{1}{2} \Rightarrow 2\alpha = 26.57° \Rightarrow \alpha = 13.28°$. Then $A' = 0.88$, $B' = 0.00$, $C' = 3.12$, $D' = 0.74$, $E' = -1.20$, and $F' = -3 \Rightarrow 0.88x'^2 + 3.12y'^2 + 0.74x' - 1.20y' - 3 = 0$, an ellipse.

29. $\tan 2\alpha = \frac{-4}{1 - 4} = \frac{4}{3} \Rightarrow 2\alpha = 53.13° \Rightarrow \alpha = 26.57°$. Then $A' = 0.00$, $B' = 0.00$, $C' = 5.00$, $D' = 0$, $E' = 0$, and $F' = -5 \Rightarrow 5.00y'^2 - 5 = 0$ or $y' = \pm 1.00$, parallel lines.

31. $\tan 2\alpha = \frac{5}{3 - 2} = 5 \Rightarrow 2\alpha = 78.69° \Rightarrow \alpha = 39.35°$. Then $A' = 5.05$, $B' = 0.00$, $C' = -0.05$, $D' = -5.07$, $E' = -6.18$, and $F' = -1 \Rightarrow 5.05x'^2 - 0.05y'^2 - 5.07x' - 6.18y' - 1 = 0$, a hyperbola.

33. a) $\frac{x'^2}{b^2} + \frac{y'^2}{a^2} = 1$

b) $\frac{y'^2}{a^2} - \frac{x'^2}{b^2} = 1$

c) $x'^2 + y'^2 = a^2$

d) $y = mx \Rightarrow y - mx = 0 \Rightarrow D = -m$, $E = 1$. $\alpha = 90°$ $\Rightarrow D' = 1, E' = m \Rightarrow my' + x' = 0 \Rightarrow y' = -\frac{1}{m}x'$

e) $y = mx + b \Rightarrow y - mx - b = 0 \Rightarrow D = -m$, $E = 1 \Rightarrow$ $D' = 1$, $E' = m$ (See part d above.) Also $F' = -b$ $\Rightarrow my' + x' - b = 0 \Rightarrow y' = -\frac{1}{m}x' + \frac{b}{m}$.

35. a) $A' = \cos 45° \sin 45° = \left(\frac{\sqrt{2}}{2}\right)\left(\frac{\sqrt{2}}{2}\right) = \frac{1}{2}$, $B' = 0$, $C' = -\cos 45° \sin 45° = -\frac{1}{2}$, $F' = -1 \Rightarrow$ $\frac{1}{2}x'^2 - \frac{1}{2}y'^2 = 1$ or $x'^2 - y'^2 = 2$.

b) $A' = \frac{1}{2}$, $C' = -\frac{1}{2}$ (See part a above.) $D' = E' = B' = 0$, $F' = -a \Rightarrow \frac{1}{2}x'^2 - \frac{1}{2}y'^2 = a$ or $x'^2 - y'^2 = 2a$

37. Let α be any angle. Then $A' = \cos^2\alpha + \sin^2\alpha = 1$, $B' = 0$, $C' = \sin^2\alpha + \cos^2\alpha = 1$, $D' = E' = 0$, $F' = -a^2$ $\Rightarrow x'^2 + y'^2 = a^2$.

39.

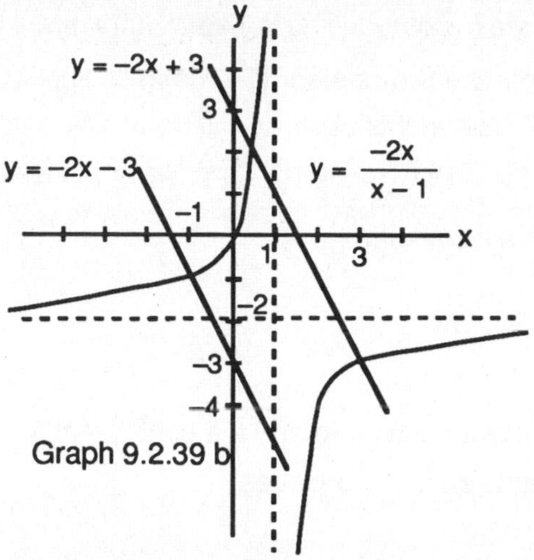

Graph 9.2.39 b

a) $B^2 - 4AC = 1 - 4(0)(0) = 1 \Rightarrow$ hyperbola

b) $xy + 2x - y = 0 \Rightarrow y(x-1) = -2x \Rightarrow y = \frac{-2x}{x-1}$

c) $y = \frac{-2x}{x-1} \Rightarrow \frac{dy}{dx} = \frac{2}{(x-1)^2}$. We want $\frac{-1}{dy/dx} = -2$, the slope of $y = -2x \Rightarrow -2 = \frac{-1}{2/(x-1)^2} \Rightarrow (x-1)^2 = 4 \Rightarrow x = 3$ or $x = -1$. $x = 3 \Rightarrow y = -3 \Rightarrow (3,-3)$ is a point on the hyperbola where the line with $m = -2$ is normal $\Rightarrow$ the line is $y + 3 = -2(x-3)$ or $y = -2x + 3$. $x = -1 \Rightarrow y = -1 \Rightarrow (-1,-1)$ is a point on the hyperbola where the line with $m = -2$ is normal $\Rightarrow$ the line is $y + 1 = -2(x+1)$ or $y = -2x - 3$.

41. Symmetric about the origin means (x,y) on the graph $\Rightarrow (-x,-y)$ is on the graph. $\therefore A(-x)^2 + B(-x)(-y) + C(-y)^2 + D(-x) + E(-y) + F = -(Ax^2 + Bxy + Cy^2 + Dx + Ey + F) \Rightarrow Ax^2 + Bxy + Cy^2 - Dx - Ey + F = -Ax^2 - Bxy - Cy^2 - Dx - Ey - F \Rightarrow 2Ax^2 + 2Bxy + 2Cy^2 + 2F = 0$ or $Ax^2 + Bxy + Cy^2 + F = 0$. If the conic passes through (1,0), then $A(1)^2 + B(1)(0) + C(0)^2 + F = 0 \Rightarrow A = -F \Rightarrow -Fx^2 + Bxy + Cy^2 + F = 0$. By implicit differentiation, $-2Fx + By + Bxy' + 2Cyy' = 0$. $y = 1$ tangent to the conic at $(-2,1) \Rightarrow \frac{dy}{dx} = 0$ at $(-2,1) \Rightarrow -2F(-2) + B(1) + B(-2)(0) + 2C(1)(0) = 0 \Rightarrow 4F + B = 0$ or $B = -4F$. Then the conic is $-Fx^2 - 4Fxy + Cy^2 + F = 0$. Also $(-2,1)$ is on the conic $\Rightarrow -F(-2)^2 - 4F(-2)(1) + C(1)^2 + F = 0 \Rightarrow C = -5F$. $\therefore$ the equation of the conic (in terms of F) is $-Fx^2 - 4Fxy - 5Fy^2 + F = 0$. Assuming $F \neq 0$, we get $-x^2 - 4xy - 5y^2 + 1 = 0$ or $x^2 + 4xy + 5y^2 - 1 = 0$. $B^2 - 4AC = -4 \Rightarrow$ Ellipse.

43. a) Let $A = C = 1$, $B = 2 \Rightarrow B^2 - 4AC = 0 \Rightarrow$ parabola

b) See part a above.

c) If $AC < 0$, then $-4AC > 0 \Rightarrow B^2 - 4AC > 0 \Rightarrow$ hyperbola $\Rightarrow$ the statement is true.

45. a) $A' + C' = \left(A\cos^2\alpha + B\cos\alpha\sin\alpha + C\sin^2\alpha\right) + \left(A\sin^2\alpha - B\cos\alpha\sin\alpha + C\sin^2\alpha\right) = A\left(\cos^2\alpha + \sin^2\alpha\right) + C\left(\sin^2\alpha + \cos^2\alpha\right) = A + C.$

b) $D'^2 + E'^2 = \left(D\cos\alpha + E\sin\alpha\right)^2 + \left(-D\sin\alpha + E\cos\alpha\right)^2 = D^2\cos^2\alpha + 2DE\cos\alpha\sin\alpha + E^2\sin^2\alpha + D^2\sin^2\alpha - 2DE\sin\alpha\cos\alpha + E^2\cos^2\alpha = D^2\left(\cos^2\alpha + \sin^2\alpha\right) + E^2\left(\sin^2\alpha + \cos^2\alpha\right) = D^2 + E^2.$

47.

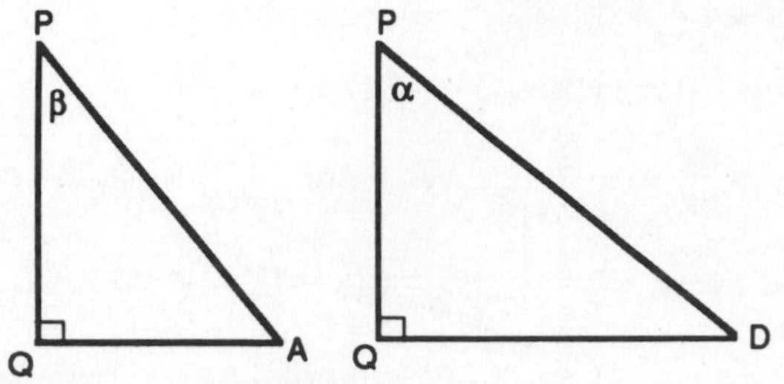

Graph 9.2.47

Using the same hypotheses as in Figure 9.29, page 647 in the text, we get the two related right triangles shown in Graph 9.2.47. But in this case, $\alpha = \beta \Rightarrow \Delta PQA$ is congruent to $\Delta PQD \Rightarrow PA = PD$. Since $PA = PF$, $PF = PD \Rightarrow e = 1 \Rightarrow$ the section is a parabola.

49. The cutting plane and the plane of the circle of tangency between the sphere and the cone are the same. $\therefore$ there is no unique line, L, a directrix, of the intersection between the two planes.

SECTION 9.3 PARAMETRIZATIONS OF CURVES

1. $x = \cos t,\ y = \sin t,\ 0 \le t \le \pi \Rightarrow$
$\cos^2 t + \sin^2 t = 1 \Rightarrow x^2 + y^2 = 1$

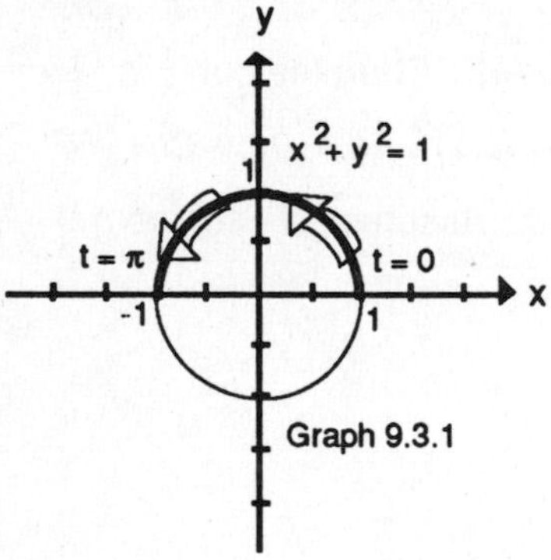

Graph 9.3.1

3. $x = \sin 2\pi t,\ y = \cos 2\pi t,\ 0 \le t \le 1 \Rightarrow$
$\sin^2 2\pi t + \cos^2 2\pi t = 1 \Rightarrow x^2 + y^2 = 1$

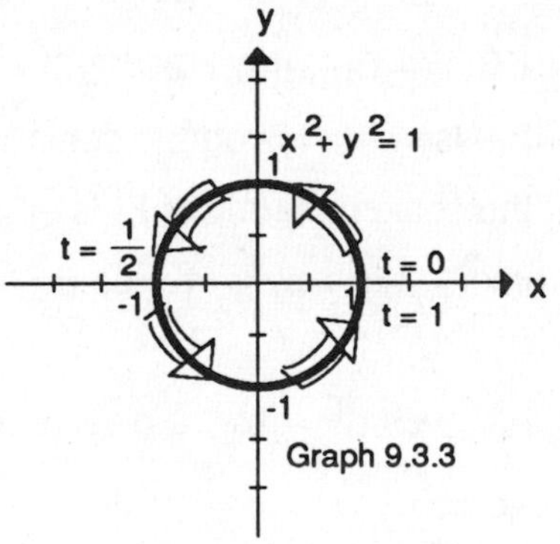

Graph 9.3.3

5. $x = 4\cos t,\ y = 2\sin t,\ 0 \le t \le 2\pi \Rightarrow$
$\frac{16\cos^2 t}{16} + \frac{4\sin^2 t}{4} = 1 \Rightarrow \frac{x^2}{16} + \frac{y^2}{4} = 1$

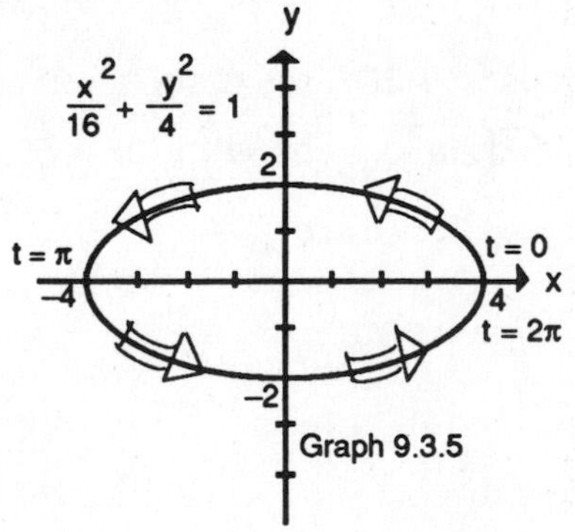

Graph 9.3.5

7. $x = 4\cos t,\ y = 5\sin t,\ 0 \le t \le 2\pi \Rightarrow$
$\frac{16\cos^2 t}{16} + \frac{25\sin^2 t}{25} = 1 \Rightarrow \frac{x^2}{16} + \frac{y^2}{25} = 1$

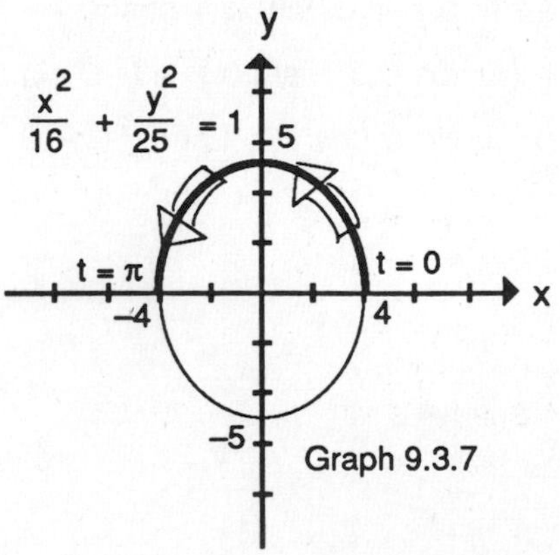

Graph 9.3.7

9. $x = 3t,\ y = 9t^2,\ -\infty < t < \infty \Rightarrow$

$y = x^2$

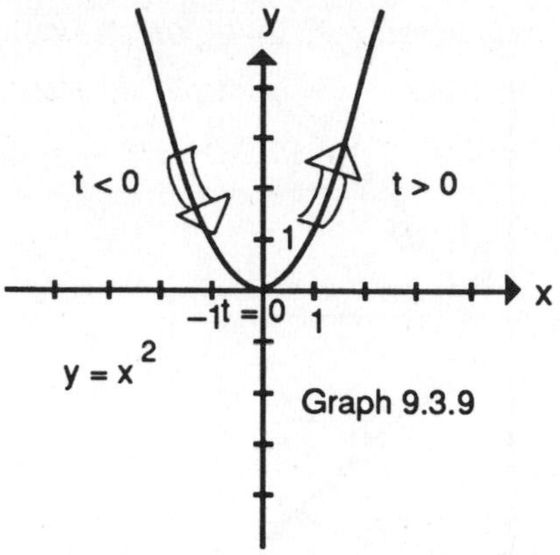

Graph 9.3.9

11. $x = t,\ y = \sqrt{t},\ t \geq 0$

$x = y^2$

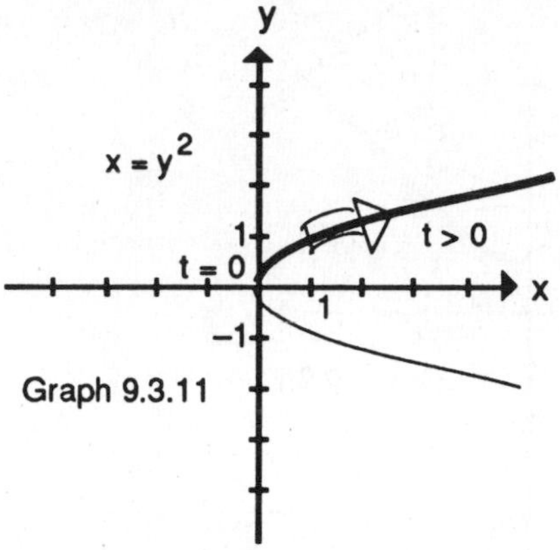

Graph 9.3.11

13. $x = -\sec t,\ y = \tan t,\ -\frac{\pi}{2} < t < \frac{\pi}{2} \Rightarrow$

$\sec^2 t - \tan^2 t = 1 \Rightarrow x^2 - y^2 = 1$

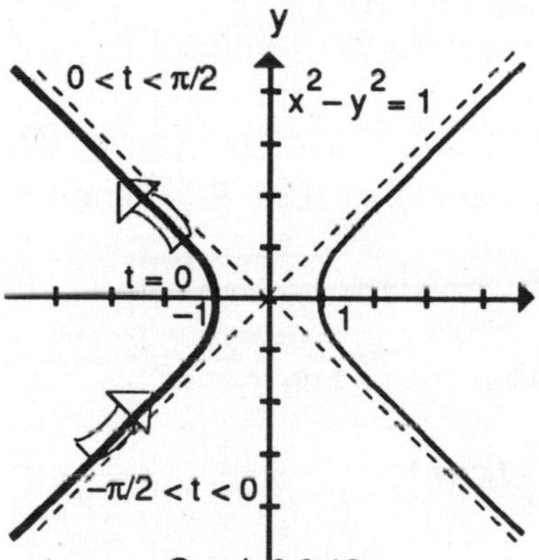

Graph 9.3.13

15. $x = 2t - 5,\ y = 4t - 7,\ -\infty < t < \infty \Rightarrow$

$x + 5 = 2t \Rightarrow 2(x + 5) = 4t \Rightarrow$

$y = 2(x + 5) - 7 \Rightarrow y = 2x + 3$

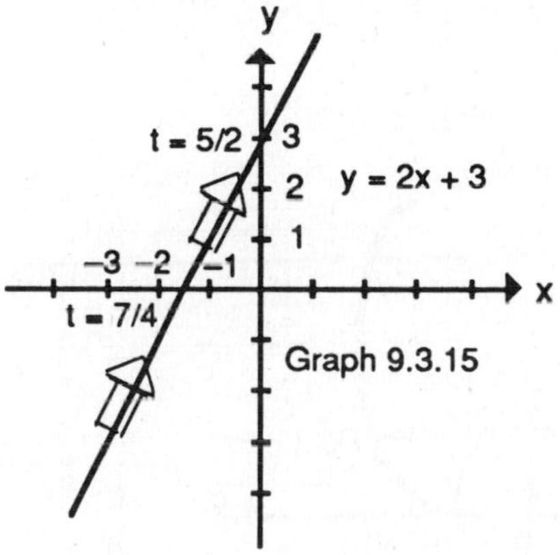

Graph 9.3.15

17. $x = t,\ y = 1 - t,\ 0 \leq t \leq 1 \Rightarrow$

$y = 1 - x$

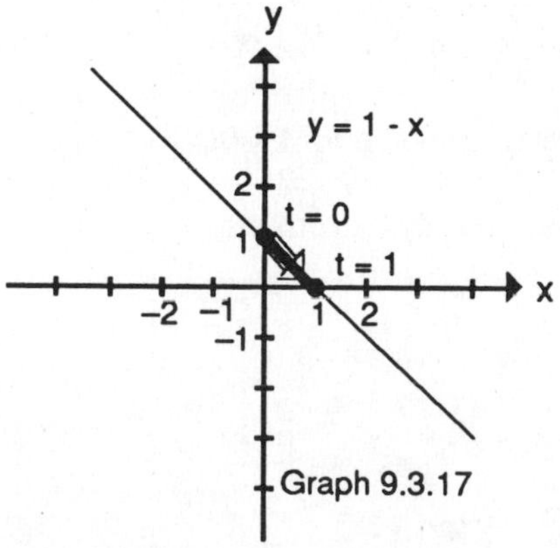

Graph 9.3.17

19. $x = t,\ y = \sqrt{1 - t^2},\ -1 \leq t \leq 1 \Rightarrow$

$y = \sqrt{1 - x^2}$

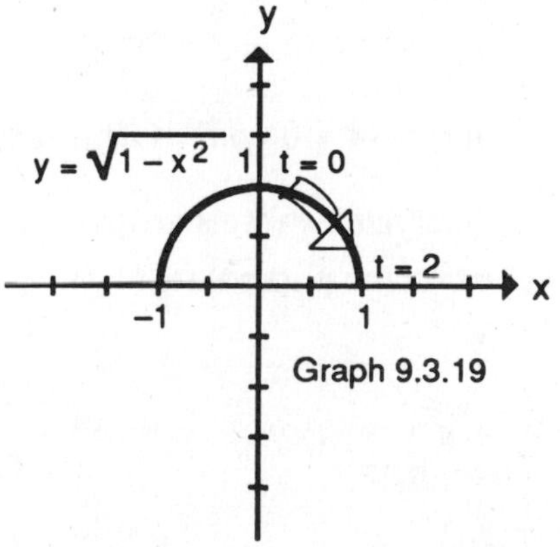

Graph 9.3.19

21. $x = t^2$, $y = \sqrt{t^4 + 1}$, $t \ge 0 \Rightarrow$
$y = \sqrt{x^2 + 1}$, $x \ge 0$

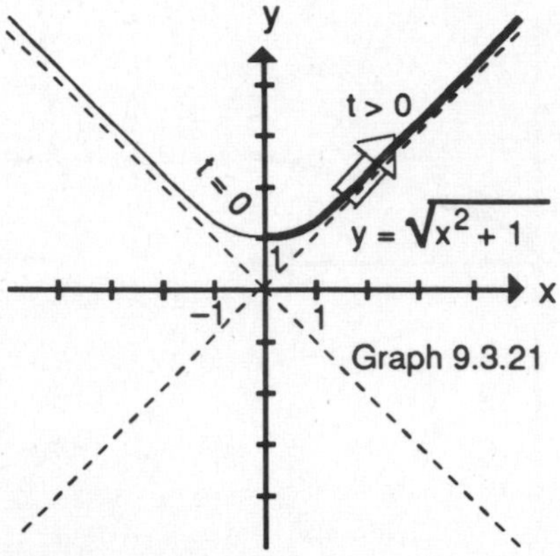

Graph 9.3.21

23. $x = \cosh t$, $y = \sinh t$, $-\infty < t < \infty \Rightarrow$
$\cosh^2 t - \sinh^2 t = 1 \Rightarrow x^2 - y^2 = 1$

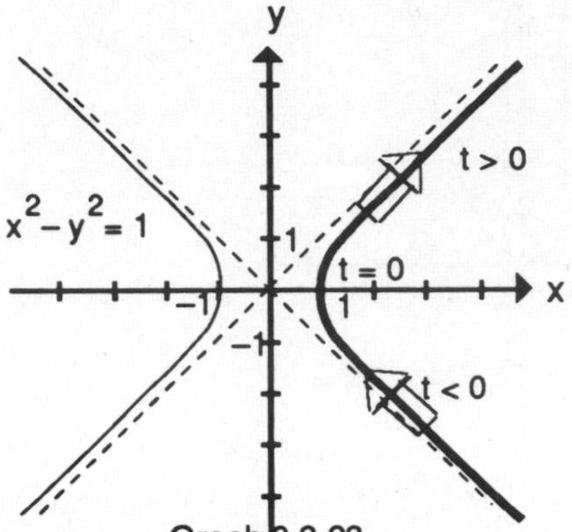

Graph 9.3.23

25. a) $x = a \cos t$, $y = -a \sin t$, $0 \le t \le 2\pi$

b) $x = a \cos t$, $y = a \sin t$, $0 \le t \le 2\pi$

c) $x = a \cos t$, $y = -a \sin t$, $0 \le t \le 4\pi$

d) $x = a \cos t$, $y = a \sin t$, $0 \le t \le 4\pi$

27.

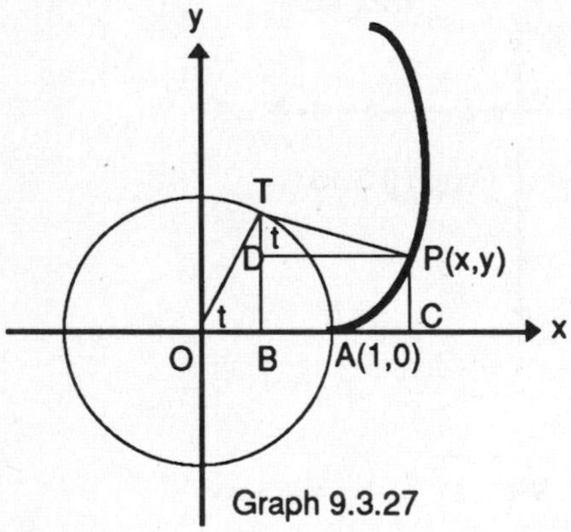

Graph 9.3.27

$\angle PTB = \angle TOB = t$. PT = arc(AT) = t since PT = length of the unwound string = length of arc(AT).

$\therefore x = OB + BC = OB + DP = \cos t + t\sin t$.

$y = PC = TB - TD = \sin t - t\cos t$

29. $d = \sqrt{(x-2)^2 + (y-\frac{1}{2})^2} \Rightarrow d^2 = (x-2)^2 + (y-\frac{1}{2})^2 = (t-2)^2 + (t^2 - \frac{1}{2})^2 \Rightarrow d^2 = t^4 - 4t + \frac{17}{4}$

$\frac{d(d^2)}{dt} = 4t^3 - 4$. Let $4t^3 - 4 = 0 \Rightarrow t = 1$ The second derivative is always positve for $t \ne 0 \Rightarrow$

$t = 1$ gives a local minimum which is an absolute minimum since it is the only extremum.

$\therefore$ the closest point on the parabola is (1,1).

31. a) $x = x_0 + (x_1 - x_0)t$, $y = y_0 + (y_1 - y_0)t \Rightarrow t = \frac{x - x_0}{x_1 - x_0} \Rightarrow y = y_0 + (y_1 - y_0)\left(\frac{x - x_0}{x_1 - x_0}\right) \Rightarrow$

$y - y_0 = \left(\frac{y_1 - y_0}{x_1 - x_0}\right)(x - x_0)$ which is the equation of the line through the points.

b) $x = x_1 t$, $y = y_1 t$ (Answer not unique)

c) $x = -1 + t$, $y = t$ or $x = -t$, $y = 1 - t$

33. a)

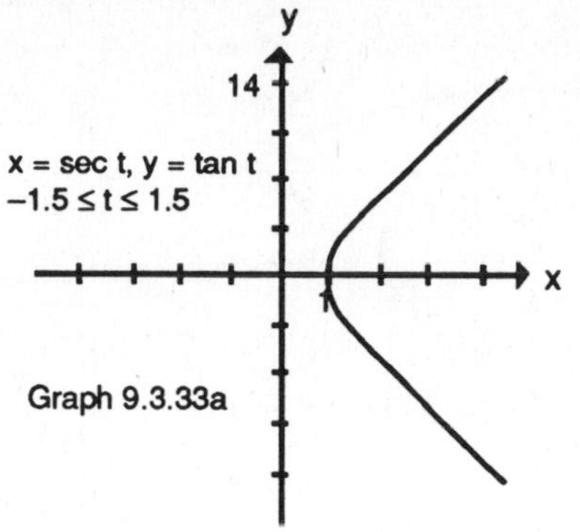

Graph 9.3.33a

b)

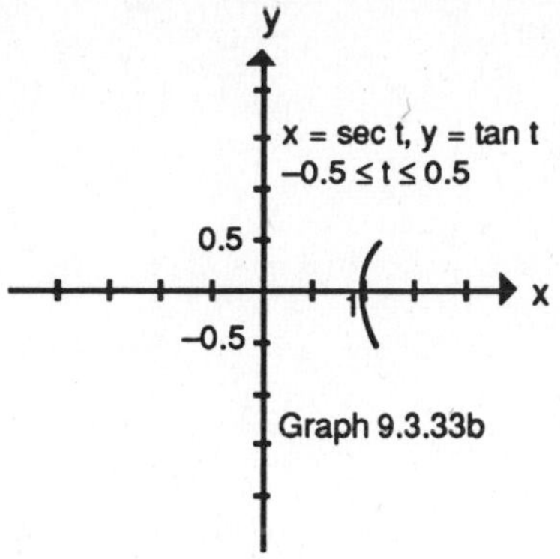

Graph 9.3.33b

c)

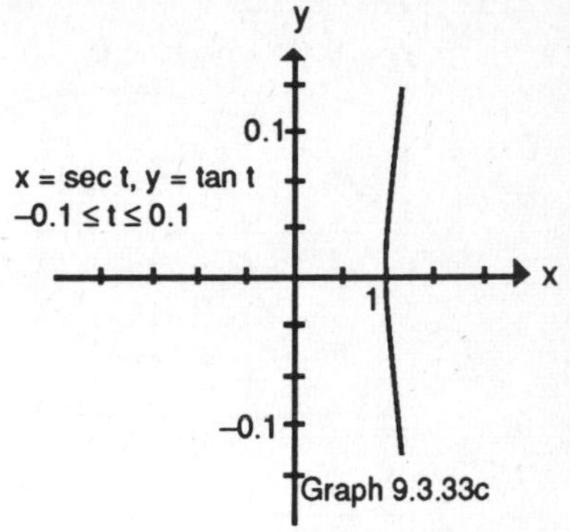

Graph 9.3.33c

35. a)

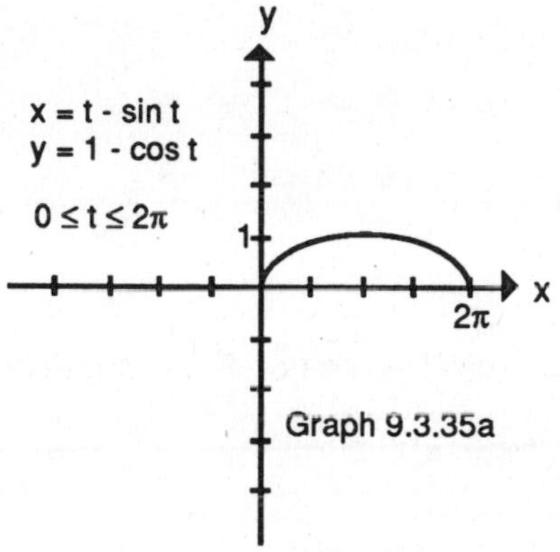

Graph 9.3.35a

b)

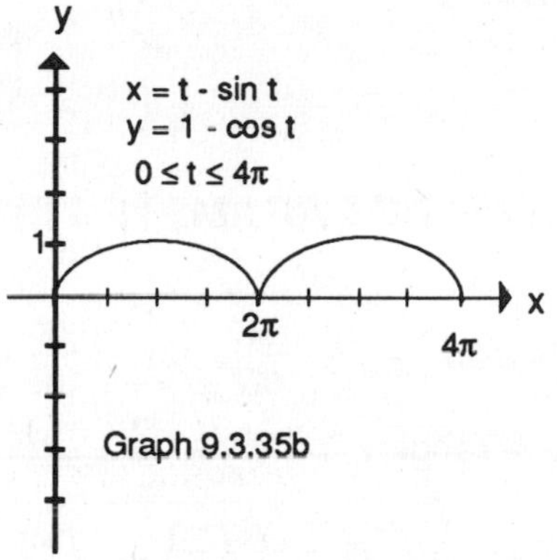

Graph 9.3.35b

c)

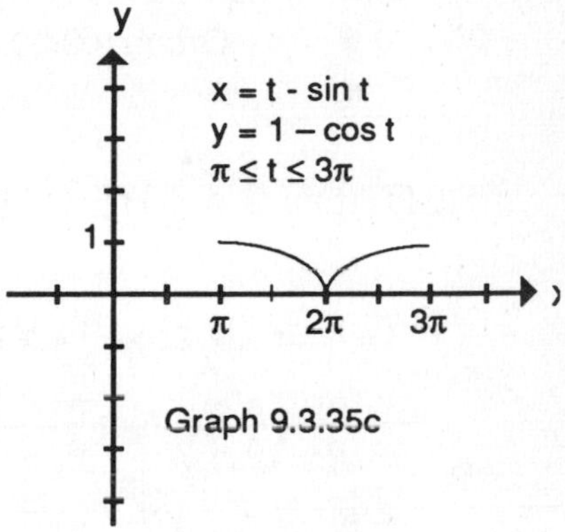

Graph 9.3.35c

37. a)

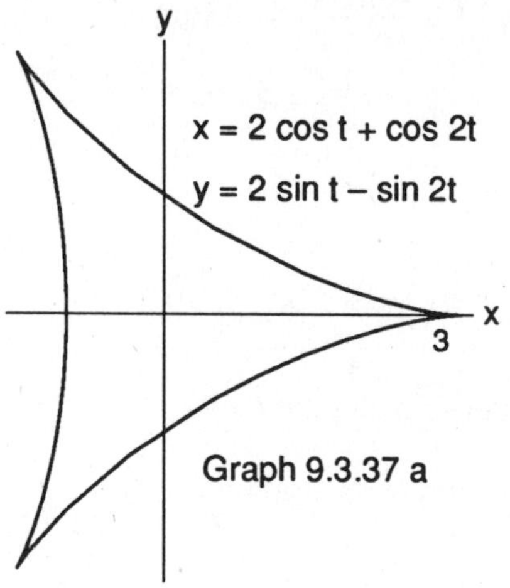

Graph 9.3.37 a

b)

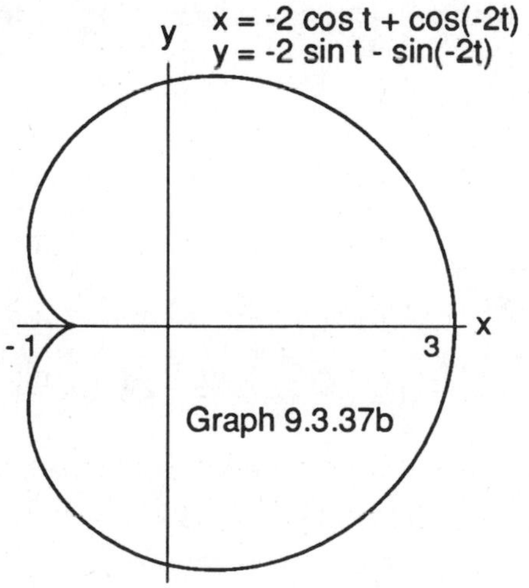

Graph 9.3.37b

39. a)

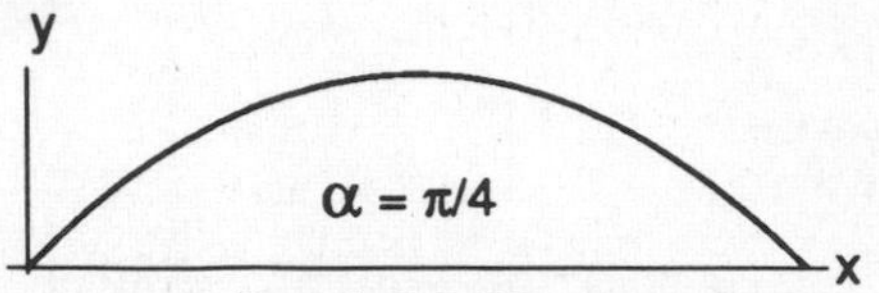

Graph 9.3.39 a

b)

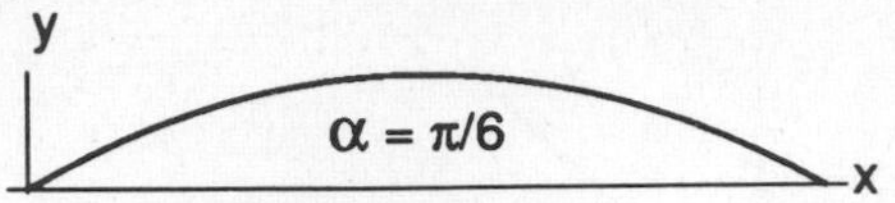

Graph 9.3.39 b

c)

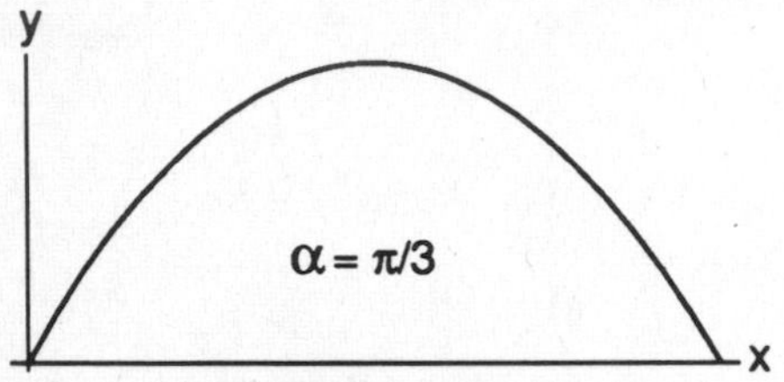

Graph 9.3.39 c

d)

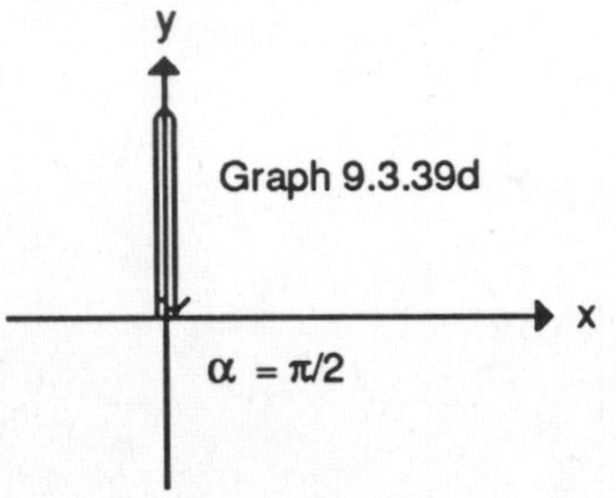

Graph 9.3.39d

SECTION 9.4 THE CALCULUS OF PARAMETRIZED CURVES

1. $t=\frac{\pi}{4}\Rightarrow x=2\cos\frac{\pi}{4}=\sqrt{2}$, $y=2\sin\frac{\pi}{4}=\sqrt{2}$. $\frac{dx}{dt}=-2\sin t$, $\frac{dy}{dt}=2\cos t\Rightarrow\frac{dy}{dx}=\frac{2\cos t}{-2\sin t}=-\cot t$.

$\therefore\ \frac{dy}{dx}\left(\frac{\pi}{4}\right)=-\cot\frac{\pi}{4}=-1$. $\therefore$ Tangent line is $y-\sqrt{2}=-1(x-\sqrt{2})\Rightarrow y=-x+2\sqrt{2}$.

$\frac{dy'}{dt}=\csc^2 t\Rightarrow\frac{d^2y}{dx^2}=\frac{\csc^2 t}{-2\sin t}=-\frac{1}{2\sin^3 t}\Rightarrow\frac{d^2y}{dx^2}\left(\frac{\pi}{4}\right)=-\sqrt{2}$

3. $t=\frac{\pi}{4}\Rightarrow x=4\sin\frac{\pi}{4}=2\sqrt{2}$, $y=2\cos\frac{\pi}{4}$. $\frac{dx}{dt}=4\cos t$, $\frac{dy}{dt}=-2\sin t\Rightarrow\frac{dy}{dx}=\frac{-2\sin t}{4\cos t}=-\frac{1}{2}\tan t\Rightarrow$

$\frac{dy}{dx}\left(\frac{\pi}{4}\right)=-\frac{1}{2}\tan\frac{\pi}{4}=-\frac{1}{2}$. $\therefore$ Tangent line is $y-\sqrt{2}=-\frac{1}{2}(x-2\sqrt{2})\Rightarrow y=-\frac{1}{2}x+2\sqrt{2}$.

$\frac{dy'}{dt}=-\frac{1}{2}\sec^2 t\Rightarrow\frac{d^2y}{dx^2}=\frac{-\frac{1}{2}\sec^2 t}{4\cos t}=-\frac{1}{8\cos^3 t}\Rightarrow\frac{d^2y}{dx^2}\left(\frac{\pi}{4}\right)=-\frac{\sqrt{2}}{4}$

5. $t=\frac{1}{4}\Rightarrow x=\frac{1}{4}$, $y=\frac{1}{2}$. $\frac{dx}{dt}=1$, $\frac{dy}{dt}=\frac{1}{2\sqrt{t}}\Rightarrow\frac{dy}{dx}=\frac{1}{2\sqrt{t}}\Rightarrow\frac{dy}{dx}\left(\frac{1}{4}\right)=\frac{1}{2\sqrt{\frac{1}{4}}}=1$.

$\therefore$ Tangent line is $y-\frac{1}{2}=1\left(x-\frac{1}{4}\right)\Rightarrow y=x+\frac{1}{4}$. $\frac{dy'}{dt}=-\frac{1}{4}t^{-3/2}\Rightarrow\frac{d^2y}{dx^2}=-\frac{1}{4}t^{-3/2}\Rightarrow\frac{d^2y}{dx^2}\left(\frac{1}{4}\right)=-2$.

7. $t=\frac{\pi}{6}\Rightarrow x=\sec\frac{\pi}{6}=\frac{2}{\sqrt{3}}$, $y=\tan\frac{\pi}{6}=\frac{1}{\sqrt{3}}$. $\frac{dx}{dt}=\sec t\tan t$, $\frac{dy}{dt}=\sec^2 t\Rightarrow\frac{dy}{dx}=\frac{\sec^2 t}{\sec t\tan t}=\csc t\Rightarrow$

$\frac{dy}{dx}\left(\frac{\pi}{6}\right)=\csc\frac{\pi}{6}=2$. $\therefore$ Tangent line is $y-\frac{1}{\sqrt{3}}=2\left(x-\frac{2}{\sqrt{3}}\right)\Rightarrow y=2x-\sqrt{3}$. $\frac{dy'}{dt}=-\csc t\cot t\Rightarrow$

$\frac{d^2y}{dx^2}=\frac{-\csc t\cot t}{\sec t\tan t}=-\cot^3 t\Rightarrow\frac{d^2y}{dx^2}\left(\frac{\pi}{6}\right)=-3\sqrt{3}$

9. $t=-1 \Rightarrow x=5,\ y=1.\ \frac{dx}{dt}=4t,\ \frac{dy}{dt}=4t^3 \Rightarrow \frac{dy}{dx}=\frac{4t^3}{4t}=t^2 \Rightarrow \frac{dy}{dx}(-1)=(-1)^2=1.$ ∴ Tangent line is $y-1=1(x-5) \Rightarrow y=x-4.\ \frac{dy'}{dt}=2t \Rightarrow \frac{d^2y}{dx^2}=\frac{2t}{4t}=\frac{1}{2} \Rightarrow \frac{d^2y}{dx^2}(-1)=\frac{1}{2}$

11. $t=\frac{\pi}{3} \Rightarrow x=\frac{\pi}{3}-\sin\frac{\pi}{3}=\frac{\pi}{3}-\frac{\sqrt{3}}{2},\ y=1-\cos\frac{\pi}{3}=1-\frac{1}{2}=\frac{1}{2}.\ \frac{dx}{dt}=1-\cos t,\ \frac{dy}{dt}=\sin t \Rightarrow \frac{dy}{dx}=\frac{\sin t}{1-\cos t}$

$\Rightarrow \frac{dy}{dx}\left(\frac{\pi}{3}\right)=\frac{\sin\frac{\pi}{3}}{1-\cos\frac{\pi}{3}}=\frac{\frac{\sqrt{3}}{2}}{\frac{1}{2}}=\sqrt{3}.$ ∴ Tangent line is $y-\frac{1}{2}=\sqrt{3}\left(x-\frac{\pi}{3}+\frac{\sqrt{3}}{2}\right) \Rightarrow y=\sqrt{3}\,x-\frac{\pi\sqrt{3}}{3}+2.$

$\frac{dy'}{dt}=\frac{(1-\cos t)\cos t-\sin t(\sin t)}{(1-\cos t)^2}=\frac{-1}{1-\cos t} \Rightarrow \frac{d^2y}{dx^2}=\frac{\frac{-1}{1-\cos t}}{1-\cos t}=\frac{-1}{(1-\cos t)^2} \Rightarrow \frac{d^2y}{dx^2}\left(\frac{\pi}{3}\right)=-4$

13. $\frac{dx}{dt}=-\sin t,\ \frac{dy}{dt}=1+\cos t \Rightarrow \sqrt{\left(\frac{dx}{dt}\right)^2+\left(\frac{dy}{dt}\right)^2}=\sqrt{(-\sin t)^2+(1+\cos t)^2}=\sqrt{2+2\cos t}.$

∴ Length $=\int_0^{\pi}\sqrt{2+2\cos t}\,dt=\sqrt{2}\int_0^{\pi}\sqrt{\frac{1-\cos t}{1-\cos t}(1+\cos t)}\,dt=\sqrt{2}\int_0^{\pi}\sqrt{\frac{\sin^2 t}{1-\cos t}}\,dt=$

$\sqrt{2}\int_0^{\pi}\frac{\sin t}{\sqrt{1-\cos t}}\,dt$ (since $\sin t\ge 0$ on $[0,\pi]$) $=\sqrt{2}\int_0^{2}u^{-1/2}\,du=\sqrt{2}\left[2u^{1/2}\right]_0^2=4.$

(Let $u=1-\cos t \Rightarrow du=\sin t\,dt;\ t=0 \Rightarrow u=0,\ t=\pi \Rightarrow u=2$)

15. $\frac{dx}{dt}=t,\ \frac{dy}{dt}=(2t+1)^{1/2} \Rightarrow \sqrt{\left(\frac{dx}{dt}\right)^2+\left(\frac{dy}{dt}\right)^2}=\sqrt{t^2+\left((2t+1)^{1/2}\right)^2}=|t+1|=t+1$ since $0\le t\le 4.$

∴ Length $=\int_0^4 (t+1)\,dt=\left[\frac{t^2}{2}+t\right]_0^4=12.$

17. $\frac{dx}{dt}=8t\cos t,\ \frac{dy}{dt}=8t\sin t \Rightarrow \sqrt{\left(\frac{dx}{dt}\right)^2+\left(\frac{dy}{dt}\right)^2}=\sqrt{(8t\cos t)^2+(8t\sin t)^2}=\sqrt{64t^2\cos^2 t+64t^2\sin^2 t}=|8t|$

$=8t$ since $0\le t\le\frac{\pi}{2}.$ ∴ Length $=\int_0^{\pi/2} 8t\,dt=\left[4t^2\right]_0^{\pi/2}=\pi^2.$

19. $\frac{dx}{dt}=-\sin t,\ \frac{dy}{dt}=\cos t \Rightarrow \sqrt{\left(\frac{dx}{dt}\right)^2+\left(\frac{dy}{dt}\right)^2}=\sqrt{(-\sin t)^2+(\cos t)^2}=1.$

∴ Area $=\int_0^{2\pi} 2\pi(2+\sin t)1\,dt=2\pi\left[2t-\cos t\right]_0^{2\pi}=8\pi^2$

21. $\frac{dx}{dt} = 1,\ \frac{dy}{dt} = t + \sqrt{2} \Rightarrow \sqrt{\left(\frac{dx}{dt}\right)^2 + \left(\frac{dy}{dt}\right)^2} = \sqrt{1^2 + (t + \sqrt{2})^2} = \sqrt{t^2 + 2\sqrt{2}\,t + 3}.$

$$\therefore \text{ Area} = \int_{-\sqrt{2}}^{\sqrt{2}} 2\pi(t + \sqrt{2})\sqrt{t^2 + 2\sqrt{2}\,t + 3}\,dt = \int_1^9 \pi\sqrt{u}\,du = \left[\frac{2}{3}\pi u^{3/2}\right]_1^9 = \frac{52}{3}\pi$$

$(\text{Let } u = t^2 + 2\sqrt{2}\,t + 3 \Rightarrow du = 2t + 2\sqrt{2};\ t = -\sqrt{2} \Rightarrow u = 1,\ t = \sqrt{2} \Rightarrow u = 9)$

23. $\frac{dx}{dt} = 2,\ \frac{dy}{dt} = 1 \Rightarrow \sqrt{\left(\frac{dx}{dt}\right)^2 + \left(\frac{dy}{dt}\right)^2} = \sqrt{2^2 + 1^2} = \sqrt{5}.\ \therefore \text{ Area} = \int_0^1 2\pi(t + 1)\sqrt{5}\,dt =$

$2\pi\sqrt{5}\left[\frac{t^2}{2} + t\right]_0^1 = 3\pi\sqrt{5}.$ The slant height is $\sqrt{5} \Rightarrow \text{Area} = \pi(1 + 2)\sqrt{5} = 3\pi\sqrt{5}.$

25. a) Let the density be $\delta = 1$. $x = \cos t + t\sin t \Rightarrow \frac{dx}{dt} = t\cos t$. $y = \sin t - t\cos t \Rightarrow \frac{dy}{dt} = \sin t$. $\therefore\ dm = 1 \cdot ds$ $= \sqrt{\left(\frac{dx}{dt}\right)^2 + \left(\frac{dy}{dt}\right)^2}\,dt = \sqrt{(t\cos t)^2 + (t\sin t)^2}\,dt = |t|\,dt = t\,dt$ since $0 \le t \le \frac{\pi}{2}$. Then the curve's mass is $M = \int_0^{\pi/2} t\,dt = \frac{\pi^2}{8}$. $M_x = \int \tilde{y}\,dm = \int_0^{\pi/2} (\sin t - t\cos t)\,t\,dt = \int_0^{\pi/2} t\sin t\,dt - \int_0^{\pi/2} t^2\cos t\,dt =$ $\left[\sin t - t\cos t\right]_0^{\pi/2}$ (By parts) $- \left[t^2\sin t - 2\sin t + 2t\cos t\right]_0^{\pi/2}$ (By parts twice) $= 3 - \frac{\pi^2}{4}$. $\therefore\ \bar{y} = \frac{M_x}{M} =$ $\frac{3 - \pi^2/4}{\pi^2/8} = \frac{24}{\pi^2} - 2$. $M_y = \int \tilde{x}\,dm = \int_0^{\pi/2} (\cos t + t\sin t)\,t\,dt = \int_0^{\pi/2} t\cos t\,dt + \int_0^{\pi/2} t^2\sin t\,dt =$ $\left[\cos t + t\sin t\right]_0^{\pi/2}$ (By parts) $+ \left[-t^2\cos t + 2\cos t + 2t\sin t\right]_0^{\pi/2}$ (By parts twice) $= \frac{3\pi}{2} - 3$. $\therefore\ \bar{x} = \frac{M_y}{M} =$ $\frac{(3\pi/2) - 3}{\pi^2/8} = \frac{12}{\pi} - \frac{24}{\pi^2}$. Then $(\bar{x}, \bar{y}) = \left(\frac{12}{\pi} - \frac{24}{\pi^2}, \frac{24}{\pi^2} - 2\right)$

b) $(\bar{x}, \bar{y}) \approx (1.4, 0.4)$

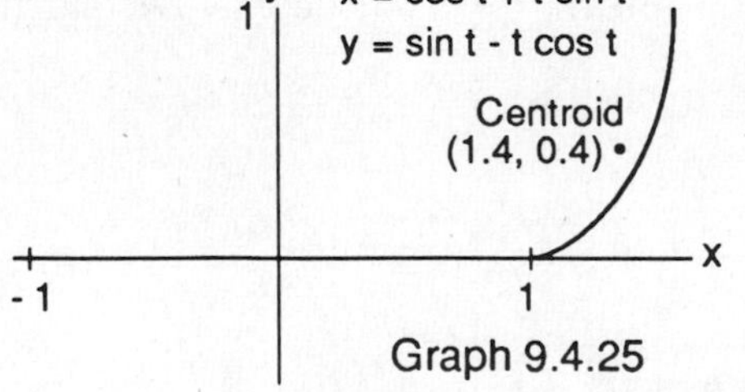

Graph 9.4.25

27. a) Let the density be $\delta = 1$. $x = \cos t \Rightarrow \frac{dx}{dt} = -\sin t$. $y = t + \sin t \Rightarrow \frac{dy}{dx} = 1 + \cos t$. $dm = 1 \cdot ds =$

$\sqrt{\left(\frac{dx}{dt}\right)^2 + \left(\frac{dy}{dt}\right)^2}\, dt = \sqrt{(-\sin t)^2 + (1 + \cos t)^2}\, dt = \sqrt{2 + 2\cos t}\, dt$. The curve's mass is $M =$

$$\int_0^{\pi} \sqrt{2 + 2\cos t}\, dt = \sqrt{2} \int_0^{\pi} \sqrt{1 + \cos t}\, dt = \sqrt{2} \int_0^{\pi} \sqrt{2\cos^2\left(\frac{1}{2}t\right)}\, dt = 2 \int_0^{\pi} \left|\cos\left(\frac{1}{2}t\right)\right| dt =$$

$$2 \int_0^{\pi} \cos\left(\frac{1}{2}t\right) dt \left(\text{since } 0 \le t \le \pi \Rightarrow 0 \le \frac{1}{2}t \le \frac{\pi}{2}\right) = 2\left[2\sin\left(\frac{1}{2}t\right)\right]_0^{\pi} = 4.\ M_x = \int \tilde{y}\, dm =$$

$$\int_0^{\pi} (t + \sin t)\left(2\cos\frac{1}{2}t\right) dt = \int_0^{\pi} 2t\cos\frac{1}{2}t\, dt + \int_0^{\pi} 2\sin t \cos\frac{1}{2}t\, dt = 2\left[4\cos\frac{1}{2}t + 2t\sin\frac{1}{2}t\right]_0^{\pi} +$$

$$2\left[-\frac{\cos\frac{3}{2}t}{3} - \frac{\cos\frac{1}{2}t}{1}\right]_0^{\pi} \text{ (By using the Tables)} = 4\pi - \frac{16}{3}.\ \therefore \bar{y} = \frac{M_x}{M} = \frac{4\pi - (16/3)}{4} = \pi - \frac{4}{3}.$$

$$M_y = \int \tilde{x}\, dm = \int_0^{\pi} \cos t\left(2\cos\frac{1}{2}t\right) dt = 2\int_0^{\pi} \cos t \cos\frac{1}{2}t\, dt = 2\left[\sin\frac{1}{2}t + \frac{\sin\frac{3}{2}t}{3}\right]_0^{\pi} = 2 - \frac{2}{3} = \frac{4}{3}$$

$$\therefore \bar{x} = \frac{M_y}{M} = \frac{4/3}{4} = \frac{1}{3}.\ \therefore (\bar{x},\bar{y}) = \left(\frac{1}{3}, \pi - \frac{4}{3}\right).$$

b) $(\bar{x},\bar{y}) \approx (0.33, 1.81)$

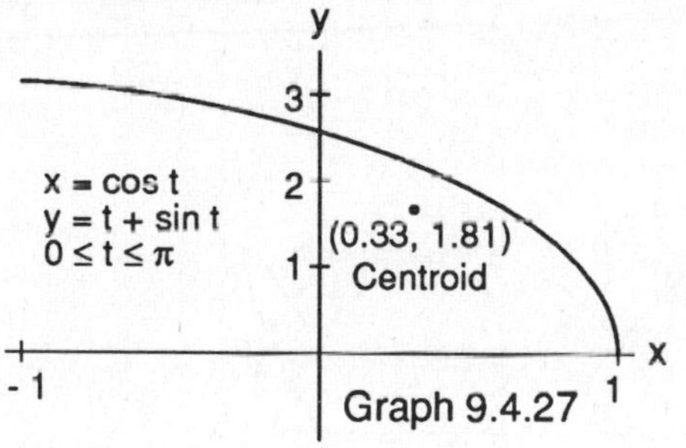

Graph 9.4.27

29. a) $\frac{dx}{dt} = -2\sin 2t,\ \frac{dy}{dt} = 2\cos 2t \Rightarrow \sqrt{\left(\frac{dx}{dt}\right)^2 + \left(\frac{dy}{dt}\right)^2} = \sqrt{(-2\sin 2t)^2 + (2\cos 2t)^2} = 2.$

$$\therefore \text{ Length} = \int_0^{\pi/2} 2\, dt = [2t]_0^{\pi/2} = \pi.$$

b) $\frac{dx}{dt} = \pi\cos \pi t,\ \frac{dy}{dt} = -\pi\sin \pi t \Rightarrow \sqrt{\left(\frac{dx}{dt}\right)^2 + \left(\frac{dy}{dt}\right)^2} = \sqrt{(\pi\cos \pi t)^2 + (-\pi\sin \pi t)^2} = \pi.$

$$\therefore \text{ Length} = \int_{-1/2}^{1/2} \pi\, dt = [\pi t]_{-1/2}^{1/2} = \pi.$$

31. $\frac{dx}{dt} = \cos t, \frac{dy}{dt} = 2\cos 2t \Rightarrow \frac{dy}{dx} = \frac{2\cos 2t}{\cos t} = \frac{2(2\cos^2 t - 1)}{\cos t}$. Let $\frac{2(2\cos^2 t - 1)}{\cos t} = 0 \Rightarrow 2\cos^2 t - 1 = 0$

$\Rightarrow \cos t = \pm\frac{1}{\sqrt{2}} \Rightarrow t = \frac{\pi}{4}, \frac{3\pi}{4}, \frac{5\pi}{4}, \frac{7\pi}{4}$. In the 1st quadrant, $t = \frac{\pi}{4} \Rightarrow x = \sin\frac{\pi}{4} = \frac{\sqrt{2}}{2}$, $y = \sin 2\left(\frac{\pi}{4}\right) = 1$

$\Rightarrow \left(\frac{\sqrt{2}}{2}, 1\right)$ is the point in the 1st quadrant where the tangent line is horizontal.

$x = 0, y = 0 \Rightarrow \sin t = 0 \Rightarrow t = 0$ or $t = \pi$ and $\sin 2t = 0 \Rightarrow t = 0, \frac{\pi}{2}, \pi, \frac{3\pi}{2}$. $\therefore$ $t = 0$ and $t = \pi$ give the tangent lines at the origin. $\frac{dy}{dx}(0) = 2 \Rightarrow y = 2x$ and $\frac{dy}{dx}(\pi) = -2 \Rightarrow y = -2x$.

33.

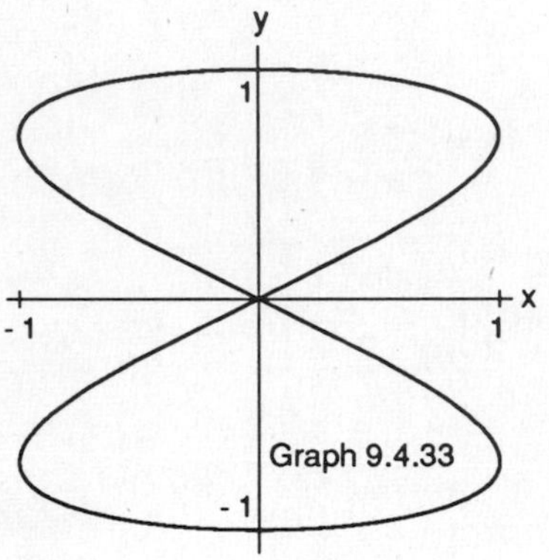

Graph 9.4.33

35.

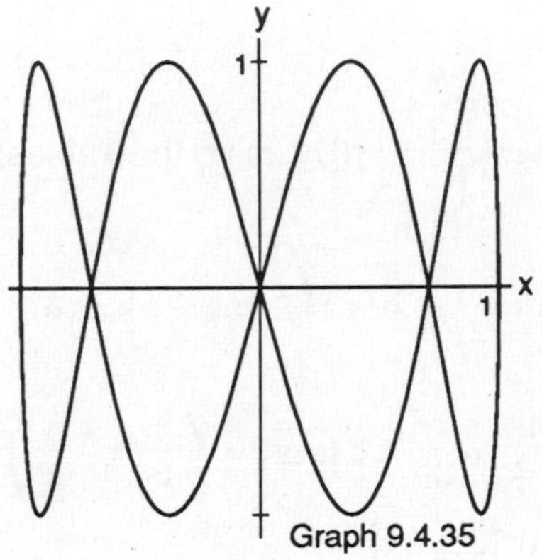

Graph 9.4.35

37.

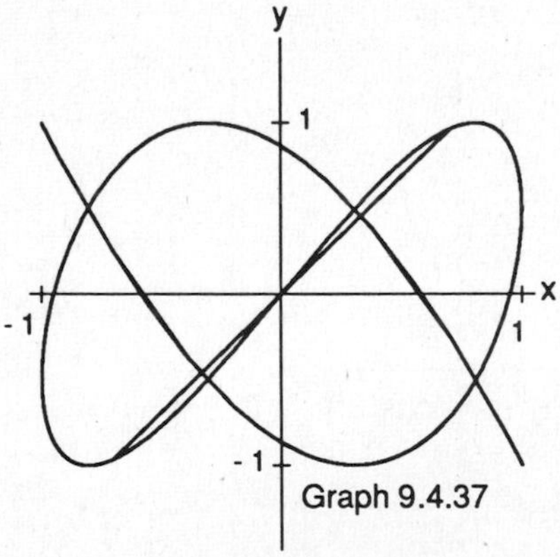

Graph 9.4.37

39.

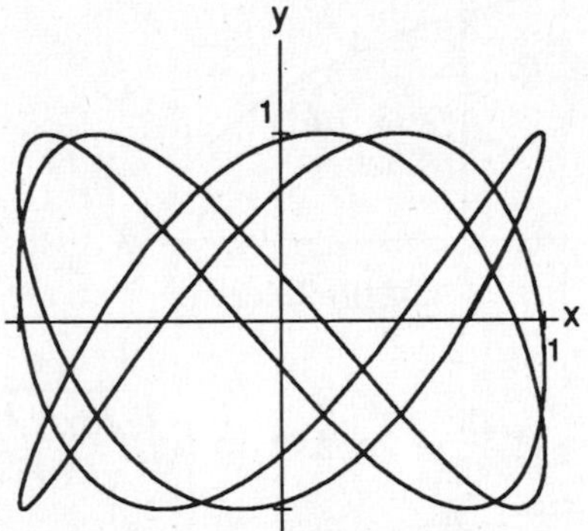

Graph 9.4.39

SECTION 9.5 POLAR COORDINATES

1. a, c; b, d; e, k; g, j; h, f; i, l; m, o; n, p

3. a) $\left(2, \frac{\pi}{2} + 2n\pi\right)$ and $\left(-2, \frac{\pi}{2} + (2n+1)\pi\right)$, n an integer

   b) $(2, 2n\pi)$ and $(-2, (2n+1)\pi)$, n an integer

   c) $\left(2, \frac{3\pi}{2} + 2n\pi\right)$ and $\left(-2, \frac{3\pi}{2} + (2n+1)\pi\right)$, n an integer

   d) $(2, (2n+1)\pi)$ and $(-2, 2n\pi)$, n an integer

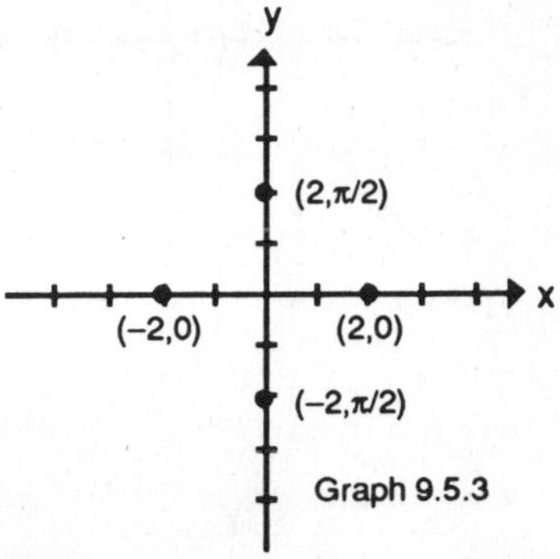

Graph 9.5.3

5. a) $x = \sqrt{2}\cos\frac{\pi}{4} = 1,\ y = \sqrt{2}\sin\frac{\pi}{4} = 1 \Rightarrow (1,1)$

   b) $x = 1\cos 0 = 1,\ y = 1\sin 0 = 1 \Rightarrow (1,0)$

   c) $x = 0\cos\frac{\pi}{2} = 0,\ y = 0\sin\frac{\pi}{2} = 0 \Rightarrow (0,0)$

   d) $x = -\sqrt{2}\cos\frac{\pi}{4} = -1,\ y = -\sqrt{2}\sin\frac{\pi}{2} = -1 \Rightarrow (-1,-1)$

   e) $x = -3\cos\frac{5\pi}{6} = \frac{3\sqrt{3}}{2},\ y = -3\sin\frac{5\pi}{6} = -\frac{3}{2} \Rightarrow \left(\frac{3\sqrt{3}}{2}, -\frac{3}{2}\right)$

   f) $x = 5\cos(\tan^{-1}\frac{4}{3}) = 3,\ y = 5\sin(\tan^{-1}\frac{4}{3}) = 4 \Rightarrow (3,4)$

   g) $x = -1\cos 7\pi = 1,\ y = -1\sin 7\pi = 0 \Rightarrow (1,0)$

   h) $x = 2\sqrt{3}\cos\frac{2\pi}{3} = -\sqrt{3},\ y = 2\sqrt{3}\sin\frac{2\pi}{3} = 3 \Rightarrow (-\sqrt{3}, 3)$

7.

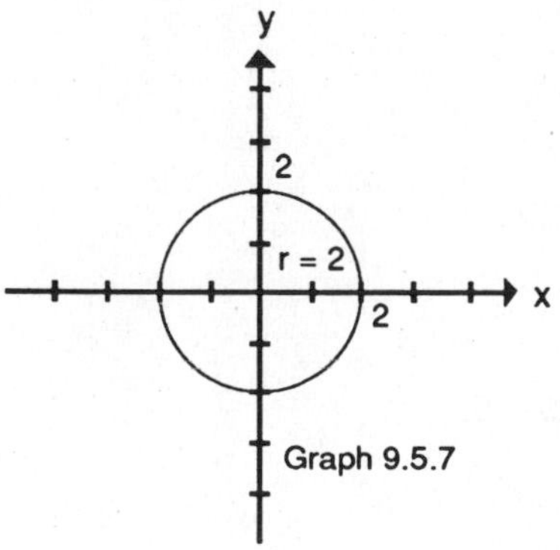

Graph 9.5.7

9.

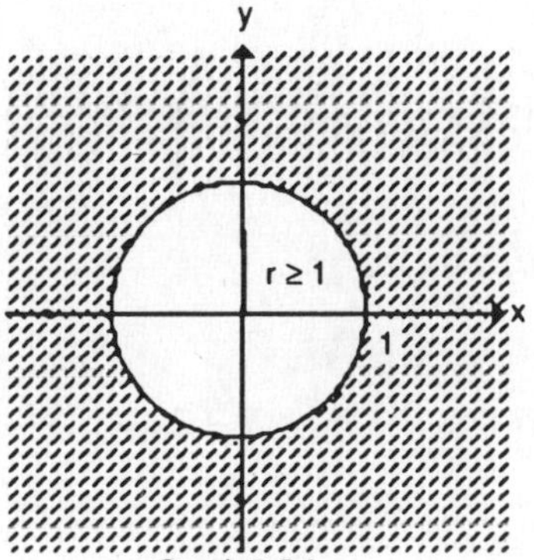

Graph 9.5.9

11.

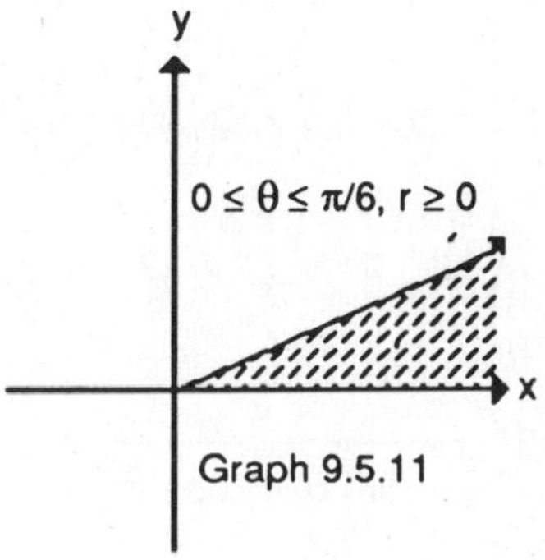

Graph 9.5.11

13.

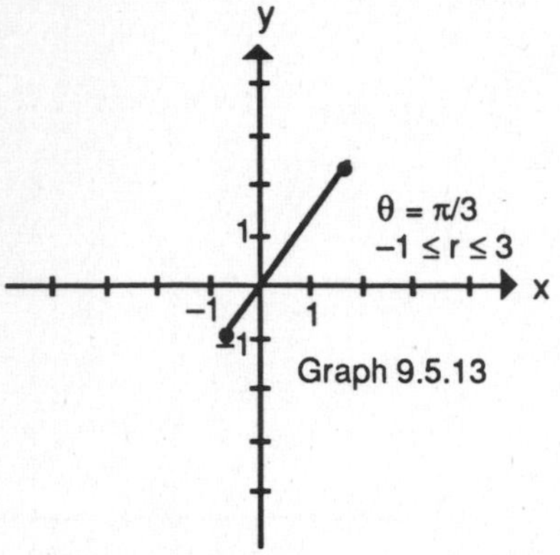

Graph 9.5.13

15.

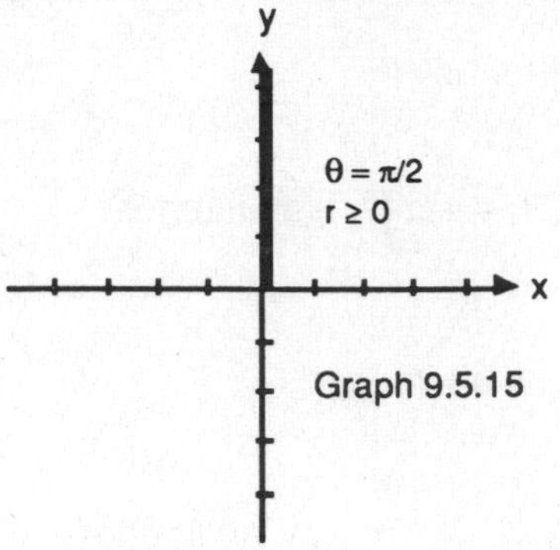

Graph 9.5.15

17.

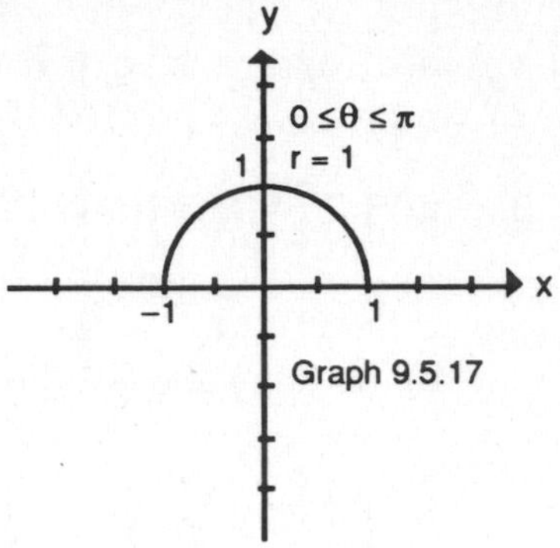

Graph 9.5.17

19.

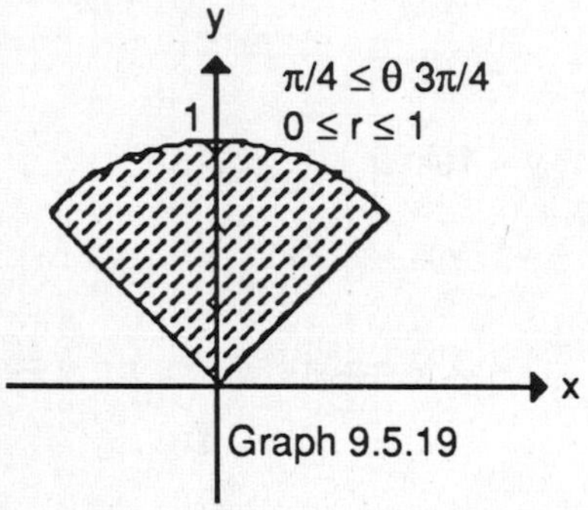

Graph 9.5.19

21.

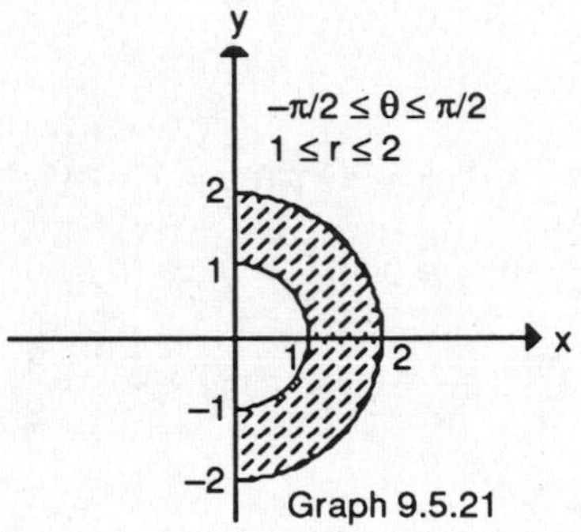

Graph 9.5.21

23. $r\cos\theta = 2 \Rightarrow x = 2$, vertical line through (2,0).

25. $r\sin\theta = 4 \Rightarrow y = 4$, horizontal line through (0,4).

27. $r\sin\theta = 0 \Rightarrow y = 0$, the x–axis.

29. $r\cos\theta + r\sin\theta = 1 \Rightarrow x + y = 1$, line, $m = -1$, $b = 1$

31. $r^2 = 1 \Rightarrow x^2 + y^2 = 1$, circle, $C = (0,0)$, $r = 1$

33. $r = \dfrac{5}{\sin\theta - 2\cos\theta} \Rightarrow r\sin\theta - 2r\cos\theta = 5 \Rightarrow y - 2x = 5$, line, $m = 2$, $b = 5$

35. $r = \cot\theta\csc\theta = \dfrac{\cos\theta}{\sin\theta}\left(\dfrac{1}{\sin\theta}\right) \Rightarrow r\sin^2\theta = \cos\theta \Rightarrow r^2\sin^2\theta = r\cos\theta \Rightarrow y^2 = x$, parabola, vertex is (0,0), opens right.

37. $r = \csc\theta\, e^{r\cos\theta} \Rightarrow r\sin\theta = e^{r\cos\theta} \Rightarrow y = e^x$, natural exponential function

39. $r^2 + 2r^2\cos\theta\sin\theta = 1 \Rightarrow x^2 + y^2 + 2xy = 1 \Rightarrow x^2 + 2xy + y^2 = 1 \Rightarrow (x + y)^2 = 1 \Rightarrow x + y = \pm 1$, two straight lines, slope of each is -1, y–intercepts are ± 1.

41. $r = 2\cos\theta + 2\sin\theta \Rightarrow r^2 = 2r\cos\theta + 2r\sin\theta \Rightarrow x^2 + y^2 = 2x + 2y \Rightarrow x^2 - 2x + y^2 - 2y = 0 \Rightarrow (x - 1)^2 + (y - 1)^2 = 2$, circle, center is (1,1), $r = \sqrt{2}$.

43. $x = 7 \Rightarrow r\cos\theta = 7$

45. $x = y \Rightarrow r\cos\theta = r\sin\theta \Rightarrow \theta = \frac{\pi}{4}$

47. $x^2 + y^2 = 4 \Rightarrow r^2 = 4$ or $r = 2$ or $r = -2$

49. $\frac{x^2}{9} + \frac{y^2}{4} = 1 \Rightarrow 4x^2 + 9y^2 = 36 \Rightarrow 4r^2\cos^2\theta + 9r^2\sin^2\theta = 36$ or $r^2\cos 2\theta = 1$

51. $y^2 = 4x \Rightarrow r^2\sin^2\theta = 4r\cos\theta$

SECTION 9.6 GRAPHING IN POLAR COORDINATES

1.

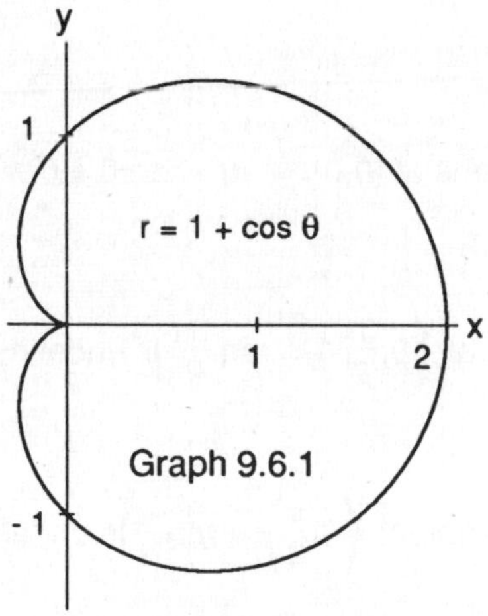

Graph 9.6.1

3.

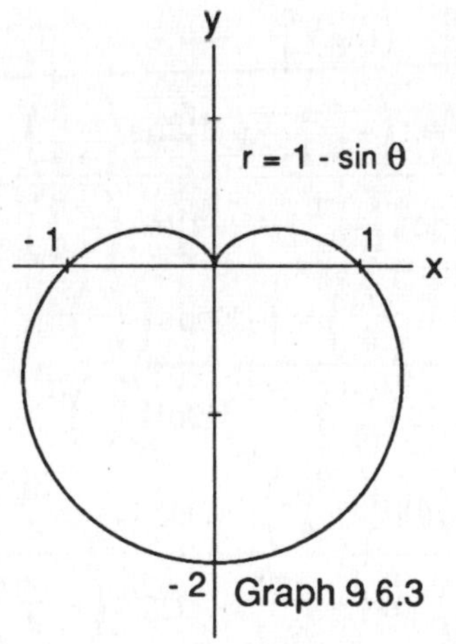

Graph 9.6.3

5.

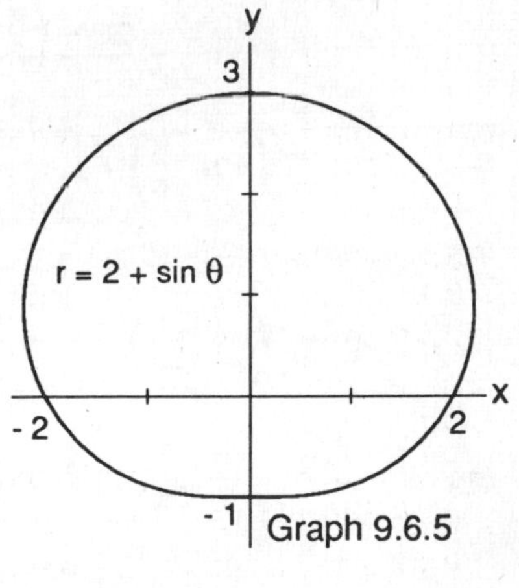

Graph 9.6.5

7.

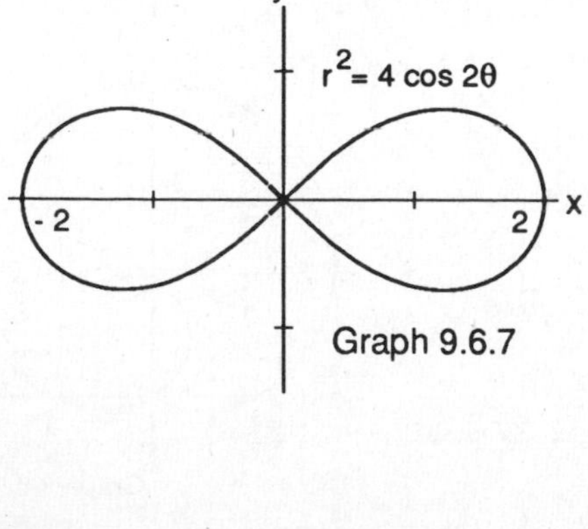

Graph 9.6.7

9.

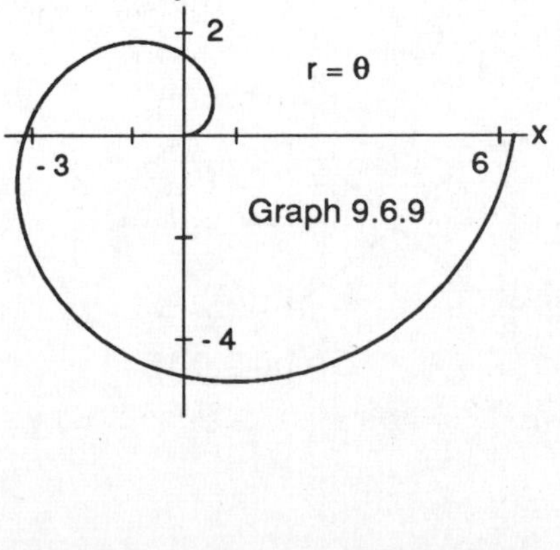

Graph 9.6.9

11. $\theta = \frac{\pi}{2} \Rightarrow r = -1 \Rightarrow \left(-1, \frac{\pi}{2}\right)$, $\theta = -\frac{\pi}{2} \Rightarrow r = -1 \Rightarrow \left(-1, -\frac{\pi}{2}\right)$

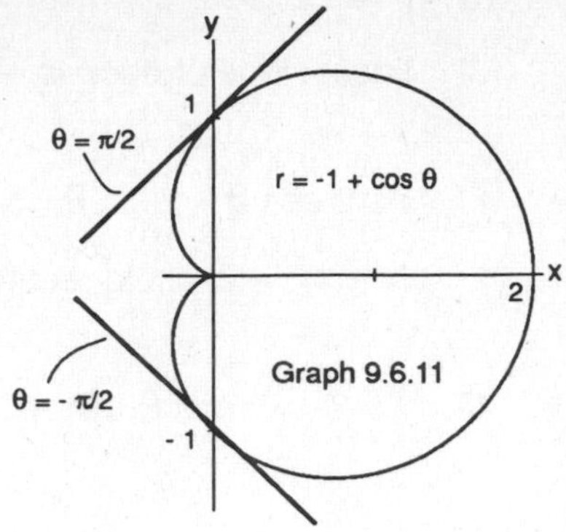

Graph 9.6.11

$r' = \frac{dr}{d\theta} = -\sin\theta$. Slope $= \frac{r'\sin\theta + r\cos\theta}{r'\cos\theta - r\sin\theta} = \frac{-\sin^2\theta + r\cos\theta}{-\sin\theta\cos\theta - r\sin\theta}$

$$\Rightarrow \text{Slope at } \left(-1, \frac{\pi}{2}\right) = \frac{-\sin^2\left(\frac{\pi}{2}\right) + (-1)\cos\frac{\pi}{2}}{-\sin\frac{\pi}{2}\cos\frac{\pi}{2} - (-1)\sin\frac{\pi}{2}} = -1. \text{ Slope at } \left(-1, -\frac{\pi}{2}\right)$$

$$= \frac{-\sin^2\left(-\frac{\pi}{2}\right) + (-1)\cos\left(-\frac{\pi}{2}\right)}{-\sin\left(-\frac{\pi}{2}\right)\cos\left(-\frac{\pi}{2}\right) - (-1)\sin\left(-\frac{\pi}{2}\right)} = 1$$

13. $\theta = \frac{\pi}{4} \Rightarrow r = 1 \Rightarrow \left(1, \frac{\pi}{4}\right)$. $\theta = -\frac{\pi}{4} \Rightarrow r = -1 \Rightarrow \left(-1, -\frac{\pi}{4}\right)$

$\theta = \frac{3\pi}{4} \Rightarrow r = -1 \Rightarrow \left(-1, \frac{3\pi}{4}\right)$. $\theta = -\frac{3\pi}{4} \Rightarrow r = 1 \Rightarrow \left(1, -\frac{3\pi}{4}\right)$

$r = 0 \Rightarrow \theta = 0, \frac{\pi}{2}, \pi, \frac{3\pi}{2}$. $r' = \frac{dr}{d\theta} = 2\cos 2\theta \Rightarrow$

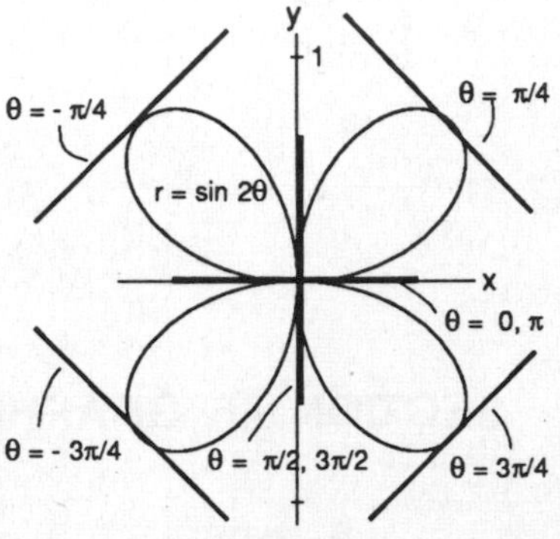

Graph 9.6.13

$$\text{Slope} = \frac{r'\sin\theta + r\cos\theta}{r'\cos\theta - r\sin\theta} = \frac{2\cos 2\theta\sin\theta + r\cos\theta}{2\cos 2\theta\cos\theta - r\sin\theta} \Rightarrow$$

$$\text{Slope at } \left(1, \frac{\pi}{4}\right) = \frac{2\cos\left(\frac{\pi}{2}\right)\sin\frac{\pi}{4} + (1)\cos\frac{\pi}{4}}{2\cos\left(\frac{\pi}{2}\right)\cos\frac{\pi}{4} - (1)\sin\frac{\pi}{4}} = -1$$

$$\text{Slope at } \left(-1, -\frac{\pi}{4}\right) = \frac{2\cos\left(-\frac{\pi}{2}\right)\sin\left(-\frac{\pi}{4}\right) + (-1)\cos\left(-\frac{\pi}{4}\right)}{2\cos\left(-\frac{\pi}{2}\right)\cos\left(-\frac{\pi}{4}\right) - (-1)\sin\left(-\frac{\pi}{4}\right)} = 1$$

Slope at $(0,0)$, $(0,\pi) = \tan 0 = 0$

$$\text{Slope at } \left(-1, \frac{3\pi}{4}\right) = \frac{2\cos\left(\frac{3\pi}{2}\right)\sin\left(\frac{3\pi}{4}\right) + (-1)\cos\left(\frac{3\pi}{4}\right)}{2\cos\left(\frac{3\pi}{2}\right)\cos\left(\frac{3\pi}{4}\right) - (-1)\sin\left(\frac{3\pi}{4}\right)} = 1$$

Slope at $\left(0, \frac{3\pi}{2}\right) = \tan\frac{3\pi}{2}$ is undefined

$$\text{Slope at } \left(1, -\frac{3\pi}{4}\right) = \frac{2\cos\left(-\frac{3\pi}{2}\right)\sin\left(-\frac{3\pi}{4}\right) + (1)\cos\left(-\frac{3\pi}{4}\right)}{2\cos\left(-\frac{3\pi}{2}\right)\cos\left(-\frac{3\pi}{4}\right) - (1)\sin\left(-\frac{3\pi}{4}\right)} = -1$$

Slope at $\left(0, \frac{\pi}{2}\right) = \tan\frac{\pi}{2}$ is undefined

15.

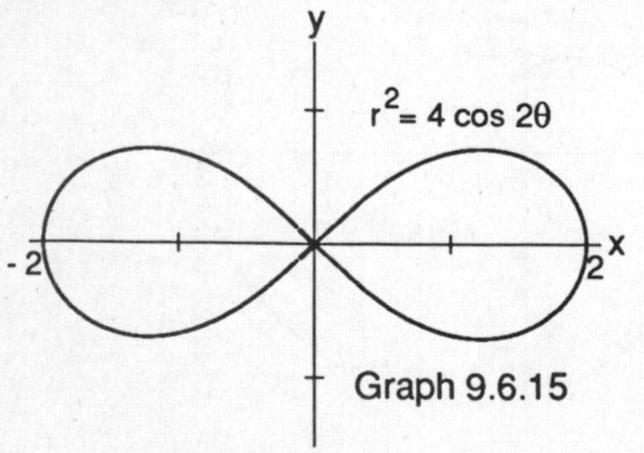

Graph 9.6.15

17. a)

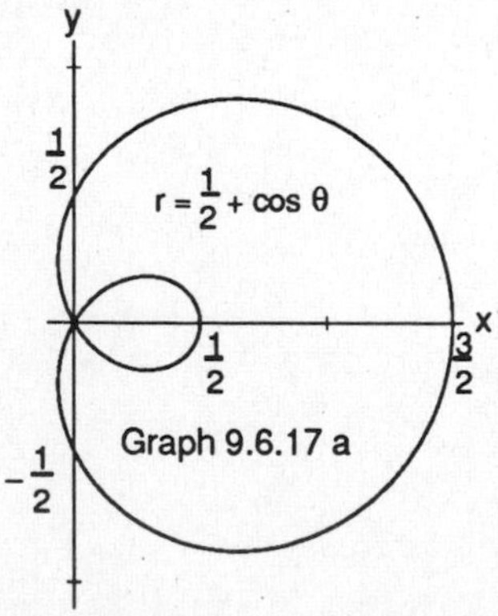

Graph 9.6.17 a

17. b)

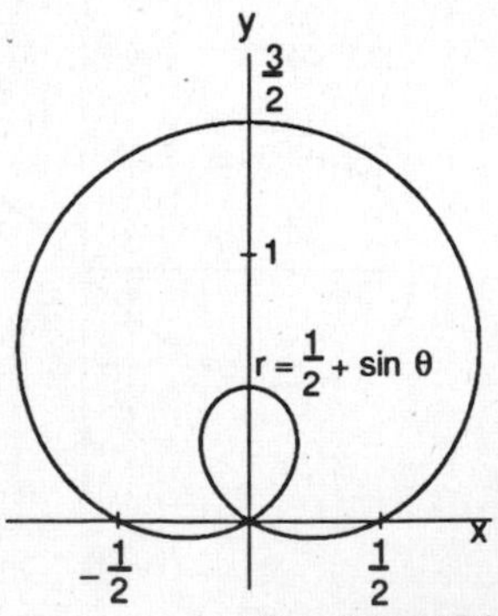

Graph 9.6.17 b

19. a)

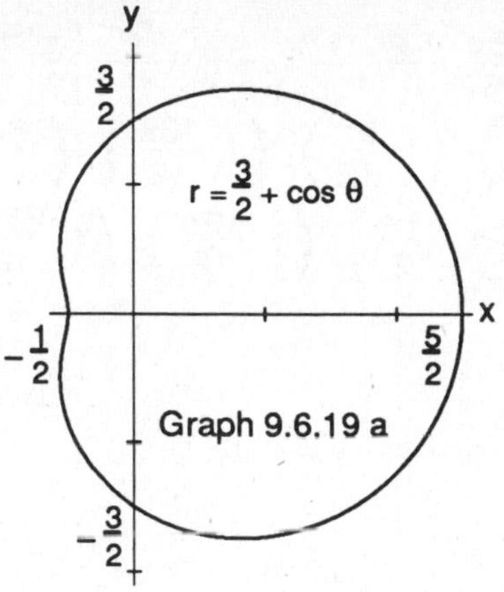

Graph 9.6.19 a

19. b)

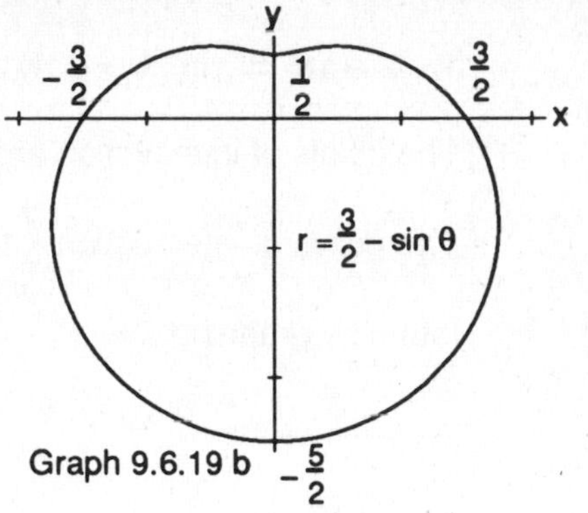

Graph 9.6.19 b

21.

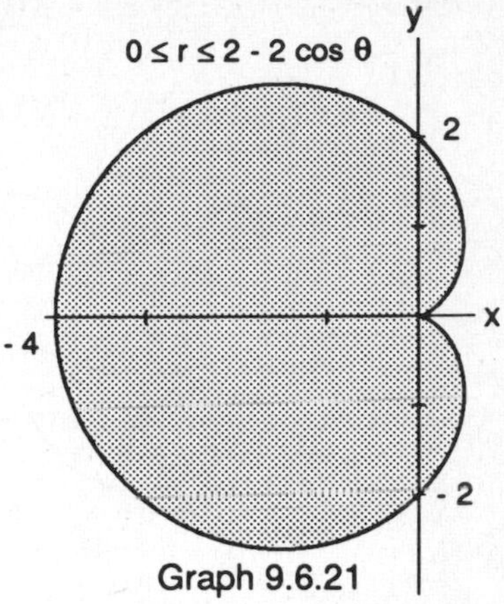

Graph 9.6.21

23. $\left(2, \frac{3\pi}{4}\right)$ is the same point as $\left(-2, -\frac{\pi}{4}\right)$. $r = 2\sin 2\left(-\frac{\pi}{4}\right) = 2\sin\left(-\frac{\pi}{2}\right) = -2 \Rightarrow \left(-2, -\frac{\pi}{4}\right)$ is on the graph $\Rightarrow \left(2, \frac{3\pi}{4}\right)$ is on the graph.

25. $1 + \cos\theta = 1 - \cos\theta \Rightarrow \cos\theta = 0 \Rightarrow$

$\theta = \frac{\pi}{2}, \frac{3\pi}{2} \Rightarrow r = 1$. Points of intersection are $\left(1, \frac{\pi}{2}\right)$ and $\left(1, \frac{3\pi}{2}\right)$. The point of intersection, (0,0), is found by graphing.

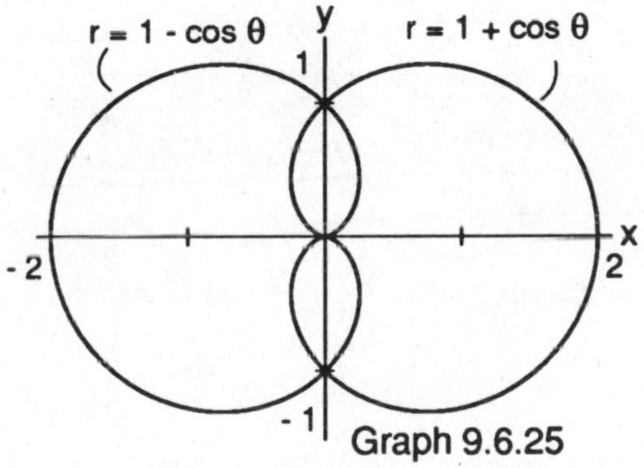

Graph 9.6.25

27. $(1 - \sin\theta)^2 = 4\sin\theta \Rightarrow 1 - 2\sin\theta + \sin^2\theta = 4\sin\theta \Rightarrow$

$1 - 6\sin\theta + \sin^2\theta = 0 \Rightarrow \sin\theta = \frac{6 \pm \sqrt{32}}{2} = 3 \pm 2\sqrt{2}.$

$\sin\theta$ cannot be $3 + 2\sqrt{2}$. $\therefore\ \theta = \sin^{-1}(3 - 2\sqrt{2}) \Rightarrow$

$r^2 = 4\sin\theta = 4(3 - 2\sqrt{2}) \Rightarrow r = \pm 2\sqrt{3 - 2\sqrt{2}} = \pm 2\sqrt{1 - 2\sqrt{2} + 2}$

$\Rightarrow r = \pm 2\left|1 - \sqrt{2}\right| = \pm 2\left(\sqrt{2} - 1\right)$. Points of intersection are $\left(2\left(\sqrt{2} - 1\right), \sin^{-1}\left(3 - 2\sqrt{2}\right)\right)$ and $\left(-2\left(\sqrt{2} - 1\right), -\sin^{-1}\left(3 - 2\sqrt{2}\right)\right)$.

Points of intersection (0,0) and $\left(2, \frac{3\pi}{2}\right)$ found by graphing.

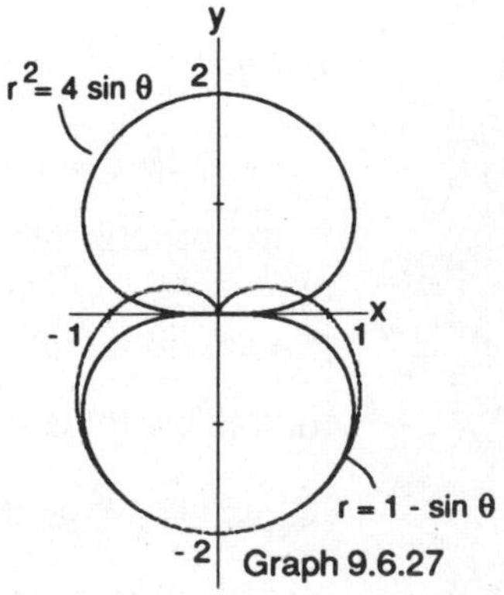

Graph 9.6.27

29.

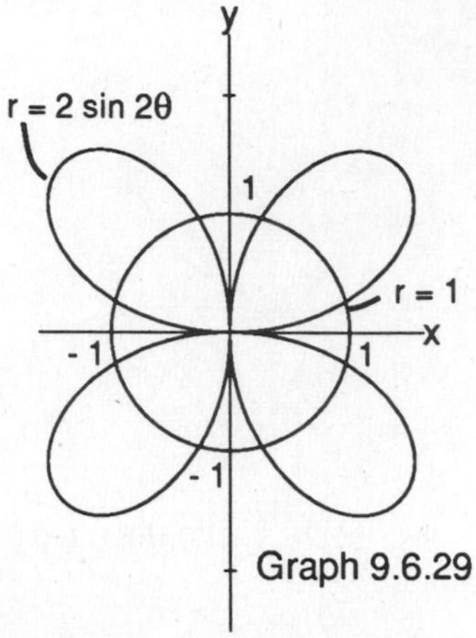

Graph 9.6.29

$1 = 2\sin\theta \Rightarrow \sin 2\theta = \frac{1}{2} \Rightarrow 2\theta = \frac{\pi}{6}, \frac{5\pi}{6}, \frac{13\pi}{6}, \frac{17\pi}{6} \Rightarrow \theta = \frac{\pi}{12}, \frac{5\pi}{12}, \frac{13\pi}{12}, \frac{17\pi}{12}$

The points of intersection are $\left(1, \frac{\pi}{12}\right), \left(1, \frac{5\pi}{12}\right), \left(1, \frac{13\pi}{12}\right), \left(1, \frac{17\pi}{12}\right)$.

The points of intersection $\left(1, \frac{7\pi}{12}\right), \left(1, \frac{11\pi}{12}\right), \left(1, \frac{19\pi}{12}\right), \left(1, \frac{23\pi}{12}\right)$ are found by graphing.

31.

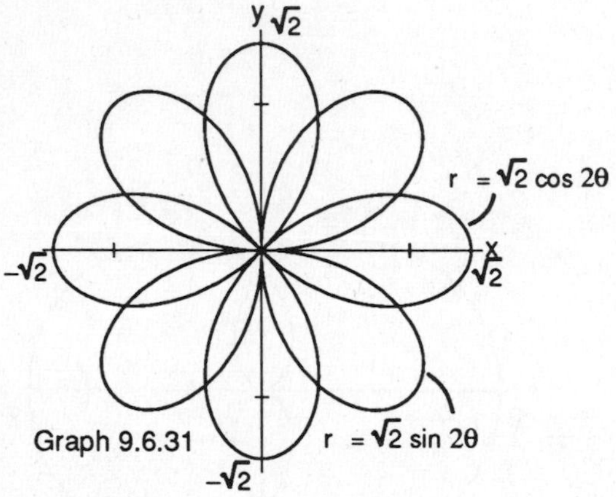

Graph 9.6.31

$\sqrt{2}\cos 2\theta = \sqrt{2}\sin 2\theta \Rightarrow \cos 2\theta = \sin 2\theta \Rightarrow 2\theta = \frac{\pi}{4}, \frac{5\pi}{4}, \frac{9\pi}{4}, \frac{13\pi}{4} \Rightarrow \theta = \frac{\pi}{8}, \frac{5\pi}{8}, \frac{9\pi}{8}, \frac{13\pi}{8}$. $\theta = \frac{\pi}{8}, \frac{9\pi}{8} \Rightarrow r = 1$. $\theta = \frac{5\pi}{8}, \frac{13\pi}{8} \Rightarrow r = -1$. The points of intersection are $\left(1, \frac{\pi}{8}\right)$, $\left(1, \frac{9\pi}{8}\right), \left(-1, \frac{5\pi}{8}\right), \left(-1, \frac{13\pi}{8}\right)$.

The points of intersection $\left(1, \frac{3\pi}{8}\right), \left(1, \frac{7\pi}{8}\right), \left(1, \frac{11\pi}{8}\right)$, $\left(1, \frac{15\pi}{8}\right)$, (0,0) are found by graphing.

33. a) $r^2 = -4\cos\theta \Rightarrow \cos\theta = -\frac{r^2}{4}$. $r = 1 - \cos\theta \Rightarrow r = 1 - \left(-\frac{r^2}{4}\right) \Rightarrow 0 = r^2 - 4r + 4 \Rightarrow (r-2)^2 = 0 \Rightarrow$ $r = 2$. $\therefore \cos\theta = -\frac{2^2}{4} = -1 \Rightarrow \theta = \pi$ $\therefore (2,\pi)$ is a point of intersection.

b) $r = 0 \Rightarrow 0^2 = -4\cos\theta \Rightarrow \cos\theta = 0 \Rightarrow \theta = \frac{\pi}{2}, \frac{3\pi}{2} \Rightarrow \left(0,\frac{\pi}{2}\right)$ or $\left(0,\frac{3\pi}{2}\right)$ is on the graph.

$r = 0 \Rightarrow 0 = 1 - \cos\theta \Rightarrow \cos\theta = 1 \Rightarrow \theta = 0 \Rightarrow (0,0)$ is on the graph. Since $(0,0) = \left(0,\frac{\pi}{2}\right)$, the graphs intersect at the origin.

35. $r^2 = \sin 2\theta$ and $r^2 = \cos 2\theta$ are generated completely for $0 \le \theta \le \frac{\pi}{2}$. Then $\sin 2\theta = \cos 2\theta$ yields $2\theta = \frac{\pi}{4}$ as the only solution on that interval $\Rightarrow$ $\theta = \frac{\pi}{8} \Rightarrow r^2 = \sin 2\left(\frac{\pi}{8}\right) = \frac{1}{\sqrt{2}} \Rightarrow r = \pm\frac{1}{\sqrt[4]{2}}$.

$\therefore$ Points of intersection are $\left(\pm\frac{1}{\sqrt[4]{2}}, \frac{\pi}{8}\right)$ The point of intersection (0,0) is found by graphing.

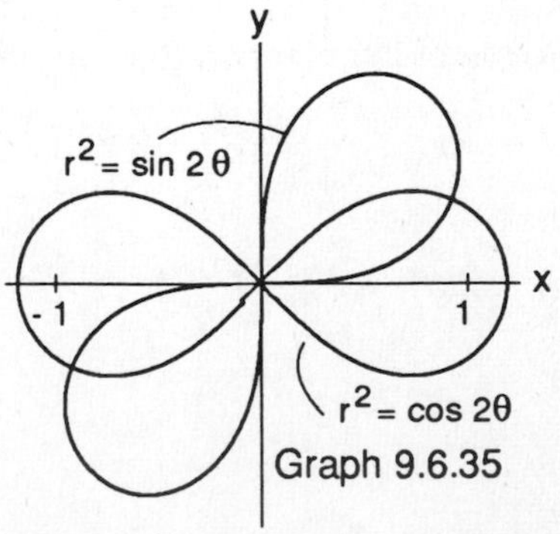

Graph 9.6.35

37. $1 = 2\sin 2\theta \Rightarrow \sin 2\theta = \frac{1}{2} \Rightarrow 2\theta = \frac{\pi}{6}, \frac{5\pi}{6}, \frac{13\pi}{6}, \frac{17\pi}{6}$ $\Rightarrow \theta = \frac{\pi}{12}, \frac{5\pi}{12}, \frac{13\pi}{12}, \frac{17\pi}{12}$. Points of intersection are $\left(1, \frac{\pi}{12}\right)$, $\left(1, \frac{5\pi}{12}\right)$, $\left(1, \frac{13\pi}{12}\right)$, and $\left(1, \frac{17\pi}{12}\right)$. Points of intersection $\left(1, \frac{7\pi}{12}\right)$, $\left(1, \frac{11\pi}{12}\right)$, $\left(1, \frac{19\pi}{12}\right)$, and $\left(1, \frac{23\pi}{12}\right)$ found by graphing and symmetry.

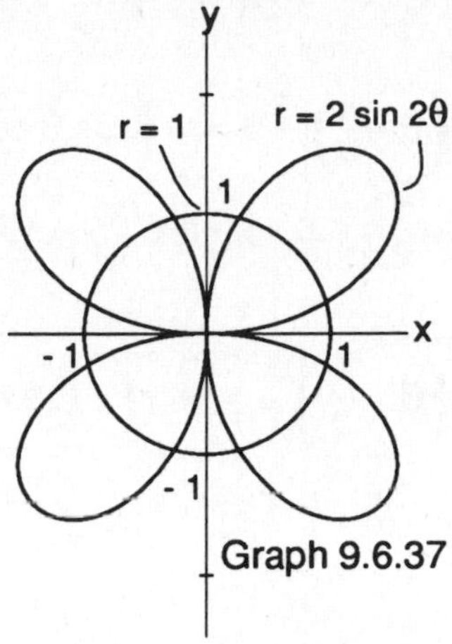

Graph 9.6.37

39.

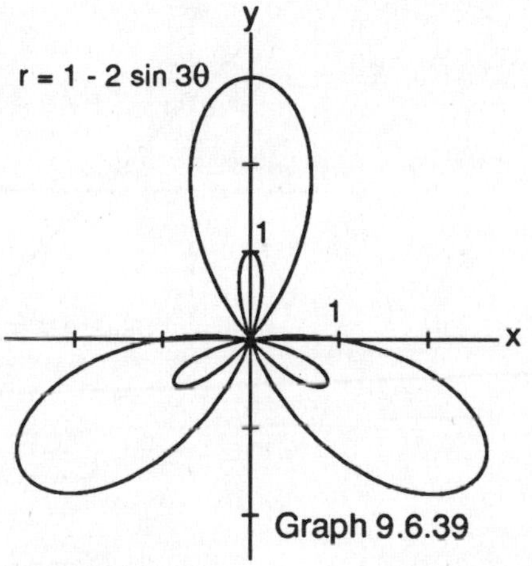

Graph 9.6.39

41. a)

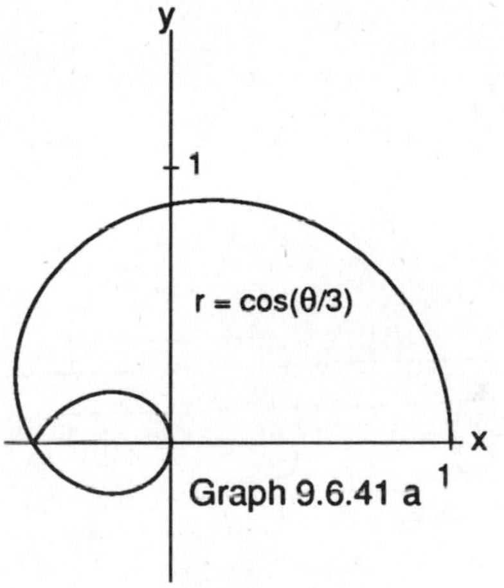

Graph 9.6.41 a

41. b)

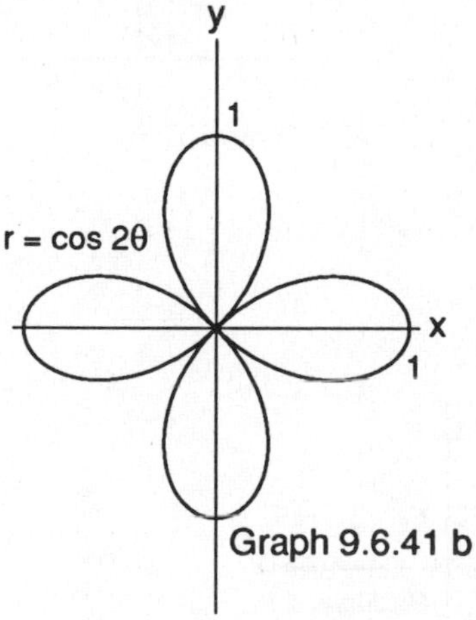

Graph 9.6.41 b

41. c)

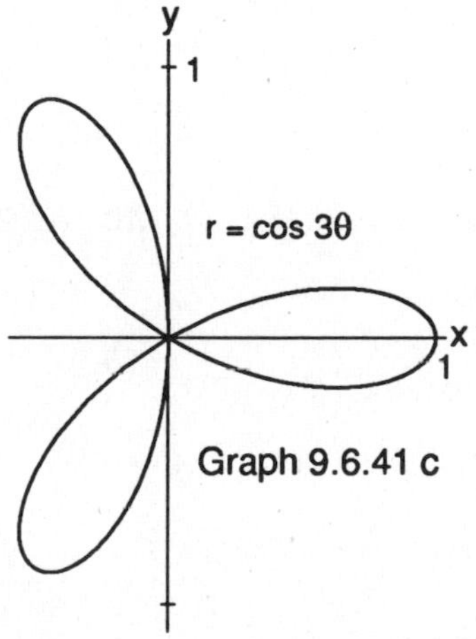

Graph 9.6.41 c

41. d)

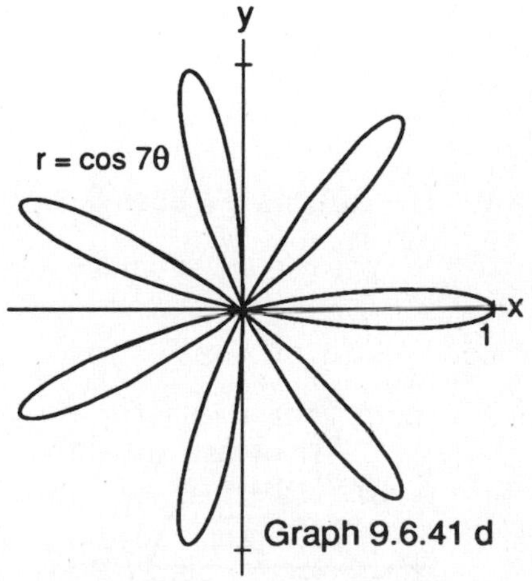

Graph 9.6.41 d

SECTION 9.7 POLAR EQUATIONS FOR CONIC SECTIONS

1. $r\cos\left(\theta-\frac{\pi}{6}\right)=5 \Rightarrow r\left(\cos\theta\cos\frac{\pi}{6}+\sin\theta\sin\frac{\pi}{6}\right)=5 \Rightarrow \frac{\sqrt{3}}{2}r\cos\theta+\frac{1}{2}r\sin\theta=5 \Rightarrow$
$\frac{\sqrt{3}}{2}x+\frac{1}{2}y=5 \Rightarrow \sqrt{3}\,x+y=10.$

3. $r\cos\left(\theta-\frac{4\pi}{3}\right)=3 \Rightarrow r\left(\cos\theta\cos\frac{4\pi}{3}+\sin\theta\sin\frac{4\pi}{3}\right)=3 \Rightarrow -\frac{1}{2}r\cos\theta-\frac{\sqrt{3}}{2}r\sin\theta=3 \Rightarrow$
$-\frac{1}{2}x-\frac{\sqrt{3}}{2}y=3 \Rightarrow x+\sqrt{3}\,y=-6$

5. $r\cos\left(\theta-\frac{\pi}{4}\right)=\sqrt{2} \Rightarrow r\left(\cos\theta\cos\frac{\pi}{4}+\sin\theta\sin\frac{\pi}{4}\right)=$
$\sqrt{2} \Rightarrow \frac{1}{\sqrt{2}}r\cos\theta+\frac{1}{\sqrt{2}}r\sin\theta=\sqrt{2} \Rightarrow$
$\frac{1}{\sqrt{2}}x+\frac{1}{\sqrt{2}}y=\sqrt{2} \Rightarrow x+y=2.$

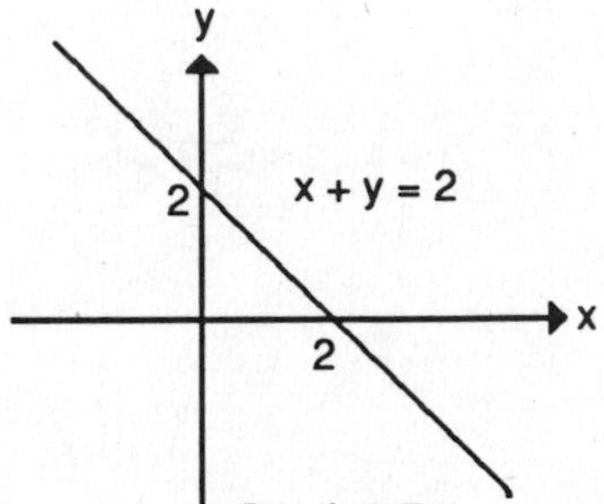

Graph 9.7.5

7. $r\cos\left(\theta-\frac{3\pi}{2}\right)=1 \Rightarrow r\left(\cos\theta\cos\frac{3\pi}{2}+\sin\theta\sin\frac{3\pi}{2}\right)=1 \Rightarrow$
$-r\sin\theta=1 \Rightarrow y=-1.$

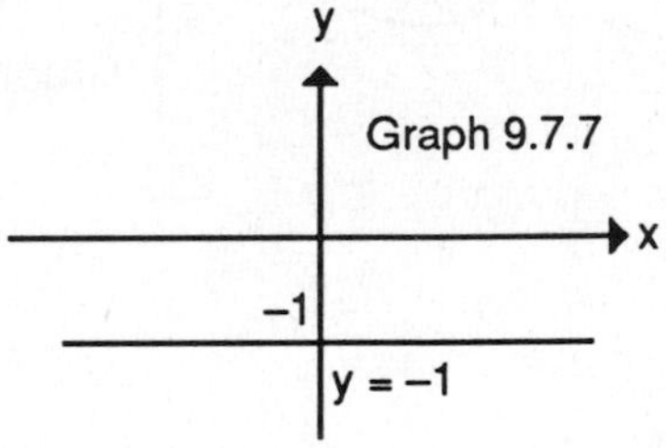

Graph 9.7.7

9. $r=2(4)\cos\theta=8\cos\theta$

11. $r=2\sqrt{2}\sin\theta$

13.

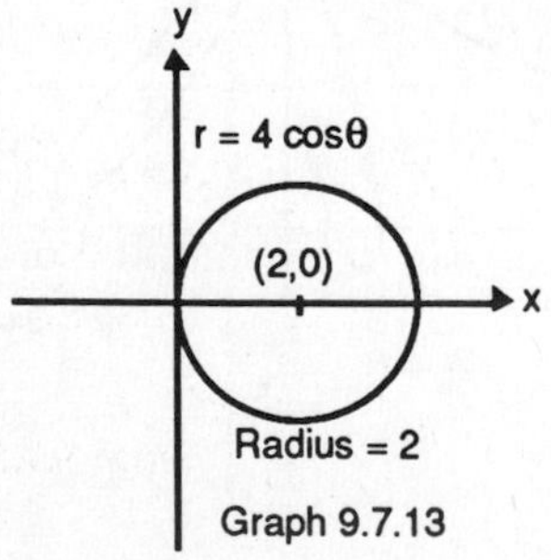

Graph 9.7.13

15.

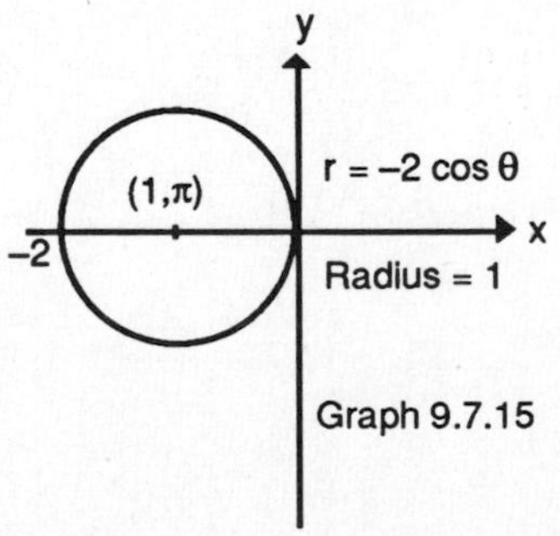

Graph 9.7.15

17. $e = 1, x = 2 \Rightarrow k = 2 \Rightarrow r = \dfrac{2(1)}{1 + (1)\cos\theta} = \dfrac{2}{1 + \cos\theta}$

19. $e = 2, x = 4 \Rightarrow k = 4 \Rightarrow r = \dfrac{4(2)}{1 + (2)\cos\theta} = \dfrac{8}{1 + 2\cos\theta}$

21. $e = \frac{1}{2}, x = 1 \Rightarrow k = 1 \Rightarrow r = \dfrac{\frac{1}{2}(1)}{1 + (1)\cos\theta} = \dfrac{1}{2 + \cos\theta}$

23. $e = \frac{1}{5}, y = -10 \Rightarrow k = 10 \Rightarrow r = \dfrac{\frac{1}{5}(10)}{1 - \frac{1}{5}\sin\theta} = \dfrac{2}{1 - \frac{1}{5}\sin\theta} = \dfrac{10}{5 - \sin\theta}$

25. $r = \dfrac{1}{1 + \cos\theta} \Rightarrow e = 1, k = 1 \Rightarrow x = 1$

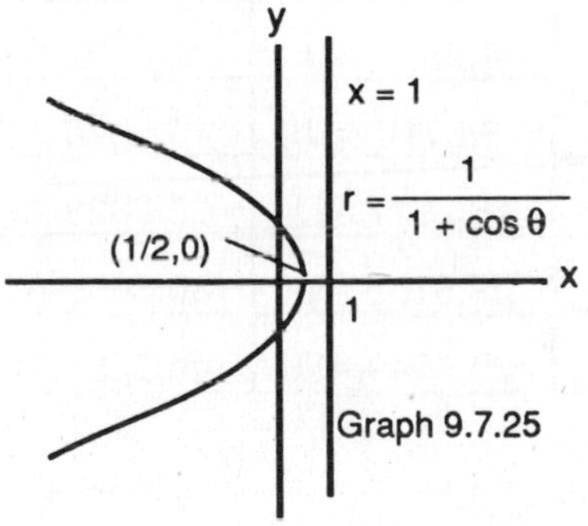

Graph 9.7.25

27. $r = \dfrac{25}{10 - 5\cos\theta} \Rightarrow r = \dfrac{\frac{25}{10}}{1 - \frac{5}{10}\cos\theta} = \dfrac{\frac{5}{2}}{1 - \frac{1}{2}\cos\theta} \Rightarrow$

$e = \frac{1}{2}, k = 5.\ a(1 - e^2) = ke \Rightarrow a\left(1 - \left(\frac{1}{2}\right)^2\right) = \frac{5}{2}$

$\Rightarrow \frac{3}{4}a = \frac{5}{2} \Rightarrow a = \frac{10}{3}.\ a - ae = \frac{10}{3} - \left(\frac{10}{3}\right)\frac{1}{2} = \frac{5}{3}$

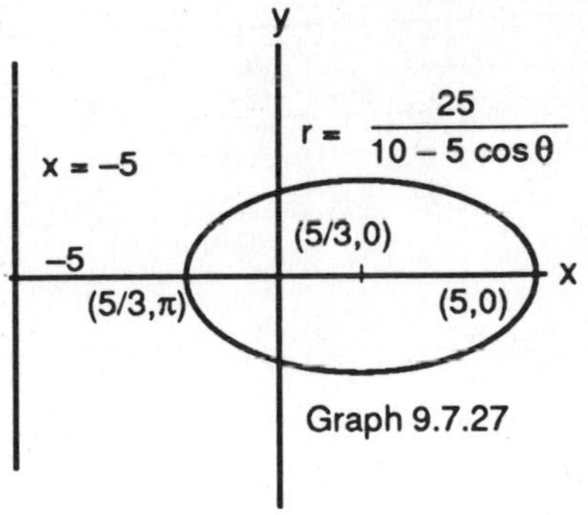

Graph 9.7.27

29. $r = \dfrac{400}{16 + 8\sin\theta} \Rightarrow r = \dfrac{\frac{400}{16}}{1 + \frac{8}{16}\sin\theta} \Rightarrow r = \dfrac{25}{1 + \frac{1}{2}\sin\theta} \Rightarrow$

$e = \frac{1}{2}, k = 50.\ a(1 - e^2) = ke \Rightarrow a\left(1 - \left(\frac{1}{2}\right)^2\right) = 25 \Rightarrow$

$\frac{3}{4}a = 25 \Rightarrow a = \frac{100}{3}.\ a - ae = \frac{100}{3} - \frac{100}{3}\left(\frac{1}{2}\right) = \frac{50}{3}.$

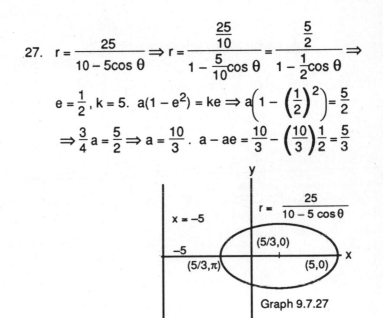

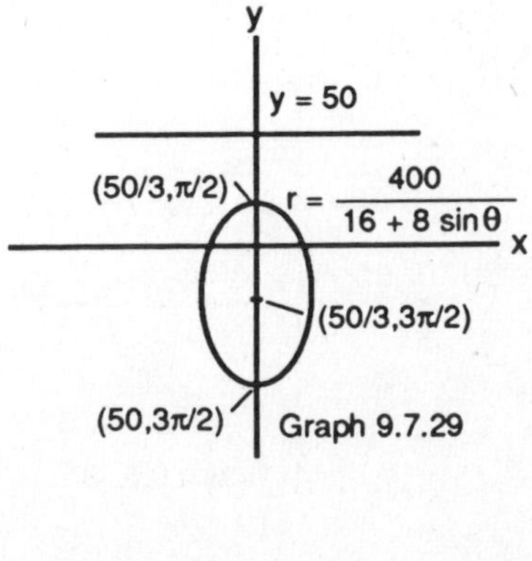

Graph 9.7.29

31. $r = \dfrac{8}{2 - 2\sin\theta} \Rightarrow r = \dfrac{4}{1 - \sin\theta} \Rightarrow e = 1,\ k = 4$

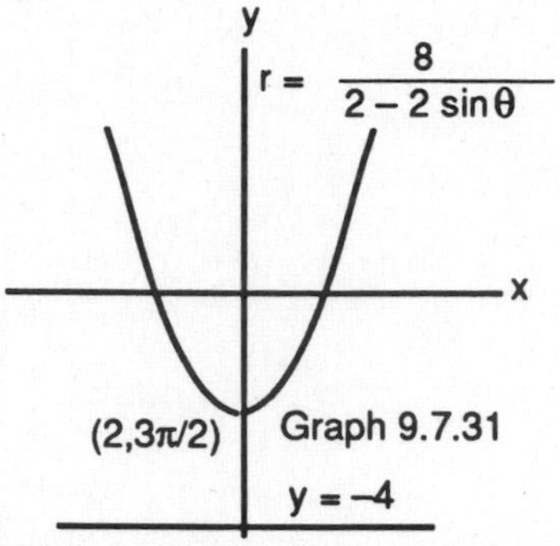

Graph 9.7.31

33.

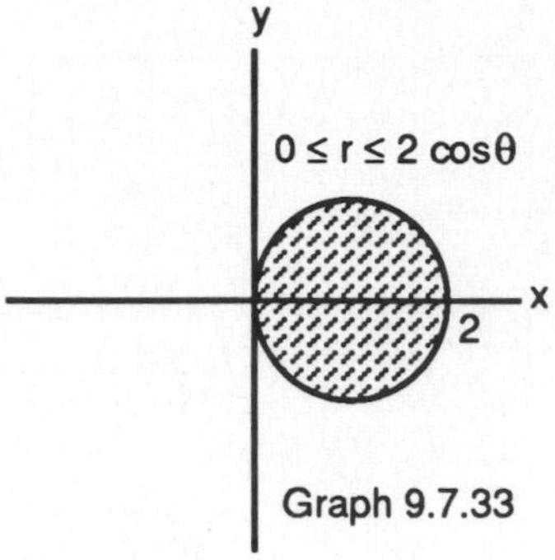

Graph 9.7.33

35. a) Perihelion $= a - ae = a(1 - e)$

Aphelion $= ea + a = a(1 + e)$

b)

Planet	Perihelion	Aphelion
Mercury	0.3075 AU	0.4667 AU
Venus	0.7184 AU	0.7282 AU
Earth	0.9833 AU	1.0167 AU
Mars	1.3817 AU	1.6663 AU
Jupiter	4.9512 AU	5.4548 AU
Saturn	9.0210 AU	10.0570 AU
Uranus	18.2977 AU	20.0623 AU
Neptune	29.8135 AU	30.3065 AU
Pluto	29.6549 AU	49.2251 AU

37. a) $r = 2\sin\theta \Rightarrow a = 1 \Rightarrow$ Center $= (0,1)$

$\Rightarrow x^2 + (y - 1)^2 = 1.$

$r = \csc\theta \Rightarrow r = \dfrac{1}{\sin\theta} \Rightarrow r\sin\theta = 1$

$\Rightarrow y = 1$

b)

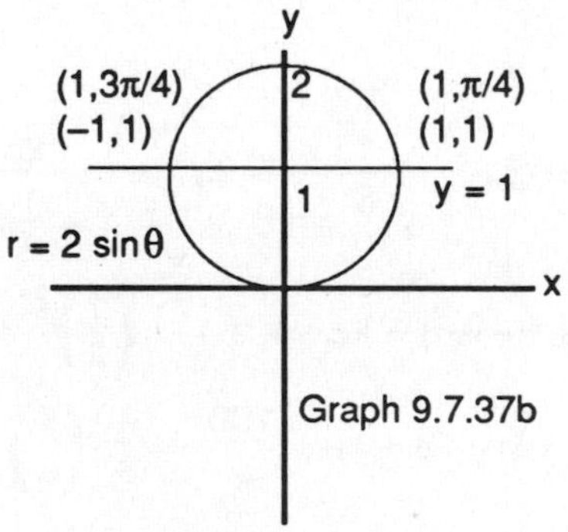

Graph 9.7.37b

39. $r\cos\theta = 4 \Rightarrow x = 4 \Rightarrow k = 4$. Parabola $\Rightarrow e = 1$. $\therefore r = \dfrac{4}{1 + \cos\theta}$

41.

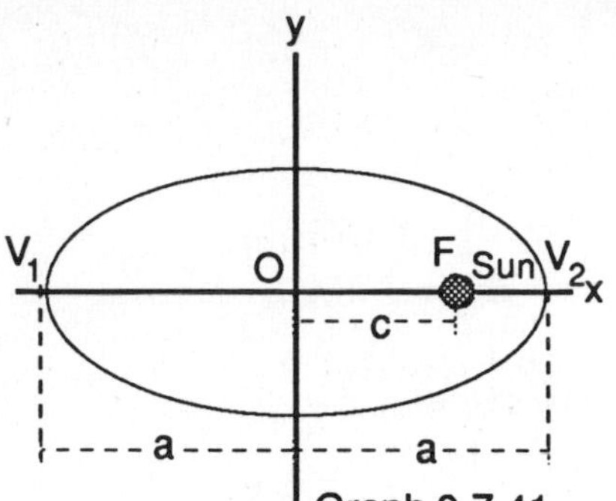

Graph 9.7.41

Let the ellipse be the orbit, with the Sun at one focus. Then $r_{max} = a + c$, $r_{min} = a - c \Rightarrow \frac{r_{max} - r_{min}}{r_{max} + r_{min}} = \frac{(a+c)-(a-c)}{(a+c)+(a-c)} = \frac{2c}{2a} = \frac{c}{a} = e$.

43. Let F_1, F_2 be the foci. Then $PF_1 + PF_2 = 10$ where P is any point on the ellipse. If P is a vertex, then $PF_1 = a + c$, $PF_2 = a - c \Rightarrow (a + c) + (a - c) = 10 \Rightarrow 2a = 10 \Rightarrow a = 5$. Since $e = \frac{c}{a}$, $0.2 = \frac{c}{5} \Rightarrow c = 1.0 \Rightarrow$ the pins are 2 inches apart.

45.

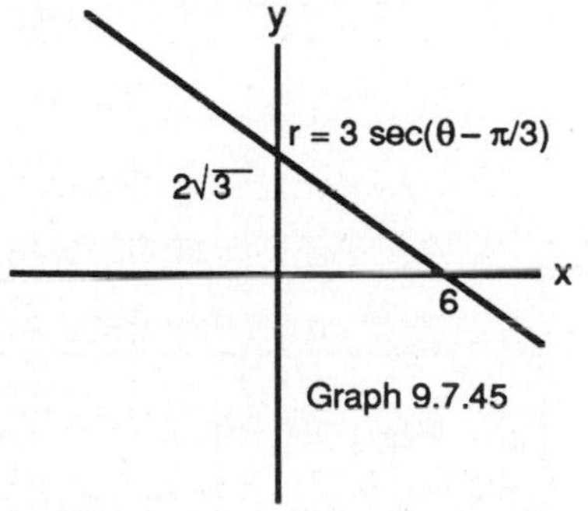

Graph 9.7.45

47.

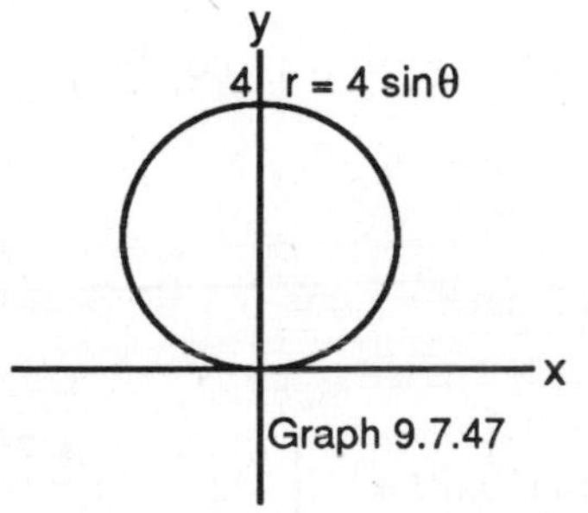

Graph 9.7.47

49.

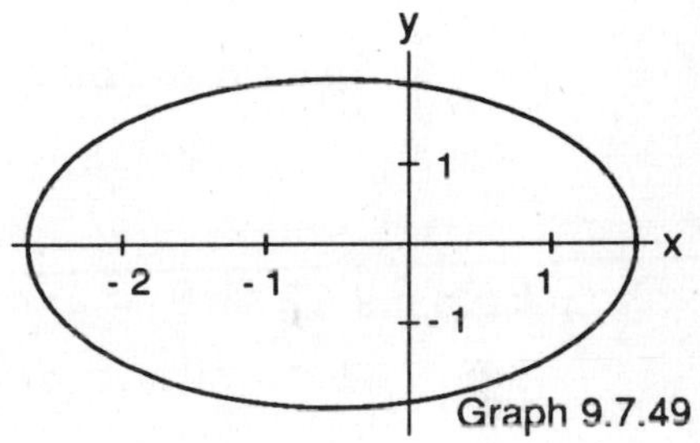

Graph 9.7.49

51.

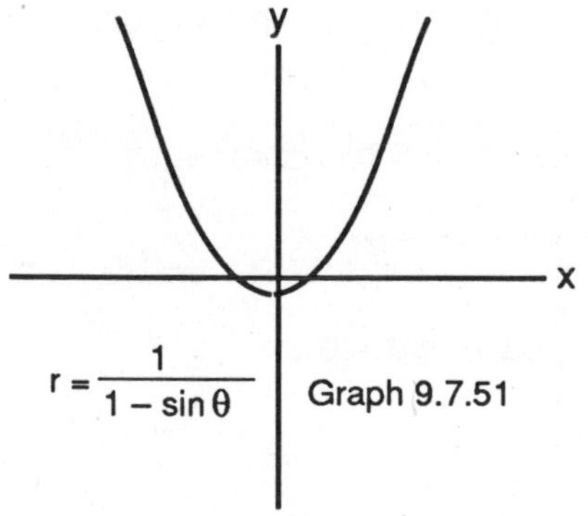

Graph 9.7.51

53.

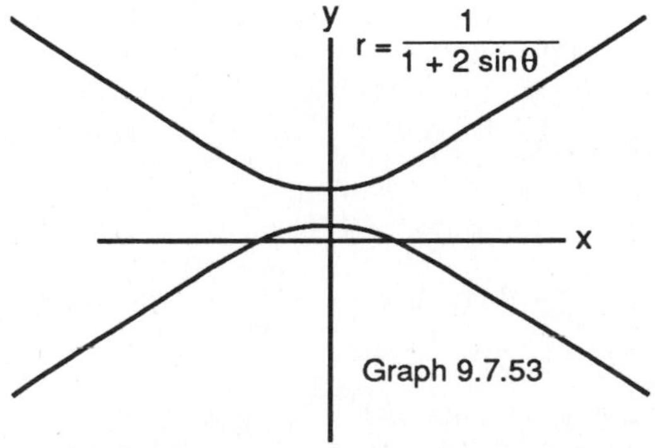

Graph 9.7.53

SECTION 9.8 INTEGRATION IN POLAR COORDINATES

1. $$A = \int_0^{2\pi} \frac{1}{2}(4 + 2\cos\theta)^2 d\theta = \int_0^{2\pi} \frac{1}{2}(16 + 16\cos\theta + 4\cos^2\theta)\, d\theta$$
$$= \int_0^{2\pi} \left(8 + 8\cos\theta + 2\left(\frac{1+\cos 2\theta}{2}\right)\right) d\theta = \int_0^{2\pi} (9 + 8\cos\theta + \cos 2\theta)\, d\theta =$$
$$\left[9\theta + 8\sin\theta + \frac{1}{2}\sin 2\theta\right]_0^{2\pi} = 18\pi$$

3. $$A = 2\int_0^{\pi/4} \frac{1}{2}\cos^2 2\theta\, d\theta = \int_0^{\pi/4} \frac{1+\cos 4\theta}{2}\, d\theta = \frac{1}{2}\left[\theta + \frac{\sin 4\theta}{4}\right]_0^{\pi/4} = \frac{\pi}{8}$$

5. $$A = \int_0^{\pi/2} \frac{1}{2}(4\sin 2\theta)\, d\theta = \int_0^{\pi/2} 2\sin 2\theta\, d\theta = \left[-\cos 2\theta\right]_0^{\pi/2} = 2$$

7. $r = 2\cos\theta$, $r = 2\sin\theta \Rightarrow 2\cos\theta = 2\sin\theta \Rightarrow \cos\theta = \sin\theta \Rightarrow \theta = \frac{\pi}{4}$
$$A = 2\int_0^{\pi/4} \frac{1}{2}(2\sin\theta)^2 d\theta = \int_0^{\pi/4} 4\sin^2\theta\, d\theta = \int_0^{\pi/4} 4\left(\frac{1-\cos 2\theta}{2}\right) d\theta = \int_0^{\pi/4} (2 - 2\cos 2\theta)\, d\theta =$$
$$\left[2\theta - \sin 2\theta\right]_0^{\pi/4} = \frac{\pi}{2} - 1$$

9. $r = 2$, $r = 2(1 - \cos\theta) \Rightarrow 2 = 2(1 - \cos\theta) \Rightarrow \cos\theta = 1 \Rightarrow \theta = 0$. Sketch a graph to see the region.
$$A = 2\int_0^{\pi/2} \frac{1}{2}\left(2(1-\cos\theta)\right)^2 d\theta + \frac{1}{2}\text{ of the area of the circle} = \int_0^{\pi/2} 4(1 - 2\cos\theta + \cos^2\theta)\, d\theta + \frac{1}{2}\pi(2)^2$$
$$= \int_0^{\pi/2} \left(4 - 8\cos\theta + \left(\frac{1+\cos 2\theta}{2}\right)\right) d\theta + 2\pi = \int_0^{\pi/2} (6 - 8\cos\theta + 2\cos 2\theta)\, d\theta + 2\pi =$$
$$\left[6\theta - 8\sin\theta + \sin 2\theta\right]_0^{\pi/2} + 2\pi = 5\pi - 8$$

11. $r = \sqrt{3}$, $r^2 = 6\cos 2\theta \Rightarrow 3 = 6\cos 2\theta \Rightarrow \cos 2\theta = \frac{1}{2} \Rightarrow \theta = \frac{\pi}{6}$ in the 1st quadrant. Use symmetry to find the area.

$$A = 4\int_0^{\pi/6}\left(\frac{1}{2}(6\cos 2\theta) - \frac{1}{2}\left(\sqrt{3}\right)^2\right)d\theta = 2\int_0^{\pi/6}(6\cos 2\theta - 3)d\theta = 2\Big[3\sin 2\theta - 3\theta\Big]_0^{\pi/6} = 3\sqrt{3} - \pi$$

13. $r = 1$, $r = -2\cos\theta \Rightarrow 1 = -2\cos\theta \Rightarrow \cos\theta = -\frac{1}{2} \Rightarrow \theta = \frac{2\pi}{3}$ in quadrant II.

$$A = 2\int_{2\pi/3}^{\pi}\frac{1}{2}\left((-2\cos\theta)^2 - 1^2\right)d\theta = \int_{2\pi/3}^{\pi}\left(4\cos^2\theta - 1\right)d\theta = \int_{2\pi/3}^{\pi}\left(2(1+\cos 2\theta) - 1\right)d\theta$$

$$= \int_{2\pi/3}^{\pi}(1 + 2\cos 2\theta)d\theta = \Big[\theta + \sin 2\theta\Big]_{2\pi/3}^{\pi} = \frac{\pi}{3} + \frac{\sqrt{3}}{2}$$

15. $r = 6$. $r = 3\csc\theta \Rightarrow r\sin\theta = 3 \therefore 6\sin\theta = 3 \Rightarrow \sin\theta = \frac{1}{2} \Rightarrow \theta = \frac{\pi}{4}, \frac{3\pi}{4}$

$$A = \int_{\pi/4}^{3\pi/4}\frac{1}{2}\left(6^2 - \left(\frac{3}{\sin\theta}\right)^2\right)d\theta = \int_{\pi/4}^{3\pi/4}(18 - 9\csc^2\theta)d\theta = \Big[18\theta + 9\cot\theta\Big]_{\pi/4}^{3\pi/4} = 9\pi - 18$$

17. a)

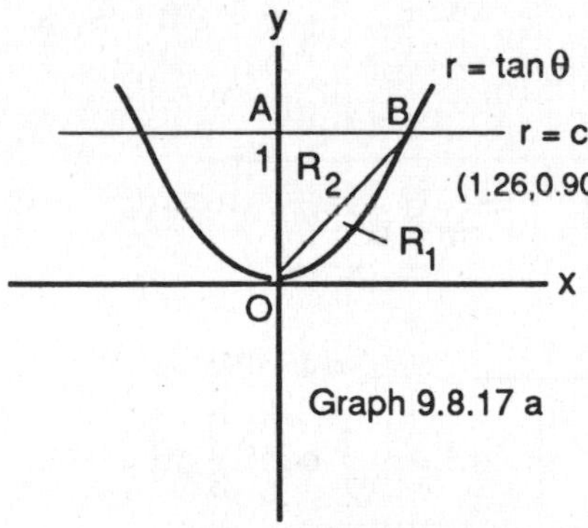

Graph 9.8.17 a

$r = \tan\theta$, $r = \csc\theta \Rightarrow \tan\theta = \csc\theta \Rightarrow \sin^2\theta = \cos\theta \Rightarrow 1 - \cos^2\theta = \cos\theta \Rightarrow \cos^2\theta + \cos\theta - 1 = 0 \Rightarrow \cos\theta = \frac{-1 \pm \sqrt{5}}{2} \Rightarrow \theta = 0.90 \Rightarrow r = \tan 0.90 = 1.26$. The area of R_1 is $A_1 = \int_0^{0.90}\frac{1}{2}\tan^2\theta\, d\theta = \frac{1}{2}\int_0^{0.90}(\sec^2\theta - 1)\, d\theta = \frac{1}{2}\Big[\tan\theta - \theta\Big]_0^{0.90} = 0.18$. AO = 1, OB = 1.26 $\Rightarrow AB = \sqrt{(1.26)^2 - 1} = 0.77 \Rightarrow$ the area of R_2 is $A_2 = \frac{1}{2}(1)(0.77) = 0.38$. $\therefore$ the area of the region shaded in the text is $2(0.38 + 0.18) = 1.12$. Note: the area must be found this way since no common interval generates the region. For example, the interval $0 \le \theta \le 0.90$ generates the arc OB of $r = \tan\theta$ but does not generate the segment AB of the line $r = \csc\theta$. Instead the interval generates the half-line from B to $+\infty$ on the line $r = \csc\theta$.

b) $\lim_{\theta \to (\pi/2)^-} \tan\theta = +\infty$. The line $x = 1$ is $r = \sec\theta$ in polar coordinates. Then $\lim_{\theta \to (\pi/2)^-}(\tan\theta - \sec\theta)$

$$= \lim_{\theta \to (\pi/2)^-}\left(\frac{\sin\theta}{\cos\theta} - \frac{1}{\cos\theta}\right) = \lim_{\theta \to (\pi/2)^-}\left(\frac{\sin\theta - 1}{\cos\theta}\right) = \lim_{\theta \to (\pi/2)^-}\left(\frac{\cos\theta}{-\sin\theta}\right) = 0 \Rightarrow r = \tan\theta \to r =$$

$\sec\theta$ as $\theta \to \frac{\pi}{2}^- \Rightarrow r = \sec\theta$ ($x = 1$) is a vertical asymptote of $r = \tan\theta$. Similarly, $r = -\sec\theta$ is the

17. b) (Continued)

polar equation of $x = -1$ and $\lim_{\theta \to (-\pi/2)^+} (\tan\theta - (-\sec\theta)) = 0 \Rightarrow r = -\sec\theta$ $(x = -1)$ is a vertical asymptote of $r = \tan\theta$.

19. $r = \theta^2,\ 0 \le \theta \le \sqrt{5} \Rightarrow \frac{dr}{d\theta} = 2\theta$. $\therefore$ Length $= \int_0^{\sqrt{5}} \sqrt{(\theta^2)^2 + (2\theta)^2}\, d\theta = \int_0^{\sqrt{5}} \sqrt{\theta^4 + 4\theta^2}\, d\theta =$

$$\int_0^{\sqrt{5}} |\theta|\sqrt{\theta^2 + 4}\, d\theta = \int_0^{\sqrt{5}} \theta\sqrt{\theta^2 + 4}\, d\theta = \int_4^9 \frac{1}{2}\sqrt{u}\, du = \frac{1}{2}\left[\frac{2}{3}u^{3/2}\right]_4^9 = \frac{19}{3}$$

Let $u = \theta^2 + 4 \Rightarrow \frac{1}{2}du = \theta d\theta$; $\theta = 0 \Rightarrow u = 4$, $\theta = \sqrt{5} \Rightarrow u = 9$

21. $r = 1 + \cos\theta \Rightarrow \frac{dr}{d\theta} = -\sin\theta$. $\therefore$ Length $= \int_0^{2\pi} \sqrt{(1 + \cos\theta)^2 + (-\sin\theta)^2}\, d\theta =$

$$2\int_0^{\pi} \sqrt{1 + 2\cos\theta + \cos^2\theta + \sin^2\theta}\, d\theta = 2\int_0^{\pi} \sqrt{2 + 2\cos\theta}\, d\theta =$$

$$2\int_0^{\pi} \sqrt{\frac{4(1 + \cos\theta)}{2}}\, d\theta = 4\int_0^{\pi} \sqrt{\frac{1 + \cos\theta}{2}}\, d\theta = 4\int_0^{\pi} \cos\frac{1}{2}\theta\, d\theta = 4\left[2\sin\frac{1}{2}\theta\right]_0^{\pi} = 8$$

23. $r = \cos^3\frac{\theta}{3} \Rightarrow \frac{dr}{d\theta} = -\sin\frac{\theta}{3}\cos^2\frac{\theta}{3}$. $\therefore$ Length $= \int_0^{\pi/4} \sqrt{\left(\cos^3\frac{\theta}{3}\right)^2 + \left(-\sin\frac{\theta}{3}\cos^2\frac{\theta}{3}\right)^2}\, d\theta =$

$$\int_0^{\pi/4} \sqrt{\cos^6\frac{\theta}{3} + \sin^2\frac{\theta}{3}\cos^4\frac{\theta}{3}}\, d\theta = \int_0^{\pi/4} \cos^2\frac{\theta}{3}\sqrt{\cos^2\frac{\theta}{3} + \sin^2\frac{\theta}{3}}\, d\theta = \int_0^{\pi/4} \cos^2\frac{\theta}{3}\, d\theta =$$

$$\int_0^{\pi/4} \frac{1 + \cos(2\theta/3)}{2}\, d\theta = \left[\frac{\theta + \frac{3}{2}\sin\frac{2\theta}{3}}{2}\right]_0^{\pi/4} = \frac{\pi}{4} + \frac{3}{8}$$

24. $r = \sqrt{1 + \sin 2\theta},\ 0 \le \theta \le \pi\sqrt{2} \Rightarrow \frac{dr}{d\theta} = \frac{1}{2}(1 + \sin 2\theta)^{-1/2}(2\cos 2\theta) = \cos 2\theta(1 + \sin 2\theta)^{-1/2} \Rightarrow$

$$r^2 + \left(\frac{dr}{d\theta}\right)^2 = \left(\sqrt{1 + \sin 2\theta}\right)^2 + \left((1 + \sin 2\theta)^{-1/2}\cos 2\theta\right)^2 = 1 + \sin 2\theta + \frac{\cos^2 2\theta}{1 + \sin 2\theta} =$$

$$\frac{1 + 2\sin 2\theta + \sin^2 2\theta + \cos^2 2\theta}{1 + \sin 2\theta} = \frac{2 + 2\sin 2\theta}{1 + \sin 2\theta} = 2.\ \therefore \text{Length} = \int_0^{\pi\sqrt{2}} \sqrt{2}\, d\theta = \left[\sqrt{2}\,\theta\right]_0^{\pi\sqrt{2}} = 2\pi$$

25. $r = \sqrt{\cos 2\theta}$, $0 \le \theta \le \frac{\pi}{6} \Rightarrow \frac{dr}{d\theta} = \frac{1}{2}(\cos 2\theta)^{-1/2}(-2\sin 2\theta) = -\sin 2\theta(\cos 2\theta)^{-1/2} \Rightarrow r^2 + \left(\frac{dr}{d\theta}\right)^2 =$

$\left(\sqrt{\cos 2\theta}\right)^2 + \left(-\sin 2\theta(\cos 2\theta)^{-1/2}\right)^2 = \cos 2\theta + \frac{\sin^2 2\theta}{\cos 2\theta} = \frac{\cos^2 2\theta + \sin^2 2\theta}{\cos 2\theta} = \frac{1}{\cos 2\theta} = \sec 2\theta.$ $\therefore$

$$\text{Length} = \int_0^{\pi/6} \sqrt{\sec 2\theta}\, d\theta = \left[\frac{1}{2}\ln\left|\sec 2\theta + \tan 2\theta\right|\right]_0^{\pi/6} = \frac{1}{2}\ln\left(2 + \sqrt{3}\right)$$

27. $r = \sqrt{\cos 2\theta}$, $0 \le \theta \le \frac{\pi}{4} \Rightarrow \frac{dr}{d\theta} = \frac{1}{2}(\cos 2\theta)^{-1/2}(-\sin 2\theta)(2) = \frac{-\sin 2\theta}{\sqrt{\cos 2\theta}}$

$$\therefore \text{Surface Area} = \int_0^{\pi/4} 2\pi r\cos\theta \sqrt{\left(\sqrt{\cos 2\theta}\right)^2 + \left(\frac{-\sin 2\theta}{\sqrt{\cos 2\theta}}\right)^2}\, d\theta =$$

$$\int_0^{\pi/4} 2\pi\sqrt{\cos 2\theta}\cos\theta\sqrt{\cos 2\theta + \frac{\sin^2 2\theta}{\cos 2\theta}}\, d\theta = \int_0^{\pi/4} 2\pi\sqrt{\cos 2\theta}\cos\theta\sqrt{\frac{1}{\cos 2\theta}}\, d\theta =$$

$$\int_0^{\pi/4} 2\pi\cos\theta\, d\theta = \left[2\pi\sin\theta\right]_0^{\pi/4} = \pi\sqrt{2}$$

29. $r^2 = \cos 2\theta \Rightarrow r = \pm\sqrt{\cos 2\theta}$. Use $r = \sqrt{\cos 2\theta}$ on $\left[0, \frac{\pi}{4}\right]$. Then $\frac{dr}{d\theta} = \frac{1}{2}(\cos 2\theta)^{-1/2}(-\sin 2\theta)(2) =$

$$\frac{-\sin 2\theta}{\sqrt{\cos 2\theta}} \quad \therefore \text{ Surface Area} = \int_0^{\pi/4} 2\pi\sqrt{\cos 2\theta}\sin\theta\sqrt{\left(\sqrt{\cos 2\theta}\right)^2 + \left(\frac{-\sin 2\theta}{\sqrt{\cos 2\theta}}\right)^2}\, d\theta =$$

$$\int_0^{\pi/4} 2\pi\sqrt{\cos 2\theta}\sin\theta\sqrt{\cos 2\theta + \frac{\sin^2 2\theta}{\cos 2\theta}}\, d\theta = \int_0^{\pi/4} 2\pi\sqrt{\cos 2\theta}\sin\theta\sqrt{\frac{1}{\cos 2\theta}}\, d\theta =$$

$$\int_0^{\pi/4} 2\pi\sin\theta\, d\theta = \left[-2\pi\cos\theta\right]_0^{\pi/4} = \pi\left(2 - \sqrt{2}\right)$$

31. a) $r = 2f(\theta)$, $\alpha \le \theta \le \beta \Rightarrow \frac{dr}{d\theta} = 2f'(\theta) \Rightarrow r^2 + \left(\frac{dr}{d\theta}\right)^2 = \left(2f(\theta)\right)^2 + \left(2f'(\theta)\right)^2 = 4\left((f(\theta))^2 + (f'(\theta))^2\right) \Rightarrow$

$$\text{Length} = \int_\alpha^\beta \sqrt{4\left((f(\theta))^2 + (f'(\theta))^2\right)}\, d\theta = 2\int_\alpha^\beta \sqrt{\left(f(\theta)\right)^2 + \left(f'(\theta)\right)^2}\, d\theta$$ or twice the length of $r = f(\theta)$,

$\alpha \le \theta \le \beta$.

31. b) Again $r = 2\,f(\theta) \Rightarrow r^2 + \left(\frac{dr}{d\theta}\right)^2 = \left(2\,f(\theta)\right)^2 + \left(2\,f'(\theta)\right)^2 = 4\left((f(\theta))^2 + (f'(\theta))^2\right) \Rightarrow$ Area =

$$\int_{\alpha}^{\beta} 2\pi(2\,f(\theta))\sin\theta\,\sqrt{4\left((f(\theta))^2 + (f'(\theta))^2\right)}\;d\theta = 4\int_{\alpha}^{\beta} 2\pi\,f(\theta)\,\sqrt{(f(\theta))^2 + (f'(\theta))^2}\;d\theta$$ or four times the area of the surface generated by revolving $r = f(\theta)$, $\alpha \le \theta \le \beta$, about the x–axis.

33. $$\bar{x} = \frac{\frac{2}{3}\int_0^{2\pi} r^3\cos\theta\,d\theta}{\int_0^{2\pi} r^2\,d\theta} = \frac{\frac{2}{3}\int_0^{2\pi} (a(1+\cos\theta))^3\cos\theta\,d\theta}{\int_0^{2\pi}(a(1+\cos\theta))^2\,d\theta} =$$

$$\frac{\frac{2}{3}a^3\int_0^{2\pi}(1 + 3\cos\theta + 3\cos^2\theta + \cos^3\theta)\cos\theta\,d\theta}{a^2\int_0^{2\pi}(1 + 2\cos\theta + \cos^2\theta)d\theta} =$$

$$\frac{\frac{2}{3}a\int_0^{2\pi}\left(\cos\theta + 3\left(\frac{1+\cos 2\theta}{2}\right) + 3(1-\sin^2\theta)\cos\theta + \left(\frac{1+\cos 2\theta}{2}\right)^2\right)d\theta}{\int_0^{2\pi}\left(1 + 2\cos\theta + \left(\frac{1+\cos 2\theta}{2}\right)\right)d\theta} =$$

(After much work using the identity $\cos A = \frac{1+\cos 2A}{2}$)

$$\frac{a\int_0^{2\pi}\left(\frac{15}{12} + \frac{8}{3}\cos\theta + \frac{4}{3}\cos 2\theta - 2\cos\theta\sin^2\theta + \frac{1}{12}\cos 4\theta\right)d\theta}{\int_0^{2\pi}\left(\frac{3}{2} + 2\cos\theta + \frac{1}{2}\cos 2\theta\right)d\theta} =$$

$$\frac{a\left[\frac{15}{12}\theta + \frac{8}{3}\sin\theta + \frac{2}{3}\sin 2\theta - \frac{2}{3}\sin^3\theta + \frac{1}{48}\sin 4\theta\right]_0^{2\pi}}{\left[\frac{3}{2}\theta + 2\sin\theta + \frac{1}{4}\sin 2\theta\right]_0^{2\pi}} = \frac{a\left(\frac{15}{6}\pi\right)}{3\pi} = \frac{5}{6}a$$

33. (Continued)

$$\bar{y} = \frac{\frac{2}{3}\int_0^{2\pi} r^3 \sin\theta\, d\theta}{\int_0^{2\pi} r^2\, d\theta} = \frac{\frac{2}{3}\int_0^{2\pi} (a(1+\cos\theta))^3 \sin\theta\, d\theta}{3\pi} = \frac{\frac{2}{3}\int_{2a}^{2a} -\frac{1}{a}u^3 du}{3\pi} = \frac{0}{3\pi} = 0 \quad \text{Centroid is } \left(\frac{5}{6}a, 0\right)$$

Let $u = a(1 + \cos\theta) \Rightarrow -\frac{1}{a}du = \sin\theta\, d\theta$; $\theta = 0 \Rightarrow u = 2a$,n $\theta = 2\pi \Rightarrow u = 2a$

SECTION 9.M MISCELLANEOUS EXERCISES

1.

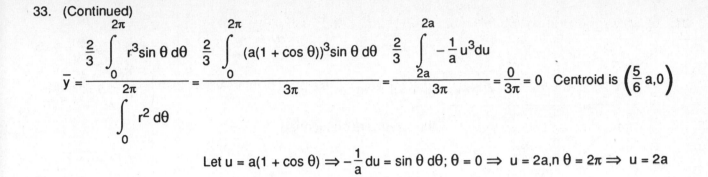

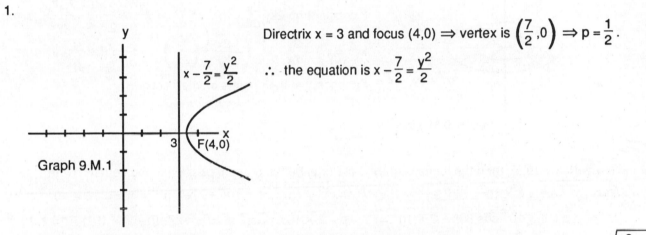

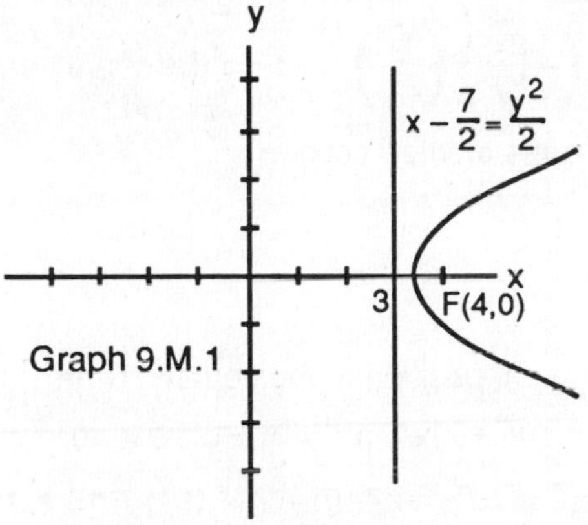

Graph 9.M.1

Directrix $x = 3$ and focus $(4,0) \Rightarrow$ vertex is $\left(\frac{7}{2}, 0\right) \Rightarrow p = \frac{1}{2}$.

$\therefore$ the equation is $x - \frac{7}{2} = \frac{y^2}{2}$

3. $x^2 = 4y \Rightarrow$ vertex is (0,0) and $p = 1 \Rightarrow$ focus is (0,1). $\therefore$ the distance from P(x,y) to the vertex is $\sqrt{x^2 + y^2}$, the distance from P to the focus is $\sqrt{x^2 + (y-1)^2} \Rightarrow \sqrt{x^2 + y^2} = 2\sqrt{x^2 + (y-2)^2} \Rightarrow x^2 + y^2 = 4\left(x^2 + (y-2)^2\right) \Rightarrow x^2 + y^2 = 4x^2 + 4y^2 - 16y + 16 \Rightarrow 3x^2 + 3y^2 - 16y + 16 = 0$, a circle.

5. Vertices = $(0,\pm 2) \Rightarrow a = 2 \Rightarrow e = \frac{c}{a} \Rightarrow 0.5 = \frac{c}{2} \Rightarrow c = 1 \Rightarrow$ Foci are $(0,\pm 1)$.

7. Let the center by (0,y)

a) Directrix $y = -1$, focus $(0,-7)$, $e = 2 \Rightarrow c - \frac{a}{e} = 6 \Rightarrow \frac{a}{e} = c - 6 \Rightarrow a = 2c - 12$. Also $c = ae = 2a$. $\therefore a = 2(2a) - 12 \Rightarrow a = 4 \Rightarrow c = 8$. $y - (-1) = \frac{a}{e} = \frac{4}{2} = 2 \Rightarrow y = 1$. $\therefore$ the center is (0,1). $c^2 = a^2 + b^2 \Rightarrow b^2 = c^2 - a^2 = 64 - 16 = 48$. $\therefore$ the equation is $\frac{(y-1)^2}{16} - \frac{x^2}{48} = 1$.

b) $e = 5 \Rightarrow c - \frac{a}{e} = 6 \Rightarrow \frac{a}{e} = c - 6 \Rightarrow a = 5c - 30$. Also, $c = ae = 5a$. $\therefore a = 5(5a) - 30 \Rightarrow 24a = 30 \Rightarrow a = \frac{5}{4} \Rightarrow c = \frac{25}{4}$. $y - (-1) = \frac{a}{e} = \frac{5/4}{5} = \frac{1}{4} \Rightarrow y = -\frac{3}{4}$. $\therefore$ the center is $\left(0, -\frac{3}{4}\right)$. $c^2 = a^2 + b^2 \Rightarrow b^2 = c^2 - a^2 = \frac{625}{16} - \frac{24}{16} = \frac{75}{2}$. $\therefore$ the equation is $\frac{\left(y + \frac{3}{4}\right)^2}{25/16} - \frac{x^2}{75/16} = 1$ or $\frac{16\left(y + \frac{3}{4}\right)^2}{25} - \frac{2x^2}{75} = 1$

9.

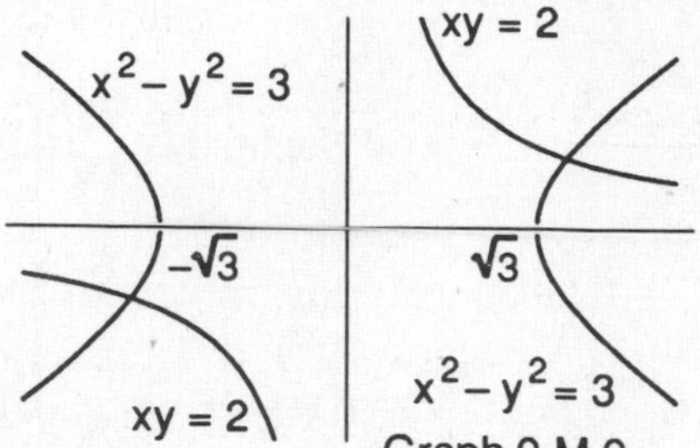

Graph 9.M.9

$xy = 2 \Rightarrow x\frac{dy}{dx} + y = 0 \Rightarrow \frac{dy}{dx} = -\frac{y}{x}$. $x^2 - y^2 = 3 \Rightarrow 2x - 2y\frac{dy}{dx} = 0 \Rightarrow \frac{dy}{dx} = \frac{x}{y}$. If (x_0,y_0) is a point of intersection, then $\left(-\frac{y_0}{x_0}\right)\left(\frac{x_0}{y_0}\right) = -1 \Rightarrow$ the curves are orthogonal.

11.

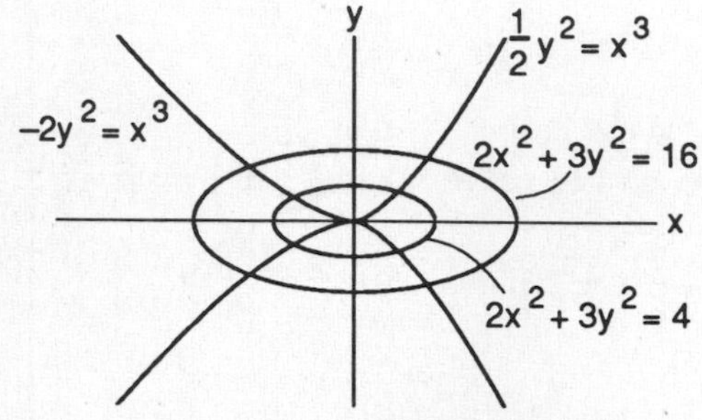

Graph 9.M.11

$2x^2 + 3y^2 = a^2 \Rightarrow 4x + 6y\frac{dy}{dx} = 0 \Rightarrow \frac{dy}{dx} = -\frac{2x}{3y}$. $ky^2 = x^3$ where k constant $\Rightarrow 2ky\frac{dy}{dx} = 3x^2 \Rightarrow \frac{dy}{dx} = \frac{3x^2}{2ky} = \frac{3x^2y}{2ky^2} = \frac{3x^2y}{2x^3}$ (since $ky^2 = x^3$) $= \frac{3y}{2x}$. $\therefore \left(-\frac{2x}{3y}\right)\left(\frac{3y}{2x}\right) = -1 \Rightarrow$ the curves are orthogonal at their points of intersection.

13. If the vertex is (0,0), then the focus is (p,0). Let P(x,y) be the present position of the comet. Then $\sqrt{(x-p)^2 + y^2} = 4 \times 10^7$. Since $y^2 = 4px$, $\sqrt{(x-p)^2 + 4px} = 4 \times 10^7 \Rightarrow (x-p)^2 + 4px = 16 \times 10^{14}$. Also, $x - p = 4 \times 10^7 \cos 60° = 2 \times 10^7 \Rightarrow x = p + 2 \times 10^7$. $\therefore (2 \times 10^7)^2 + 4p(p + 2 \times 10^7) = 16 \times 10^{14}$ $\Rightarrow 4 \times 10^{14} + 4p^2 + 8p \times 10^7 = 16 \times 10^{14} \Rightarrow 4p^2 + 8p \times 10^7 - 12 \times 10^{14} = 0 \Rightarrow p^2 + 2p \times 10^7 - 3 \times 10^{14}$ $= 0 \Rightarrow (p + 3 \times 10^7)(p - 10^7) = 0 \Rightarrow p = -3 \times 10^7$ or $p = 10^7$. Since p is positive, $p = 10^7$ miles.

15. a) $Q_{circle} = \int_0^a \sqrt{a^2 - x^2}\,dx$ b) $Q_{ellipse} = \int_0^a \frac{b}{a}\sqrt{a^2 - x^2}\,dx$

c) From part a and part b above, $Q_{ellipse} = \frac{b}{a}\int_0^a \sqrt{a^2 - x^2}\,dx = \frac{b}{a}Q_{circle} = \frac{b}{a}(\pi a^2) = \pi ab$

17.

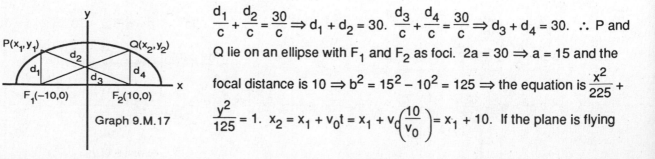

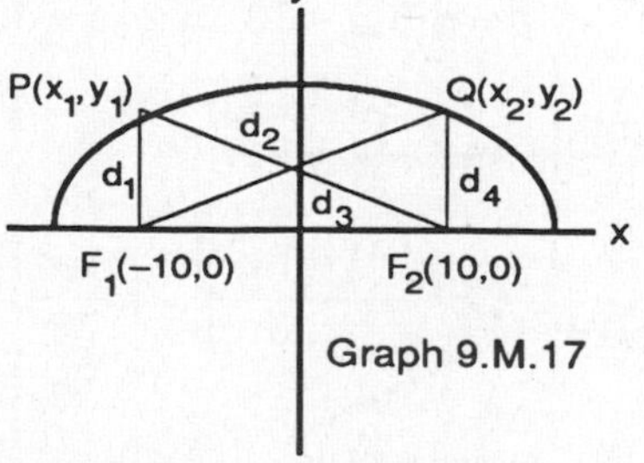

Graph 9.M.17

$\frac{d_1}{c} + \frac{d_2}{c} = \frac{30}{c} \Rightarrow d_1 + d_2 = 30$. $\frac{d_3}{c} + \frac{d_4}{c} = \frac{30}{c} \Rightarrow d_3 + d_4 = 30$. $\therefore$ P and Q lie on an ellipse with F_1 and F_2 as foci. $2a = 30 \Rightarrow a = 15$ and the focal distance is $10 \Rightarrow b^2 = 15^2 - 10^2 = 125 \Rightarrow$ the equation is $\frac{x^2}{225} + \frac{y^2}{125} = 1$. $x_2 = x_1 + v_0t = x_1 + v_0\left(\frac{10}{v_0}\right) = x_1 + 10$. If the plane is flying

17. (Continued)

level, P and Q must be symmetric to the y–axis $\Rightarrow x_1 = -x_2 \Rightarrow x_2 = -x_2 + 10 \Rightarrow x_2 = 5 \Rightarrow \frac{5^2}{225} + \frac{y_2^{\,2}}{125} = 1$ $\Rightarrow y_2^{\,2} = \frac{1000}{9} \Rightarrow y_2 = \frac{10\sqrt{10}}{3}$ since y_2 must be positive. ∴ the position is $\left(5, \frac{10\sqrt{10}}{3}\right)$ where (0,0) is midway between the two stations.

19. The time for the bullet to hit the target remains constant, say $t = t_0$. Let the time it takes for sound to travel from the target to the listener be t_2. Since the listener hears the sounds simultaneously, $t_1 = t_0 + t_2$. If v is the velocity of sound, $vt_1 = vt_0 + vt_2$ or $vt_1 - vt_2 = vt_0$. vt_1 is the distance from the rifle to the listener, vt_2 is the distance from the target to the listener. ∴ the difference of the distances is constant since vt_0 is constant ⇒ the listener is on a branch of a hyperbola with foci at the rifle and the target. The branch is the one with the target as focus.

21. $y^2 = 4px \Rightarrow 2y\frac{dy}{dx} = 4p \Rightarrow \frac{dy}{dx} = \frac{2p}{y} \Rightarrow m_{tan} = \frac{2p}{y_1}$ at $P(x_1,y_1) \Rightarrow$ the tangent line is $y - y_1 = \frac{2p}{y_1}(x - x_1)$. The axis of symmetry $y = 0 \Rightarrow -y_1 = \frac{2p}{y_1}(x - x_1) \Rightarrow -\frac{y_1^{\,2}}{2p} + x_1 = x \Rightarrow -\frac{4px_1}{2p} + x_1 = x \Rightarrow x = -x_1$ or x_1 units to the left.

23.

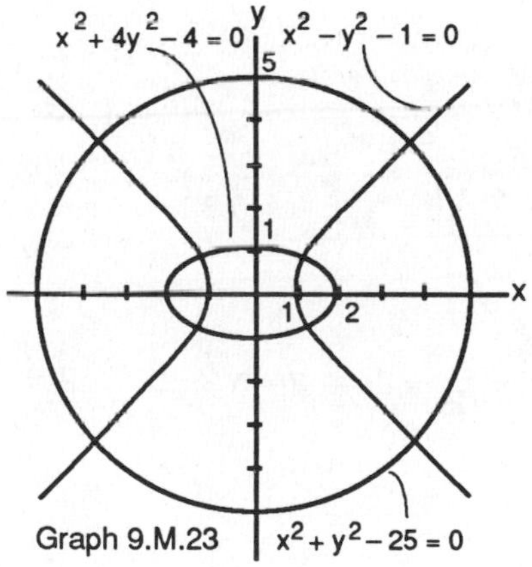

Graph 9.M.23

25.

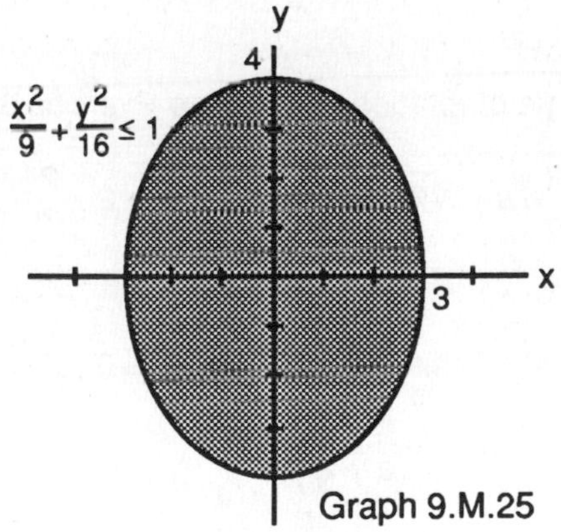

Graph 9.M.25

27. $(9x^2 + 4y^2 - 36)(4x^2 + 9y^2 - 16) \le 0 \Rightarrow$
$9x^2 + 4y^2 - 36 \le 0$ and $4x^2 + 9y^2 - 16 \ge 0$ or
$9x^2 + 4y^2 - 36 \ge 0$ and $4x^2 + 9y^2 - 16 \le 0$

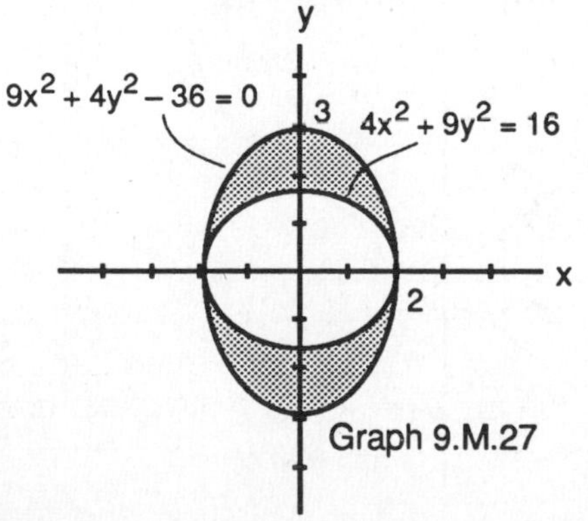

Graph 9.M.27

29. $x^4 - (y^2 - 9)^2 = 0 \Rightarrow x^2 - (y^2 - 9) = 0$ or
$x^2 + (y^2 - 9) = 0 \Rightarrow y^2 - x^2 = 9$ or $x^2 + y^2 = 9$

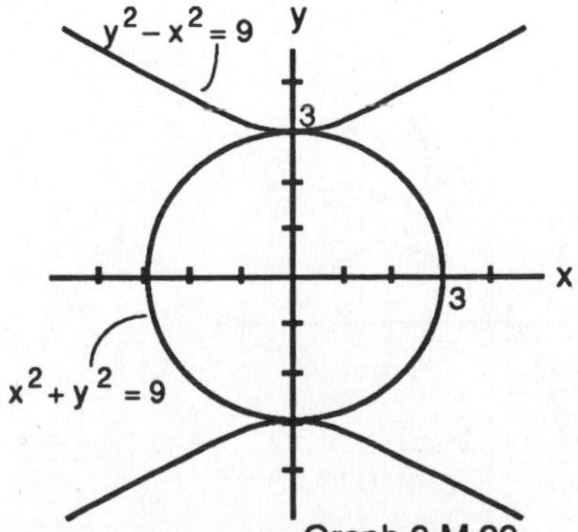

Graph 9.M.29

31. $B^2 - 4AC = 1 - 4(1)(1) = -3 < 0 \Rightarrow$ Ellipse

33. $B^2 - 4AC = 4^2 - 4(1)(4) = 0 \Rightarrow$ Parabola

35. $B^2 - 4AC = 1^2 - 4(2)(2) = -15 < 0 \Rightarrow$ Ellipse. $\cot 2\alpha = \frac{A - C}{B} = 0 \Rightarrow 2\alpha = \frac{\pi}{2} \Rightarrow \alpha = \frac{\pi}{4}$

$x = \frac{\sqrt{2}}{2}x' - \frac{\sqrt{2}}{2}y'$, $y = \frac{\sqrt{2}}{2}x' + \frac{\sqrt{2}}{2}y' \Rightarrow 2\left(\frac{\sqrt{2}}{2}x' - \frac{\sqrt{2}}{2}y'\right)^2 + \left(\frac{\sqrt{2}}{2}x' - \frac{\sqrt{2}}{2}y'\right)\left(\frac{\sqrt{2}}{2}x' + \frac{\sqrt{2}}{2}y'\right) +$
$2\left(\frac{\sqrt{2}}{2}x' + \frac{\sqrt{2}}{2}y'\right)^2 - 15 = 0 \Rightarrow 5x'^2 + 3y'^2 - 30 = 0.$

37. $x = 2t,\ y = t^2 \Rightarrow y = \frac{x^2}{4}$. Let $d = \sqrt{(x-0)^2 + \left(\frac{x^2}{4} - 3\right)^2} = \sqrt{x^2 + \frac{x^4}{16} - \frac{3}{2}x^2 + 9} = \sqrt{\frac{x^4}{16} - \frac{1}{2}x^2 + 9} =$
$\frac{1}{4}\sqrt{x^4 - 8x^2 + 144}$ be the distance from any point on the parabola to (0,3). We want to minimize d. Then
$\frac{dd}{dx} = \frac{1}{8}(x^4 - 8x^2 + 144)^{-1/2}(4x^3 - 16x) = \frac{\frac{1}{2}x^3 - 2x}{\sqrt{x^4 - 8x^2 + 144}}$. The critical numbers are only where $\frac{dd}{dx} = 0 \Rightarrow$
$\frac{1}{2}x^3 - 2x = 0 \Rightarrow x^3 - 4x = 0 \Rightarrow x = 0$ or $x = \pm 2$. $x = 0 \Rightarrow y = 0$, $x = \pm 2 \Rightarrow y = 1$. The distance from (0,0) to (0,3) is $d = 3$. The distance from $(\pm 2,1)$ to (0,3) is $d = \sqrt{(\pm 2)^2 + (1-3)^2} = 2\sqrt{2}$ which is less than 3. $\therefore$ the points closest to (0,3) are $(\pm 2,1)$.

39. The angle of rotation is 45° (See Exercise 30). Then $A' = \frac{3}{2}$, $B' = 0$, $C' = \frac{1}{2} \Rightarrow \frac{3}{2}x'^2 + \frac{1}{2}y'^2 = 1 \Rightarrow$
$b = \sqrt{\frac{2}{3}}$, $a = \sqrt{2} \Rightarrow c^2 = a^2 - b^2 = 2 - \frac{2}{3} = \frac{4}{3} \Rightarrow c = \frac{2}{\sqrt{3}}$. $\therefore e = \frac{c}{a} = \frac{2/\sqrt{3}}{\sqrt{2}} = \sqrt{\frac{2}{3}} \approx 0.82.$

41. $\alpha = \frac{\pi}{4} \Rightarrow A' = \sin\frac{\pi}{4}\cos\frac{\pi}{4} = \frac{1}{2}$, $B' = 0$, $C' = -\sin\frac{\pi}{4}\cos\frac{\pi}{4} = -\frac{1}{2} \Rightarrow \frac{x'^2}{2} - \frac{y'^2}{2} = 1 \Rightarrow a = \sqrt{2}$, $b = \sqrt{2} \Rightarrow c^2 =$
$4 \Rightarrow c = 2 \Rightarrow e = \frac{c}{a} = \frac{2}{\sqrt{2}} = \sqrt{2}$.

43. $x = \frac{t}{2},\ y = t + 1 \Rightarrow 2x = t \Rightarrow y = 2x + 1$

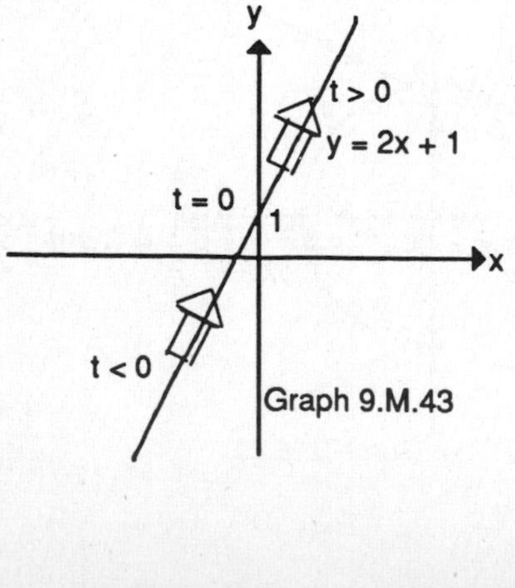

Graph 9.M.43

45. $x = \frac{1}{2}\tan t,\ y = \frac{1}{2}\sec t \Rightarrow x^2 = \frac{1}{4}\tan^2 t,$
$y = \frac{1}{4}\sec^2 t \Rightarrow 4x^2 = \tan^2 t,\ 4y^2 = \sec^2 t \Rightarrow$
$4x^2 + 1 = 4y^2 \Rightarrow 1 = 4y^2 - 4x^2$

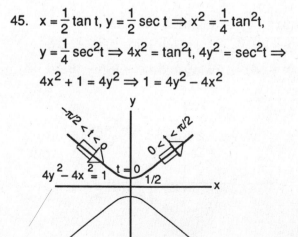

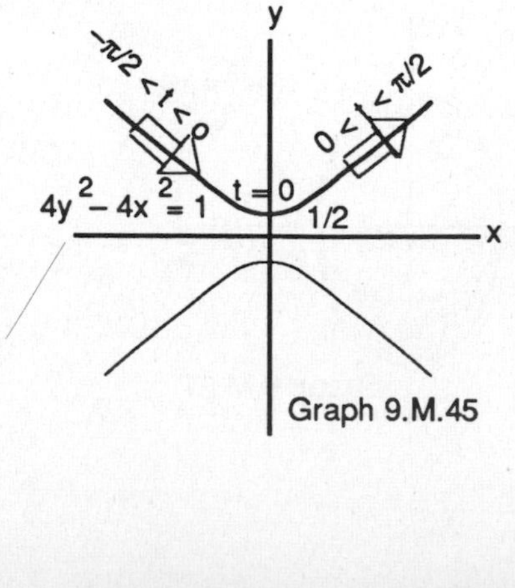

Graph 9.M.45

47. $x = -\cos t,\ y = \cos^2 t \Rightarrow y = (-x)^2 = x^2$

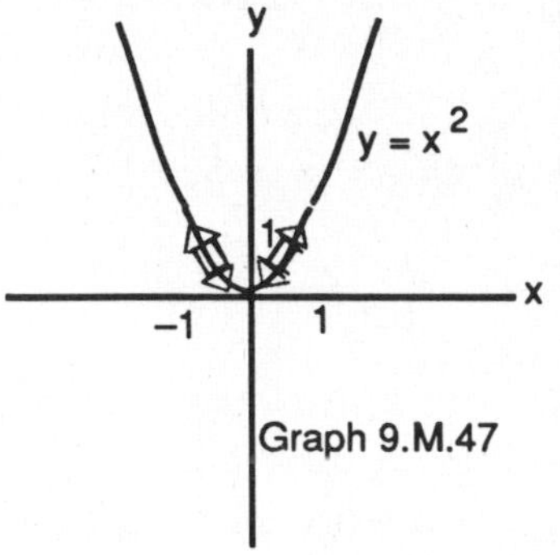

Graph 9.M.47

49. $16x^2 + 9y^2 = 144 \Rightarrow \frac{x^2}{9} + \frac{y^2}{16} = 1 \Rightarrow$

$a = 3,\ b = 4 \Rightarrow x = 3\cos t,\ y = 4\sin t,\ 0 \le t \le 2\pi$

51.

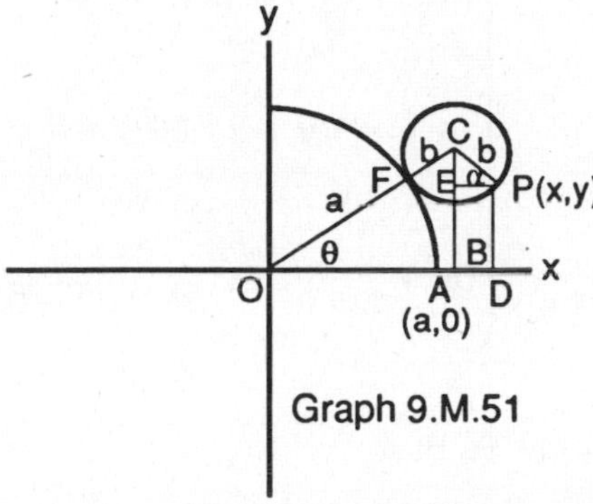

Graph 9.M.51

Arc PF = Arc AF since each is the distance rolled. $\angle PCF = \frac{\text{Arc PF}}{b}$ $\Rightarrow$ Arc PF = b($\angle$PCF). $\theta = \frac{\text{Arc AF}}{a} \Rightarrow$ Arc AF = $a\theta$. $\therefore a\theta =$ $b(\angle PCF) \Rightarrow \angle PCF = \frac{a}{b}\theta$. $\angle OCB = \frac{\pi}{2} - \theta$ and $\angle OCB = \angle PCF -$ $\angle PCE = \angle PCF - \left(\frac{\pi}{2} - \alpha\right) = \frac{a}{b}\theta - \left(\frac{\pi}{2} - \alpha\right) \Rightarrow \frac{\pi}{2} - \theta = \frac{a}{b}\theta -$ $\left(\frac{\pi}{2} - \alpha\right) \Rightarrow \frac{\pi}{2} - \theta = \frac{a}{b}\theta - \frac{\pi}{2} + \alpha \Rightarrow \alpha = \pi - \theta - \frac{a}{b}\theta \Rightarrow \alpha = \pi -$ $\left(\frac{a+b}{b}\theta\right)$. Now $x = OB + BD = OB + EP = (a + b)\cos\theta + b\cos\alpha$

$= (a + b)\cos\theta + b\cos\left(\pi - \left(\frac{a+b}{b}\theta\right)\right) = (a + b)\cos\theta + b\cos\pi\cos\left(\frac{a+b}{b}\theta\right) + b\sin\pi\sin\left(\frac{a+b}{b}\theta\right) =$

(a + b) cos θ − b cos\b\bc\((\f(a + b,b) θ). $y = PD = CB - CE = (a + b)\sin\theta - b\sin\alpha = (a + b)\sin\theta - b$ $\sin\left(\pi - \frac{a+b}{b}\theta\right)$

$= (a + b)\sin\theta - b\sin\pi\cos\left(\frac{a+b}{b}\theta\right) + b\cos\pi\sin\left(\frac{a+b}{b}\theta\right) = (a + b)\sin\theta - b\sin\left(\frac{a+b}{b}\theta\right)$.

$\therefore x = (a + b)\cos\theta - b\cos\left(\frac{a+b}{b}\theta\right),\ y = (a + b)\sin\theta - b\sin\left(\frac{a+b}{b}\theta\right)$

53. P traces a hypocycloid where the larger radius is 2a and the smaller is a $\Rightarrow x = (2a - a)\cos\theta +$ $a\cos\left(\frac{2a-a}{a}\theta\right) = 2a\cos\theta,\ 0 \le \theta \le 2\pi$. $y = (2a - a)\sin\theta - a\sin\left(\frac{2a-a}{a}\theta\right) = a\sin\theta - a\sin\theta = 0$.

$\therefore$ P traces the diameter of the circle back and forth as θ goes from 0 to 2π.

55. $x = \frac{1}{2}\tan t,\ y = \frac{1}{2}\sec t \Rightarrow \frac{dy}{dx} = \frac{dy/dt}{dx/dt} = \frac{\frac{1}{2}\sec t\tan t}{\frac{1}{2}\sec^2 t} = \frac{\tan t}{\sec t} = \sin t \Rightarrow \frac{dy}{dx}\left(\frac{\pi}{3}\right) = \sin\frac{\pi}{3} = \frac{\sqrt{3}}{2}$. $t = \frac{\pi}{3} \Rightarrow$ $x = \frac{1}{2}\tan\frac{\pi}{3} = \frac{\sqrt{3}}{2}$ and $y = \frac{1}{2}\sec t = 1 \Rightarrow y = \frac{\sqrt{3}}{2}x + \frac{1}{4}$. $\frac{d^2y}{dx^2} = \frac{\cos t}{\frac{1}{2}\sec^2 t} = 2\cos^3 t \Rightarrow \frac{d^2y}{dx^2}\left(\frac{\pi}{3}\right) = 2\cos^3\frac{\pi}{3}$ $= \frac{1}{4}$

57. $x = e^{2t} - \frac{t}{8}$, $y = e^t$, $0 \le t \le \ln 2 \Rightarrow \frac{dx}{dt} = 2e^{2t} - \frac{1}{8}$, $\frac{dy}{dt} = e^t \Rightarrow$ Length $= \int_0^{\ln 2} \sqrt{\left(2e^{2t} - \frac{1}{8}\right)^2 + \left(e^t\right)^2}\, dt =$

$$\int_0^{\ln 2} \sqrt{4e^{4t} + \frac{1}{2}e^{2t} + \frac{1}{16}}\, dt = \int_0^{\ln 2} \sqrt{\left(2e^{2t} + \frac{1}{8}\right)^2}\, dt = \int_0^{\ln 2} \left(2e^{2t} + \frac{1}{8}\right) dt =$$

$$\left[e^{2t} + \frac{t}{8}\right]_0^{\ln 2} = 3 + \frac{\ln 2}{8}$$

59. $x = \frac{t^2}{2}$, $y = 2t$, $0 \le t \le \sqrt{5} \Rightarrow \frac{dx}{dt} = t^2$, $\frac{dy}{dt} = 2 \Rightarrow$ Area $= \int_0^{\sqrt{5}} 2\pi(2t)\sqrt{t^2 + 4}\, dt = \int_4^9 2\pi\, u^{1/2}\, du =$

$2\pi\left[\frac{2}{3}u^{3/2}\right]_4^9 = \frac{76\pi}{3}$ Let $u = t^2 + 4 \Rightarrow du = 2t\, dt$. $x = 0 \Rightarrow u = 4$, $x = \sqrt{5} \Rightarrow u = 9$

61. $x = a(t - \sin t) \Rightarrow \frac{dx}{dt} = a(1 - \cos t)$. Let $\delta = 1$. Then $dm = dA = y\, dx = y\left(\frac{dx}{dt}\right)dt = a(1 - \cos t)a(1 - \cos t)\, dt$

$= a^2(1 - \cos t)^2 dt$. $\therefore$ $A = \int_0^{2\pi} a^2(1 - \cos t)^2 dt = a^2 \int_0^{2\pi} (1 - 2\cos t + \cos^2 t)\, dt =$

$a^2 \int_0^{2\pi} \left(1 - 2\cos t + \frac{1}{2} + \frac{1}{2}\cos 2t\right) dt = a^2\left[\frac{3}{2}t - 2\sin t + \frac{\sin 2t}{2}\right]_0^{2\pi} = 3\pi a^2$. $\tilde{x} = x = a(t - \sin t)$, $\tilde{y} = \frac{1}{2}y =$

$\frac{1}{2}a(1 - \cos t)$. $M_x = \int \tilde{y}\, dm = \int \tilde{y}\, \delta\, dA = \int_0^{2\pi} \frac{1}{2}a(1 - \cos t)a^2(1 - \cos t)^2\, dt = \frac{1}{2}a^3 \int_0^{2\pi} (1 - \cos t)^3\, dt =$

$\frac{a^3}{2} \int_0^{2\pi} (1 - 3\cos t + 3\cos^2 t - \cos^3 t)\, dt = \frac{a^3}{2} \int_0^{2\pi} \left(1 - 3\cos t + \frac{3}{2} + \frac{3\cos 2t}{2} - (1 - \sin^2 t)\cos t\right) dt =$

$\frac{a^3}{2}\left[\frac{5}{2}t - 3\sin t + \frac{3\sin 2t}{4} - \sin t + \frac{\sin^3 t}{3}\right]_0^{2\pi} = \frac{5\pi a^3}{2}$. $\therefore$ $\bar{y} = \frac{M_x}{M} = \frac{5\pi a^3/2}{3\pi a^2} = \frac{5}{6}a$. $M_y = \int \tilde{x}\, dm = \int \tilde{x}\, \delta\, dA$

$= \int_0^{2\pi} a(t - \sin t)a^2(1 - \cos t)^2\, dt = a^3 \int_0^{2\pi} (t - 2t\cos t + t\cos^2 t - \sin t + 2\sin t\cos t - \sin t\cos^2 t)\, dt =$

$a^3\left[\frac{t^2}{2} - 2\cos t - 2t\sin t + \frac{1}{4}t^2 + \frac{1}{8}\cos 2t + \frac{t}{4}\sin 2t + \cos t + \sin^2 t + \frac{\cos^3 t}{3}\right]_0^{2\pi} = 3\pi^2 a^3$. $\therefore$ $\bar{x} = \frac{M_y}{M} =$

$\frac{3\pi^2 a^3}{3\pi a^2} = \pi a$. $\Rightarrow \left(\pi a, \frac{5}{6}a\right)$ is the center of mass.

63. a) $\frac{dx}{dt} = \frac{1}{t+2} \Rightarrow x = \int \frac{1}{t+2}\,dt = \ln|t+2| + C$. $x = \ln 2$ when $t = 0 \Rightarrow \ln 2 = \ln|0+2| + C \Rightarrow C = 0$.

$\therefore x = \ln|t+2| = \ln(t+2)$ since $t \ge 0$. $\frac{dy}{dt} = 2t \Rightarrow y = \int 2t\,dt = t^2 + C$. $y = 1$ when $t = 0 \Rightarrow 1 = 0 + C \Rightarrow$ $C = 1$. $\therefore y = t^2 + 1$.

b) $x = \ln(t+2) \Rightarrow e^x = t + 2 \Rightarrow t = e^x - 2$. $\therefore y = (e^x - 2)^2 + 1 = e^{2x} - 4e^x + 5$.

c) $y = t^2 + 1 \Rightarrow y - 1 = t^2 \Rightarrow \sqrt{y-1} = t$ since $t \ge 0$. $\therefore x = \ln\left(\sqrt{y-1} + 2\right)$.

d) $t = 0 \Rightarrow y = 1, t = 2 \Rightarrow y = 5 \Rightarrow \Delta y = 5 - 1 = 4$. $t = 0 \Rightarrow x = \ln 2, t = 2 \Rightarrow x = \ln 4 \Rightarrow \Delta x = \ln 4 - \ln 2 =$ $\ln 2$. $\therefore \frac{\Delta y}{\Delta x} = \frac{4}{\ln 2}$.

e) $\frac{dy}{dx} = \frac{dy/dt}{dx/dt} = \frac{2t}{1/(t+2)} = 2t(t+2) \Rightarrow \left.\frac{dy}{dx}\right|_{t=1} = 6$.

65. $x = \frac{2}{3}t^{3/2} \Rightarrow \frac{dx}{dt} = t^{1/2}$, $y = 2t^{1/2} \Rightarrow \frac{dy}{dt} = t^{-1/2}$. Let $\delta = 1 \Rightarrow dm = dA = y\,dx = y\left(\frac{dx}{dt}\right)dt \Rightarrow (2t^{1/2})(t^{1/2})dt =$ $2t\,dt$. $\tilde{x} = x = \frac{2}{3}t^{3/2}$, $\tilde{y} = \frac{1}{2}y = t^{1/2}$. $M_x = \int \tilde{y}\,dm = \int_0^{\sqrt{3}} t^{1/2}(2t\,dt) = \int_0^{\sqrt{3}} 2t^{3/2}\,dt = \left[\frac{4}{5}t^{5/2}\right]_0^{\sqrt{3}} = \frac{36}{5}\sqrt{3}$.

$M_y = \int \tilde{x}\,dm = \int \tilde{x}\,dA = \int_0^{\sqrt{3}} \frac{2}{3}t^{3/2}(2t\,dt) = \int_0^{\sqrt{3}} \frac{4}{3}t^{5/2}\,dt = \left[\frac{8}{21}t^{7/2}\right]_0^{\sqrt{3}} = \frac{72}{7}\sqrt{3}$.

67. d

69. l

71. k

73. i

75.

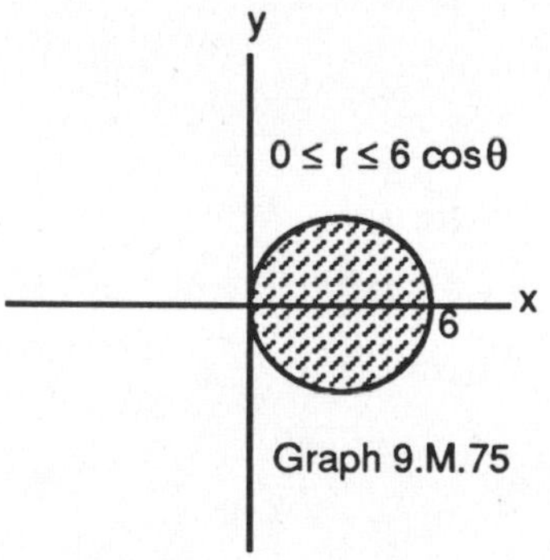

Graph 9.M.75

77. $r = \sin\theta$, $r = 1 + \sin\theta \Rightarrow \sin\theta = 1 + \sin\theta \Rightarrow \varnothing$. There are no points of intersection found by solving the system. The point of intersection (0,0) is found by graphing.

79. $r = 1 + \sin\theta$ and $r = -1 + \sin\theta$ intersect at all points of $r = 1 + \sin\theta$. This can be seen by graphing them.

81. $r = a$ and $r = a(1 - \sin\theta)$, $a > 0 \Rightarrow a = a(1 - \sin\theta) \Rightarrow \sin\theta = 0 \Rightarrow \theta = 0$ or $\pi \Rightarrow$ the points of intersection are $(a,0)$ and (a,π). No points are found by graphing.

83. $r = a\cos\theta$, $r = a(1 + \cos\theta)$, $a > 0 \Rightarrow a\cos\theta = a(1 + \cos\theta) \Rightarrow \cos\theta = 1 + \cos\theta \Rightarrow$ no points of intersection are found by solving. The point (0,0) is found by graphing.

85. $r^2 = \cos 2\theta \Rightarrow r = 0$ when $\cos 2\theta = 0 \Rightarrow 2\theta = \frac{\pi}{2}, \frac{3\pi}{2} \Rightarrow \theta = \frac{\pi}{4}, \frac{3\pi}{4}$. $\theta_1 = \frac{\pi}{4} \Rightarrow m_1 = \tan\frac{\pi}{4} = 1 \Rightarrow$ $y = x$ is one tangent line. $\theta_2 = \frac{3\pi}{4} \Rightarrow m_2 = \tan\frac{3\pi}{4} = -1 \Rightarrow y = -x$ is other tangent line.

87. Tips of the petals are at $\theta = \frac{\pi}{4}, \frac{3\pi}{4}, \frac{5\pi}{4}, \frac{7\pi}{4}$, $r = 1$ at those values of π. Then for $\theta = \frac{\pi}{4}$, the line is $r\cos\left(\theta - \frac{\pi}{4}\right) = 1$; for $\theta = \frac{3\pi}{4}$, $r\cos\left(\theta - \frac{3\pi}{4}\right) = 1$; for $\theta = \frac{5\pi}{4}$, $r\cos\left(\theta - \frac{5\pi}{4}\right) = 1$; and for $\theta = \frac{7\pi}{4}$, $r\cos\left(\theta - \frac{7\pi}{4}\right) = 1$.

89. $r = \dfrac{2}{1+\cos\theta} \Rightarrow e = 1 \Rightarrow$ Parabola

Vertex = (1,0)

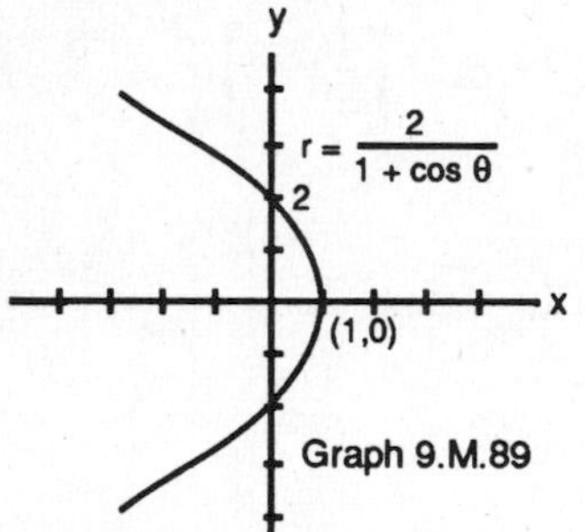

91. $r = \dfrac{6}{1-2\cos\theta} \Rightarrow e = 2 \Rightarrow$ Hyperbola

$ke = 6 \Rightarrow 2k = 6 \Rightarrow k = 3 \Rightarrow$ Vertices are $(2,\pi)$ and $(6,\pi)$.

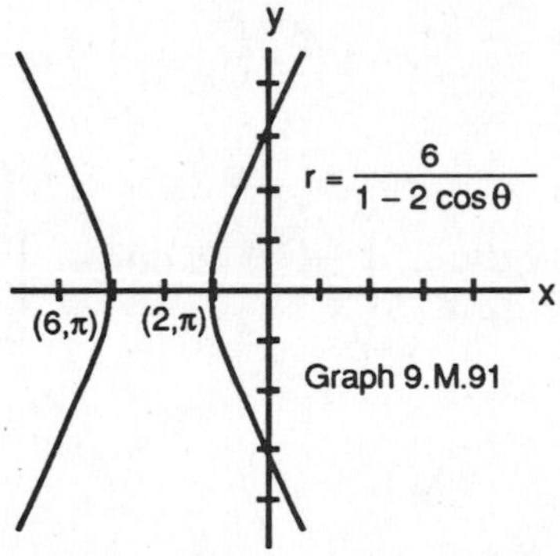

93. $e = 2$, $r\cos\theta = 2 \Rightarrow x = 2$ is directrix $\Rightarrow k = 2$ The conic is a hyperbola. $r = \dfrac{ke}{1+e\cos\theta} \Rightarrow$

$$r = \frac{2(2)}{1+2\cos\theta} \Rightarrow r = \frac{4}{1+2\cos\theta}$$

95. $e = \frac{1}{2}$, $r\sin\theta = 2$ Y $y = 2$ is directrix $\Rightarrow k = 2$. The conic is an ellipse. $r = \dfrac{ke}{1+e\sin\theta} \Rightarrow$

$$r = \frac{2\left(\frac{1}{2}\right)}{1+\frac{1}{2}\sin\theta} \Rightarrow r = \frac{2}{2+\sin\theta}$$

97. a) The equation of a parabola with focus (0,0) and vertex (a,0) is $r = \dfrac{2a}{1+\cos\theta}$. To rotate this parabola 45°: $r = \dfrac{2a}{1+\cos\left(\theta - \frac{\pi}{4}\right)}$.

b) Foci at (0,0) and (2,0) $\Rightarrow$ the center is (1,0) $\Rightarrow a = 3$ and $c = 1$ since a vertex is (4,0). Then $e = \frac{c}{a} = \frac{1}{3}$.

For ellipses with one focus at the origin, major axis along the x–axis, $r = \dfrac{a(1-e^2)}{1-e\cos\theta} = \dfrac{3\left(1-\frac{1}{9}\right)}{1-\frac{1}{3}\cos\theta} =$

97. b) (Continued)

$$\frac{8}{3 - \cos\theta}.$$

c) Center at $(2, \frac{\pi}{2})$ and focus at (0,0) $\Rightarrow$ c = 2. Center at $\left(1, \frac{\pi}{2}\right)$ and vertex at $\left(1, \frac{\pi}{2}\right) \Rightarrow a = 1 \therefore e = \frac{2}{1}$

$= 2$. Then $k = ae - \frac{a}{e} = (1)(2) - \frac{1}{2} = \frac{3}{2}$. $\therefore\ r = \dfrac{ke}{1 + e\sin\theta} = \dfrac{\frac{3}{2}(2)}{1 + 2\sin\theta} = \dfrac{3}{1 + 2\sin\theta}$.

99. $A = 2\displaystyle\int_0^{\pi} \frac{1}{2} r^2\, d\theta = \int_0^{\pi} (2 - \cos\theta)^2\, d\theta = \int_0^{\pi} \left(4 - 2\cos\theta + \cos^2\theta\right) d\theta =$

$$\int_0^{\pi} \left(4 - 2\cos\theta + \frac{1 + \cos 2\theta}{2}\right) d\theta = \int_0^{\pi} \left(\frac{9}{2} - 2\cos\theta + \frac{\cos 2\theta}{2}\right) d\theta =$$

$$\left[\frac{9}{2}\theta - 2\sin\theta + \frac{\sin 2\theta}{4}\right]_0^{\pi} = \frac{9}{2}\pi$$

101. $r = 1 + \cos 2\theta$, $r = 1 \Rightarrow 1 = 1 + \cos 2\theta \Rightarrow 0 = \cos 2\theta \Rightarrow 2\theta = \frac{\pi}{2} \Rightarrow \theta = \frac{\pi}{4}$.

$$\therefore\ A = 4\int_0^{\pi/4} \frac{1}{2}\left((1 + \cos 2\theta)^2 - 1^2\right) d\theta = 2\int_0^{\pi/4} (1 + 2\cos 2\theta + \cos^2 2\theta - 1)\, d\theta =$$

$$2\int_0^{\pi/4} \left(2\cos 2\theta + \frac{1}{2} + \frac{\cos 4\theta}{2}\right) d\theta = 2\left[\sin 2\theta + \frac{1}{2}\theta + \frac{\sin 4\theta}{8}\right]_0^{\pi/4} = 2 + \frac{\pi}{4}$$

103. $r = \sqrt{\cos 2\theta} \Rightarrow \dfrac{dr}{d\theta} = \dfrac{-\sin 2\theta}{\sqrt{\cos 2\theta}}$

$$\text{Surface Area} = \int_0^{\pi/4} 2\pi r \sin\theta \sqrt{r^2 + \left(\frac{dr}{d\theta}\right)^2}\, d\theta =$$

$$\int_0^{\pi/4} 2\pi\sqrt{\cos 2\theta}\, \sin\theta \sqrt{\left(\sqrt{\cos 2\theta}\right)^2 + \left(\frac{-\sin 2\theta}{\sqrt{\cos 2\theta}}\right)^2}\, d\theta =$$

$$\int_0^{\pi/4} 2\pi\sqrt{\cos 2\theta}\, \sin\theta \sqrt{\cos 2\theta + \frac{\sin^2 2\theta}{\cos 2\theta}}\, d\theta =$$

103. (Continued)

$$\int_0^{\pi/4} 2\pi\sqrt{\cos 2\theta}\,\sin\theta\sqrt{\frac{1}{\cos 2\theta}}\,d\theta = \int_0^{\pi/4} 2\pi \sin\theta\, d\theta = \Big[2\pi(-\cos\theta)\Big]_0^{\pi/4} = 2\pi\left(1-\frac{\sqrt{2}}{2}\right)$$

105. a) $x = e^{2t}\cos t$, $y = e^{2t}\sin t \Rightarrow x^2 + y^2 = e^{4t}\cos^2 t + e^{4t}\sin^2 t = e^{4t}$. Also $\frac{y}{x} = \frac{e^{2t}\sin t}{e^{2t}\cos t} = \tan t \Rightarrow t = \arctan\frac{y}{x}$. $\therefore\ x^2 + y^2 = e^{4(\arctan(y/x))}$ is the Cartesian equation. Since $r^2 = x^2 + y^2$ and $\theta = \arctan\frac{y}{x}$, the polar equation is $r^2 = e^{4\theta}$ or $r = e^{2\theta}$.

b)

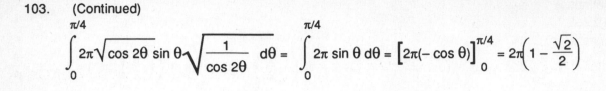

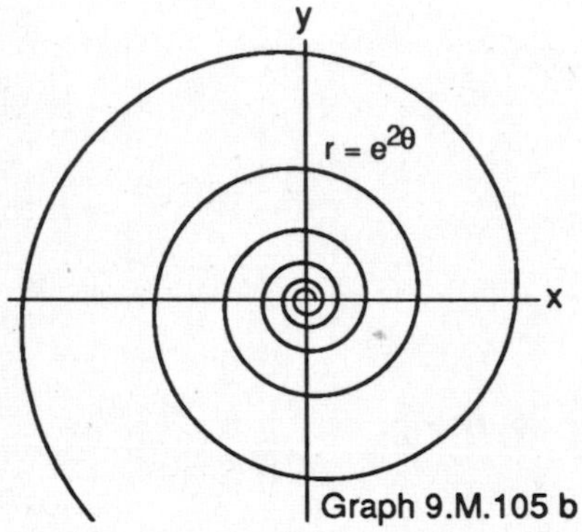

Graph 9.M.105 b

c) $ds^2 = r^2 d\theta^2 + dr^2$. $r = e^{2\theta} \Rightarrow dr = 2e^{2\theta}\,d\theta \Rightarrow$
$ds^2 = r^2\,d\theta^2 + \left(2e^{2\theta}\,d\theta\right)^2 = \left(e^{2\theta}\right)^2 d\theta^2 + 4e^{4\theta}\,d\theta^2 = 5\,e^{4\theta}\,d\theta^2 \Rightarrow ds = \sqrt{5}\,e^{2\theta}\,d\theta \Rightarrow$

$$L = \int_0^{2\pi} \sqrt{5}\,e^{2\theta}\,d\theta = \left[\frac{\sqrt{5}\,e^{2\theta}}{2}\right]_0^{2\pi} = \frac{\sqrt{5}\,e^{4\pi}}{2} - \frac{\sqrt{5}}{2}$$

107. $r = 1 + \cos\theta$. $S = \int 2\pi\rho\,ds$ where $\rho = y = r\sin\theta$ and $ds = \sqrt{r^2\,d\theta^2 + dr^2} =$

$$\sqrt{(1+\cos\theta)^2\,d\theta^2 + \sin^2\theta\,d\theta^2}\ \sqrt{1 + 2\cos\theta + \cos^2\theta + \sin^2\theta}\ d\theta = \sqrt{2 + 2\cos\theta}\ d\theta = \sqrt{4\cos^2\left(\frac{\theta}{2}\right)}\ d\theta$$

$= 2\cos\left(\frac{\theta}{2}\right)d\theta$ since $0 \le \theta \le \frac{\pi}{2}$. Then $S = \int_0^{\pi/2} 2\pi(r\sin\theta)\,2\cos\left(\frac{\theta}{2}\right)d\theta =$

$$\int_0^{\pi/2} 4\pi(1+\cos\theta)\sin\theta\cos\left(\frac{\theta}{2}\right)d\theta = \int_0^{\pi/2} 4\pi\left(2\cos^2\left(\frac{\theta}{2}\right)\right)\left(2\sin\left(\frac{\theta}{2}\right)\cos\left(\frac{\theta}{2}\right)\cos\left(\frac{\theta}{2}\right)\right)d\theta =$$

$$\int_0^{\pi/2} 16\pi\cos^4\left(\frac{\theta}{2}\right)\sin\left(\frac{\theta}{2}\right)d\theta = \left[\frac{-32\pi\cos^5(\theta/2)}{5}\right]_0^{\pi/2} = \frac{32\pi - 4\pi\sqrt{2}}{5}.$$

109.

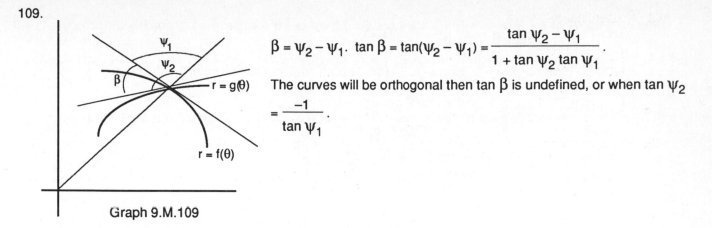

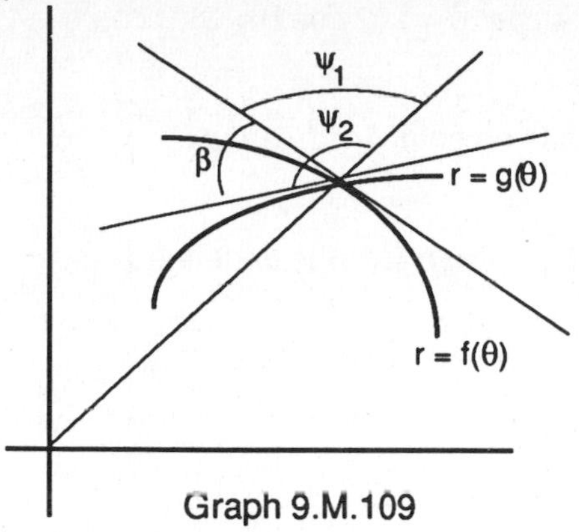

Graph 9.M.109

$\beta = \psi_2 - \psi_1$. $\tan \beta = \tan(\psi_2 - \psi_1) = \dfrac{\tan \psi_2 - \psi_1}{1 + \tan \psi_2 \tan \psi_1}$.

The curves will be orthogonal then $\tan \beta$ is undefined, or when $\tan \psi_2 = \dfrac{-1}{\tan \psi_1}$.

111. $r = 2a \sin 3\theta \Rightarrow \dfrac{dr}{d\theta}$ $6a \cos 3\theta \Rightarrow \tan \psi = \dfrac{r}{dr/d\theta} = \dfrac{2a \sin 3\theta}{6a \cos 3\theta} = \dfrac{1}{3} \tan 3\theta$. When $\theta = \dfrac{\pi}{3}$, $\tan \psi = \dfrac{1}{3} \tan \pi = 0$ so $\psi = 0$.

113. $\tan \psi_1 = \dfrac{\sqrt{3} \cos \theta}{-\sqrt{3} \sin \theta} = -\cot \theta$ is $-\dfrac{1}{\sqrt{3}}$ at $\theta = \dfrac{\pi}{3}$. $\tan \psi_2 = \dfrac{\sin \theta}{\cos \theta} = \tan \theta$ is $\sqrt{3}$ at $\theta = \dfrac{\pi}{3}$. Since the product of these slopes is -1, the tangents are perpendicular.

115. $r_1 = \dfrac{1}{1 - \cos \theta} \Rightarrow \dfrac{dr_1}{d\theta} = -\dfrac{\sin \theta}{(1 - \cos \theta)^2}$. $r_2 = \dfrac{3}{1 + \cos \theta} \Rightarrow \dfrac{dr_2}{d\theta} = \dfrac{3 \sin \theta}{(1 + \cos \theta)^2}$. $\dfrac{1}{1 - \cos \theta} = \dfrac{3}{1 + \cos \theta} \Rightarrow$ $1 + \cos \theta = 3 - 3 \cos \theta \Rightarrow 4 \cos \theta = 2 \Rightarrow \cos \theta = \dfrac{1}{2} \Rightarrow \theta = \pm \dfrac{\pi}{3} \Rightarrow r_1 = r_2 = 2 \Rightarrow$ the curves intersect at $\left(2, \pm \dfrac{\pi}{3}\right)$. $\tan \psi_1 = \dfrac{\dfrac{1}{1 - \cos \theta}}{\dfrac{-\sin \theta}{(1 - \cos \theta)^2}} = -\dfrac{1 - \cos \theta}{\sin \theta}$ is $-\dfrac{1}{\sqrt{3}}$ at $\theta = \dfrac{\pi}{3}$. $\tan \psi_2 = \dfrac{\dfrac{3}{1 + \cos \theta}}{\dfrac{3 \sin \theta}{(1 + \cos \theta)^2}} = \dfrac{1 + \cos \theta}{\sin \theta}$ is $\sqrt{3}$ at $\theta = \dfrac{\pi}{3}$. $\therefore \tan \beta$ is undefined at $\theta = \dfrac{\pi}{3}$ since $1 + \tan \psi_1 \tan \psi_2 = 1 + \left(-\dfrac{1}{\sqrt{3}}\right)\left(\sqrt{3}\right) = 0 \Rightarrow \beta = \dfrac{\pi}{2}$.

$\tan \psi_1 \Big|_{\theta = -\pi/3} = -\dfrac{1 - \cos(-\pi/3)}{\sin(-\pi/3)} = \dfrac{1}{\sqrt{3}}$, $\tan \psi_2 \Big|_{\theta = -\pi/3} = \dfrac{1 + \cos(-\pi/3)}{\sin(-\pi/3)} = -\sqrt{3}$. Then $\tan \beta$ is undefined at $\theta = -\dfrac{\pi}{3}$ also $\Rightarrow \beta = \dfrac{\pi}{2}$.

117. $r_1 = \dfrac{a}{1 + \cos \theta} \Rightarrow \dfrac{dr_1}{d\theta} = \dfrac{a \sin \theta}{(1 + \cos \theta)^2}$. $r_2 = \dfrac{b}{1 - \cos \theta} \Rightarrow \dfrac{dr_2}{d\theta} = -\dfrac{b \sin \theta}{(1 - \cos \theta)^2}$. Then $\tan \psi_1 = \dfrac{\dfrac{a}{1 + \cos \theta}}{\dfrac{a \sin \theta}{(1 + \cos \theta)^2}} = \dfrac{1 + \cos \theta}{\sin \theta}$. $\tan \psi_2 = \dfrac{\dfrac{b}{1 - \cos \theta}}{\dfrac{-b \sin \theta}{(1 - \cos \theta)^2}} = \dfrac{1 - \cos \theta}{-\sin \theta}$. Then $1 + \tan \psi_1 \tan \psi_2 = 1 +$ $\left(\dfrac{1 + \cos \theta}{\sin \theta}\right)\left(\dfrac{1 - \cos \theta}{-\sin \theta}\right) = 1 - \dfrac{1 - \cos^2 \theta}{\sin^2 \theta} = 0 \Rightarrow \beta$ is undefined $\Rightarrow$ the parabolas are orthogonal at each point of intersection.

119. $r = 3\sec\theta \Rightarrow r = \frac{3}{\cos\theta}$. $\frac{3}{\cos\theta} = 4 + 4\cos\theta \Rightarrow 3 = 4\cos\theta + 4\cos^2\theta \Rightarrow (2\cos\theta + 3)(2\cos\theta - 1) = 0$

$\Rightarrow \cos\theta = \frac{1}{2}$ or $\cos\theta = -\frac{3}{2} \Rightarrow \theta = \frac{\pi}{3}, \frac{5\pi}{3}$ (the second equation has no solutions) $\tan\psi_2 = \frac{4(1+\cos\theta)}{-4\sin\theta} =$

$-\frac{1+\cos\theta}{\sin\theta}$ is $-\sqrt{3}$ at $\frac{\pi}{3}$. $\tan\psi_1 = \frac{3\sec\theta}{3\sec\theta\tan\theta} = \cot\theta$ is $\frac{1}{\sqrt{3}}$ at $\frac{\pi}{3}$. Then $\tan\beta$ is undefined since

$1 - \tan\psi_1\tan\psi_2 = 1 + \left(\frac{1}{\sqrt{3}}\right)(-\sqrt{3}) = 0 \Rightarrow \beta = \frac{\pi}{2}$.

121. $\frac{1}{1-\cos\theta} = \frac{1}{1-\sin\theta} \Rightarrow 1 - \cos\theta = 1 - \sin\theta \Rightarrow \cos\theta = \sin\theta \Rightarrow \theta = \frac{\pi}{4}$. $\tan\psi_1 = \dfrac{\frac{1}{1-\cos\theta}}{\frac{-\sin\theta}{(1-\cos\theta)^2}} =$

$\frac{1-\cos\theta}{-\sin\theta}$. $\tan\psi_2 = \dfrac{\frac{1}{1-\sin\theta}}{\frac{\cos\theta}{(1-\sin\theta)^2}} = \frac{1-\sin\theta}{\cos\theta}$. At $\theta = \frac{\pi}{4}$, $\tan\psi_1 = \frac{1-\cos(\pi/4)}{-\sin(\pi/4)} = 1 - \sqrt{2}$ and $\tan\psi_2 =$

$\frac{1-\sin(\pi/4)}{\cos(\pi/4)} = \sqrt{2} - 1$. Then $\tan\beta = \frac{(\sqrt{2}-1) - (1-\sqrt{2})}{1 + (\sqrt{2}-1)(1-\sqrt{2})} = \frac{2\sqrt{2}-2}{2\sqrt{2}-2} = 1 \Rightarrow \beta = \frac{\pi}{4}$.

123. a) $\tan\alpha = \frac{r}{dr/d\theta} \Rightarrow \frac{dr}{r} = \frac{d\theta}{\tan\alpha} \Rightarrow \ln r = \frac{1}{\tan\alpha}\theta + C$ (by integration) $\Rightarrow r = B\,e^{\theta/\tan\alpha}$.

$$A = \frac{1}{2}\int_{\theta_1}^{\theta_2} B^2 e^{2\theta/\tan\alpha}\, d\theta = \left[\frac{1}{4}B^2 \tan\alpha\, e^{2\theta/\tan\alpha}\right]_{\theta_1}^{\theta_2} = \frac{1}{4}\tan\alpha\left[B^2 e^{2\theta_2/\tan\alpha} - B^2 e^{2\theta_1/\tan\alpha}\right] =$$

$\frac{\tan\alpha}{4}\left(r_2^2 - r_1^2\right)$ since $r_2^2 = B^2 e^{2\theta_2/\tan\alpha}$ and $r_1^2 = B^2 e^{2\theta_1/\tan\alpha}$. $K = \frac{\tan\alpha}{4}$.

b) $\tan\alpha = \frac{r}{dr/d\theta} \Rightarrow dr = \frac{r\,d\theta}{\tan\alpha} \Rightarrow dr^2 = \frac{r^2 d\theta^2}{\tan^2\alpha}$. $ds^2 = r^2 d\theta^2 + dr^2 = r^2 d\theta^2 + \frac{r^2 d\theta^2}{\tan^2\alpha} = r^2 d\theta^2\left(\frac{\tan^2\alpha + 1}{\tan^2\alpha}\right)$

$= r^2 d\theta^2\left(\frac{\sec^2\alpha}{\tan^2\alpha}\right) \Rightarrow ds = r\,d\theta\left(\frac{\sec\alpha}{\tan\alpha}\right)$ $S = \int_{\theta_1}^{\theta_2} B\,e^{\theta/\tan\alpha}\frac{\sec\alpha}{\tan\alpha}\, d\theta = \left[B\frac{\sec\alpha}{\tan\alpha}\tan\alpha\, e^{\theta/\tan\alpha}\right]_{\theta_1}^{\theta_2} =$

$\left[B\sec\alpha\, e^{\theta/\tan\alpha}\right]_{\theta_1}^{\theta_2} = \sec\alpha\left[B\,e^{\theta_2/\tan\alpha} - B\,e^{\theta_1/\tan\alpha}\right] = K(r_2 - r_1)$ where $K = \sec\alpha$.

CHAPTER 10

VECTORS AND ANALYTIC GEOMETRY IN SPACE

10.1 Vectors in the Plane

1.

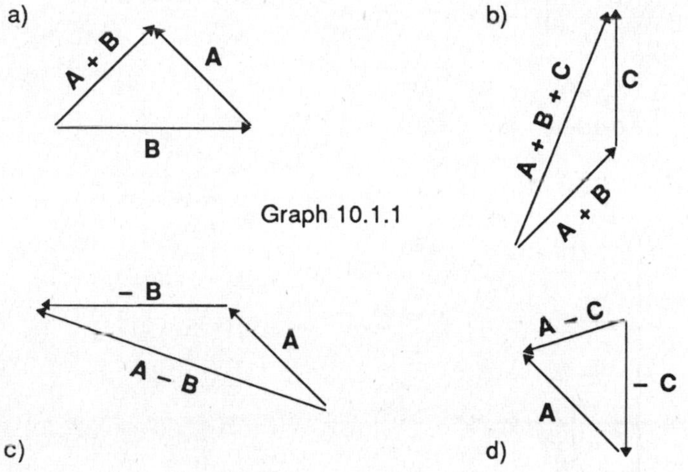

Graph 10.1.1

3. $(2\mathbf{i} - 7\mathbf{j}) + (\mathbf{i} + 6\mathbf{j}) = 3\mathbf{i} - \mathbf{j}$

5. $(-2\mathbf{i} + 6\mathbf{j}) - 2(\mathbf{i} + \mathbf{j}) + 3\mathbf{i} - 4\mathbf{j} = -\mathbf{i}$

7. $2\Big((\ln 2)\mathbf{i} + \mathbf{j}\Big) - \Big((\ln 8)\mathbf{i} + \pi\mathbf{j}\Big) = -\ln 2\,\mathbf{i} + (2 - \pi)\mathbf{j}$

9. $\overrightarrow{P_1P_2} = \mathbf{i} - 4\mathbf{j}$

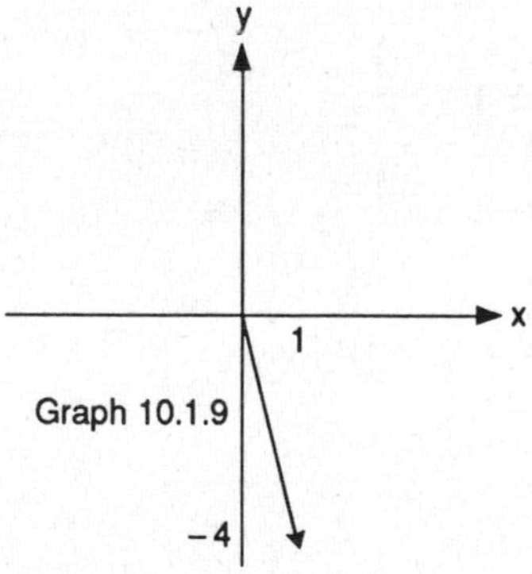

Graph 10.1.9

11. $\overrightarrow{AO} = -2\mathbf{i} - 3\mathbf{j}$

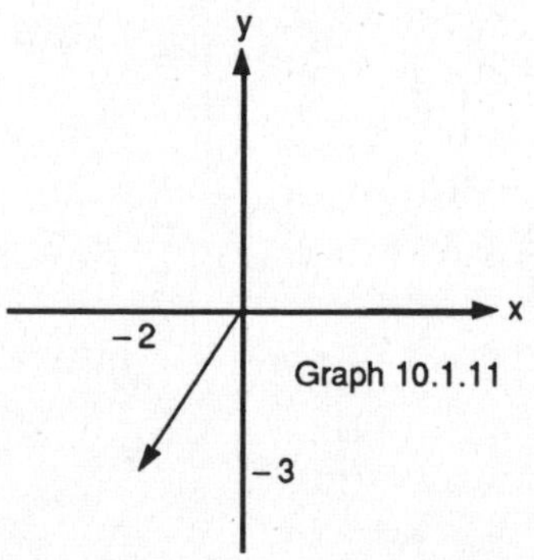

Graph 10.1.11

13. $\mathbf{u} = \frac{\sqrt{3}}{2}\mathbf{i} + \frac{1}{2}\mathbf{j}$, when $\theta = \frac{\pi}{6}$

$\mathbf{u} = -\frac{1}{2}\mathbf{i} + \frac{\sqrt{3}}{2}\mathbf{j}$, when $\theta = \frac{2\pi}{3}$

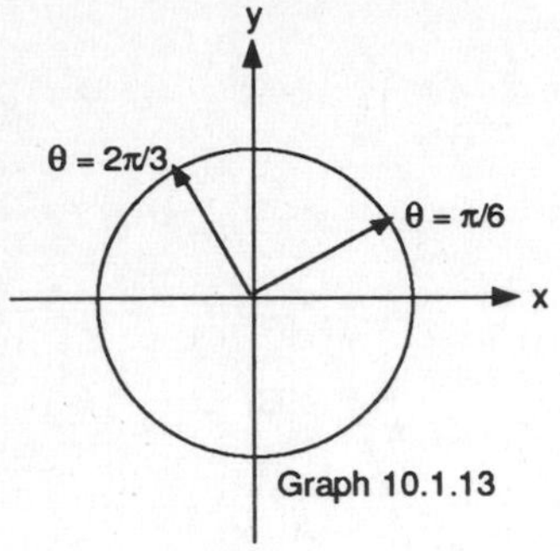

Graph 10.1.13

15. $\mathbf{u} = \frac{\sqrt{3}}{2}\mathbf{i} - \frac{1}{2}\mathbf{j}$

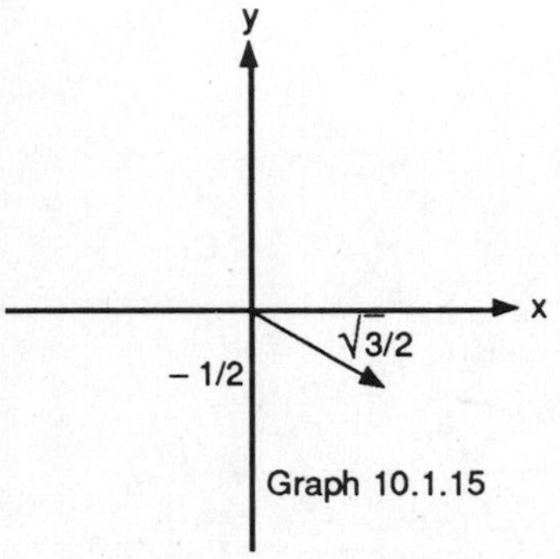

Graph 10.1.15

17. $\mathbf{u} = \frac{1}{\sqrt{17}}\mathbf{i} + \frac{4}{\sqrt{17}}\mathbf{j}$, $-\mathbf{u} = -\frac{1}{\sqrt{17}}\mathbf{i} - \frac{4}{\sqrt{17}}\mathbf{j}$,

$\mathbf{n} = \frac{4}{\sqrt{17}}\mathbf{i} - \frac{1}{\sqrt{17}}\mathbf{j}$, $-\mathbf{n} = -\frac{4}{\sqrt{17}}\mathbf{i} + \frac{1}{\sqrt{17}}\mathbf{j}$

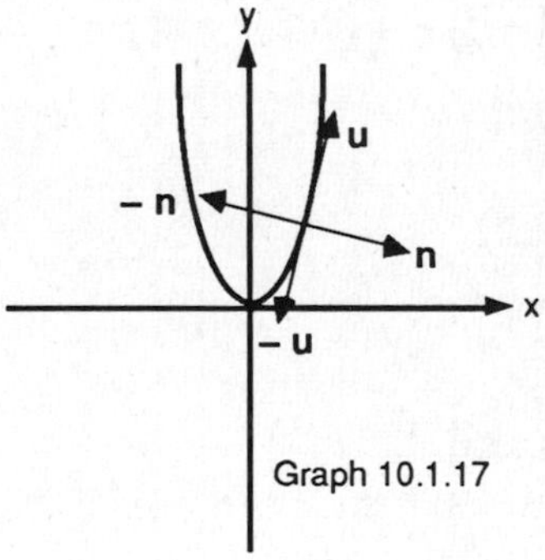

Graph 10.1.17

19. $\mathbf{u} = \frac{1}{\sqrt{5}}(2\mathbf{i} + \mathbf{j})$, $-\mathbf{u} = \frac{1}{\sqrt{5}}(-2\mathbf{i} - \mathbf{j})$

$\mathbf{n} = \frac{1}{\sqrt{5}}(-\mathbf{i} + 2\mathbf{j})$, $-\mathbf{n} = \frac{1}{\sqrt{5}}(\mathbf{i} - 2\mathbf{j})$

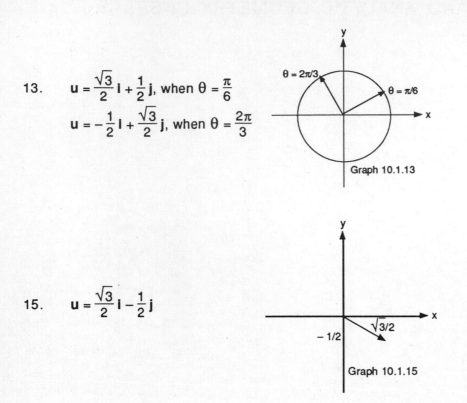

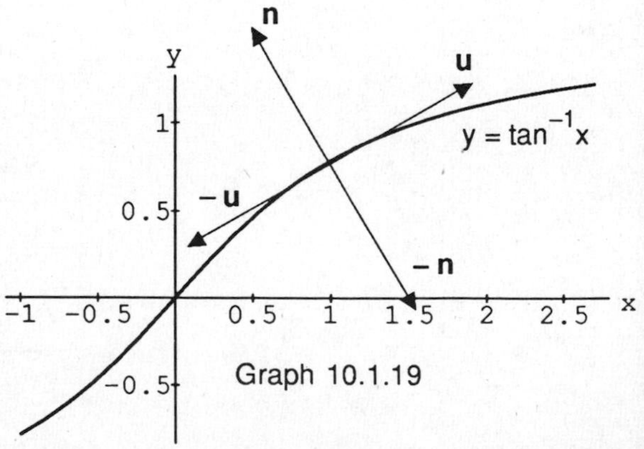

Graph 10.1.19

21. $3x^2 + 8xy + 2y^2 - 3 = 0 \Rightarrow m = \frac{-6x - 8y}{8x + 4y}$ at $(1,0) = -\frac{3}{4}$ $\therefore$ $\mathbf{u} = \frac{\pm 1}{5}(-4\mathbf{i} + 3\mathbf{j})$, $\mathbf{v} = \frac{\pm 1}{5}(3\mathbf{i} + 4\mathbf{j})$

23. $y = \int_0^x \sqrt{3 + t^4}\,dt \Rightarrow m = \sqrt{3 + x^4}$ at $(0,0) = \sqrt{3}$ $\therefore$ $\mathbf{u} = \frac{\pm 1}{2}\left(\mathbf{i} + \sqrt{3}\mathbf{j}\right)$, $\mathbf{v} = \frac{\pm 1}{2}\left(-\sqrt{3}\mathbf{i} + \mathbf{j}\right)$

25. $|\mathbf{i} + \mathbf{j}| = \sqrt{1^2 + 1^2} = \sqrt{2}$, $\mathbf{i} + \mathbf{j} = \sqrt{2}\left[\frac{1}{\sqrt{2}}\mathbf{i} + \frac{1}{\sqrt{2}}\mathbf{j}\right]$

27. $\left|\sqrt{3}\,\mathbf{i} + \mathbf{j}\right| = \sqrt{\left(\sqrt{3}\right)^2 + 1^2} = 2$, $2\left[\frac{\sqrt{3}}{2}\mathbf{i} + \frac{1}{2}\mathbf{j}\right]$

29. $|5\mathbf{i} + 12\mathbf{j}| = 13$, $13\left[\frac{5}{13}\mathbf{i} + \frac{12}{13}\mathbf{j}\right]$

31. $|\mathbf{A}| = |3\mathbf{i} + 6\mathbf{j}| = \sqrt{3^2 + 6^2} = 3\sqrt{5} \Rightarrow \mathbf{A} = 3\sqrt{5}\left[\frac{1}{\sqrt{5}}\mathbf{i} + \frac{2}{\sqrt{5}}\mathbf{j}\right]$;

$|\mathbf{B}| = |-\mathbf{i} - 2\mathbf{j}| = \sqrt{5} \Rightarrow \mathbf{B} = \sqrt{5}\left[-\frac{1}{\sqrt{5}}\mathbf{i} - \frac{2}{\sqrt{5}}\mathbf{j}\right]$

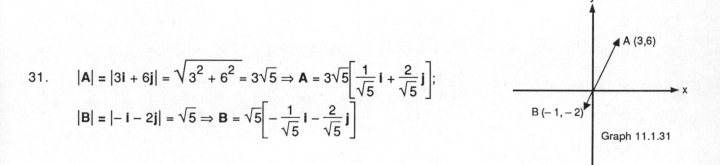

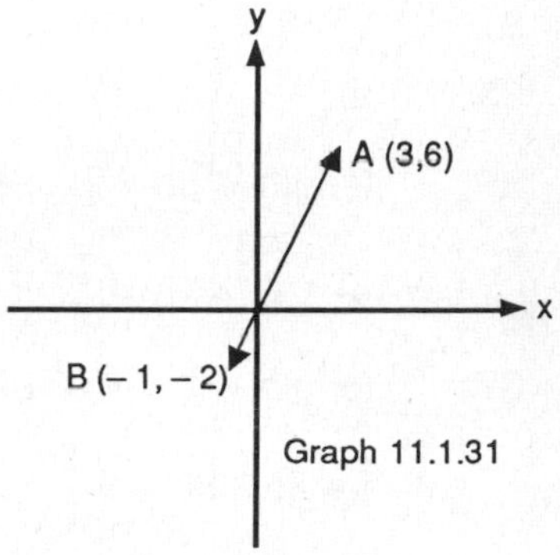

Graph 11.1.31

33. $\frac{2}{\sqrt{2}}(-\mathbf{i} - \mathbf{j})$, one

35. The same, if $\mathbf{v} = a\mathbf{i} + b\mathbf{j}$ where $a \neq 0$, then the slope is $\frac{b}{a}$ while the slope of $-\mathbf{v} = -a\mathbf{i} - b\mathbf{j}$ is $\frac{-b}{-a} = \frac{b}{a}$ also.

10.2 CARTESIAN (RECTANGULAR) COORDINATES AND VECTORS IN SPACE

1. a line through the point (2,3,0) parallel to the z–axis
3. the x–axis
5. the circle, $x^2 + y^2 = 4$ in the xy–plane
7. the circle, $x^2 + z^2 = 4$ in the xz–plane
9. the circle, $y^2 + z^2 = 1$ in the yz–plane
11. the circle, $x^2 + y^2 = 16$ in the xy–plane
13. a) the first quadrant of the xy–plane b) the fourth quadrant of the xy–plane
15. a) a solid sphere of radius 1 centered at the origin

    b) all points which are greater than 1 unit from the origin

17. a) the upper hemisphere of radius 1 centered at the origin

    b) the solid upper hemisphere of radius 1 centered at the origin

19. a) $x = 3$ b) $y = -1$ c) $z = -2$

21. a) $z = 1$ b) $x = 3$ c) $y = -1$

23. a) $x^2 + (y-2)^2 = 4$ b) $(y-2)^2 + z^2 = 4$ c) $x^2 + z^2 = 4$

25. a) $y = 3, z = -1$ b) $x = 1, z = -1$ c) $x = 1, y = 3$

27. $x^2 + y^2 + z^2 = 25, z = 3$

29. $0 \le z \le 1$

31. $z \le 0$

33. a) $(x-1)^2 + (y-1)^2 + (z-1)^2 < 1$ b) $(x-1)^2 + (y-1)^2 + (z-1)^2 > 1$

35. length $= |2\mathbf{i} + \mathbf{j} - 2\mathbf{k}| = \sqrt{2^2 + 1^2 + (-2)^2} = 3$, the direction is $\frac{2}{3}\mathbf{i} + \frac{1}{3}\mathbf{j} - \frac{2}{3}\mathbf{k} \Rightarrow$
$2\mathbf{i} + \mathbf{j} - 2\mathbf{k} = 3\left[\frac{2}{3}\mathbf{i} + \frac{1}{3}\mathbf{j} - \frac{2}{3}\mathbf{k}\right]$

37. length $= |\mathbf{i} + 4\mathbf{j} - 8\mathbf{k}| = \sqrt{1 + 16 + 64} = 9$, the direction is $\frac{1}{9}\mathbf{i} + \frac{4}{9}\mathbf{j} - \frac{8}{9}\mathbf{k} \Rightarrow$
$\mathbf{i} + 4\mathbf{j} - 8\mathbf{k} = 9\left[\frac{1}{9}\mathbf{i} + \frac{4}{9}\mathbf{j} - \frac{8}{9}\mathbf{k}\right]$

39. length $= |5\mathbf{k}| = \sqrt{25} = 5$, the direction is $\mathbf{k} \Rightarrow 5\mathbf{k} = 5\,[\mathbf{k}]$

41. length $= |-4\mathbf{j}| = \sqrt{(-4)^2} = \sqrt{16} = 4$, the direction is $-\mathbf{j} \Rightarrow 4[-\mathbf{j}]$

43. length $= \left|-\frac{1}{3}\mathbf{j} + \frac{1}{4}\mathbf{k}\right| = \sqrt{\left(-\frac{1}{3}\right)^2 + \left(\frac{1}{4}\right)^2} = \frac{5}{12}$, the direction is $-\frac{4}{5}\mathbf{j} + \frac{3}{5}\mathbf{k} \Rightarrow$
$-\frac{1}{3}\mathbf{j} + \frac{1}{4}\mathbf{k} = \frac{5}{12}\left[-\frac{4}{5}\mathbf{j} + \frac{3}{5}\mathbf{k}\right]$

45. length $= \left|\frac{1}{\sqrt{6}}\mathbf{i} - \frac{1}{\sqrt{6}}\mathbf{j} - \frac{1}{\sqrt{6}}\mathbf{k}\right| = \sqrt{3\left(\frac{1}{\sqrt{6}}\right)^2} = \sqrt{\frac{1}{2}}$, the direction is $\frac{1}{\sqrt{3}}\mathbf{i} - \frac{1}{\sqrt{3}}\mathbf{j} - \frac{1}{\sqrt{3}}\mathbf{k} \Rightarrow$
$\frac{1}{\sqrt{6}}\mathbf{i} - \frac{1}{\sqrt{6}}\mathbf{j} - \frac{1}{\sqrt{6}}\mathbf{k} = \sqrt{\frac{1}{2}}\left[\frac{1}{\sqrt{3}}\mathbf{i} - \frac{1}{\sqrt{3}}\mathbf{j} - \frac{1}{\sqrt{3}}\mathbf{k}\right]$

47. the length $= \left|\overrightarrow{P_1P_2}\right| = |2\mathbf{i} + 2\mathbf{j} - \mathbf{k}| = \sqrt{2^2 + 2^2 + (-1)^2} = 3$, $2\mathbf{i} + 2\mathbf{j} - \mathbf{k} = 3\left[\frac{2}{3}\mathbf{i} + \frac{2}{3}\mathbf{j} - \frac{1}{3}\mathbf{k}\right] \Rightarrow$
the direction is $\frac{2}{3}\mathbf{i} + \frac{2}{3}\mathbf{j} - \frac{1}{3}\mathbf{k}$, the midpoint is (2,2,1/2)

49. the length $= \left|\overrightarrow{P_1P_2}\right| = |(3\mathbf{i} - 6\mathbf{j} + 2\mathbf{k})| = \sqrt{9 + 36 + 4} = 7$, $3\mathbf{i} - 6\mathbf{j} + 2\mathbf{k} = 7\left[\frac{3}{7}\mathbf{i} - \frac{6}{7}\mathbf{j} + \frac{2}{7}\mathbf{k}\right] \Rightarrow$
the direction is $\frac{3}{7}\mathbf{i} - \frac{6}{7}\mathbf{j} + \frac{2}{7}\mathbf{k}$, the midpoint is (5/2,1,6)

51. the length $= \left|\overrightarrow{P_1P_2}\right| = |2\mathbf{i} - 2\mathbf{j} - 2\mathbf{k}| = \sqrt{3 \cdot 2^2} = 2\sqrt{3}$, $2\mathbf{i} - 2\mathbf{j} - 2\mathbf{k} = 2\sqrt{3}\left[\frac{1}{\sqrt{3}}\mathbf{i} - \frac{1}{\sqrt{3}}\mathbf{j} - \frac{1}{\sqrt{3}}\mathbf{k}\right] \Rightarrow$
the direction is $\frac{1}{\sqrt{3}}\mathbf{i} - \frac{1}{\sqrt{3}}\mathbf{j} - \frac{1}{\sqrt{3}}\mathbf{k}$, the midpoint is (1,–1,–1)

53. a) $2\mathbf{i}$ b) $4\mathbf{j}$ c) $\sqrt{3}\mathbf{k}$
d) $\frac{3}{10}\mathbf{j} + \frac{2}{5}\mathbf{k}$ e) $6\mathbf{i} + 2\mathbf{j} + 3\mathbf{k}$ f) $au_1\mathbf{i} + au_2\mathbf{j} + au_3\mathbf{k}$

55. $|\mathbf{A}| = |\mathbf{i} + \mathbf{j} + \mathbf{k}| = \sqrt{3}$, $\mathbf{U} = \sqrt{3}\left[\frac{1}{\sqrt{3}}\mathbf{i} + \frac{1}{\sqrt{3}}\mathbf{j} + \frac{1}{\sqrt{3}}\mathbf{k}\right] \Rightarrow$ the desired vector is $7\left[\frac{1}{\sqrt{3}}\mathbf{i} + \frac{1}{\sqrt{3}}\mathbf{j} + \frac{1}{\sqrt{3}}\mathbf{k}\right]$

57. a) center $(-2,0,2)$, radius $2\sqrt{2}$ b) center $(-1/2,-1/2,-1/2)$, radius $\frac{\sqrt{21}}{2}$
c) center $(\sqrt{2},\sqrt{2},-\sqrt{2})$, radius $\sqrt{2}$ d) center $(0,-1/3,1/3)$, radius $\frac{\sqrt{29}}{3}$

59. $x^2 + y^2 + z^2 + 4x - 4z + 0 \Rightarrow (x+2)^2 + y^2 + (z-2)^2 = \left(\sqrt{8}\right)^2 \Rightarrow$ the center is at $(-2,0,2)$ and the radius is $\sqrt{8}$

61. $x^2 + y^2 + z^2 - 2az = 0 \Rightarrow x^2 + y^2 + (z-a)^2 = a^2 \Rightarrow$ the center is at $(0,0,a)$ and the radius is a

63. a) $\sqrt{y^2 + z^2}$ b) $\sqrt{x^2 + z^2}$ c) $\sqrt{x^2 + y^2}$

65. a) the midpoint of AB is $(5/2,5/2,0)$ and $\overrightarrow{CM} = (5/2-1)\mathbf{i} + (5/2-1)\mathbf{j} + (0-3)\mathbf{k} = \frac{3}{2}\mathbf{i} + \frac{3}{2}\mathbf{j} - 3\mathbf{k}$

b) the desired vector is $\left(\frac{2}{3}\right)\overrightarrow{CM} = \frac{2}{3}\left(\frac{3}{2}\mathbf{i} + \frac{3}{2}\mathbf{j} - 3\mathbf{k}\right) = \mathbf{i} + \mathbf{j} - 2\mathbf{k}$

c) the vector whose sum is the vector from the origin to C and the result of part b) will terminate at the center of mass $\therefore$ the terminal point of $(\mathbf{i} + \mathbf{j} + 3\mathbf{k}) + (\mathbf{i} + \mathbf{j} - 2\mathbf{k}) = 2\mathbf{i} + 2\mathbf{j} + \mathbf{k}$ is the $(2,2,1)$ the location of the center of mass

67. Without loss of generality we can place the quadrilateral such that $A(0,0,0)$, $B(x_b,0\ 0)$, $C(x_c,y_c,0)$ and $D(x_d,y_d,z_d) \Rightarrow$ the midpoint of AB,$M_{AB}\left(\frac{x_b}{2},0,0\right)$, the midpoint of BC, $M_{BC}\left(\frac{x_b+x_c}{2},\frac{y_c}{2},0\right)$, the midpoint of CD, $M_{CD}\left(\frac{x_c+x_d}{2},\frac{y_c+y_d}{2},\frac{z_d}{2}\right)$ and the midpoint of AD, $M_{AD}\left(\frac{x_d}{2},\frac{y_d}{2},\frac{z_d}{2}\right)$ $\therefore$ the midpoint of $M_{AB}M_{CD}$ is $\left(\frac{\frac{x_b}{2}+\frac{x_c+x_d}{2}}{2},\frac{y_c+y_d}{4},\frac{z_d}{4}\right)$ and the same as the midpoint of $M_{AD}M_{BC}$ is $\left(\frac{\frac{x_b+x_c}{2}+\frac{x_d}{2}}{2},\frac{y_c+y_d}{4},\frac{z_d}{4}\right)$

69. Without loss of generality we can place the triangle such that $A(0,0)$, $B(b,0)$ and $C(x_c, y_c) \Rightarrow$ a is located at $\left(\frac{b+x_c}{2},\frac{y_c}{2}\right)$, b is at $\left(\frac{x_c}{2},\frac{y_c}{2}\right)$ and c is at $\left(\frac{b}{2},0\right)$ $\therefore$ $\overrightarrow{Aa} = \left(\frac{b}{2}+\frac{x_c}{2}\right)\mathbf{i} + \left(\frac{y_c}{2}\right)\mathbf{j}$, $\overrightarrow{Bb} = \left(\frac{x_c}{2}-b\right)\mathbf{i} + \left(\frac{y_c}{2}\right)\mathbf{j}$, $\overrightarrow{Cc} = \left(\frac{b}{2}-x_c\right)\mathbf{i} + \left(-y_c\right)\mathbf{k}$ and $\overrightarrow{Aa} + \overrightarrow{Bb} + \overrightarrow{Cc} = \mathbf{0}$

10.3 DOT PRODUCTS

	A·B	$\lvert A\rvert$	$\lvert B\rvert$	cos θ	$\lvert B\rvert$ cos θ	Proj_A B
1.	10	$\sqrt{13}$	$\sqrt{26}$	$\frac{10}{13\sqrt{2}}$	$\frac{10}{\sqrt{13}}$	$\frac{10}{13}[3\mathbf{i} + 2\mathbf{j}]$
3.	4	$\sqrt{14}$	2	$\frac{2}{\sqrt{14}}$	$\frac{4}{\sqrt{14}}$	$\frac{2}{7}[3\mathbf{i} - 2\mathbf{j} - \mathbf{k}]$
5.	2	$\sqrt{34}$	$\sqrt{3}$	$\frac{2}{\sqrt{3}\sqrt{34}}$	$\frac{2}{\sqrt{34}}$	$\frac{1}{17}[5\mathbf{j} - 3\mathbf{k}]$
7.	$\sqrt{3} - \sqrt{2}$	$\sqrt{2}$	3	$\frac{\sqrt{3} - \sqrt{2}}{3\sqrt{2}}$	$\frac{\sqrt{3} - \sqrt{2}}{\sqrt{2}}$	$\frac{\sqrt{3} - \sqrt{2}}{2}[-\mathbf{i} + \mathbf{j}]$
9.	− 25	5	5	− 1	− 5	$-2\mathbf{i} + 4\mathbf{j} - \sqrt{5}\mathbf{k}$
11.	25	15	5	$\frac{1}{3}$	$\frac{5}{3}$	$\frac{1}{9}[10\mathbf{i} + 11\mathbf{j} - 2\mathbf{k}]$

13. $\mathbf{B} = \left(\frac{\mathbf{A}\cdot\mathbf{B}}{\mathbf{A}\cdot\mathbf{A}}\mathbf{A}\right) + \left(\mathbf{B} - \frac{\mathbf{A}\cdot\mathbf{B}}{\mathbf{A}\cdot\mathbf{A}}\mathbf{A}\right) = \frac{3}{2}[\mathbf{i} + \mathbf{j}] + \left[(3\mathbf{j} + 4\mathbf{k}) - \frac{3}{2}(\mathbf{i} + \mathbf{j})\right] = \left[\frac{3}{2}\mathbf{i} + \frac{3}{2}\mathbf{j}\right] + \left[-\frac{3}{2}\mathbf{i} + \frac{3}{2}\mathbf{j} + 4\mathbf{k}\right]$, where $\mathbf{A}\cdot\mathbf{B} = 3$ and $\mathbf{A}\cdot\mathbf{A} = 2$

15. $\mathbf{B} = \left(\frac{\mathbf{A}\cdot\mathbf{B}}{\mathbf{A}\cdot\mathbf{A}}\mathbf{A}\right) + \left(\mathbf{B} - \frac{\mathbf{A}\cdot\mathbf{B}}{\mathbf{A}\cdot\mathbf{A}}\mathbf{A}\right) = \frac{14}{3}[\mathbf{i} + 2\mathbf{j} - \mathbf{k}] + \left[(8\mathbf{i} + 4\mathbf{j} - 12\mathbf{k}) - \left(\frac{14}{3}\mathbf{i} + \frac{28}{3}\mathbf{j} - \frac{14}{3}\mathbf{k}\right)\right] =$ $\left[\frac{14}{3}\mathbf{i} + \frac{28}{3}\mathbf{j} - \frac{14}{3}\mathbf{k}\right] + \left[\frac{10}{3}\mathbf{i} - \frac{16}{3}\mathbf{j} - \frac{22}{3}\mathbf{k}\right]$, where $\mathbf{A}\cdot\mathbf{B} = 28$ and $\mathbf{A}\cdot\mathbf{A} = 6$

17. $(\mathbf{i} + 2\mathbf{j})\cdot\left((x - 2)\mathbf{i} + (y - 1)\mathbf{j}\right) = 0 \Rightarrow x + 2y = 4$

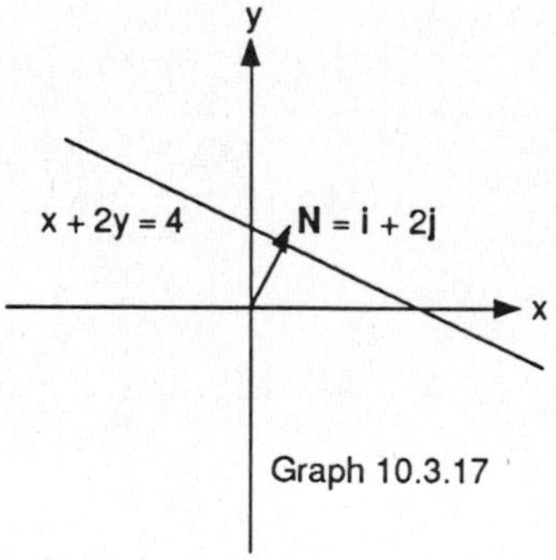

Graph 10.3.17

19. $(-2\mathbf{i} - \mathbf{j})\cdot\left((x + 1)\mathbf{i} + (y - 2)\mathbf{j}\right) = 0 \Rightarrow -2x - y = 0$

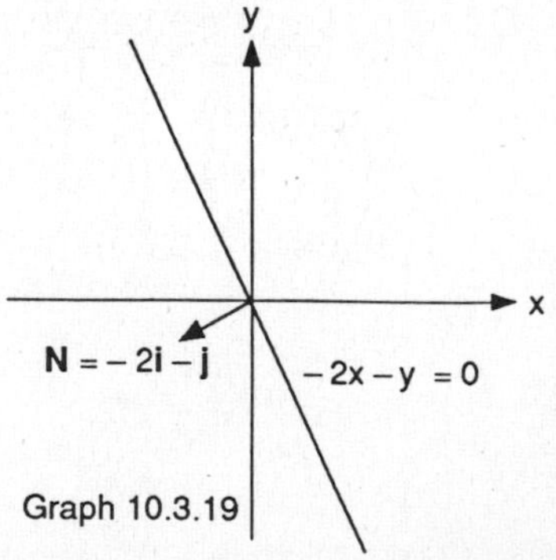

Graph 10.3.19

21. distance $= \left|\text{proj}_{\mathbf{N}}\overrightarrow{PS}\right| = \left|\frac{\mathbf{N}\cdot\overrightarrow{PS}}{|\mathbf{N}|}\right| = \left|\frac{(\mathbf{i}+3\mathbf{j})\cdot(2\mathbf{i}+6\mathbf{j})}{\sqrt{1^2+3^2}}\right| = \left|\frac{20}{\sqrt{10}}\right| = 2\sqrt{10}$, where S(2,8), P(0,2) and $\mathbf{N} + \mathbf{i} + 3\mathbf{j}$

23. distance $= \left|\text{proj}_{\mathbf{N}}\overrightarrow{PS}\right| = \left|\frac{\mathbf{N}\cdot\overrightarrow{PS}}{|\mathbf{N}|}\right| = \left|\frac{(\mathbf{i}+\mathbf{j})\cdot(\mathbf{i}+\mathbf{j})}{\sqrt{1+1}}\right| = \sqrt{2}$, where S(2,1), P(1,0) and $\mathbf{N} = \mathbf{i} + \mathbf{j}$

25. $\mathbf{A}\cdot\mathbf{B} = \left(\frac{1}{\sqrt{3}}\right)(0) + \left(-\frac{1}{\sqrt{3}}\right)\left(\frac{1}{\sqrt{2}}\right) + \left(\frac{1}{\sqrt{3}}\right)\left(\frac{1}{\sqrt{2}}\right) = 0$, $\mathbf{A}\cdot\mathbf{C} = \left(\frac{1}{\sqrt{3}}\right)\left(-\frac{2}{\sqrt{6}}\right) + \left(-\frac{1}{\sqrt{3}}\right)\left(-\frac{1}{\sqrt{6}}\right) + \left(\frac{1}{\sqrt{3}}\right)\left(\frac{1}{\sqrt{6}}\right) = 0$ and $\mathbf{B}\cdot\mathbf{C} = (0)\left(-\frac{2}{\sqrt{6}}\right) + \left(\frac{1}{\sqrt{2}}\right)\left(-\frac{1}{\sqrt{6}}\right) + \left(\frac{1}{\sqrt{2}}\right)\left(\frac{1}{\sqrt{6}}\right) = 0 \Rightarrow$ that **A**, **B** and **C** are orthogonal

27. If $\mathbf{A} = (\mathbf{i}+\mathbf{j}+\mathbf{k})$, $\mathbf{B}_1 = \mathbf{i}$, and $\mathbf{B}_2 = \mathbf{j}$, then $\mathbf{A}\cdot\mathbf{B}_1 = \mathbf{A}\cdot\mathbf{B}_2$, but $\mathbf{B}_1 \neq \mathbf{B}_2$

29. Let ABCD be a rhombus. Clearly, $\overrightarrow{AC} = \overrightarrow{AB} + \overrightarrow{BC}$ and $\overrightarrow{BD} = \overrightarrow{CD} + \overrightarrow{BC}$. $\overrightarrow{AC}\cdot\overrightarrow{BD} = \left(\overrightarrow{AB} + \overrightarrow{BC}\right)\cdot\left(\overrightarrow{CD} + \overrightarrow{BC}\right) = \left(\overrightarrow{AB} + \overrightarrow{BC}\right)\cdot\left(-\overrightarrow{AB} + \overrightarrow{BC}\right) = \left|\overrightarrow{AB}\right|^2 - \left|\overrightarrow{BC}\right|^2 = 0.$

$\therefore$ the diagonals of a rhombus are perpendicular.

31. $\overrightarrow{AB} = 3\mathbf{i} + \mathbf{j} - 3\mathbf{k}$, $\overrightarrow{AC} = 2\mathbf{i} - 2\mathbf{j}$, $\overrightarrow{BA} = -3\mathbf{i} - \mathbf{j} + 3\mathbf{k}$, $\overrightarrow{CA} = -2\mathbf{i} + 2\mathbf{j}$, $\overrightarrow{CB} = \mathbf{i} + 3\mathbf{j} - 3\mathbf{k}$, $\overrightarrow{BC} = -\mathbf{i} - 3\mathbf{j} + 3\mathbf{k} \Rightarrow \angle A = \cos^{-1}\left(\frac{\overrightarrow{AB}\cdot\overrightarrow{AC}}{\left|\overrightarrow{AB}\right|\left|\overrightarrow{AC}\right|}\right) = \cos^{-1}\left(\frac{4}{\sqrt{152}}\right) \approx 71.1°$,

$\angle B = \cos^{-1}\left(\frac{\overrightarrow{BA}\cdot\overrightarrow{BC}}{\left|\overrightarrow{BA}\right|\left|\overrightarrow{BC}\right|}\right) = \cos^{-1}\left(\frac{15}{19}\right) \approx 37.9°$, $\angle C = \cos^{-1}\left(\frac{\overrightarrow{CA}\cdot\overrightarrow{CB}}{\left|\overrightarrow{CA}\right|\left|\overrightarrow{CB}\right|}\right) = \cos^{-1}\left(\frac{4}{\sqrt{152}}\right) \approx 71.1°$

33. $\theta = \cos^{-1}\left(\frac{\mathbf{A}\cdot\mathbf{B}}{|\mathbf{A}|\,|\mathbf{B}|}\right) = \cos^{-1}\left(\frac{2}{\sqrt{2}\sqrt{3}}\right) \approx 35.3°$, where $\mathbf{A} = \mathbf{i} + \mathbf{k}$ and $\mathbf{B} = \mathbf{i} + \mathbf{j} + \mathbf{k}$

35. P(0,0,0) , Q(1,1,1) and $\mathbf{F} = -5\mathbf{k} \Rightarrow \overrightarrow{PQ} = \mathbf{i} + \mathbf{j} + \mathbf{k}$, $W = \mathbf{F}\cdot\overrightarrow{PQ} = (-5\mathbf{k})\cdot(\mathbf{i}+\mathbf{j}+\mathbf{k}) = -5$ N · m

37. $W = |\mathbf{F}|\left|\overrightarrow{PQ}\right|\cos\theta = (200)(20)\left(\cos\frac{\pi}{6}\right) = 20000\sqrt{3} \approx 3464.10$ N · m

39. The angle between the corresponding normals is equal to the angle between the corresponding tangents. $\theta = \cos^{-1}\left(\frac{\mathbf{N}_1\cdot\mathbf{N}_2}{|\mathbf{N}_1|\,|\mathbf{N}_2|}\right) = \cos^{-1}\left(\frac{1}{\sqrt{2}}\right) = 45°$, where $\mathbf{N}_1 = 3\mathbf{i} + \mathbf{j}$, $\mathbf{N}_2 = 2\mathbf{i} - \mathbf{j}$; the angle is either 45° or 135°

41. The curve $y = \sqrt{\frac{3}{4} + x}$ has $y = \frac{1}{\sqrt{3}}x + \frac{\sqrt{3}}{2}$ as its tangent line at $(0, \sqrt{3}/2)$. The normal vector for $y = \sqrt{\frac{3}{4} + x}$ at $(0, \sqrt{3}/2)$ is $\mathbf{N}_1 = \frac{1}{\sqrt{3}}\mathbf{i} - \mathbf{j}$. The curve $y = \sqrt{\frac{3}{4} - x}$ has $y = -\frac{1}{\sqrt{3}}x + \frac{\sqrt{3}}{2}$ as its tangent line at $(0, \sqrt{3}/2)$. The normal vector for $y = \sqrt{\frac{3}{4} + x}$ at $(0, \sqrt{3}/2)$ is $\mathbf{N}_2 = -\frac{1}{\sqrt{3}}\mathbf{i} - \mathbf{j}$.

$\theta = \cos^{-1}\left(\frac{\mathbf{N}_1 \cdot \mathbf{N}_2}{|\mathbf{N}_1|\,|\mathbf{N}_2|}\right) = \cos^{-1}\left(\frac{1}{2}\right) = \frac{\pi}{3}$; the angle is either $\frac{\pi}{3}$ or $\frac{2\pi}{3}$. See exercise 39 for additional details.

43. The points of intersection for the curves $y = x^2$ and $y = \sqrt[3]{x}$ are (0,0) and (1,1). At (0,0) the tangent line for $y = x^2$ is $y = 0$ and the tangent line for $y = \sqrt{x}$ is $x = 0$. Therefore, the angle of intersection at (0,0) is $\frac{\pi}{2}$. At (1,1) the tangent line for $y = x^2$ is $y = 2x - 1$ and the tangent line for $y = \sqrt[3]{x}$ is $y = \frac{1}{3}x + \frac{2}{3}$. The corresponding normal vectors are: $\mathbf{N}_1 = 2\mathbf{i} - \mathbf{j}$, $\mathbf{N}_2 = \frac{1}{3}\mathbf{i} - \mathbf{j}$.

$\theta = \cos^{-1}\left(\frac{\mathbf{N}_1 \cdot \mathbf{N}_2}{|\mathbf{N}_1|\,|\mathbf{N}_2|}\right) = \cos^{-1}\frac{1}{\sqrt{2}} = \frac{\pi}{4}$; the angle is either 45° or 135°

10.4 CROSS PRODUCTS

1. $\mathbf{A} \times \mathbf{B} = \begin{vmatrix} \mathbf{i} & \mathbf{j} & \mathbf{k} \\ 2 & -2 & -1 \\ 1 & 0 & -1 \end{vmatrix} = 3\left[\frac{2}{3}\mathbf{i} + \frac{1}{3}\mathbf{j} + \frac{2}{3}\mathbf{k}\right] \Rightarrow$ length = 3 and the direction is $\frac{2}{3}\mathbf{i} + \frac{1}{3}\mathbf{j} + \frac{2}{3}\mathbf{k}$

$\mathbf{B} \times \mathbf{A} = \begin{vmatrix} \mathbf{i} & \mathbf{j} & \mathbf{k} \\ 1 & 0 & -1 \\ 2 & -2 & -1 \end{vmatrix} = -3\left[\frac{2}{3}\mathbf{i} + \frac{1}{3}\mathbf{j} + \frac{2}{3}\mathbf{k}\right] \Rightarrow$ length = 3 and the direction is $-\frac{2}{3}\mathbf{i} - \frac{1}{3}\mathbf{j} - \frac{2}{3}\mathbf{k}$

$\mathbf{B} \times \mathbf{A} = \begin{vmatrix} \mathbf{i} & \mathbf{j} & \mathbf{k} \\ -1 & 1 & 0 \\ 2 & 3 & 0 \end{vmatrix} = -5\,[\mathbf{k}] \Rightarrow$ length = 5 and the direction is $-\mathbf{k}$

3. $\mathbf{A} \times \mathbf{B} = \begin{vmatrix} \mathbf{i} & \mathbf{j} & \mathbf{k} \\ 2 & -2 & 4 \\ -1 & 1 & -2 \end{vmatrix} = \mathbf{0} \Rightarrow$ length = 0 and has no direction

$\mathbf{B} \times \mathbf{A} = \begin{vmatrix} \mathbf{i} & \mathbf{j} & \mathbf{k} \\ -1 & 1 & -2 \\ 2 & -2 & 4 \end{vmatrix} = \mathbf{0} \Rightarrow$ length = 0 and has no direction

$\mathbf{B} \times \mathbf{A} = \begin{vmatrix} \mathbf{i} & \mathbf{j} & \mathbf{k} \\ 0 & 0 & 0 \\ 1 & 1 & -1 \end{vmatrix} = \mathbf{0} \Rightarrow$ length = 0 and has no direction

5. $\mathbf{A} \times \mathbf{B} = \begin{vmatrix} \mathbf{i} & \mathbf{j} & \mathbf{k} \\ 2 & 0 & 0 \\ 0 & -3 & 0 \end{vmatrix} = -6\,[\mathbf{k}] \Rightarrow$ length = 6 and the direction is $-\mathbf{k}$

$\mathbf{B} \times \mathbf{A} = \begin{vmatrix} \mathbf{i} & \mathbf{j} & \mathbf{k} \\ 0 & -3 & 0 \\ 2 & 0 & 0 \end{vmatrix} = 6\,[\mathbf{k}] \Rightarrow$ length = 6 and the direction is $\mathbf{k}$

7. $\mathbf{A} \times \mathbf{B} = \begin{vmatrix} \mathbf{i} & \mathbf{j} & \mathbf{k} \\ -8 & -2 & -4 \\ 2 & 2 & 1 \end{vmatrix} = [6\mathbf{i} - 12\mathbf{k}] \Rightarrow$ length $= 6\sqrt{5}$ and the direction is $\frac{1}{\sqrt{5}}\mathbf{i} - \frac{2}{\sqrt{5}}\mathbf{k}$

$\mathbf{B} \times \mathbf{A} = \begin{vmatrix} \mathbf{i} & \mathbf{j} & \mathbf{k} \\ 2 & 2 & 1 \\ -8 & -2 & -4 \end{vmatrix} = -[6\mathbf{i} - 12\mathbf{k}] \Rightarrow$ length $= 6\sqrt{5}$ and the direction is $-\frac{1}{\sqrt{5}}\mathbf{i} + \frac{2}{\sqrt{5}}\mathbf{k}$

9. $\mathbf{A} \times \mathbf{B} = \begin{vmatrix} \mathbf{i} & \mathbf{j} & \mathbf{k} \\ 1 & 0 & 0 \\ 0 & 1 & 0 \end{vmatrix} = \mathbf{k}$

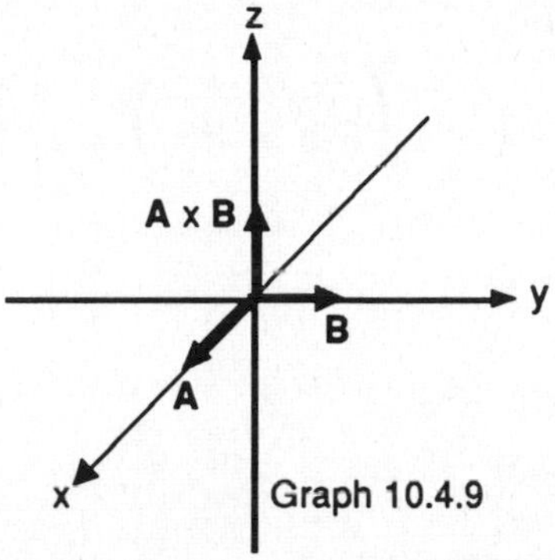

Graph 10.4.9

11. $\mathbf{A} \times \mathbf{B} = \begin{vmatrix} \mathbf{i} & \mathbf{j} & \mathbf{k} \\ 1 & 0 & -1 \\ 0 & 1 & 1 \end{vmatrix} = \mathbf{i} - \mathbf{j} + \mathbf{k}$

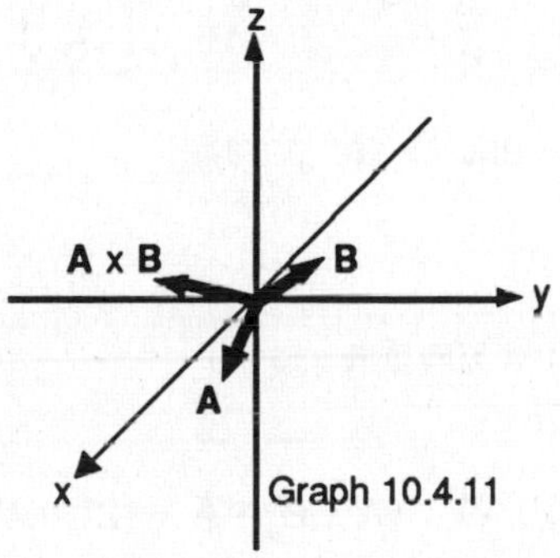

Graph 10.4.11

13. $\mathbf{A} \times \mathbf{B} = \begin{vmatrix} \mathbf{i} & \mathbf{j} & \mathbf{k} \\ 1 & 3 & 2 \\ 0 & 0 & 1 \end{vmatrix} = 3\mathbf{i} - \mathbf{j}$

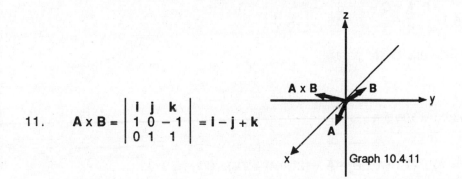

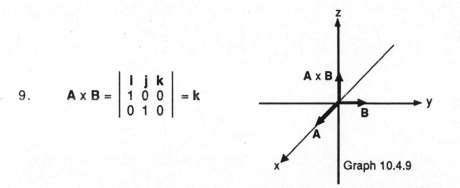

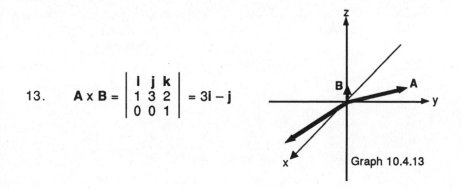

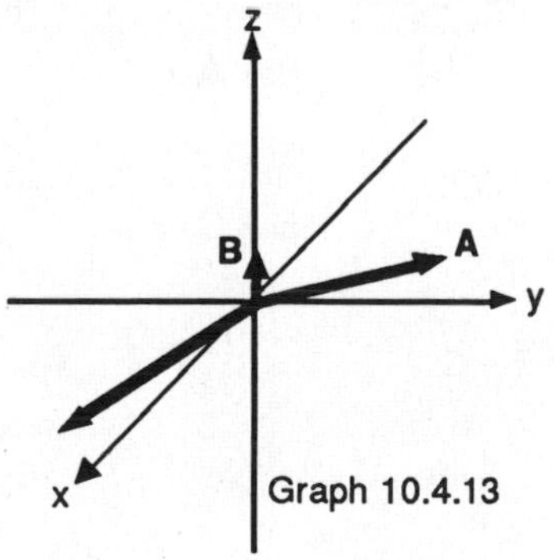

Graph 10.4.13

15. a) $\overrightarrow{PQ} = \mathbf{i} + \mathbf{j} - 3\mathbf{k},\ \overrightarrow{PR} = -\mathbf{i} + 3\mathbf{j} - \mathbf{k} \Rightarrow \pm\left(\overrightarrow{PQ} \times \overrightarrow{PR}\right) = \pm\begin{vmatrix} \mathbf{i} & \mathbf{j} & \mathbf{k} \\ 1 & 1 & -3 \\ -1 & 3 & -1 \end{vmatrix} = \pm(8\mathbf{i} + 4\mathbf{j} + 4\mathbf{k})$

b) $\dfrac{\left|\overrightarrow{PQ} \times \overrightarrow{PR}\right|}{2} = \dfrac{\sqrt{64 + 16 + 16}}{2} = 2\sqrt{6}$

c) $\pm\left[\dfrac{2}{\sqrt{6}}\mathbf{i} + \dfrac{1}{\sqrt{6}}\mathbf{j} + \dfrac{1}{\sqrt{6}}\mathbf{k}\right]$

17. a) $\overrightarrow{PQ} = \mathbf{i} + \mathbf{j} + \mathbf{k},\ \overrightarrow{PR} = \mathbf{i} + \mathbf{j} \Rightarrow \pm\left(\overrightarrow{PQ} \times \overrightarrow{PR}\right) = \pm\begin{vmatrix} \mathbf{i} & \mathbf{j} & \mathbf{k} \\ 1 & 1 & 1 \\ 1 & 1 & 0 \end{vmatrix} = \pm(-\mathbf{i} + \mathbf{j})$

b) $\dfrac{\left|\overrightarrow{PQ} \times \overrightarrow{PR}\right|}{2} = \dfrac{\sqrt{2}}{2}$

c) $\pm\left[\dfrac{-1}{\sqrt{2}}\mathbf{i} + \dfrac{1}{\sqrt{2}}\mathbf{j}\right]$

19. a) $\mathbf{A} \cdot \mathbf{B} = -6,\ \mathbf{A} \cdot \mathbf{C} = -81,\ \mathbf{B} \cdot \mathbf{C} = 18 \Rightarrow$ none

b) $\mathbf{A} \times \mathbf{B} = \begin{vmatrix} \mathbf{i} & \mathbf{j} & \mathbf{k} \\ 5 & -1 & 1 \\ 0 & 1 & -5 \end{vmatrix} \neq \mathbf{0},\ \mathbf{A} \times \mathbf{C} = \begin{vmatrix} \mathbf{i} & \mathbf{j} & \mathbf{k} \\ 5 & -1 & 1 \\ -15 & 3 & -3 \end{vmatrix} = \mathbf{0},$

$\mathbf{B} \times \mathbf{C} = \begin{vmatrix} \mathbf{i} & \mathbf{j} & \mathbf{k} \\ 0 & 1 & -5 \\ -15 & 3 & -3 \end{vmatrix} \neq \mathbf{0} \Rightarrow$ **A** and **C** are parallel

21. $\left|\overrightarrow{PQ} \times \mathbf{F}\right| = \left|\overrightarrow{PQ}\right| |\mathbf{F}| \sin(60°) = 10\sqrt{3}$ ft · lb

23. $\mathbf{A} \times \mathbf{B} = \begin{vmatrix} \mathbf{i} & \mathbf{j} & \mathbf{k} \\ 2 & -1 & 0 \\ 1 & 3 & -2 \end{vmatrix} = 2\mathbf{i} + 4\mathbf{j} + 7\mathbf{k},\ (\mathbf{A} \times \mathbf{B}) \cdot \mathbf{A} = (2\mathbf{i} + 4\mathbf{j} + 7\mathbf{k}) \cdot (2\mathbf{i} - \mathbf{j}) = 0,$

$(\mathbf{A} \times \mathbf{B}) \cdot \mathbf{B} = (2\mathbf{i} + 4\mathbf{j} + 7\mathbf{k}) \cdot (\mathbf{i} + 3\mathbf{j} - 2\mathbf{k}) = 0$

25. a) $\text{proj}_{\mathbf{B}}\,\mathbf{A} = \dfrac{\mathbf{A} \cdot \mathbf{B}}{\mathbf{B} \cdot \mathbf{B}}\mathbf{B}$ b) $(\pm)(\mathbf{A} \times \mathbf{B})$ c) $\sqrt{\mathbf{A} \cdot \mathbf{A}}\dfrac{\mathbf{B}}{\sqrt{\mathbf{B} \cdot \mathbf{B}}}$

d) $(\pm)(\mathbf{A} \times \mathbf{B}) \times \mathbf{C}$ e) $(\pm)(\mathbf{B} \times \mathbf{C}) \times \mathbf{A}$

27. $\dfrac{1}{2}\left|\overrightarrow{AB} \times \overrightarrow{AC}\right| = \dfrac{\sqrt{4 + 16 + 16}}{2} = 3$

10.5 LINES AND PLANES IN SPACE

1 the direction $\mathbf{i} + \mathbf{j} + \mathbf{k}$ and $P(3,-4,-1) \Rightarrow x = 3 + t,\ y = -4 + t,\ z = -1 + t$

3. the direction $\overrightarrow{PQ} = 5\mathbf{i} + 5\mathbf{j} - 5\mathbf{k}$ and $P(-2,0,3) \Rightarrow x = -2 + 5t,\ y = 5t,\ z = 3 - 5t$

5. the direction $2\mathbf{j} + \mathbf{k}$ and $P(0,0,0) \Rightarrow x = 0,\ y = 2t,\ z = t$

7. the direction $\mathbf{k}$ and $P(1,1,1) \Rightarrow x = 1,\ y = 1,\ z = 1 + t$

9. the direction $\mathbf{i} + 2\mathbf{j} + 2\mathbf{k}$ and $(0,-7,0) \Rightarrow x = t,\ y = -7 + 2t,\ z = 2t$

11. the direction $\mathbf{i}$ and $P(0,0,0) \Rightarrow x = t,\ y = 0,\ z = 0$

13. the direction $\overrightarrow{PQ} = \mathbf{i} + \mathbf{j} + \mathbf{k}$ and $P(0,0,0) \Rightarrow$
$x = t,\ y = t,\ z = t$, where $0 \le t \le 1$

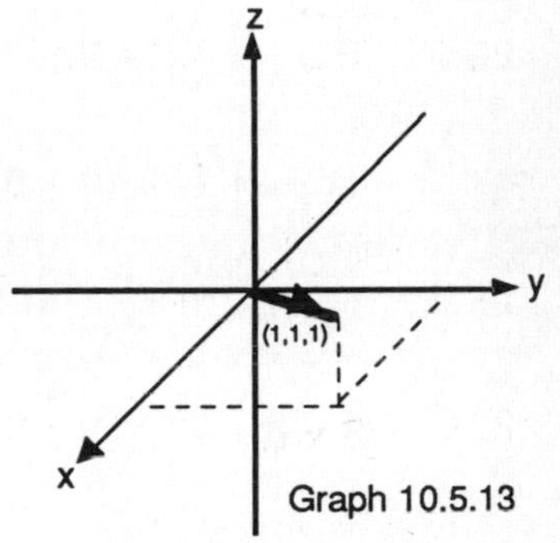

Graph 10.5.13

15. the direction $\overrightarrow{PQ} = \mathbf{j}$ and $P(1,0,0) \Rightarrow$
$x = 1,\ y = 1 + t,\ z = 0$, where $-1 \le t \le 0$

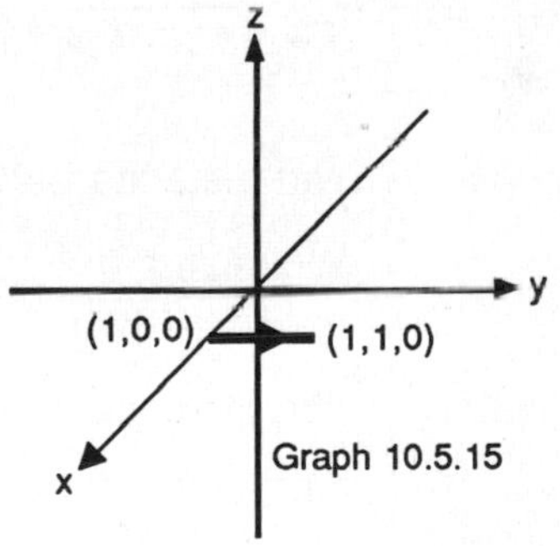

Graph 10.5.15

17. the direction $\overrightarrow{PQ} = 2\mathbf{j}$ and $P(0,-1,1) \Rightarrow$
$x = 0,\ y = -1 + 2t,\ z = 1$, where $0 \le t \le 1$

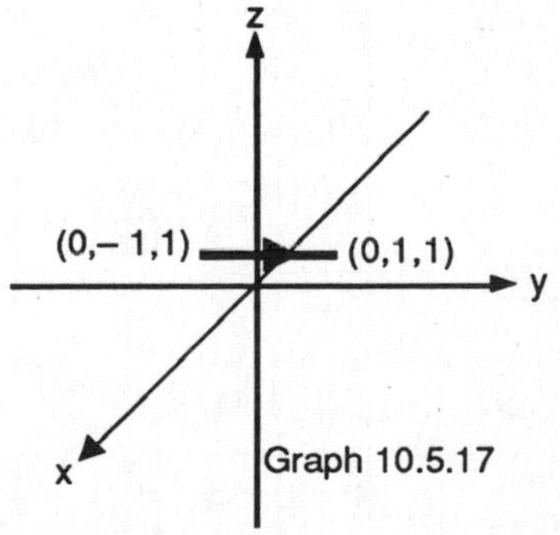

Graph 10.5.17

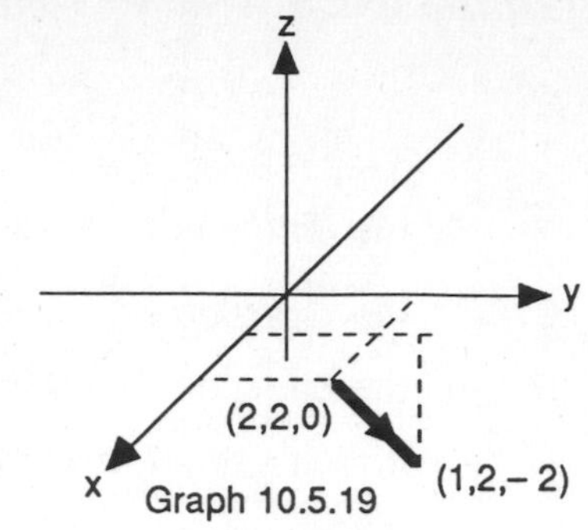

19. the direction $\overrightarrow{PQ} = -\mathbf{i} - 2\mathbf{k}$ and $P(2,2,0) \Rightarrow$
$x = 2 - t,\ y = 2,\ z = -2t$, where $0 \le t \le 1$

21. $3(x) + (-2)(y - 2) + (-1)(z + 1) = 0 \Rightarrow 3x - 2y - z = -3$

23. $\overrightarrow{PQ} = \mathbf{i} - \mathbf{j} + 3\mathbf{k},\ \overrightarrow{PS} = -\mathbf{i} - 3\mathbf{j} + 2\mathbf{k} \Rightarrow \overrightarrow{PQ} \times \overrightarrow{PS} = \begin{vmatrix} \mathbf{i} & \mathbf{j} & \mathbf{k} \\ 1 & -1 & 3 \\ -1 & -3 & 2 \end{vmatrix} = 7\mathbf{i} - 5\mathbf{j} - 4\mathbf{k}$, the normal;

$(x - 2)(7) + (y - 0)(-5) + (z - 2)(-4) = 0 \Rightarrow 7x - 5y - 4z = 6$

25. $\mathbf{N} = \mathbf{i} + 3\mathbf{j} + 4\mathbf{k},\ P(2,4,5) \Rightarrow (x - 2)(1) + (y - 4)(3) + (z - 5)(4) = 0 \Rightarrow x + 3y + 4z = 34$

27. $x = 2 + 2t,\ y = -4 - t,\ z = 7 + 3t;\ x = -2 - t,\ y = -2 + \frac{1}{2}t,\ z = 1 - \frac{3}{2}t$

29. The distance between $(4t, -2t, 2t)$ and $(0,0,12)$ is $d = \sqrt{(4t)^2 + (-2t)^2 + (2t - 12)^2}$. If $f(t) = (4t)^2 + (-2t)^2 + (2t - 12)^2$ is minimized, then d is minimized. $f'(t) = 0 \Rightarrow t = 1 \Rightarrow$
$d = \sqrt{16 + 4 + 100} = 2\sqrt{30}$

31. The distance between $(2 + 2t, 1 + 6t, 3)$ and $(2,1,3)$ is $d = \sqrt{(2t)^2 + (6t)^2}$. If $f(t) = (2t)^2 + (6t)^2$ is minimized, then d is minimized. $f'(t) = 0 \Rightarrow t = 0 \Rightarrow d = 0$

33. The distance between $(4 - t, 3 + 2t, -5 + 3t)$ and $(3, -1, 4)$ is $d = \sqrt{(1 - t)^2 + (4 + 2t)^2 + (3t - 9)^2}$. If $f(t) = (1 - t)^2 + (4 + 2t)^2 + (3t - 9)^2$ is minimized, then d is minimized. $f'(t) = 0 \Rightarrow t = \frac{10}{7} \Rightarrow$

$$d = \sqrt{\left(\frac{3}{7}\right)^2 + \left(\frac{48}{7}\right)^2 + \left(\frac{33}{7}\right)^2} = \frac{\sqrt{3402}}{7} = \frac{9\sqrt{42}}{7}$$

35. $S(2, -3, 4),\ x + 2y + 2z = 13$ and $P(13,0,0)$ is on the plane $\Rightarrow \overrightarrow{PS} = -11\mathbf{i} - 3\mathbf{j} + 4\mathbf{k},\ \mathbf{N} = \mathbf{i} + 2\mathbf{j} + 2\mathbf{k}$;
$$d = \left|\overrightarrow{PS} \cdot \frac{\mathbf{N}}{|\mathbf{N}|}\right| = \left|\frac{-11 - 6 + 8}{\sqrt{1 + 4 + 4}}\right| = 3$$

37. $S(0,1,1),\ 4y + 3z = -12$ and $P(0, -3, 0)$ is on the plane $\Rightarrow \overrightarrow{PS} = 4\mathbf{j} + \mathbf{k},\ \mathbf{N} = 4\mathbf{j} + 3\mathbf{k}$;
$$d = \left|\overrightarrow{PS} \cdot \frac{\mathbf{N}}{|\mathbf{N}|}\right| = \left|\frac{16 + 3}{\sqrt{16 + 9}}\right| = \frac{19}{5}$$

39. $S(0,-1,0)$, $2x + y + 2z = 4$ and $P(2,0,0)$ is on the plane $\Rightarrow \overrightarrow{PS} = -2\mathbf{i} - \mathbf{j}$, $\mathbf{N} = 2\mathbf{i} + \mathbf{j} + 2\mathbf{k}$;

$$d = \left|\overrightarrow{PS} \cdot \frac{\mathbf{N}}{|\mathbf{N}|}\right| = \left|\frac{-4-1+0}{\sqrt{4+1+4}}\right| = \frac{5}{3}$$

41. $2(1-t) - (3t) + 3(1+t) = 6 \Rightarrow t = -\ 1/2 \Rightarrow (3/2, -3/2, 1/2)$

43. $1(1+2t) + 1(1+5t) + 1(3t) = 2 \Rightarrow t = 0 \Rightarrow (1,1,0)$

45. $\mathbf{N}_1 = \mathbf{i} + \mathbf{j}$, $\mathbf{N}_2 = 2\mathbf{i} + \mathbf{j} - 2\mathbf{k} \Rightarrow \theta = \cos^{-1}\left(\frac{\mathbf{N}_1 \cdot \mathbf{N}_2}{|\mathbf{N}_1|\,|\mathbf{N}_2|}\right) = \cos^{-1}\left(\frac{2+1}{\sqrt{2}\sqrt{9}}\right) = \cos^{-1}\frac{1}{\sqrt{2}} = \frac{\pi}{4}$

47. $\mathbf{N}_1 = 2\mathbf{i} + 2\mathbf{j} + 2\mathbf{k}$, $\mathbf{N}_2 = 2\mathbf{i} - 2\mathbf{j} - \mathbf{k} \Rightarrow \theta = \cos^{-1}\left(\frac{\mathbf{N}_1 \cdot \mathbf{N}_2}{|\mathbf{N}_1|\,|\mathbf{N}_2|}\right) = \cos^{-1}\left(\frac{4-4-2}{\sqrt{12}\sqrt{9}}\right) =$

$\cos^{-1}\left(\frac{-1}{3\sqrt{3}}\right) \approx 101.1°$

49. $\mathbf{N}_1 = 2\mathbf{i} + 2\mathbf{j} - \mathbf{k}$, $\mathbf{N}_2 = \mathbf{i} + 2\mathbf{j} + \mathbf{k} \Rightarrow \theta = \cos^{-1}\left(\frac{\mathbf{N}_1 \cdot \mathbf{N}_2}{|\mathbf{N}_1|\,|\mathbf{N}_2|}\right) = \cos^{-1}\left(\frac{2+4-1}{\sqrt{9}\sqrt{6}}\right) =$

$\cos^{-1}\left(\frac{5}{3\sqrt{6}}\right) \approx 47.1°$

51. $\mathbf{N}_1 = \mathbf{i} + \mathbf{j} + \mathbf{k}$, $\mathbf{N}_2 = \mathbf{i} + \mathbf{j} \Rightarrow \mathbf{N}_1 \times \mathbf{N}_2 = \begin{vmatrix} \mathbf{i} & \mathbf{j} & \mathbf{k} \\ 1 & 1 & 1 \\ 1 & 1 & 0 \end{vmatrix} = -\mathbf{i} + \mathbf{j}$, the direction of the desired line; $(1,1,-1)$ is

on both planes; the desired line is $x = 1 - t$, $y = 1 + t$, $z = -1$

53. $\mathbf{N}_1 = \mathbf{i} - 2\mathbf{j} + 4\mathbf{k}$, $\mathbf{N}_2 = \mathbf{i} + \mathbf{j} - 2\mathbf{k} \Rightarrow \mathbf{N}_1 \times \mathbf{N}_2 = \begin{vmatrix} \mathbf{i} & \mathbf{j} & \mathbf{k} \\ 1 & -2 & 4 \\ 1 & 1 & -2 \end{vmatrix} = 6\mathbf{j} + 3\mathbf{k}$, the direction of the desired line;

$(4,3,1)$ is on both planes; the desired line is $x = 4$, $y = 3 + 6t$, $z = 1 + 3t$

55. $x = 0 \Rightarrow t = -\frac{1}{2}, y = -\frac{1}{2}, z = -\frac{3}{2} \Rightarrow \left(0, -\frac{1}{2}, -\frac{3}{2}\right)$; $y = 0 \Rightarrow t = -1, x = -1, z = -3 \Rightarrow (-1,0,-3)$;

$z = 0 \Rightarrow t = 0 \Rightarrow x = 1, y = -1 \Rightarrow (1,-1,0)$

57. a) $\overrightarrow{EP} = \overrightarrow{EP_1} \Rightarrow -x_o\mathbf{i} + y\mathbf{j} + z\mathbf{k} = (x_1 - x_o)\mathbf{i} + y_1\mathbf{j} + z_1\mathbf{k} \Rightarrow -x_o = c(x_1 - x_o)$, $y = cy_1$, $z = cz_1$,

where c is a positive real number

b) at $x_1 = 0 \Rightarrow c = 1$, $y = y_1$ and $z = z_1$; at $x_1 = x_o \Rightarrow x_o = 0$, $y = 0$, $z = 0$; $\lim_{x_o \to \infty} c =$

$\lim_{x_o \to \infty} \frac{-x_o}{x_1 - x_o} = \lim_{x_o \to \infty} \frac{-1}{-1} = 1 \Rightarrow c \to 1, y \to y_1$ and $z \to z_1$

59. $\left|\overrightarrow{OA}\right| = 1.41421356$; $\left|\overrightarrow{OB}\right| = 2$; $\left|\overrightarrow{AB}\right| = 2.44948974$; the midpoint of AB is (.5,1,– .5); $\mathbf{A}\cdot\mathbf{B} = 0$; the angle between $\overrightarrow{OA}$ and $\overrightarrow{OB}$ is 1.5707965 radians ≈ 90°; $\mathbf{A} \times \mathbf{B} = 2\mathbf{i} + 0\mathbf{j} + 2\mathbf{k}$; the line through A and B: $x = 1 - t$, $y = 2t$, $z = -1 + t$; the line through C parallel to AB: $x = -1 - t$, $y = 2t$, $z = 1 + t$; the distance from C to AB is 2.30940108; the equation of the ABC plane is $-4x - 4z = 0$; the distance from S to the ABC plane is 2.12132034

61. $\left|\overrightarrow{OA}\right| = 4.47213595$; $\left|\overrightarrow{OB}\right| = 8.24621126$; $\left|\overrightarrow{AB}\right| = 9.79795897$; the midpoint of AB is (0,4,2); $\mathbf{A}\cdot\mathbf{B} = -4$; the angle between $\overrightarrow{OA}$ and $\overrightarrow{OB}$ is 1.62032412 radians ≈ 92.8377235°; $\mathbf{A} \times \mathbf{B} = -32\mathbf{i} + 8\mathbf{j} - 16\mathbf{k}$; the line through A and B: $x = -2 + 4t$, $y = 8t$, $z = 4 - 4t$; the line through C parallel to AB: $x = 2 + 4t$, $y = 4 + 8t$, $z = -2 - 4t$; the distance from C to AB is 3.74165739; the equation of the ABC plane is $32x - 8y + 16z = 0$; the distance from S to the ABC plane is 0

10.6 PRODUCTS OF THREE VECTORS OR MORE

1. If $\mathbf{A} = a_1\mathbf{i} + a_2\mathbf{j} + a_3\mathbf{k}$, $\mathbf{B} = b_1\mathbf{i} + b_2\mathbf{j} + b_3\mathbf{k}$, and $\mathbf{C} = c_1\mathbf{i} + c_2\mathbf{j} + c_3\mathbf{k}$, then $\mathbf{A}\cdot(\mathbf{B}\times\mathbf{C}) = \begin{vmatrix} a_1 & a_2 & a_3 \\ b_1 & b_2 & b_3 \\ c_1 & c_2 & c_3 \end{vmatrix}$, $\mathbf{B}\cdot(\mathbf{C}\times\mathbf{A}) = \begin{vmatrix} b_1 & b_2 & b_3 \\ c_1 & c_2 & c_3 \\ a_1 & a_2 & a_3 \end{vmatrix}$ and $\mathbf{C}\cdot(\mathbf{A}\times\mathbf{B}) = \begin{vmatrix} c_1 & c_2 & c_3 \\ a_1 & a_2 & a_3 \\ b_1 & b_2 & b_3 \end{vmatrix}$ which all have the same value, since the interchanging of two rows, in a determinant, does not change its value. The volume is $\left|(\mathbf{A}\times\mathbf{B})\cdot\mathbf{C}\right| = \begin{vmatrix} 2 & 0 & 0 \\ 0 & 2 & 0 \\ 0 & 0 & 2 \end{vmatrix} = 8$, $(\mathbf{A}\times\mathbf{B})\times\mathbf{C} = \mathbf{0}$, $\mathbf{A}\times(\mathbf{B}\times\mathbf{C}) = \mathbf{0}$

3. $\left|(\mathbf{A}\times\mathbf{B})\cdot\mathbf{C}\right| = \begin{vmatrix} 2 & 1 & 0 \\ 2 & -1 & 1 \\ 1 & 0 & 2 \end{vmatrix} = 7$, $(\mathbf{A}\times\mathbf{B})\times\mathbf{C} = -4\mathbf{i} - 6\mathbf{j} + 2\mathbf{k}$, $\mathbf{A}\times(\mathbf{B}\times\mathbf{C}) = \mathbf{i} - 2\mathbf{j} - 4\mathbf{k}$ For details about verification, see exercise 1.

5.

a)	true	b)	false, $(2\mathbf{i})\cdot(2\mathbf{i}) = 4$ while $\lvert 2\mathbf{i}\rvert = \sqrt{2^2} = 2$	c)	true
d)	true	e)	false, they have opposite directions	f)	true
g)	true	h)	true	i)	true

7. a) $(\mathbf{A} \times \mathbf{B}) \times \mathbf{C} = 0 \Rightarrow \mathbf{A}$ and $\mathbf{C}$ are perpendicular $\Rightarrow \mathbf{A}\cdot\mathbf{C} = 0$

b) $(\mathbf{A} \times \mathbf{B}) \cdot \mathbf{C} = |\mathbf{A} \times \mathbf{B}|\,|\mathbf{C}| \cos\theta = |\mathbf{A} \times \mathbf{B}|\,|\mathbf{C}| \cos 0 = |\mathbf{A} \times \mathbf{B}|\,|\mathbf{C}|$, $|\mathbf{A} \times \mathbf{B}| = |\mathbf{A}|\,|\mathbf{B}| \sin\left(\frac{\pi}{2}\right) =$ $|\mathbf{A}|\,|\mathbf{B}|$ $\therefore\ 8 = (\mathbf{A} \times \mathbf{B}) \cdot \mathbf{C} = |\mathbf{A}|\,|\mathbf{B}|\,|\mathbf{C}| = 4|\mathbf{C}| \Rightarrow |\mathbf{C}| = 2$, since $|\mathbf{A}| = |\mathbf{B}| = 2$

c) From part a) we have $\mathbf{B} \times \mathbf{C} = t\mathbf{A}$ for some t. $(\mathbf{A} \times \mathbf{B})\cdot\mathbf{C} = \mathbf{A}\cdot(\mathbf{B}\times\mathbf{C}) = \mathbf{A}\cdot t\mathbf{A} = t|\mathbf{A}|^2$

$\therefore\ 4t = 8$ and $|\mathbf{B} \times \mathbf{C}| = |2\mathbf{A}| = 2|\mathbf{A}| = 4$

9. a) $\mathbf{A}\cdot(\mathbf{B}\times\mathbf{C}) = \begin{vmatrix} a_1 & a_2 & a_3 \\ b_1 & b_2 & b_3 \\ c_1 & c_2 & c_3 \end{vmatrix} = - \begin{vmatrix} a_1 & a_2 & a_3 \\ c_1 & c_2 & c_3 \\ b_1 & b_2 & b_3 \end{vmatrix} - \mathbf{A}\cdot(\mathbf{C}\times\mathbf{B}) =$

b) $\mathbf{A}\cdot(\mathbf{A}\times\mathbf{B}) = \begin{vmatrix} a_1 & a_2 & a_3 \\ a_1 & a_2 & a_3 \\ c_1 & c_2 & c_3 \end{vmatrix} = 0$

c) $(\mathbf{A} + \mathbf{B})\cdot(\mathbf{B}\times\mathbf{C}) = \begin{vmatrix} a_1 + d_1 & a_2 + d_2 & a_3 + d_3 \\ b_1 & b_2 & b_3 \\ c_1 & c_2 & c_3 \end{vmatrix} = \begin{vmatrix} a_1 & a_2 & a_3 \\ b_1 & b_2 & b_3 \\ c_1 & c_2 & c_3 \end{vmatrix} + \begin{vmatrix} d_1 & d_2 & d_3 \\ b_1 & b_2 & b_3 \\ c_1 & c_2 & c_3 \end{vmatrix} =$

$\mathbf{A}\cdot(\mathbf{B}\times\mathbf{C}) + \mathbf{D}\cdot(\mathbf{B}\times\mathbf{C})$

11. a) $\overrightarrow{PQ} = 2\mathbf{i} - 3\mathbf{j} + 5\mathbf{k}$; $\overrightarrow{SR} = t((2 - x)\mathbf{i} + (6 - y)\mathbf{j} + (2 - z)\mathbf{k})$, where t may be 1, since $\overrightarrow{PQ}$ and $\overrightarrow{SR}$ are parallel $\therefore$ $x = 0,\ y = 9,\ z = -3$ and S is located at $(0,9,-3)$

b) $\overrightarrow{PQ} \times \overrightarrow{PS} = \begin{vmatrix} \mathbf{i} & \mathbf{j} & \mathbf{k} \\ 2 & -3 & 5 \\ -1 & 7 & -2 \end{vmatrix} = -29\mathbf{i} - \mathbf{j} + 11\mathbf{k}$; the area is $\sqrt{(-29)^2 + (-1)^2 + (11)^2} = \sqrt{963}$

c)

plane	the area of projection
xy	$\lvert(-29\mathbf{i} - \mathbf{j} + 11\mathbf{k}) \cdot \mathbf{k}\rvert = 11$
yz	$\lvert(-29\mathbf{i} - \mathbf{j} + 11\mathbf{k}) \cdot \mathbf{i}\rvert = 29$
xz	$\lvert(-29\mathbf{i} - \mathbf{j} + 11\mathbf{k}) \cdot \mathbf{j}\rvert = 1$

13. Without loss of generality we can place the parallelogram such that $P(0,0,0)$, $Q(x_q,0,0)$, $R(x_r,y_r,0)$ and $S(x_s, y_s, z_s) \Rightarrow \overrightarrow{PQ} \times \overrightarrow{PR} = 0\mathbf{i} + 0\mathbf{j} + (x_q y_r)\mathbf{k}$. The area of the parallelogram $= \left|\overrightarrow{PQ} \times \overrightarrow{PR}\right| =$ $\left\| \begin{matrix} \mathbf{i} & \mathbf{j} & \mathbf{k} \\ x_q & 0 & 0 \\ x_r & y_r & 0 \end{matrix} \right\| = \sqrt{0^2 + 0^2 + (x_q y_r)^2} = |x_q y_r|$, while the square root of the squares of the areas of the parallelogram's orthogonal projections is

$$\sqrt{\left(\left(\overrightarrow{PQ} \times \overrightarrow{PR}\right) \cdot \mathbf{i}\right)^2 + \left(\left(\overrightarrow{PQ} \times \overrightarrow{PR}\right) \cdot \mathbf{j}\right)^2 + \left(\left(\overrightarrow{PQ} \times \overrightarrow{PR}\right) \cdot \mathbf{k}\right)^2} = |x_q y_r|.$$

10.7 SURFACES IN SPACE

1. $x^2 + y^2 = 4$

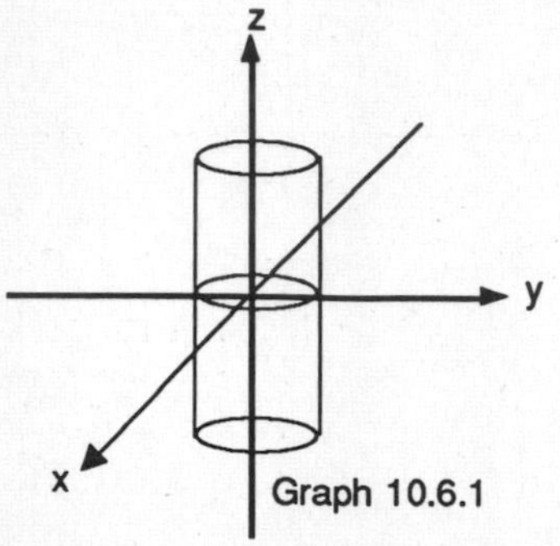

Graph 10.6.1

3. $z = y^2 - 1$

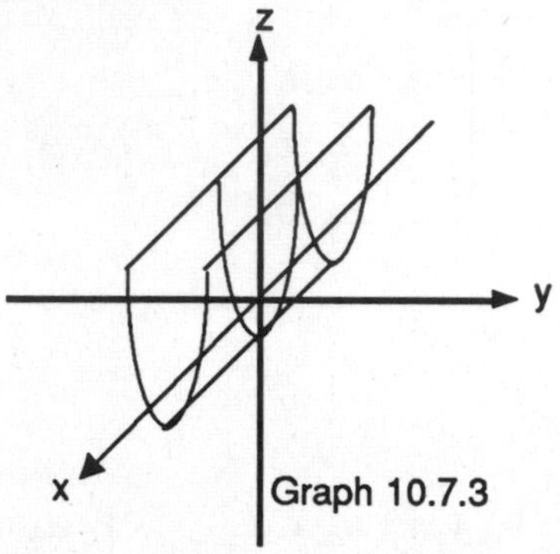

Graph 10.7.3

5. $x^2 + 4z^2 = 16$

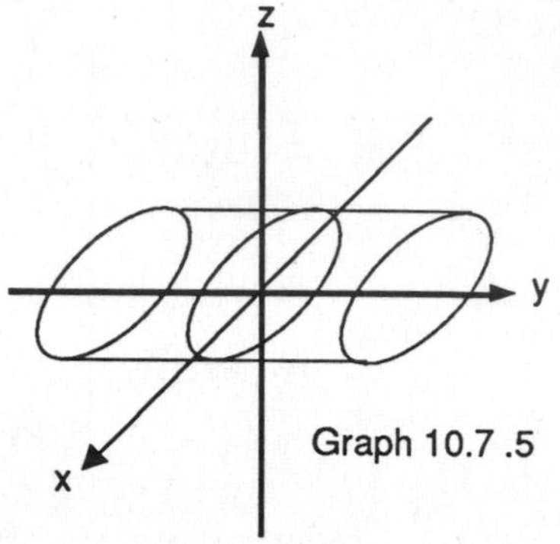

Graph 10.7 .5

7. $z^2 - y^2 = 1$

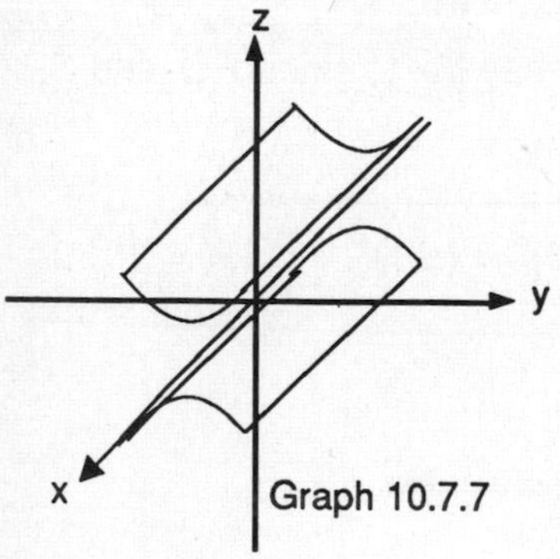

Graph 10.7.7

9. $9x^2 + y^2 + z^2 = 9$

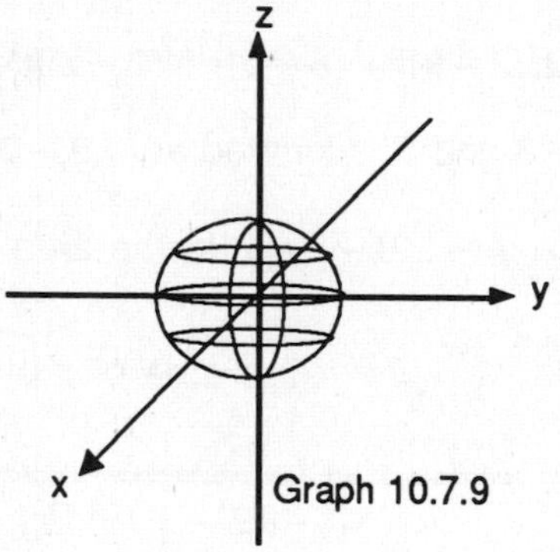

Graph 10.7.9

11. $4x^2 + 9y^2 + 4z^2 = 36$

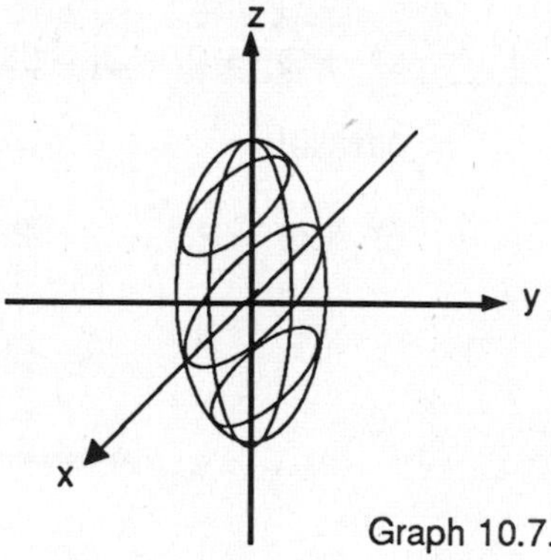

Graph 10.7.11

13. $x^2 + 4y^2 = z$

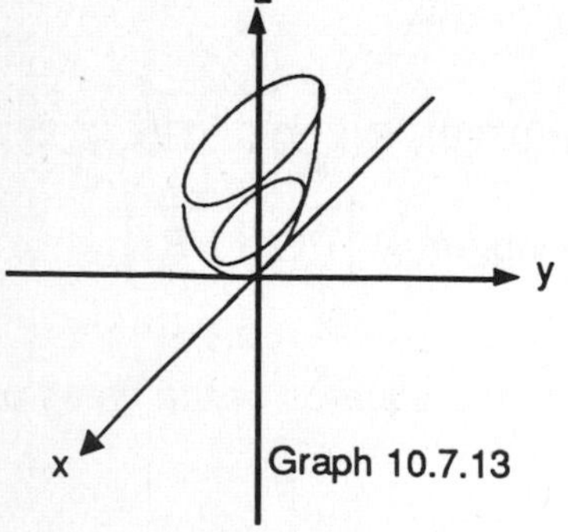

Graph 10.7.13

15. $z = 8 - x^2 - y^2$

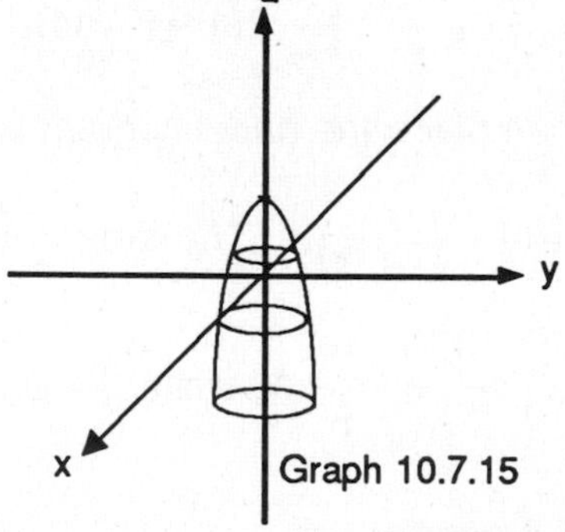

Graph 10.7.15

17. $x = 4 - 4y^2 - z^2$

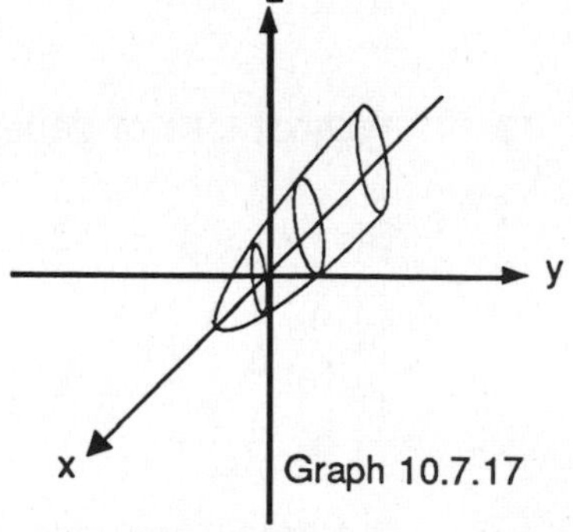

Graph 10.7.17

19. $x^2 + y^2 = z^2$

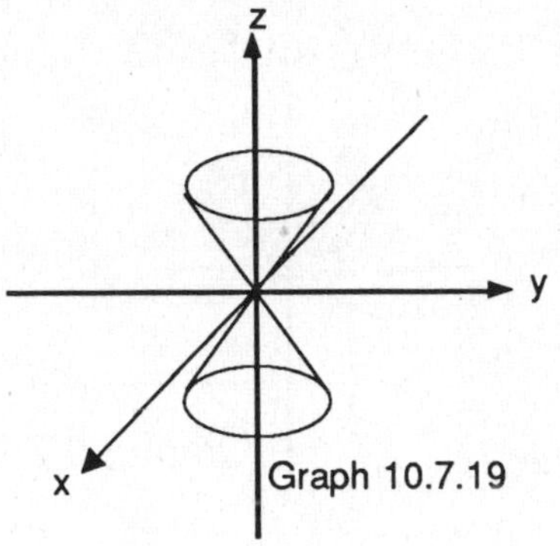

21. $4x^2 + 9z^2 = 9y^2$

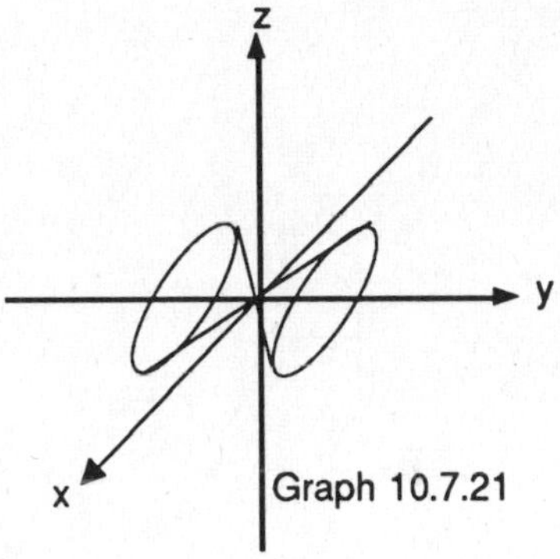

23. $x^2 + y^2 - z^2 = 1$

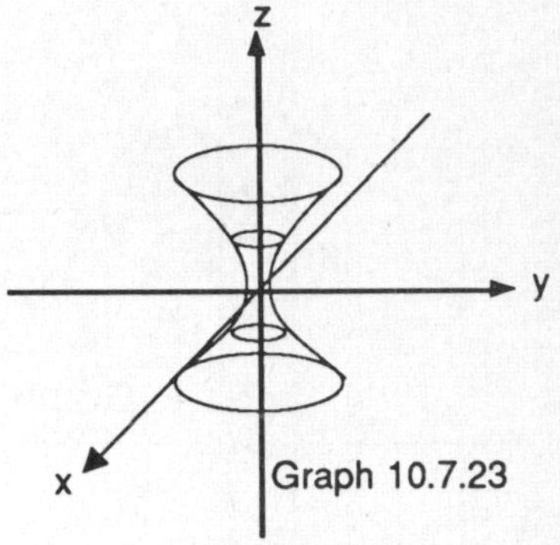

25. $\left(y^2/4\right) + \left(z^2/9\right) - \left(x^2/4\right) = 1$

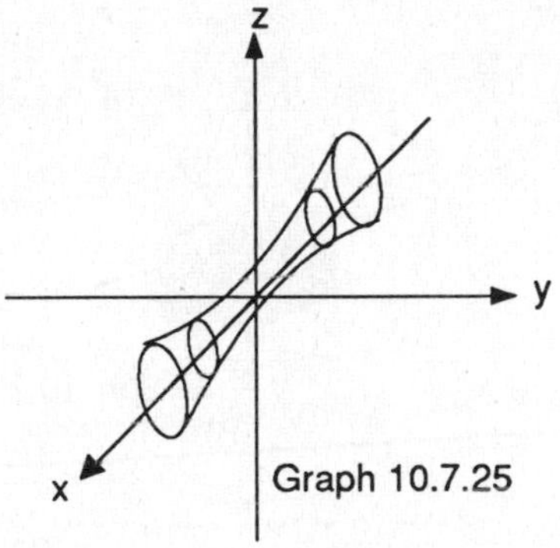

27. $z^2 - x^2 - y^2 = 1$

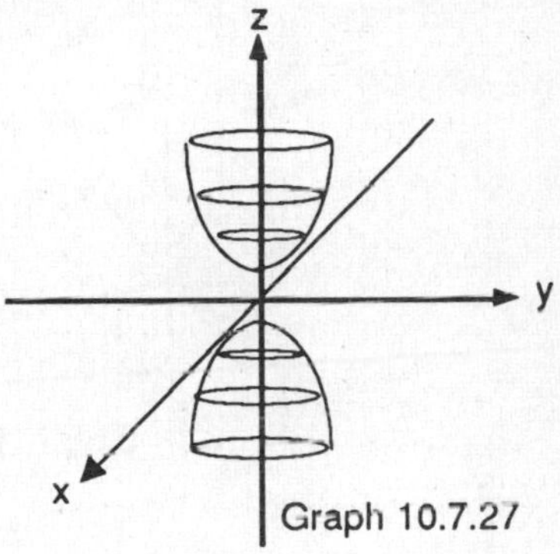

29. $x^2 - y^2 - \left(z^2/4\right) = 1$

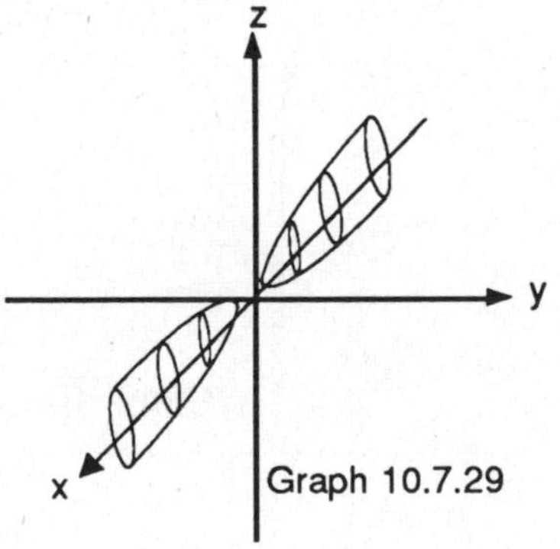

31. $y^2 - x^2 = z$

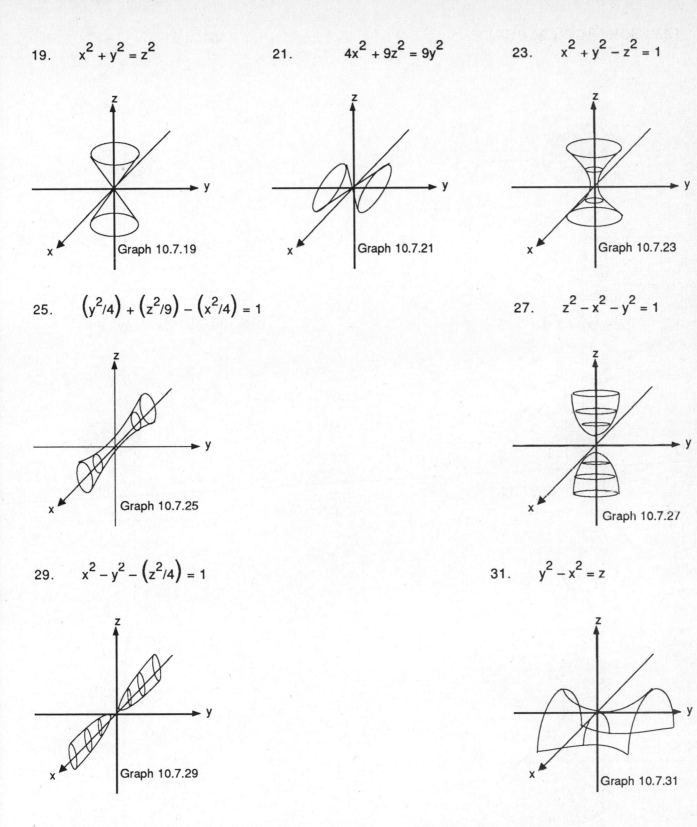

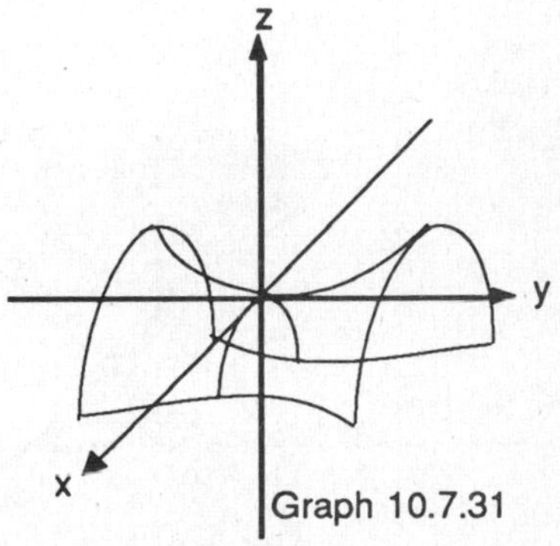

33. $x^2 + y^2 + z^2 = 4$

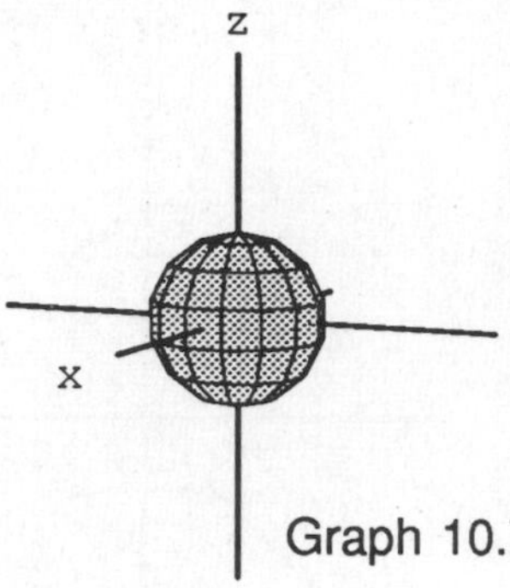

Graph 10.7.33

35. $z = 1 + y^2 - x^2$

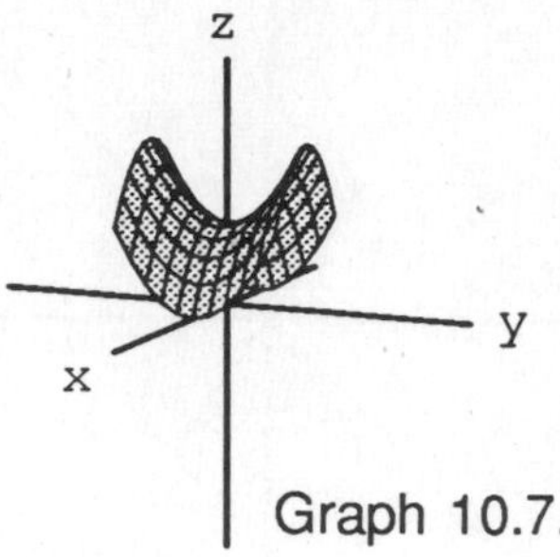

Graph 10.7.35

37. $y = -\left(x^2 + z^2\right)$

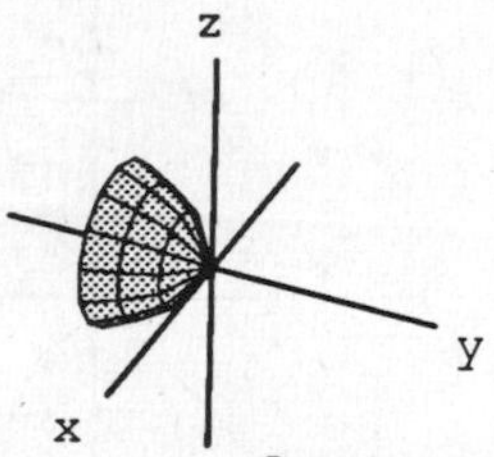

Graph 10.7.37

39. $16x^2 + 4y^2 = 1$

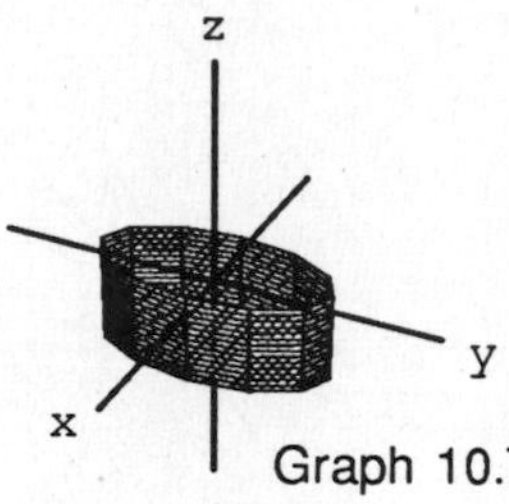

Graph 10.7.39

41. $x^2 + y^2 - z^2 = 4$

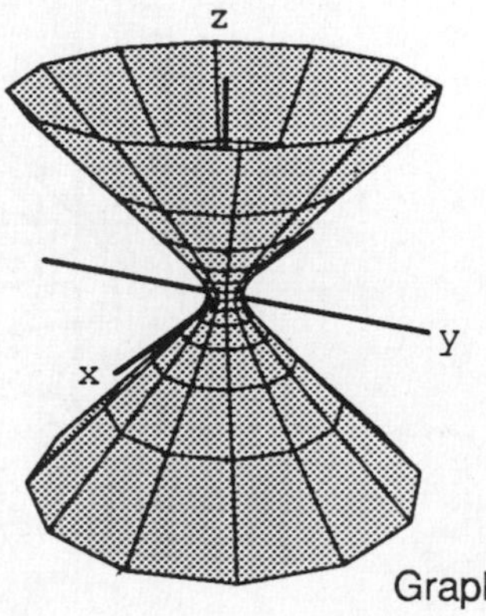

Graph 10.7.41

43. $x^2 + z^2 = y$

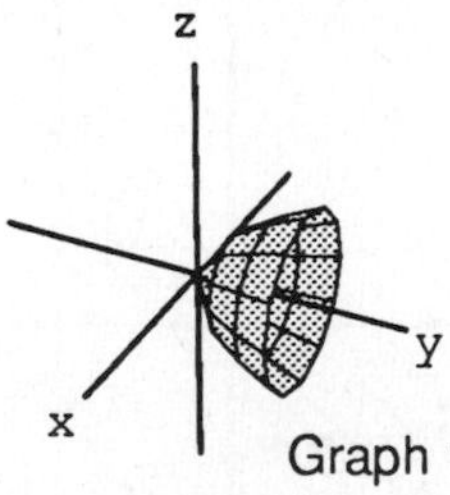

Graph 10.7.43

45. $x^2 + z^2 = 1$

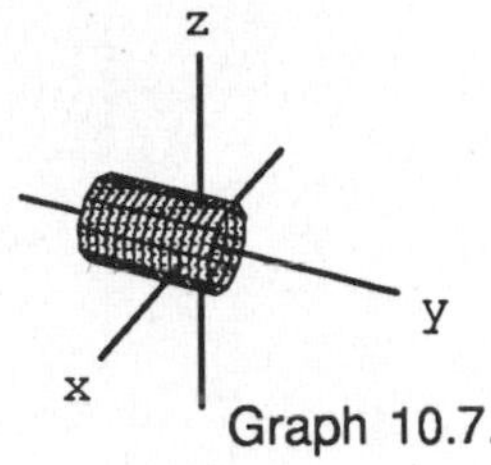

Graph 10.7.45

47. $16y^2 + 9z^2 = 4x^2$

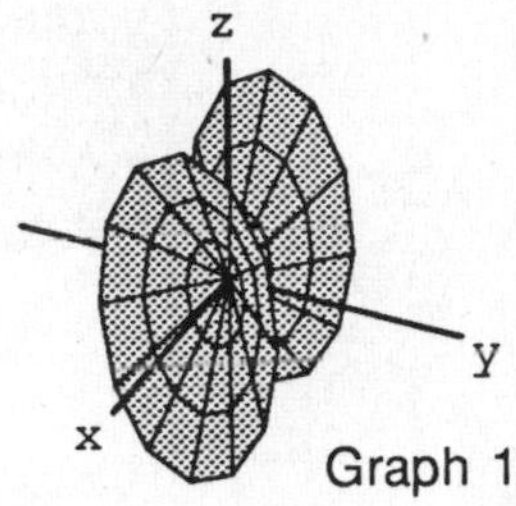

Graph 10.7.47

49. $9x^2 + 4y^2 + z^2 = 36$

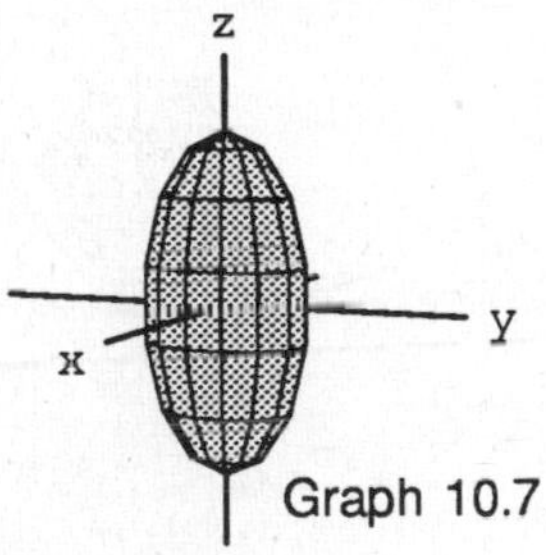

Graph 10.7.49

51. $x^2 + y^2 - 16z^2 = 16$

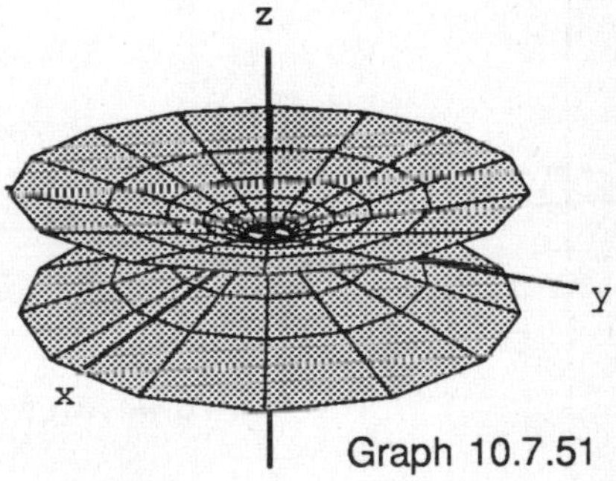

Graph 10.7.51

53. $z = -\left(x^2 + y^2\right)$

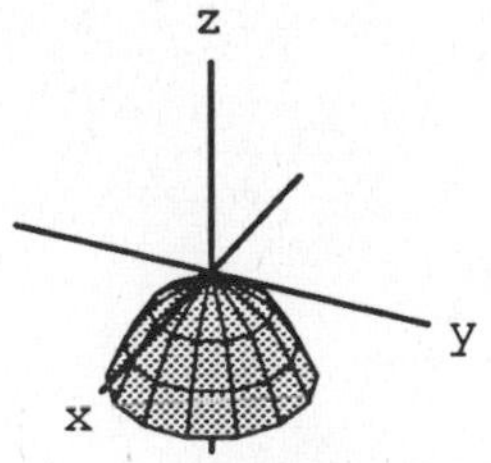

Graph 10.7.53

55. $x^2 - 4y^2 = 1$

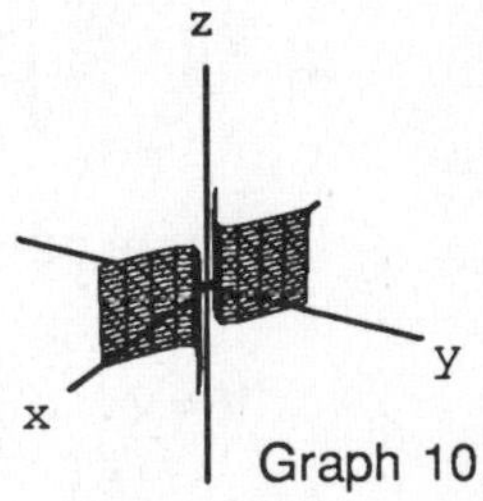

Graph 10.7.55

57. $4y^2 + z^2 - 4x^2 = 4$

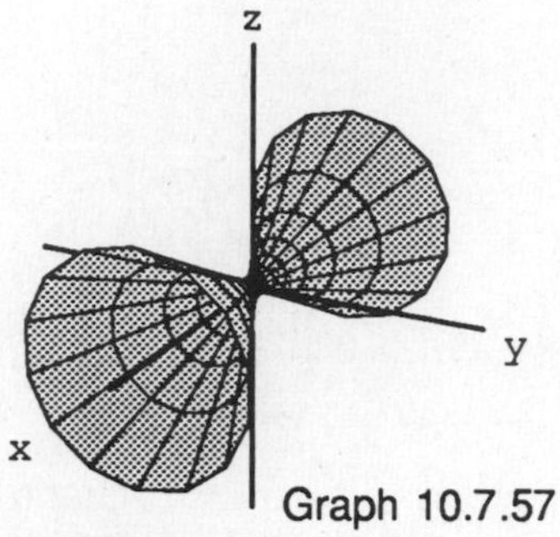

Graph 10.7.57

59. $x^2 + y^2 = z$

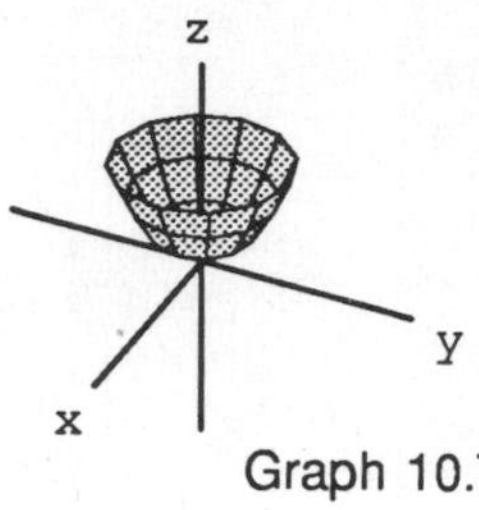

Graph 10.7.59

61. $yz = 1$

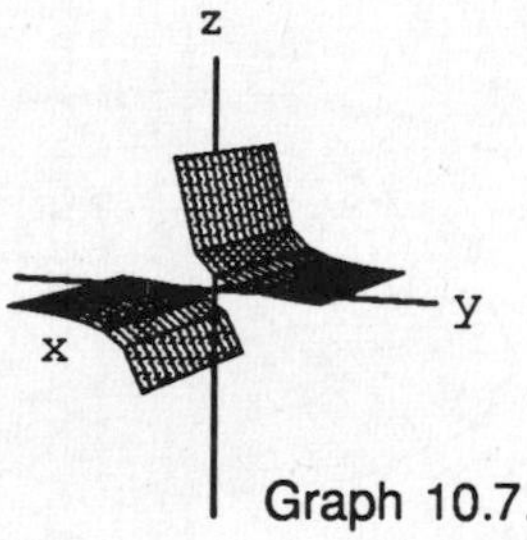

Graph 10.7.61

63. $9x^2 + 16y^2 = 4z^2$

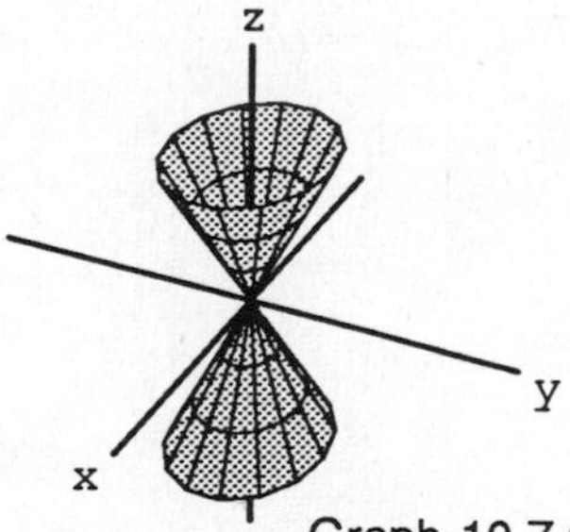

Graph 10.7.63

65. a) If $x^2 + \frac{y^2}{4} + \frac{z^2}{9} = 1$ and $z = c$, then $x^2 + \frac{y^2}{4} = \frac{9-c^2}{9} \Rightarrow \frac{x^2}{\frac{9-c^2}{9}} + \frac{y^2}{\frac{4(9-c^2)}{9}} = 1 \Rightarrow A = ab\pi =$

$$\pi\left(\frac{\sqrt{9-c^2}}{3}\right)\left(\frac{2\sqrt{9-c^2}}{3}\right) = \frac{2\pi\left(9-c^2\right)}{9}$$

b) From part a) each slice has the area $\frac{2\pi\left(9-z^2\right)}{9}$ where $-3 \le z \le 3$.

$$\therefore\ V = 2\int_0^3 \frac{2\pi}{9}\left(9-z^2\right) dz = \frac{4\pi}{9}\int_0^3 9 - z^2\, dz = \frac{4\pi}{9}\left[9z - \frac{z^3}{3}\right]_0^3 = 8\pi$$

c) $\frac{x^2}{a^2} + \frac{y^2}{b^2} + \frac{z^2}{c^2} = 1 \Rightarrow \frac{x^2}{\frac{a^2\left(c^2 - z^2\right)}{c^2}} + \frac{y^2}{\frac{a^2\left(c^2 - z^2\right)}{c^2}} = 1$ and $V = 2\int_0^c \frac{\pi ab}{c^2}\left(c^2 - z^2\right)\,dz =$

$\frac{4\pi abc}{3}$. If $r = a = b = c$, then $V = \frac{4\pi r^3}{3}$ the volume of a sphere.

67. If z be fixed in $\frac{x^2}{a^2} + \frac{y^2}{b^2} = \frac{z}{c} \Rightarrow \frac{x^2}{\frac{za^2}{c}} + \frac{y^2}{\frac{zb^2}{c}} = 1$, then the cross–sectional area of the ellipse is $\frac{\pi abz}{c}$.

If $z = h$, then the base is $\frac{\pi abh}{c}$. $V = \int_0^h \frac{\pi ab}{c} z\,dz = \frac{\pi abh^2}{2c} = \frac{1}{2}h \cdot \frac{\pi abh}{c} = \frac{1}{2}$ (base)(height)

69. If $x = 0$, then $Hy^2 + Iz^2 + Kyz + Ny + Pz + Q = 0$ the general form of a conic in terms of y and z. Likewise when $y = 0$ or $z = 0$ we also have a conic.

71. $z = y^2$

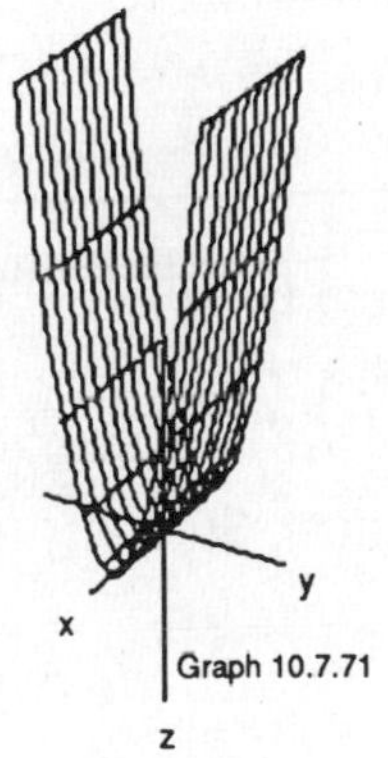

Graph 10.7.71

73. $z = x^2 + y^2$

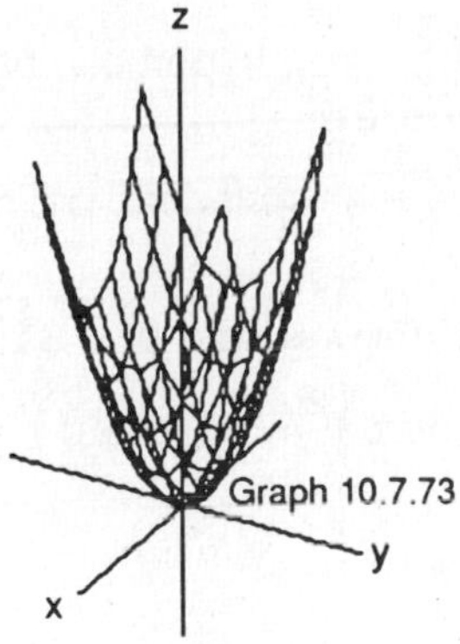

Graph 10.7.73

75. $z = \sqrt{1 - x^2}$

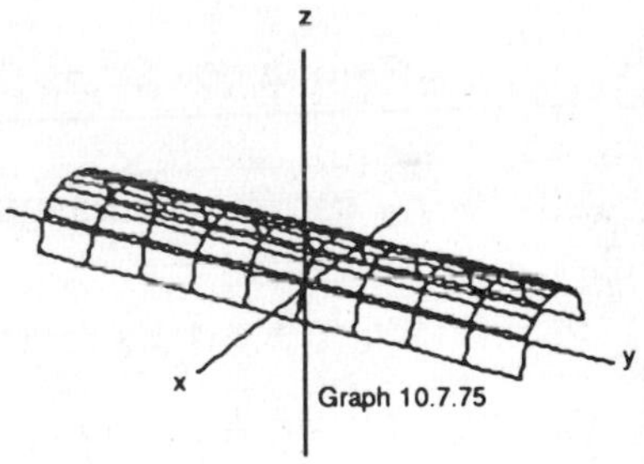

Graph 10.7.75

77. $z = \sqrt{x^2 + 2y^2 + 4}$

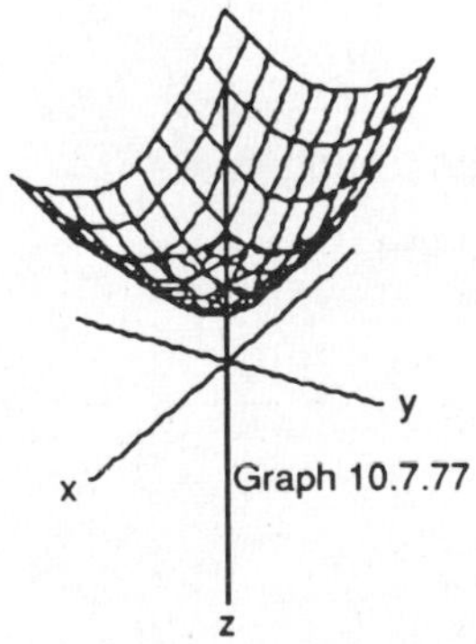

Graph 10.7.77

10.8 CYLINDRICAL AND SPHERICAL COORDINATES

	Rectangular	Cylindrical	Spherical
1.	(0,0,0)	(0,0,0)	(0,0,0)
3.	(0,1,0)	$(1,\pi/2,0)$	$(1,\pi/2,\pi/2)$
5.	(1,0,0)	(1,0,0)	$(1,\pi/2,0)$
7.	(0,1,1)	$(1,\pi/2,1)$	$(\sqrt{2},\pi/4,\pi/2)$
9.	$(0,-2\sqrt{2},0)$	$(2\sqrt{2},3\pi/2,0)$	$(2\sqrt{2},\pi/2,3\pi/2)$

* any angle may be used

11. $r = 0 \Rightarrow$ rectangular, $x^2 + y^2 = 0$; spherical, $\phi = 0$ and $\phi = \pi$; the z–axis

13. $z = 0 \Rightarrow$ cylindrical, $z = 0$; spherical, $\phi = \frac{\pi}{2}$; the xy–plane

15. $\rho \cos \phi = 3 \Rightarrow$ rectangular, $z = 3$; cylindrical, $z = 3$; the plane $z = 3$

17. $\rho \sin \phi \cos \phi = 0 \Rightarrow$ rectangular, $x = 0$; cylindrical $\theta = \frac{\pi}{2}$; the yz–plane

19. $x^2 + y^2 + z^2 = 4 \Rightarrow$ cylindrical, $r^2 + z^2 = 4$; spherical, $\rho = 2$; a sphere centered at the origin with a radius of 2

21. $z = r^2 \cos 2\theta \Rightarrow$ rectangular, $z = r^2\left(\cos^2\theta - \sin^2\theta\right) \Rightarrow z = x^2 - y^2$; spherical, $\rho \cos^2\phi = \rho^2\sin^2\phi \cos 2\theta \Rightarrow \rho = \dfrac{\cos \phi}{\sin^2\phi \cos 2\theta}$; hyperbolic paraboloid

23. $r = \csc \theta \Rightarrow$ rectangular, $r = \frac{r}{y} \Rightarrow y = 1$; spherical, $\rho \sin \phi = \csc \theta \Rightarrow \rho = \dfrac{1}{\sin \phi \sin \theta}$; the plane $y = 1$

25. $3 \tan^2\phi = 1 \Rightarrow$ rectangular, $3\left(\sin^2\phi\right) = \cos^2\phi \Rightarrow 3\left(\rho^2\sin^2\phi\right)\left(\sin^2\theta + \cos^2\theta\right) = \rho^2\cos^2\phi \Rightarrow 3\left(\rho \sin \phi \sin \theta\right)^2 + 3\left(\rho \sin \phi \cos \theta\right)^2 = \left(\rho \cos \phi\right)^2 \Rightarrow 3x^2 + 3y^2 = z^2$; cylindrical, $3r^2 = z^2$, a cone

27. a right circular cylinder whose generating curve is a circle of radius 4 in the $r\theta$–plane

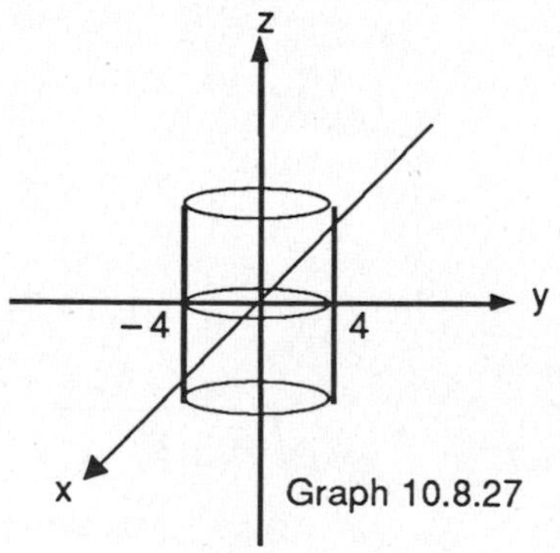

Graph 10.8.27

29. a cylinder whose generation curve is a cardioid in the $r\theta$–plane

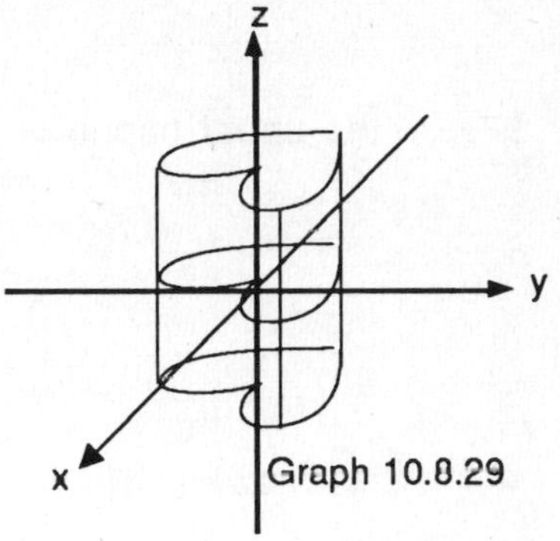

Graph 10.8.29

31. a circle contained in the plane $z = 3$ having a radius of 2 and center at (0,0,3)

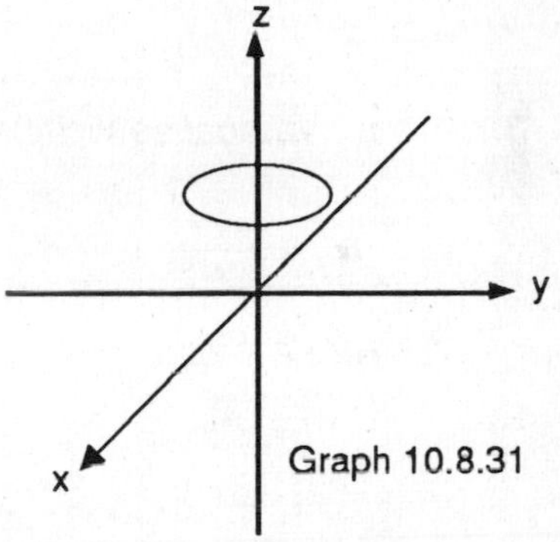

Graph 10.8.31

33. a space curve called a helix

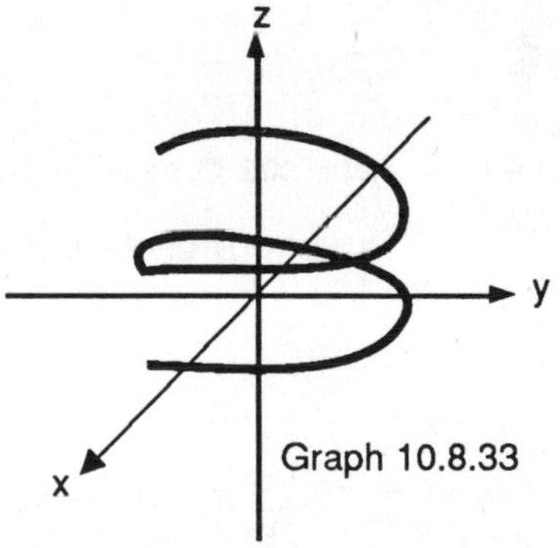

Graph 10.8.33

35. $r^2 + z^2 = 4r\cos\theta + 6r\sin\theta + 2z \Rightarrow x^2 + y^2 + z^2 = 4x + 6y + 2z \Rightarrow$
$\left(x^2 - 4x + 4\right) + \left(y^2 - 6y + 9\right) + \left(z^2 - 2z + 1\right) = 14 \Rightarrow (x-2)^2 + (y-3)^2 + (z-1)^2 = 14 \Rightarrow$
the center is located at (2,3,1)

37. the upper nappe of a cone

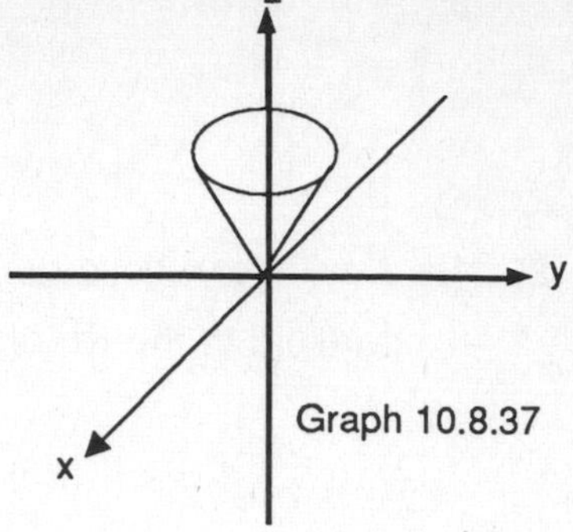

Graph 10.8.37

39. a " vertical semicircle " in the $y = x$ plane

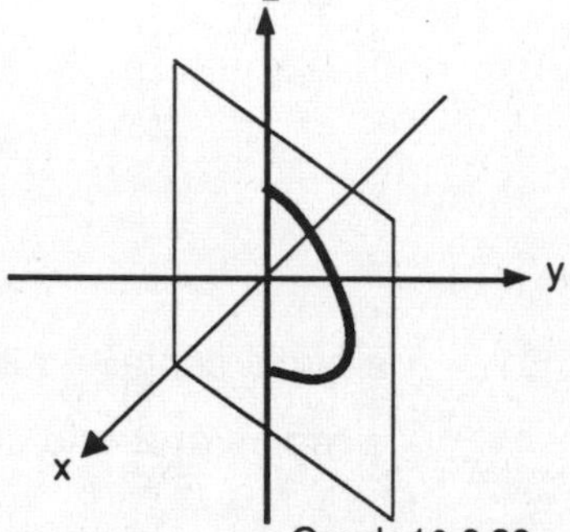

Graph 10.8.39

41. $\rho = \cos\phi \Rightarrow \rho^2 - z = 0 \Rightarrow x^2 + y^2 + (z - 1/2)^2 = (1/2)^2$

a sphere centered at (0,0,1/2) with a radius of 1/2

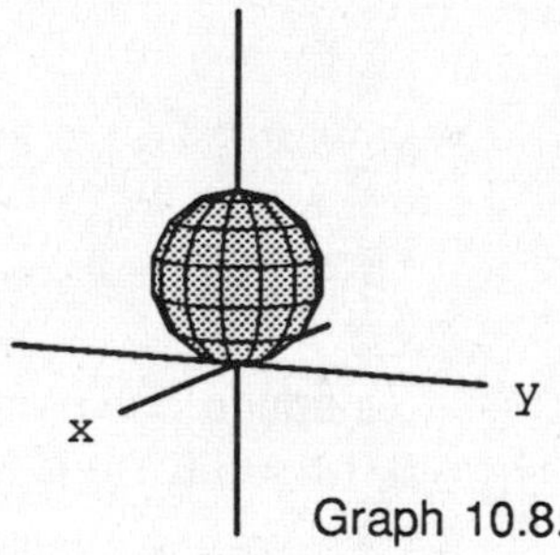

Graph 10.8.41

43. $\rho = \sin\phi$, a torus like object centered at the origin

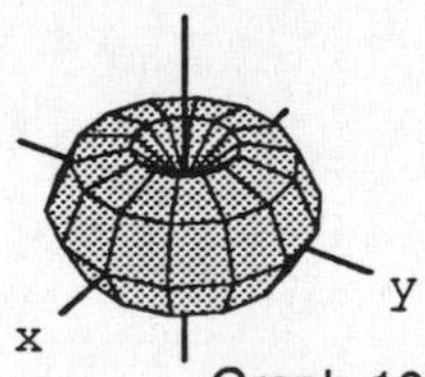

Graph 10.8.43

10.M MISCELLANEOUS EXERCISES

1. $\theta = 0 \Rightarrow \mathbf{u} = \mathbf{i}$; $\theta = \frac{\pi}{2} \Rightarrow \mathbf{u} = \mathbf{j}$; $\theta = \frac{2\pi}{3} \Rightarrow \mathbf{u} = -\frac{1}{2}\mathbf{i} + \frac{\sqrt{3}}{2}\mathbf{j}$;
$\theta = \frac{5\pi}{4} \Rightarrow \mathbf{u} = -\frac{1}{\sqrt{2}}\mathbf{i} - \frac{1}{\sqrt{2}}\mathbf{j}$; $\theta = \frac{5\pi}{3} \Rightarrow \mathbf{u} = \frac{1}{2}\mathbf{i} - \frac{\sqrt{3}}{2}\mathbf{j}$

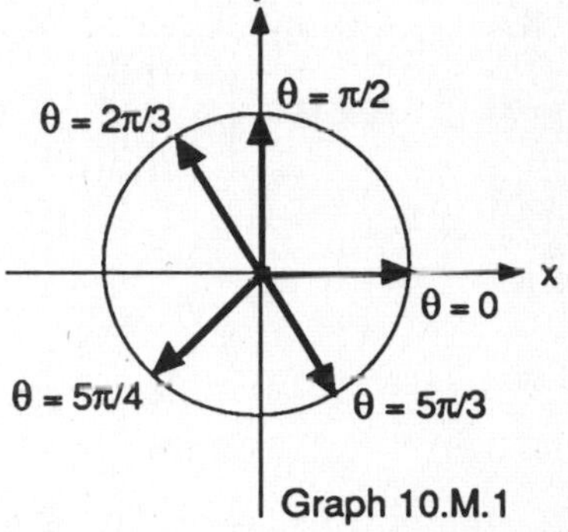

Graph 10.M.1

3. $y = \tan x \Rightarrow [y']_{\pi/4} = \left[\sec^2 x\right]_{\pi/4} = 2 = \frac{2}{1} \Rightarrow \mathbf{T} = \mathbf{i} + 2\mathbf{j} \Rightarrow$ the unit tangents are $\pm\left(\frac{1}{\sqrt{5}}\mathbf{i} + \frac{2}{\sqrt{5}}\mathbf{j}\right)$ and the unit normals are $\pm\left(-\frac{2}{\sqrt{5}}\mathbf{i} + \frac{1}{\sqrt{5}}\mathbf{j}\right)$

5. length $= \left|\sqrt{2}\mathbf{i} + \sqrt{2}\mathbf{j}\right| = \sqrt{2+2} = 2$, $\sqrt{2}\mathbf{i} + \sqrt{2}\mathbf{j} = 2\left[\frac{1}{\sqrt{2}}\mathbf{i} + \frac{1}{\sqrt{2}}\mathbf{j}\right] \Rightarrow$
the direction is $\frac{1}{\sqrt{2}}\mathbf{i} + \frac{1}{\sqrt{2}}\mathbf{j}$

7. length $= |2\mathbf{i} - 3\mathbf{j} + 6\mathbf{k}| = \sqrt{4+9+36} = 7$, $2\mathbf{i} - 3\mathbf{j} + 6\mathbf{k} = 7\left[\frac{2}{7}\mathbf{i} - \frac{3}{7}\mathbf{j} + \frac{6}{7}\mathbf{k}\right] \Rightarrow$
the direction is $= \frac{2}{7}\mathbf{i} - \frac{3}{7}\mathbf{j} + \frac{6}{7}\mathbf{k}$

9. If $\mathbf{A} = 1\mathbf{i} + b\mathbf{j}$ and $\mathbf{B} = c\mathbf{i} + d\mathbf{j}$, then $\mathbf{A}\cdot\mathbf{B} = |\mathbf{A}|\,|\mathbf{B}|\cos\theta \Rightarrow ac + bd = \sqrt{a^2+b^2}\sqrt{c^2+d^2}\cos\theta \Rightarrow$
$(ac+bd)^2 = (a^2+b^2)(c^2+d^2)\cos^2\theta \Rightarrow (ac+bd)^2 \le (a^2+b^2)(c^2+d^2)$ since $c0s^2\theta \le 1$.

11. The equation of the line through (2,3) and (4,1) is $x = 2 + 2t$, $y = 3 - 2t$. If (5,y) is to lie on this line, then $5 = 2 + 2t \Rightarrow t = \frac{3}{2}$. $\therefore\ y = 3 - 2\left(\frac{3}{2}\right) = 0$.

13. $|\mathbf{A}| = \sqrt{2}$, $|\mathbf{B}| = 3$, $\mathbf{A}\cdot\mathbf{B} = 3$, $\mathbf{B}\cdot\mathbf{A} = 3$, $\mathbf{A}\times\mathbf{B} = \begin{vmatrix} \mathbf{i} & \mathbf{j} & \mathbf{k} \\ 1 & 1 & 0 \\ 2 & 1 & -2 \end{vmatrix} = -2\mathbf{i} + 2\mathbf{j} - \mathbf{k}$,

$\mathbf{B}\times\mathbf{A} = \begin{vmatrix} \mathbf{i} & \mathbf{j} & \mathbf{k} \\ 2 & 1 & -2 \\ 1 & 1 & 0 \end{vmatrix} = 2\mathbf{i} - 2\mathbf{j} + \mathbf{k}$, $|\mathbf{A}\times\mathbf{B}| = \sqrt{4+4+1} = 3$, $\theta = \cos^{-1}\left(\frac{\mathbf{A}\cdot\mathbf{B}}{|\mathbf{A}|\,|\mathbf{B}|}\right) = \cos^{-1}\left(\frac{1}{\sqrt{2}}\right) = \frac{\pi}{4}$,

$\text{comp}_{\mathbf{A}}\,\mathbf{B} = \frac{\mathbf{A}\cdot\mathbf{B}}{|\mathbf{A}|} = \frac{3}{\sqrt{2}}$, $\text{proj}_{\mathbf{A}}\,\mathbf{B} = \frac{\mathbf{A}\cdot\mathbf{B}}{\mathbf{A}\cdot\mathbf{A}}\mathbf{A} = \frac{3}{2}[\mathbf{i} + \mathbf{j}]$

15. $\mathbf{B} = \left(\frac{\mathbf{A}\cdot\mathbf{B}}{\mathbf{A}\cdot\mathbf{A}}\mathbf{A}\right) + \left(\mathbf{B} - \frac{\mathbf{A}\cdot\mathbf{B}}{\mathbf{A}\cdot\mathbf{A}}\mathbf{A}\right) = \frac{4}{3}[2\mathbf{i} + \mathbf{j} - \mathbf{k}] + \left[(\mathbf{i} + \mathbf{j} - 5\mathbf{k}) - \frac{4}{3}(2\mathbf{i} + \mathbf{j} - \mathbf{k})\right] =$
$\frac{4}{3}[2\mathbf{i} + \mathbf{j} - \mathbf{k}] - \frac{1}{3}[5\mathbf{i} + \mathbf{j} + 11\mathbf{k}]$, where $\mathbf{A}\cdot\mathbf{B} = 8$ and $\mathbf{A}\cdot\mathbf{A} = 6$

17. $\mathbf{A} \times \mathbf{B} = \begin{vmatrix} \mathbf{i} & \mathbf{j} & \mathbf{k} \\ 1 & 0 & 0 \\ 1 & 1 & 0 \end{vmatrix} = \mathbf{k}$

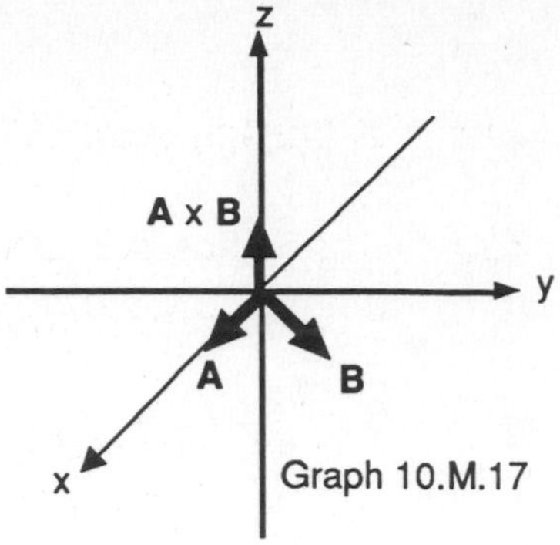

Graph 10.M.17

19. $W = \mathbf{F} \cdot \overrightarrow{PQ} = |\mathbf{F}|\left|\overrightarrow{PQ}\right| \cos\theta = [70 \text{ lb}](800 \text{ ft}) \cos 30° = 28000\sqrt{3} \text{ ft} \cdot \text{lb}$

21. $|\mathbf{A}+\mathbf{B}| = (\mathbf{A}+\mathbf{B}) \cdot (\mathbf{A}+\mathbf{B}) = \mathbf{A} \cdot \mathbf{A} + 2\mathbf{A} \cdot \mathbf{B} + \mathbf{B} \cdot \mathbf{B} \le |\mathbf{A}|^2 + 2\,|\mathbf{A}|\,|\mathbf{B}| + |\mathbf{B}|^2 = \left(|\mathbf{A}| + |\mathbf{B}|\right)^2$

$\therefore\ |\mathbf{A}+\mathbf{B}| \le |\mathbf{A}| + |\mathbf{B}|$

23. Let C divide the angle between **A** and **B** into α and β. Then $\cos\alpha = \dfrac{\mathbf{C} \cdot \mathbf{A}}{|\mathbf{C}|\,|\mathbf{A}|} = \dfrac{(a\,\mathbf{B} + b\,\mathbf{A}) \cdot \mathbf{A}}{|\mathbf{C}|\,|\mathbf{A}|} =$

$\dfrac{(a\,\mathbf{B} \cdot \mathbf{A} + b\,\mathbf{A} \cdot \mathbf{A})}{|\mathbf{C}|\,|\mathbf{A}|} = \dfrac{(a\,\mathbf{B} \cdot \mathbf{A} + b\,\mathbf{A} \cdot \mathbf{A})}{|\mathbf{C}|\,|\mathbf{A}|} = \dfrac{(a\,\mathbf{B} \cdot \mathbf{A} + b\,a^2)}{|\mathbf{C}|\,a} = \dfrac{\mathbf{B} \cdot \mathbf{A} + b\,a}{|\mathbf{C}|}$ where $a = |\mathbf{A}|$ and

$b = |\mathbf{B}|$. Likewise, $\cos\beta = \dfrac{\mathbf{A} \cdot \mathbf{B} + b\,a}{|\mathbf{C}|}$. Since, the angle between A and B is always $\le \frac{\pi}{2}$ and

$\cos\alpha = \cos\beta \Rightarrow \alpha = \beta$.

25. If $\mathbf{A} = a\mathbf{i} + b\mathbf{j} + c\mathbf{k}$, then $\mathbf{A} \cdot \mathbf{A} = a^2 + b^2 + c^2 \ge 0$ and $\mathbf{A} \cdot \mathbf{A} = 0$ iff $a = b = 0$.

27. $d = \dfrac{|(3)(3) + (4)(2) - 2|}{\sqrt{3^2 + 4^2}} = 3$

29. The desired distance $d(t) = \sqrt{(-t-2)^2 + (t-2)^2 + (t-1)^2}$ is minimized when $f(t) = (-t-2)^2 + (t-2)^2 + (t-1)^2$ is minimized. $f'(t) = 2(t+2) + 2(t-2) + 2(t-1) = 0 \Rightarrow t = \frac{1}{3}$. $\therefore$ the distance is

$d(1/3) = \sqrt{\left(\frac{7}{3}\right)^2 + \left(\frac{-5}{3}\right)^2 + \left(\frac{-2}{3}\right)^2} = \frac{\sqrt{78}}{3}$.

31. $x = 1 - 3t,\ y = 2,\ z = 3 + 7t$

33. $S(6,0,-6)$, $P(4,0,0)$ is on $x - y = 4 \Rightarrow \overrightarrow{PS} = 2\mathbf{i} - 6\mathbf{k}$ and $\mathbf{N} = \mathbf{i} - \mathbf{j} \Rightarrow d = \left|\dfrac{\mathbf{N} \cdot \overrightarrow{PS}}{|\mathbf{N}|}\right| = \left|\dfrac{2}{\sqrt{2}}\right| = \sqrt{2}$

35. $(3,-2,1)$ and $\mathbf{N} = 2\mathbf{i} + \mathbf{j} - \mathbf{k} \Rightarrow 2(x-3) + 1(y+2) + (-1)(z-1) = 0 \Rightarrow 2x + y - z = 3$

37. Let $\mathbf{N} = n_1\mathbf{i} + n_2\mathbf{j} + n_3\mathbf{k}$ and $\overrightarrow{P_oP} = (x - p_1)\mathbf{i} + (y - p_2)\mathbf{j} + (z - p_3)\mathbf{k}$. Now $\mathbf{N} \cdot \overrightarrow{P_oP} = n_1(x - p_1) + n_2(y - p_2) + n_3(z - p_3) > 0 \Rightarrow n_1x + n_2y + n_3z > n_1p_1 + n_2p_2 + n_3p_3$. $\therefore$ the set is all planes that are perpendicular to **N** as one moves away from the given plane in the direction of **N**.

39. $\mathbf{N}_1 = \mathbf{i} + \mathbf{j}$ and $\mathbf{N}_2 = \mathbf{j} + \mathbf{k} \Rightarrow$ the desired angle is $\arccos\left(\dfrac{\mathbf{N}_1 \cdot \mathbf{N}_2}{|\mathbf{N}_1|\,|\mathbf{N}_2|}\right) = \arccos\left(\frac{1}{2}\right) = \frac{\pi}{3}$

41. If A(0,0,0) and P(2,2,3), then the distance is $\left|\dfrac{\overrightarrow{AP}\cdot \mathbf{N}}{|\mathbf{N}|}\right| = \left|\dfrac{(2\mathbf{i}+2\mathbf{j}+3\mathbf{k})\cdot(2\mathbf{i}+3\mathbf{j}+5\mathbf{k})}{\sqrt{4+9+25}}\right| = \dfrac{25}{\sqrt{38}}$

43. The vector, $\mathbf{A}\times(\mathbf{B}\times\mathbf{C}) = (\mathbf{A}\cdot\mathbf{C})\mathbf{B} - (\mathbf{A}\cdot\mathbf{B})\mathbf{C} = (2-1-2)\mathbf{B} - (2-2+1)\mathbf{C} = -\mathbf{B}-\mathbf{C} = -2\mathbf{i}-3\mathbf{j}+\mathbf{k}$, clearly is in the plane of **B** and **C** and orthogonal to **A**. The desired unit vector is $\dfrac{1}{\sqrt{14}}(-2\mathbf{i}-3\mathbf{j}+\mathbf{k})$.

45. The line containing (0,0,0) normal to the plane is $x = 2t$, $y = -t$, $z = -t$. This line intersects the plane $3x - 5y + 2z = 6$ when $3(2t) - 5(-t) + 2(-t) = 6 \Rightarrow t = \dfrac{2}{3}$. The point is $\left(\dfrac{4}{3}, -\dfrac{2}{3}, -\dfrac{2}{3}\right)$.

47. The direction of the intersection, $\mathbf{N}_1 \times \mathbf{N}_2 = \begin{vmatrix} \mathbf{i} & \mathbf{j} & \mathbf{k} \\ 1 & 2 & -2 \\ 5 & -2 & -1 \end{vmatrix} = -6\mathbf{i} - 9\mathbf{j} - 12\mathbf{k} = -3(2\mathbf{i}+3\mathbf{j}+4\mathbf{k})$ is the same as the direction of the given line.

49. The direction of the intersection is $\mathbf{A} = \mathbf{N}_1 \times \mathbf{N}_2 = \begin{vmatrix} \mathbf{i} & \mathbf{j} & \mathbf{k} \\ 2 & 1 & -1 \\ 1 & 1 & 2 \end{vmatrix} = 3\mathbf{i} - 5\mathbf{j} + \mathbf{k}$. $\therefore\ \theta = \cos^{-1}\left(\dfrac{\mathbf{A}\cdot\mathbf{i}}{|\mathbf{A}|\,|\mathbf{i}|}\right) = \arccos\left(\dfrac{3}{\sqrt{35}}\right) \approx 59.5°$

51. If P(a,b,c) is a point on the line of intersedction, then P is in both planes $\Rightarrow a - 2b + c + 3 = 0$ and $2a - b - c + 1 = 0$. $\therefore\ (a - 2b + c + 3) + k(2a - b - c + 1) = 0$.

53. Information from ship A indicates the submarine is on the line L_1: $x = 4 + 2t$, $y = 3t$, $z = -\dfrac{1}{3}t$ now; information from ship B indicates the submarine is on the line L_2: $x = 18s$, $y = 5 - 6s$, $z = -s$ now. The current position of the sub is $\left(6,3,-\dfrac{1}{3}\right)$ and occurs when $t = 1$ or $s = \dfrac{1}{3}$. The path of the sub contains $P\left(2,-1,-\dfrac{1}{3}\right)$ and $Q\left(6,3,-\dfrac{1}{3}\right)$; the line representing this path is L: $x = 2 + 4t$, $y = -1 + 4t$, $z = -\dfrac{1}{3}$. The distance between P and Q the sub traveled in 4 minutes $\Rightarrow$ a speed of $\sqrt{2}$ thousand ft/min. In 20 minutes ythe sub will move $20\sqrt{2}$ thousand ft from Q long line L $\Rightarrow$ $20\sqrt{2} = \sqrt{(2+4t-6)^2 + (-1+4t-3)^2 + 0^2} \Rightarrow t = 6$. $\therefore$ in 20 minutes the sub will be at $\left(26,23,-\dfrac{1}{3}\right)$.

55. $\overrightarrow{AB} = -2\mathbf{i} + \mathbf{j} + \mathbf{k}$, $\overrightarrow{CD} = \mathbf{i} + 4\mathbf{j} - \mathbf{k}$ and $\overrightarrow{AC} = 2\mathbf{i} + \mathbf{j}$. $\mathbf{N} = \begin{vmatrix} \mathbf{i} & \mathbf{j} & \mathbf{k} \\ -2 & 1 & 1 \\ 1 & 4 & -1 \end{vmatrix} = -5\mathbf{i} - \mathbf{j} - 9\mathbf{k}$. The distance is $\left|\dfrac{(2\mathbf{i}+\mathbf{j})\cdot(-5\mathbf{i}-\mathbf{j}-9\mathbf{k})}{\sqrt{25+1+81}}\right| = \dfrac{11}{\sqrt{107}}$.

57. a) The line through A and B is: $x = 1 + t$, $y = -t$, $z = -1 + 5t$; the line through C and D must be parallel and is L_1: $x = 1 + t$, $y = 2 - t$, $z = 3 + 5t$. The line through B and C is: $x = 1$, $y = 2 + 2s$, $z = 3 + 4s$; the line through A and D must be parallel and is L_2: $x = 2$, $y = -1 + 2s$, $z = 4 + 4s$. The lines L_1 and L_2 intersect at D(2,1,8).

b) $\cos\theta = \dfrac{(2\mathbf{j} + 4\mathbf{k}) \cdot (\mathbf{i} - \mathbf{j} + 5\mathbf{k})}{\sqrt{20}\sqrt{27}} = \dfrac{3}{\sqrt{15}}$

c) $\left(\dfrac{\overrightarrow{BA} \cdot \overrightarrow{BC}}{\overrightarrow{BC} \cdot \overrightarrow{BC}}\right)\overrightarrow{BC} = \dfrac{9}{10}\overrightarrow{BC} = \dfrac{9}{5}(\mathbf{i} + 2\mathbf{k})$

d) $A = |(2\mathbf{j} + 4\mathbf{k}) \times (\mathbf{i} - \mathbf{j} + 5\mathbf{k})| = |14\mathbf{i} + 4\mathbf{j} - 2\mathbf{k}| = 6\sqrt{6}$

e) From part d) we have $\mathbf{N} = 14\mathbf{i} + 4\mathbf{j} - 2\mathbf{k}$. $\therefore$ the plane is $(14\mathbf{i} + 4\mathbf{k} - 2\mathbf{k}) \cdot ((x - 1)\mathbf{i} + y\mathbf{j} + (z + 1)\mathbf{k} = 0 \Rightarrow 7x + 2y - z = 8$

f) $\overrightarrow{BC} \times \overrightarrow{BA} = (2\mathbf{j} + 4\mathbf{k}) \times (-\mathbf{i} + \mathbf{j} - \mathbf{k}) = -14\mathbf{i} - 4\mathbf{j} + 2\mathbf{k}$.
The area of the projection on the yz–plane is $|(-14\mathbf{i} - 4\mathbf{j} + 2\mathbf{k}) \cdot (\mathbf{i})| = 14$
The area of the projection on the xz–plane is $|(-14\mathbf{i} - 4\mathbf{j} + 2\mathbf{k}) \cdot (\mathbf{j})| = 4$
The area of the projection on the xy–plane is $|(-14\mathbf{i} - 4\mathbf{j} + 2\mathbf{k}) \cdot (\mathbf{k})| = 2$

59. a) If P(x,y,z) is a point in the plane determined by $P_i(x_i,y_i,z_i)$, where i = 1, 2, 3, then the given determinant, $\overrightarrow{PP_1} \cdot \left(\overrightarrow{PP_2} \times \overrightarrow{PP_3}\right)$, is the equation of the plane containing $P_i(x_i,y_i,z_i)$'s.

b) Subtract row 1 from rows 2, 3 and 4. Expand by the minors of column 4. Now we have the deterninant equation of part a. $\therefore$ we have a plane described by the points (x_1,y_1,z_1), (x_2,y_2,z_2), and (x_3,y_3,z_3).

61. a) Let (x,y,z) be any point on $Ax + By + Cz - D = 0$. Let $\overrightarrow{QP} = (x - x_1)\mathbf{i} + (y - y_1)\mathbf{j} + (z - z_1)\mathbf{k}$, and $\mathbf{N} = \dfrac{A\mathbf{i} + B\mathbf{j} + C\mathbf{k}}{\sqrt{A^2 + B^2 + C^2}}$. The distance is $|\text{proj}_{\mathbf{N}}\mathbf{P}| =$

$$\left|\left((x - x_1)\mathbf{i} + (y - y_1)\mathbf{j} + (z - z_1)\mathbf{k}\right) \cdot \left(\frac{A\mathbf{i} + B\mathbf{j} + C\mathbf{k}}{\sqrt{A^2 + B^2 + C^2}}\right)\right| = \frac{|Ax_1 + By_1 + Cz_1 - (Ax + By + Cz)|}{\sqrt{a^2 + b^2 + c^2}} = \frac{|Ax_1 + By_1 + Cz_1 - D|}{\sqrt{A^2 + B^2 + C^2}}$$

b) Since, both tangent planes are parallel one half of the distance between them is equal to the radius of the sphere i.e. $r = \dfrac{1}{2}\dfrac{|3 - 9|}{\sqrt{1 + 1 + 1}} = \sqrt{3}$. Clearly, the points (1,2,3) and (– 1,– 2,– 3) are on the line containing the sphere's center. Hence, the line containing the center is $x = 1 + 2t$, $y = 2 + 4t$, $z = 3 + 6t$. The distance from the plane, $x + y + z - 3 = 0$, and the center is $\sqrt{3} \Rightarrow$ $\dfrac{|(1 + 2t) + (2 + 4t) + (3 + 6t) - 3|}{\sqrt{1 + 1 + 1}} = \sqrt{3} \Rightarrow t = 0 \Rightarrow$ the center is at (1,2,3). $\therefore$ the equation of the sphere is $(x - 1)^2 + (y - 2)^2 + (z - 3)^2 = 3$.

63. a) If (x_1,y_1,z_1) is on the plane $Ax + By + Cz = D_1$, then the distance between the planes,d, is $\frac{|Ax_1 + By_1 + Cz_1 - D_2|}{\sqrt{A^2 + B^2 + C^2}} = \frac{|D_1 - D_2|}{|A\mathbf{i} + B\mathbf{j} + C\mathbf{k}|}$, since $Ax_1 + By_1 + Cz_1 = D_1$.

b) $d = \frac{|12 - 6|}{\sqrt{4 + 9 + 1}} = \frac{6}{\sqrt{14}}$

c) $\frac{|2(3) + (-1)(2) + 2(-1) + 4|}{\sqrt{14}} = \frac{|2(3) + (-1)(2) + 2(-1) + D|}{\sqrt{14}} \Rightarrow D = -8 \text{ or } 4.$

$\therefore$ the desired plane is $2x - y + 2x = 8$

d) Choose the point (2,0,1) on the plane. Then $\frac{|3 - D|}{\sqrt{6}} = 5 \Rightarrow D = 3 \pm 5\sqrt{6} \Rightarrow$ the desired planes are $x - 2y + z = 3 + 5\sqrt{6}$ and $x - 2y + z = 3 - 5\sqrt{6}$

65. Let $\mathbf{N} = \overrightarrow{AB} \times \overrightarrow{BC}$ and P(x,y,x) be any point in the plane determined by A, B and C. Then $\mathbf{N} \cdot \overrightarrow{AP} = 0$ is the equation of the plane. Therefore, point D lies in this plane if and only if $\overrightarrow{AD} \cdot \mathbf{N} = 0 \Leftrightarrow \overrightarrow{AD} \cdot (\overrightarrow{AB} \times \overrightarrow{BC})$.

67. the y–axis in the xy–plane and the yz–plane in three dimensional space

69. a circle centered at (0,0) with a radius of 2 in the xy–plane; a cylinder parallel with the z–axis in three dimensional space with the circle as its generating curve

71. a horizontal parabola opening to the right with its vertex at (0,0) in the xy–plane; a cylinder parallel with the z–axis in three dimensional space with the parabola as the generating curve

73. a horizontal cardioid in the $r\theta$–plane; a cylinder parallel with the z–axis in three dimensional space with the cardioid as the generating curve

75. a horizontal lemniscate of length $2\sqrt{2}$ in the $r\theta$–plane; a cylinder parallel with the z–axis in three dimensional space with the lemniscate as the generating curve

77. a sphere with a radius of 2 centered at the origin

79. the upper nappe of a cone whose surface makes a $\frac{\pi}{6}$ angle with the z–axis

81. the upper hemisphere of a sphere with a radius 1 centered at the origin

83. $z = 2 \Rightarrow$ cylindrical, $z = 2$; spherical, $\rho \cos \phi = 2$; a plane parallel with the xy–plane

85. $z = r^2 \cos 2\theta \Rightarrow$ rectangular, $z = r^2(\cos^2\theta - \sin^2\theta) \Rightarrow z = x^2 - y^2$; spherical, $\rho \cos \phi = \rho^2 \sin^2\phi \cos 2\theta \Rightarrow \rho = \frac{\cos \phi}{\sin^2\phi \cos 2\theta}$; hyperbolic parabloid

87. $\rho = 4 \sec \phi \Rightarrow$ rectangular, $\rho \cos \phi = 4 \Rightarrow z = 4$; cylinderical, $z = 4$, the plane $z = 4$

89. Let $\overrightarrow{AB} = (0 - 1)\mathbf{i} + (1 - 0)\mathbf{j} = -\mathbf{i} + \mathbf{j}$; $\overrightarrow{AD} = (0 - 1)\mathbf{i} + (-1 - 0)\mathbf{j} = -\mathbf{i} - \mathbf{j}$

$\overrightarrow{AB} \times \overrightarrow{AD} = (-\mathbf{i} + \mathbf{j}) \times (-\mathbf{i} - \mathbf{j}) = 2\mathbf{k}$. Area $= |2\mathbf{k}| = 2$.

91. Let $\overrightarrow{AB} = (2 + 1)\mathbf{i} + (0 - 2)\mathbf{j} = 3\mathbf{i} - 2\mathbf{j}$; $\overrightarrow{AD} = (4 + 1)\mathbf{i} + (3 - 2)\mathbf{j} = 5\mathbf{i} + \mathbf{j}$

$\overrightarrow{AB} \times \overrightarrow{AD} = (3\mathbf{i} - 2\mathbf{j}) \times (5\mathbf{i} + \mathbf{j}) = 2\mathbf{k}$. Area $= |13\mathbf{k}| = 13$.

93. Let $\overrightarrow{AB} = (-2 - 0)\mathbf{i} + (3 - 0)\mathbf{j} = -2\mathbf{i} + 3\mathbf{j}$; $\overrightarrow{AD} = (3 - 0)\mathbf{i} + (1 - 0)\mathbf{j} = 3\mathbf{i} + \mathbf{j}$

$\overrightarrow{AB} \times \overrightarrow{AD} = (-2\mathbf{i} + 3\mathbf{j}) \times (3\mathbf{i} + \mathbf{j}) = -11\mathbf{k}$. Area $= \frac{1}{2}|-11\mathbf{k}| = \frac{11}{2}$.

95. Let $\overrightarrow{AB} = (1 + 5)\mathbf{i} + (-2 - 3)\mathbf{j} = 6\mathbf{i} - 5\mathbf{j}$; $\overrightarrow{AD} = (6 + 5)\mathbf{i} + (-2 - 3)\mathbf{j} = 11\mathbf{i} - 5\mathbf{j}$

$\overrightarrow{AB} \times \overrightarrow{AD} = (6\mathbf{i} - 5\mathbf{j}) \times (11\mathbf{i} - 5\mathbf{j}) = 25\mathbf{k}$. Area $= \frac{1}{2}|25\mathbf{k}| = \frac{25}{2}$.

97. If $\mathbf{A} = a_1\mathbf{i} + a_2\mathbf{j}$ and $\mathbf{B} = b_1\mathbf{i} + b_2\mathbf{j}$, then $\mathbf{A} \times \mathbf{B} = \begin{vmatrix} \mathbf{i} & \mathbf{j} & \mathbf{k} \\ a_1 & a_2 & 0 \\ b_1 & b_2 & 0 \end{vmatrix} = \begin{vmatrix} a_1 & a_2 \\ b_1 & b_2 \end{vmatrix} \mathbf{k}$ and the area is

$\frac{1}{2}|\mathbf{A} \times \mathbf{B}| = \pm\frac{1}{2}\begin{vmatrix} a_1 & a_2 \\ b_1 & b_2 \end{vmatrix}$. The sign is controlled by the directions of **A** and **B**.

CHAPTER 11

VECTOR–VALUED FUNCTIONS AND MOTION IN SPACE

SECTION 11.1 VECTOR–VALUED FUNCTIONS AND SPACE CURVES

1. $\mathbf{r} = (2\cos t)\,\mathbf{i} + (3\sin t)\,\mathbf{j} + 4t\,\mathbf{k} \Rightarrow \mathbf{v} = \frac{d\mathbf{r}}{dt} = (-2\sin t)\,\mathbf{i} + (3\cos t)\,\mathbf{j} + 4\,\mathbf{k}$

   $\mathbf{a} = \frac{d^2\mathbf{r}}{dt^2} = (-2\cos t)\,\mathbf{i} - (3\sin t)\,\mathbf{j}$. Speed: $\left|\mathbf{v}\left(\frac{\pi}{2}\right)\right| = \sqrt{\left(-2\sin\frac{\pi}{2}\right)^2 + \left(3\cos\frac{\pi}{2}\right)^2 + 4^2} = 2\sqrt{5}$

   Direction: $\dfrac{\mathbf{v}\left(\frac{\pi}{2}\right)}{\left|\mathbf{v}\left(\frac{\pi}{2}\right)\right|} = \left(-\frac{2}{2\sqrt{5}}\sin\frac{\pi}{2}\right)\mathbf{i} + \left(\frac{3}{2\sqrt{5}}\cos\frac{\pi}{2}\right)\mathbf{j} + \frac{4}{2\sqrt{5}}\mathbf{k} = -\frac{1}{\sqrt{5}}\mathbf{i} + \frac{2}{\sqrt{5}}\mathbf{k}$

   $\mathbf{v}\left(\frac{\pi}{2}\right) = 2\sqrt{5}\left[-\frac{1}{\sqrt{5}}\mathbf{i} + \frac{2}{\sqrt{5}}\mathbf{k}\right]$

3. $\mathbf{r} = (\cos 2t)\,\mathbf{j} + (2\sin t)\,\mathbf{k} \Rightarrow \mathbf{v} = (-2\sin 2t)\,\mathbf{j} + (2\cos t)\,\mathbf{k}$, $\mathbf{a} = (-4\cos 2t)\,\mathbf{j} - (2\sin t)\,\mathbf{k}$. Speed: $|\mathbf{v}(0)| = \sqrt{(-2\sin 2(0))^2 + (2\cos 0)^2} = 2$ Direction: $\frac{\mathbf{v}(0)}{|\mathbf{v}(0)|} = \frac{-2\sin 2(0)\,\mathbf{j} + 2\cos 0\,\mathbf{k}}{2} = \mathbf{k}$. $\mathbf{v}(0) = 2\,\mathbf{k}$

5. $\mathbf{r} = (\sec t)\,\mathbf{i} + (\tan t)\,\mathbf{j} + \frac{4}{3}t\,\mathbf{k} \Rightarrow \mathbf{v} = \frac{d\mathbf{r}}{dt} = (\sec t\tan t)\,\mathbf{i} + (\sec^2 t)\,\mathbf{j} + \frac{4}{3}\mathbf{k}$, $\mathbf{a} = \frac{d^2\mathbf{r}}{dt^2} = (\sec t\tan^2 t + \sec^3 t)\,\mathbf{i} + (2\sec^2 t\tan t)\,\mathbf{j}$. Speed: $\left|\mathbf{v}\left(\frac{\pi}{6}\right)\right| = \sqrt{\left(\sec\frac{\pi}{6}\tan\frac{\pi}{6}\right)^2 + \left(\sec^2\frac{\pi}{6}\right)^2 + \left(\frac{4}{3}\right)^2} = 2$. Direction: $\dfrac{\mathbf{v}\left(\frac{\pi}{6}\right)}{\left|\mathbf{v}\left(\frac{\pi}{6}\right)\right|}$

   $= \dfrac{\sec\frac{\pi}{6}\tan\frac{\pi}{6}\,\mathbf{i} + \sec^2\frac{\pi}{6}\,\mathbf{j} + \frac{4}{3}\mathbf{k}}{2} = \frac{1}{3}\mathbf{i} + \frac{2}{3}\mathbf{j} + \frac{2}{3}\mathbf{k}$. $\mathbf{v}\left(\frac{\pi}{6}\right) = 2\left(\frac{1}{3}\mathbf{i} + \frac{2}{3}\mathbf{j} + \frac{2}{3}\mathbf{k}\right)$

7. $\mathbf{v} = 3\,\mathbf{i} + \sqrt{3}\,\mathbf{j} + 2t\,\mathbf{k}$, $\mathbf{a} = 2\,\mathbf{k}$. $\mathbf{v}(0) = 3\,\mathbf{i} + \sqrt{3}\,\mathbf{j}$, $\mathbf{a}(0) = 2\,\mathbf{k} \Rightarrow |\mathbf{v}(0)| = \sqrt{3^2 + \left(\sqrt{3}\right)^2 + 0^2} = \sqrt{12}$, $|\mathbf{a}(0)| = \sqrt{2^2} = 2$. $\mathbf{v}(0)\cdot\mathbf{a}(0) = 0 \Rightarrow \cos\theta = \frac{0}{2\sqrt{12}} = 0 \Rightarrow \theta = \frac{\pi}{2}$

9. $\mathbf{v} = \frac{2t}{t^2+1}\mathbf{i} + \frac{1}{t^2+1}\mathbf{j} + t\left(t^2+1\right)^{-1/2}\mathbf{k}$, $\mathbf{a} = \frac{-2t^2+2}{\left(t^2+1\right)^2}\mathbf{i} - \frac{2t}{\left(t^2+1\right)^2}\mathbf{j} + \frac{1}{\left(t^2+1\right)^{3/2}}\mathbf{k}$. $\mathbf{v}(0) = \mathbf{j}$, $\mathbf{a}(0) = 2\,\mathbf{i} + \mathbf{k} \Rightarrow |\mathbf{v}(0)| = 1$, $|\mathbf{a}(0)| = \sqrt{2^2 + 1^2} = \sqrt{5}$. $\mathbf{v}(0)\cdot\mathbf{a}(0) = 0 \Rightarrow \cos\theta = \frac{0}{1\sqrt{5}} = 0 \Rightarrow \theta = \frac{\pi}{2}$

11. $\mathbf{v} = (1-\cos t)\,\mathbf{i} + (\sin t)\,\mathbf{j}$, $\mathbf{a} = (\sin t)\,\mathbf{i} + (\cos t)\,\mathbf{j} \Rightarrow \mathbf{v}\cdot\mathbf{a} = \sin t(1-\cos t) + \sin t(\cos t) = \sin t$.

    $\mathbf{v}\cdot\mathbf{a} = 0 \Rightarrow \sin t = 0 \Rightarrow t = 0, \pi, 2\pi$

13. $\displaystyle\int_0^1 \left(t^3\,\mathbf{i} + 7\,\mathbf{j} + (t+1)\,\mathbf{k}\right)dt = \left[\frac{t^4}{4}\right]_0^1\mathbf{i} + \left[7t\right]_0^1\mathbf{j} + \left[\frac{t^2}{2} + t\right]_0^1\mathbf{k} = \frac{1}{4}\mathbf{i} + 7\,\mathbf{j} + \frac{3}{2}\mathbf{k}$

15. $\int_{-\pi/4}^{\pi/4} \left((\sin t)\,\mathbf{i} + (1+\cos t)\,\mathbf{j} + (\sec^2 t)\,\mathbf{k}\right) dt = \left[-\cos t\right]_{-\pi/4}^{\pi/4}\mathbf{i} + \left[t+\sin t\right]_{-\pi/4}^{\pi/4}\mathbf{j} + \left[\tan t\right]_{-\pi/4}^{\pi/4}\mathbf{k} =$
$\left(\frac{\pi+2\sqrt{2}}{2}\right)\mathbf{j} + 2\,\mathbf{k}$

17. $\mathbf{v} = (\cos t)\,\mathbf{i} - (\sin t)\,\mathbf{j},\ \mathbf{a} = -(\sin t)\,\mathbf{i} - (\cos t)\,\mathbf{j} \Rightarrow$ For $t = \frac{\pi}{4}$, $\mathbf{v}\left(\frac{\pi}{4}\right) = \frac{\sqrt{2}}{2}\mathbf{i} - \frac{\sqrt{2}}{2}\mathbf{j}$, $\mathbf{a}\left(\frac{\pi}{4}\right) = -\frac{\sqrt{2}}{2}\mathbf{i} - \frac{\sqrt{2}}{2}\mathbf{j}$;
For $t = \frac{\pi}{2}$, $\mathbf{v}\left(\frac{\pi}{2}\right) = -\mathbf{j}$, $\mathbf{a}\left(\frac{\pi}{2}\right) = -\mathbf{i}$

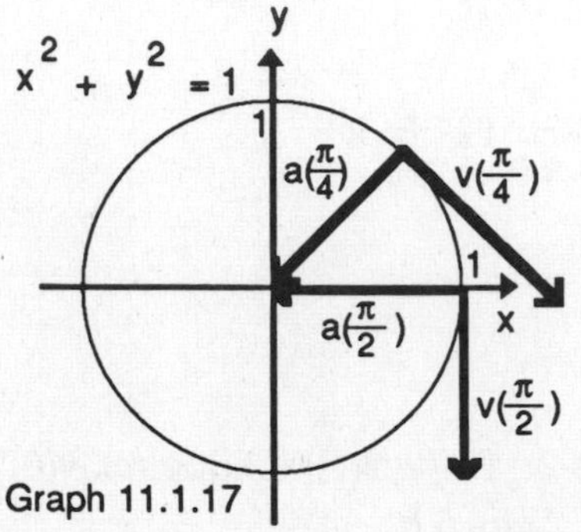

Graph 11.1.17

19. $\mathbf{v} = (1-\cos t)\,\mathbf{i} + (\sin t)\,\mathbf{j},\ \mathbf{a} = (\sin t)\,\mathbf{i} + (\cos t)\,\mathbf{j} \Rightarrow$ For $t = \pi$, $\mathbf{v}(\pi) = 2\,\mathbf{i}$, $\mathbf{a}(\pi) = -\mathbf{j}$; For $t = \frac{3\pi}{2}$,
$\mathbf{v}\left(\frac{3\pi}{2}\right) = \mathbf{i} - \mathbf{j}$, $\mathbf{a}\left(\frac{3\pi}{2}\right) = -\mathbf{i}$

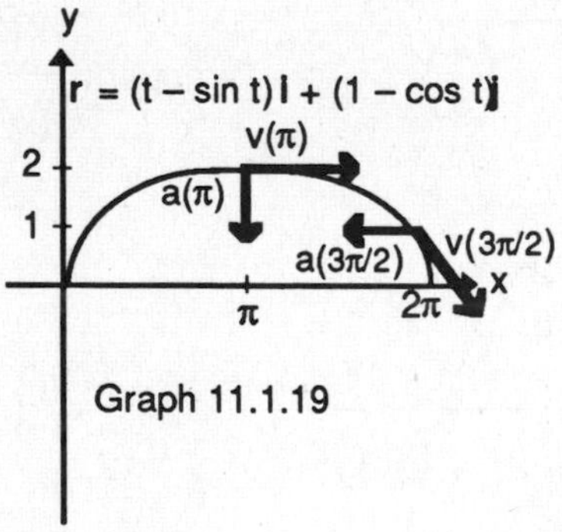

Graph 11.1.19

21. $\mathbf{r} = \int (-t\,\mathbf{i} - t\,\mathbf{j} - t\,\mathbf{k})\,dt = -\frac{t^2}{2}\mathbf{i} - \frac{t^2}{2}\mathbf{j} - \frac{t^2}{2}\mathbf{k} + \mathbf{C}.$ $\mathbf{r}(0) = 0\,\mathbf{i} - 0\,\mathbf{j} - 0\,\mathbf{k} + \mathbf{C} = \mathbf{i} + 2\,\mathbf{j} + 3\,\mathbf{k} \Rightarrow$
$\mathbf{C} = \mathbf{i} + 2\,\mathbf{j} + 3\,\mathbf{k}.\ \therefore\ \mathbf{r} = \left(-\frac{t^2}{2} + 1\right)\mathbf{i} + \left(-\frac{t^2}{2} + 2\right)\mathbf{j} + \left(-\frac{t^2}{2} + 3\right)\mathbf{k}$

23. $\mathbf{r} = \int \left(\frac{3}{2}(t+1)^{1/2}\,\mathbf{i} + e^{-t}\,\mathbf{j} + \frac{1}{t+1}\mathbf{k}\right) dt = (t+1)^{3/2}\,\mathbf{i} - e^{-t}\,\mathbf{j} + \ln(t+1)\,\mathbf{k} + \mathbf{C}.$ $\mathbf{r}(0) = \mathbf{k} \Rightarrow$
$(0+1)^{3/2}\,\mathbf{i} - e^{-0}\,\mathbf{j} + \ln(0+1)\,\mathbf{k} + \mathbf{C} = \mathbf{k} \Rightarrow \mathbf{C} = -\mathbf{i} + \mathbf{j} + \mathbf{k}.$
$\therefore\ \mathbf{r} = \left((t+1)^{3/2} - 1\right)\mathbf{i} + \left(1 - e^{-t}\right)\mathbf{j} + (1 + \ln(t+1))\,\mathbf{k}$

25. $\frac{d\mathbf{r}}{dt} = \int (-32\,\mathbf{k})\,dt = -32t\,\mathbf{k} + \mathbf{C}_1$. $\frac{d\mathbf{r}}{dt}(0) = 8\,\mathbf{i} + 8\,\mathbf{j} \Rightarrow -32(0)\,\mathbf{k} + \mathbf{C}_1 = 8\,\mathbf{i} + 8\,\mathbf{j} \Rightarrow \mathbf{C}_1 = 8\,\mathbf{i} + 8\,\mathbf{j}$

$\therefore\ \frac{d\mathbf{r}}{dt} = 8\,\mathbf{i} + 8\,\mathbf{j} - 32t\,\mathbf{k}$. $\mathbf{r} = \int (8\,\mathbf{i} + 8\,\mathbf{j} - 32t\,\mathbf{k})\,dt = 8t\,\mathbf{i} + 8t\,\mathbf{j} - 16t^2\,\mathbf{k} + \mathbf{C}_2$. $\mathbf{r}(0) = 100\,\mathbf{k} \Rightarrow$ $8(0)\,\mathbf{i} + 8(0)\,\mathbf{j} - 16(0)^2\,\mathbf{k} + \mathbf{C}_2 = 100\,\mathbf{k} \Rightarrow \mathbf{C}_2 = 100\,\mathbf{k}$. $\therefore\ \mathbf{r} = 8t\,\mathbf{i} + 8t\,\mathbf{j} + (100 - 16t^2)\,\mathbf{k}$

27. $\mathbf{v} = (1 - \cos t)\,\mathbf{i} + (\sin t)\,\mathbf{j}$, $\mathbf{a} = (\sin t)\,\mathbf{i} + (\cos t)\,\mathbf{j}$. $|\mathbf{v}|^2 = (1 - \cos t)^2 + \sin^2 t = 2 - 2\cos t$. $|\mathbf{v}|^2$ is at a max when $\cos t = -1 \Rightarrow t = \pi, 3\pi, 5\pi$, etc. At these values of t, $|\mathbf{v}|^2 = 4 \Rightarrow \max|\mathbf{v}| = \sqrt{4} = 2$. $|\mathbf{v}|^2$ is at a min when $\cos t = 1 \Rightarrow t = 0, 2\pi, 4\pi$, etc. At these values of t, $|\mathbf{v}|^2 = 0 \Rightarrow \min|\mathbf{v}| = 0$.
$|\mathbf{a}|^2 = \sin^2 t + \cos^2 t = 1$ for every $t \Rightarrow \max|\mathbf{a}| = \min|\mathbf{a}| = \sqrt{1} = 1$.

29. The length of **r** is constant (it equals the radius of the sphere). Hence $\mathbf{r} \cdot (d\mathbf{r}/dt) = 0$.

31. a) $\frac{d}{dt}(\mathbf{u} \cdot \mathbf{v} \times \mathbf{w}) = \frac{d\mathbf{u}}{dt} \cdot (\mathbf{v} \cdot \mathbf{w}) + \mathbf{u} \cdot \frac{d}{dt}(\mathbf{u} \times \mathbf{w}) = \frac{d\mathbf{u}}{dt} \cdot (\mathbf{v} \times \mathbf{w}) + \mathbf{u} \cdot \left(\frac{d\mathbf{v}}{dt} \times \mathbf{w} + \mathbf{v} \times \frac{d\mathbf{w}}{dt}\right) = \frac{d\mathbf{u}}{dt} \cdot (\mathbf{v} \times \mathbf{w}) +$ $\mathbf{u} \cdot \frac{d\mathbf{v}}{dt} \times \mathbf{w} + \mathbf{v} \times \frac{d\mathbf{w}}{dt}$

b) Each of the determinants is equivalent to each quantity in Equation 15. See Equation 4, Section 10.6.

33. Let $\mathbf{f} = \mathbf{C}$, a constant vector. Then $\mathbf{f} = c_1\,\mathbf{i} + c_2\,\mathbf{j} + c_3\,\mathbf{k}$ where c_1, c_2, c_3 are Real Numbers.
$\frac{d\mathbf{f}}{dt} = 0\,\mathbf{i} + 0\,\mathbf{j} + 0\,\mathbf{k} = 0$

35. Let $\mathbf{u} = f_1(t)\,\mathbf{i} + f_2(t)\,\mathbf{j} + f_3(t)\,\mathbf{k}$, $\mathbf{v} = g_1(t)\,\mathbf{i} + g_2(t)\,\mathbf{j} + g_3(t)\,\mathbf{k}$. Then $\mathbf{u} + \mathbf{v} = (f_1(t) + g_1(t))\,\mathbf{i} + (f_2(t) + g_2(t))\,\mathbf{j} +$ $(f_3(t) + g_3(t))\,\mathbf{k} \Rightarrow \frac{d}{dt}(\mathbf{u} + \mathbf{v}) = (f'_1(t) + g'_1(t))\,\mathbf{i} + (f'_2(t) + g'_2(t))\,\mathbf{j} + (f'_3(t) + g'_3(t))\,\mathbf{k} =$
$(f'_1(t)\,\mathbf{i} + f'_2(t)\,\mathbf{j} + f'_3(t)\,\mathbf{k}) + (g'_1(t)\,\mathbf{i} + g'_2(t)\,\mathbf{j} + g'_3(t)\,\mathbf{k} = \frac{d\mathbf{u}}{dt} + \frac{d\mathbf{v}}{dt}$.
$\mathbf{u} - \mathbf{v} = (f_1(t) - g_1(t))\,\mathbf{i} + (f_2(t) - g_2(t))\,\mathbf{j} + (f_3(t) - g_3(t))\,\mathbf{k} \Rightarrow \frac{d}{dt}(\mathbf{u} - \mathbf{v}) = (f'_1(t) - g'_1(t))\,\mathbf{i} + (f'_2(t) - g'_2(t))\,\mathbf{j} +$ $(f'_3(t) - g'_3(t))\,\mathbf{k} = (f'_1(t)\,\mathbf{i} + f'_2(t)\,\mathbf{j} + f'_3(t)\,\mathbf{k}) - (g'_1(t)\,\mathbf{i} + g'_2(t)\,\mathbf{j} + g'_3(t)\,\mathbf{k}) = \frac{d\mathbf{u}}{dt} - \frac{d\mathbf{v}}{dt}$.

37. $$\lim_{t\to t_0} (\mathbf{f}(t) \times \mathbf{g}(t)) = \lim_{t\to t_0} \begin{vmatrix} \mathbf{i} & \mathbf{j} & \mathbf{k} \\ f_1(t) & f_2(t) & f_3(t) \\ g_1(t) & g_2(t) & g_3(t) \end{vmatrix} = \begin{vmatrix} \mathbf{i} & \mathbf{j} & \mathbf{k} \\ \lim_{t\to t_0} f_1(t) & \lim_{t\to t_0} f_2(t) & \lim_{t\to t_0} f_3(t) \\ \lim_{t\to t_0} g_1(t) & \lim_{t\to t_0} g_2(t) & \lim_{t\to t_0} g_3(t) \end{vmatrix} =$$

$$\lim_{t\to t_0} \mathbf{f}(t) \times \lim_{t\to t_0} \mathbf{g}(t)$$

39. a) $\int_a^b k\,\mathbf{f}(t)\,dt = \int_a^b (kf(t)\,\mathbf{i} + kg(t)\,\mathbf{j} + kh(t)\,\mathbf{k})\,dt \quad \int_a^b k\,f(t)\,\mathbf{i}\,dt + \int_a^b k\,g(t)\,\mathbf{j}\,dt + \int_a^b k\,h(t)\,\mathbf{k}\,dt =$

$k\left(\int_a^b f(t)\,\mathbf{i}\,dt + \int_a^b g(t)\,\mathbf{j}\,dt + \int_a^b h(t)\,\mathbf{k}\,dt\right) = k\int_a^b \mathbf{f}(t)\,dt$

b) $\int_a^b (\mathbf{f}(t) + \mathbf{g}(t))\,dt = \int_a^b (f_1(t)\,\mathbf{i} + g_1(t)\,\mathbf{j} + h_1(t)\,\mathbf{k} + f_2(t)\,\mathbf{i} + g_2(t)\,\mathbf{j} + h_2(t)\,\mathbf{k})\,dt =$

$\int_a^b \big((f_1(t) + f_2(t))\,\mathbf{i} + (g_1(t) + g_2(t))\,\mathbf{j} + (h_1(t) + h_2(t))\,\mathbf{k}\big)\,dt = \int_a^b (f_1(t) + f_2(t))\,dt\,\mathbf{i} + \int_a^b (g_1(t) + g_2(t))\,dt\,\mathbf{j}$

$+ \int_a^b (h_1(t) + h_2(t))\,dt\,\mathbf{k} = \int_a^b f_1(t)\,dt\,\mathbf{i} + \int_a^b f_2(t)\,dt\,\mathbf{i} + \int_a^b g_1(t)\,dt\,\mathbf{j} + \int_a^b g_2(t)\,dt\,\mathbf{j} + \int_a^b h_1(t)\,dt\,\mathbf{k} +$

$\int_a^b h_2(t)\,dt\,\mathbf{k} = \int_a^b \mathbf{f}(t)\,dt + \int_a^b \mathbf{g}(t)\,dt$

c) $\int_a^b \mathbf{C}\cdot\mathbf{f}(t)\,dt = \int_a^b (c_1\,f(t) + c_2 g(t) + c_3 h(t))\,dt = c_1\int_a^b f(t)\,dt + c_2\int_a^b g(t)\,dt + c_3\int_a^b h(t)\,dt =$

$\mathbf{C}\cdot\int_a^b \mathbf{f}(t)\,dt$

$\int_a^b \mathbf{C}\times\mathbf{f}(t)\,dt = \int_a^b \big[c_2h(t) - c_3f(t)\big]\,\mathbf{i} + \big[c_3(t) - c_1h(t)\big]\,\mathbf{j} + \big[c_1g(t) - c_2f(t)\big]\,\mathbf{k}\,dt = c_2\int_a^b h(t)\,dt\,\mathbf{i} -$

$c_3\int_a^b f(t)\,dt\,\mathbf{i} + c_3\int_a^b f(t)\,dt\,\mathbf{j} - c_1\int_a^b h(t)\,dt\,\mathbf{j} + c_1\int_a^b g(t)\,dt\,\mathbf{k} - c_2\int_a^b f(t)\,dt\,\mathbf{k} = \mathbf{C}\times\int_a^b \mathbf{f}(t)\,dt$

41. a) If $\mathbf{F}_1(t)$ and $\mathbf{F}_2(t)$ have identical derivatives on I then $\frac{d\mathbf{F}_1}{dt} = \frac{df_1}{dt}\mathbf{i} + \frac{dg_1}{dt}\mathbf{j} + \frac{dh_1}{dt} = \frac{df_2}{dt} + \frac{dg_2}{dt} + \frac{dh_2}{dt} = \frac{d\mathbf{F}_2}{dt} \Rightarrow$
$\frac{df_1}{dt} = \frac{df_2}{dt}, \frac{dg_1}{dt} = \frac{dg_2}{dt}, \frac{dh_1}{dt} = \frac{dh_2}{dt} \Rightarrow f_1(t) = f_2(t) + c_1,\ g_1(t) = g_2(t) + c_2,\ h_1(t) = h_2(t) + c_3 \Rightarrow$
$f_1(t)\,\mathbf{i} + g_1(t)\,\mathbf{j} + h_1(t)\,\mathbf{k} = (f_2(t) + c_1)\,\mathbf{i} + (g_2(t) + c_2)\,\mathbf{j} + (h_2(t) + c_3)\,\mathbf{k} \Rightarrow \mathbf{F}_1(t) = \mathbf{F}_2(t) + \mathbf{C}$ where
$\mathbf{C} = c_1\,\mathbf{i} + c_2\,\mathbf{j} + c_3\,\mathbf{k}$.

b) Let $\mathbf{F}(t)$ be an antiderivative of $\mathbf{f}(t)$ on I. Then $\mathbf{F}'(t) = \mathbf{f}(t)$. If $\mathbf{G}(t)$ is an antiderivative of $\mathbf{f}(t)$ on I, then $\mathbf{G}'$(t0 $= \mathbf{f}(t)$. Thus $\mathbf{G}'(t) = \mathbf{F}'(t)$ on I $\Rightarrow \mathbf{G}(t) = \mathbf{F}(t) + \mathbf{C}$.

SECTION 11.2 MODELING PROJECTILE MOTION

1. $x = \left(v_0 \cos \alpha\right)t \Rightarrow (21 \text{ km})\left(\frac{1000 \text{ m}}{1 \text{ km}}\right) = 840 \text{ m/s}(\cos 60°)t \Rightarrow t = \frac{21\,000 \text{ m}}{(840 \text{ km/s})(\cos 60°)} = 50$ seconds

3. a) $t = \frac{2v_0 \sin \alpha}{g} = \frac{2(500 \text{ m/s})\sin 45°}{9.8 \text{ m/s}^2} = 72.2$ seconds. $R = \frac{v_0^2}{g}\sin 2\alpha = \frac{(500 \text{ m/s})^2}{9.8 \text{ m/s}^2}(\sin 2(45°)) =$ 25 510.2 m

b) $x = (v_0 \cos \alpha)t \Rightarrow 5000 \text{ m} = (500 \text{ m/s})(\cos 45°)t \Rightarrow t = \frac{5000 \text{ m}}{(500 \text{ m/s})\cos 45°} = 14.14$ s

$y = (v_0 \sin \alpha)t - \frac{1}{2}gt^2 \Rightarrow y = (500 \text{ m/s})(\sin 45°)(14.14 \text{ s}) - \frac{1}{2}\left(9.8 \text{ m/s}^2\right)(14.14 \text{ s})^2 = 4020.3$ m

c) $y_{max} = \frac{\left(v_0 \sin \alpha\right)^2}{2g} = \frac{((500 \text{ m/s})\sin 45°)^2}{2\left(9.8 \text{ m/s}^2\right)} = 6377.6$ m

5. $R = \frac{v_0^2}{g}\sin 2\alpha = \frac{v_0^2}{g}\left(2 \sin \alpha \cos \alpha\right) = \frac{v_0^2}{g}\left(2 \cos(90° - \alpha) \sin(90° - \alpha)\right) = \frac{v_0^2}{g}\left(\sin 2(90° - \alpha)\right)$

7. $R = \frac{v_0^2}{g}\sin 2\alpha \Rightarrow 10 \text{ m} = \frac{v_0^2}{9.8 \text{ m/s}^2}\sin 2(45°) \Rightarrow v_0^2 = 98 \text{ m}^2/\text{s}^2 \Rightarrow v_0 = 9.9$ m/s.

$6 \text{ m} = \frac{(9.9 \text{ m/s})^2}{9.8 \text{ m/s}^2}\sin 2\alpha \Rightarrow \sin 2\alpha = 0.59999 \Rightarrow 2\alpha = 36.87° \text{ or } 143.12° \Rightarrow \alpha = 18.44° \text{ or } 71.56°$

9. a) $R = \frac{v_0^2}{g}\sin 2\alpha \Rightarrow (746.4 \text{ ft}) = \frac{v_0^2}{32 \text{ ft/sec}^2}\sin 2(9°) \Rightarrow v_0^2 = 77\,292.84 \text{ ft}^2/\text{sec}^2 \Rightarrow v_0 = 278.01 \text{ ft/sec} =$ 189.6 mph

b) Let $v_1 = 0$ ft/sec, $v_2 = 278.01$ ft/sec. The weight of the golf ball is $\frac{1.6}{16}$ lb $= 0.1$ lb $\Rightarrow$ the mass is $0.1\left(\frac{1}{32}\right)$ slug $= 0.003125$ slug. Then the work is $w = \frac{1}{2}mv_2^2 - \frac{1}{2}mv_1^2 = \frac{1}{2}(0.003125)(278.01)^2 - \frac{1}{2}(0.003125)(0)^2 = 120.8$ ft–lbs.

11. $y_{max} = \frac{\left(v_0 \sin \alpha\right)^2}{2g} \Rightarrow \frac{3}{4}y_{max} = \frac{3\left(v_0 \sin \alpha\right)^2}{8g}$. $y = \left(v_0 \sin \alpha\right)t - \frac{1}{2}gt^2 \Rightarrow \frac{3\left(v_0 \sin \alpha\right)^2}{8g} = \left(v_0 \sin \alpha\right)t - \frac{1}{2}gt^2 \Rightarrow 3\left(v_0 \sin \alpha\right)^2 = \left(8gv_0 \sin \alpha\right)t - 4g^2t^2 \Rightarrow$

$4g^2t^2 - \left(8gv_0 \sin \alpha\right)t + 3\left(v_0 \sin \alpha\right)^2 = 0 \Rightarrow 2gt - 3v_0 \sin \alpha = 0 \text{ or } 2gt - v_0 \sin \alpha = 0 \Rightarrow t = \frac{3v_0 \sin \alpha}{2g}$ or $t = \frac{v_0 \sin \alpha}{2g}$. Since the time it takes to reach y_{max} is $t_{max} = \frac{v_0 \sin \alpha}{g}$, then the time it takes the projectile to reach $\frac{3}{4}$ of y_{max} is $t = \frac{v_0 \sin \alpha}{2g}$ or $\frac{1}{2}t_{max}$.

13. $x = \left(v_0 \cos \alpha\right)t \Rightarrow 135 \text{ ft} = (90 \text{ ft/sec})(\cos 30°)t \Rightarrow t = 1.732 \text{ sec}.$ $y = \left(v_0 \sin \alpha\right)t - \frac{1}{2}gt^2 \Rightarrow$

$y = (90 \text{ ft/sec})(\sin 30°)(1.732 \text{ sec}) - \frac{1}{2}\left(32 \text{ ft/sec}^2\right)(1.732 \text{ sec})^2 \Rightarrow y = 29.94 \text{ ft}.$ The golf ball will clip the leaves at the top.

15. $x = 0 + (44 \cos 40°)\,t = 33.706\,t,\ y = 8 + (44 \sin 40°)\,t - 16t^2 = 8 + 28.283\,t - 16t^2.$ $y = 0 \Rightarrow$

$t = \frac{28.283 + \sqrt{(28.283)^2 + 512}}{32} = 2.016$ sec, the positive answer. Then $x = 33.706(2.016) = 67.95 \text{ ft} \Rightarrow$

the difference in distances is $67.95 - 67.64 = 0.31$ ft or 3.72 inches.

17. $x = \left(v_0 \cos \alpha\right) t \Rightarrow 315 \text{ ft} = (v_0 \cos 20°)t \Rightarrow v_0 = \frac{315}{t \cos 20°}.$ $y = \left(v_0 \sin \alpha\right)t - \frac{1}{2}gt^2 \Rightarrow$

$34 \text{ ft} = \frac{315}{t \cos 20°}(t \sin 20°) - \frac{1}{2}(32)t^2 \Rightarrow 34 = 315 \tan 20° - 16t^2 \Rightarrow t^2 = 5.04 \text{ sec}^2 \Rightarrow t = 2.25 \text{ sec}$

$t = 2.25 \text{ sec} \Rightarrow v_0 = \frac{315}{(2.25)\cos 20°} = 148.98 \text{ ft/sec}$

19. Height of the Marble A, R units downrange: $x = \left(v_0 \cos \alpha\right) t$ and $x = R \Rightarrow R = \left(v_0 \cos \alpha\right) t \Rightarrow$

$t = \frac{R}{v_0 \cos \alpha}.$ $y = \left(v_0 \sin \alpha\right)t - \frac{1}{2}gt^2 \Rightarrow y = \left(v_0 \sin \alpha\right)\left(\frac{R}{v_0 \cos \alpha}\right) - \frac{1}{2}g\left(\frac{R}{v_0 \cos \alpha}\right)^2 \Rightarrow$

$y = R \tan \alpha - \frac{1}{2}g\left(\frac{R^2}{v_0^2 \cos^2\alpha}\right)$ is the height of Marble A after $t = \frac{R}{v_0 \cos \alpha}$ seconds.

Height of the Marble B, at $t = \frac{R}{v_0 \cos \alpha}$ seconds: $y = R \tan \alpha - \frac{1}{2}gt^2 = R \tan \alpha - \frac{1}{2}g\left(\frac{R}{v_0 \cos \alpha}\right)^2 =$

$R \tan \alpha - \frac{1}{2}g\left(\frac{R^2}{v_0^2 \cos^2\alpha}\right)$ which is the height of Marble A. $\therefore$ They collide regardless of the initial velocity.

21. $\mathbf{a}(t) = -g\,\mathbf{k} \Rightarrow \mathbf{v}(t) = -gt\,\mathbf{k} + \mathbf{C}.$ $\mathbf{v}(0) = \mathbf{v}_0 \Rightarrow \mathbf{C} = \mathbf{v}_0.$ Then $\mathbf{v}(t) = -gt\,\mathbf{k} + \mathbf{v}_0 \Rightarrow \mathbf{r}(t) = -\frac{1}{2}gt^2\,\mathbf{k} + \mathbf{v}_0 t + \mathbf{C}.$

$\mathbf{r}(0) = \mathbf{0} \Rightarrow \mathbf{C} = \mathbf{0}.$ $\therefore\ \mathbf{r}(t) = -\frac{1}{2}gt^2\,\mathbf{k} + \mathbf{v}_0 t$

23. a)

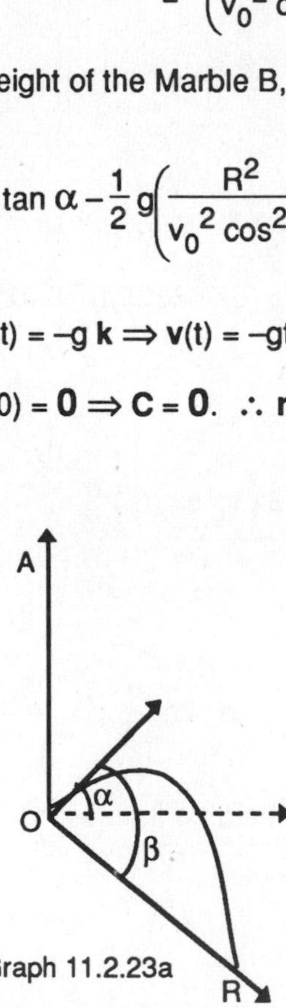

Graph 11.2.23a

$\tan \beta = \frac{y}{x}.$ $x = \left(v_0 \cos \alpha\right) t$ and $y = \left(v_0 \sin \alpha\right)t - \frac{1}{2}gt^2 \Rightarrow \tan \beta =$

$$\frac{\left(v_0 \sin \alpha\right)t - \frac{1}{2}gt^2}{\left(v_0 \cos \alpha\right) t} = \frac{(v_0 \sin \alpha) - \frac{1}{2}gt}{\left(v_0 \cos \alpha\right)} \Rightarrow v_0 \cos \alpha \tan \beta = v_0 \sin \alpha - \frac{1}{2}gt \Rightarrow$$

$t = \frac{2v_0 \sin \alpha - 2v_0 \cos \alpha \tan \beta}{g}.$ This is the time when the projectile will hit the

downhill slope. $\therefore\ x = v_0 \cos \alpha\left(\frac{2v_0 \sin \alpha - 2v_0 \cos \alpha \tan \beta}{g}\right) =$

23. a) (Continued)

$\frac{2v_0^2}{g}\left(\sin\alpha\cos\alpha - \cos^2\alpha\tan\beta\right)$. If x is maximized, then OR is maximized.

$\frac{dx}{d\alpha} = \frac{2v_0^2}{g}(\cos 2\alpha + \sin 2\alpha\tan\beta)$. $\frac{dx}{d\alpha} = 0 \Rightarrow \cos 2\alpha + \sin 2\alpha\tan\beta = 0 \Rightarrow \cot 2\alpha + \tan\beta = 0 \Rightarrow$

$\cot 2\alpha = -\tan\beta = \tan(-\beta) \Rightarrow 2\alpha = 90° - (-\beta) = 90° + \beta \Rightarrow \alpha = \frac{1}{2}\left(90° + \beta\right) = \frac{1}{2}$ of $\angle$ AOR.

b)

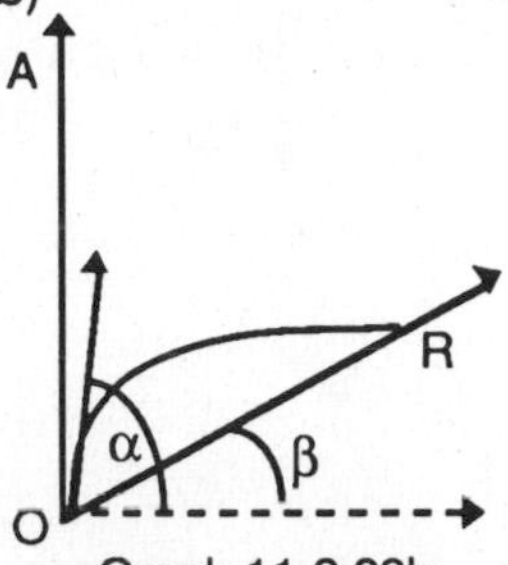

Graph 11.2.23b

By the same reasoning in 18 a), $\tan\beta = \frac{y}{x}$. $x = \left(v_0\cos\alpha\right)t$ and

$y = \left(v_0\sin\alpha\right)t - \frac{1}{2}gt^2 \Rightarrow \tan\beta = \frac{\left(v_0\sin\alpha\right)t - \frac{1}{2}gt^2}{\left(v_0\cos\alpha\right)t} = \frac{(v_0\sin\alpha) - \frac{1}{2}gt}{\left(v_0\cos\alpha\right)} \Rightarrow$

$v_0\cos\alpha\tan\beta = v_0\sin\alpha - \frac{1}{2}gt \Rightarrow t = \frac{2v_0\sin\alpha - 2v_0\cos\alpha\tan\beta}{g}$.

This is the time when the projectile will hit the uphill slope.

$\therefore x = v_0\cos\alpha\left(\frac{2v_0\sin\alpha - 2v_0\cos\alpha\tan\beta}{g}\right) =$

$\frac{2v_0^2}{g}\left(\sin\alpha\cos\alpha - \cos^2\alpha\tan\beta\right)$. If x is maximized, then OR is maximized.

$\frac{dx}{d\alpha} = \frac{2v_0^2}{g}(\cos 2\alpha + \sin 2\alpha\tan\beta)$. $\frac{dx}{d\alpha} = 0 \Rightarrow \cos 2\alpha + \sin 2\alpha\tan\beta = 0 \Rightarrow \cot 2\alpha + \tan\beta = 0 \Rightarrow$

$\cot 2\alpha = -\tan\beta = \tan(-\beta) \Rightarrow 2\alpha = 90° - (-\beta) = 90° + \beta \Rightarrow \alpha = \frac{1}{2}\left(90° + \beta\right) = \frac{1}{2}(90°) + \frac{1}{2}\beta =$

$\frac{1}{2}(90°) + \beta - \frac{1}{2}\beta = \beta + \frac{1}{2}(90° - \beta)$. Now $\frac{1}{2}(90° - \beta) =$

$\frac{1}{2}$ of $\angle$ AOR. $\therefore$ v_0 would bisect $\angle$ AOR for maximum range uphill.

SECTION 11.3 DIRECTED DISTANCE AND THE UNIT TANGENT VECTOR T

1. $\mathbf{r} = (2\cos t)\,\mathbf{i} + (2\sin t)\,\mathbf{j} + \sqrt{5}\,t\,\mathbf{k} \Rightarrow \mathbf{v} = (-2\sin t)\,\mathbf{i} + (2\cos t)\,\mathbf{j} + \sqrt{5}\,\mathbf{k} \Rightarrow$

$|\mathbf{v}| = \sqrt{(-2\sin t)^2 + (2\cos t)^2 + \left(\sqrt{5}\right)^2} = \sqrt{4\sin^2 t + 4\cos^2 t + 5} = 3$. $\mathbf{T} = \frac{\mathbf{v}}{|\mathbf{v}|} = \left(-\frac{2}{3}\sin t\right)\mathbf{i} +$

$\left(\frac{2}{3}\cos t\right)\mathbf{j} + \frac{\sqrt{5}}{3}\mathbf{k}$. Length $= \int_0^{\pi} |\mathbf{v}|\,dt = \int_0^{\pi} 3\,dt = [3t]_0^{\pi} = 3\pi$

3. $\mathbf{r} = t\,\mathbf{i} + \frac{2}{3}t^{3/2}\,\mathbf{k} \Rightarrow \mathbf{v} = \mathbf{i} + t^{1/2}\,\mathbf{k} \Rightarrow |\mathbf{v}| = \sqrt{1^2 + \left(t^{1/2}\right)^2} = \sqrt{1+t}$. $\mathbf{T} = \frac{\mathbf{v}}{|\mathbf{v}|} = \frac{1}{\sqrt{1+t}}\mathbf{i} + \frac{\sqrt{t}}{\sqrt{1+t}}\mathbf{k}$

Length $= \int_0^{8} \sqrt{1+t}\,dt = \left[\frac{2}{3}(1+t)^{3/2}\right]_0^8 = \frac{52}{3}$

5. $\mathbf{r} = (2 + t)\,\mathbf{i} - (t + 1)\,\mathbf{j} + t\,\mathbf{k} \Rightarrow \mathbf{v} = \mathbf{i} - \mathbf{j} + \mathbf{k} \Rightarrow |\mathbf{v}| = \sqrt{1^2 + (-1)^2 + 1^2} = \sqrt{3}$. $\mathbf{T} = \frac{\mathbf{v}}{|\mathbf{v}|} = \frac{1}{\sqrt{3}}\mathbf{i} - \frac{1}{\sqrt{3}}\mathbf{j} + \frac{1}{\sqrt{3}}\mathbf{k}$

Length $= \int_0^3 \sqrt{3}\,dt = \left[\sqrt{3}\,t\right]_0^3 = 3\sqrt{3}$

7. $\mathbf{r} = (t\cos t)\,\mathbf{i} + (t\sin t)\,\mathbf{j} + \frac{2\sqrt{2}}{3}t^{3/2}\,\mathbf{k} \Rightarrow \mathbf{v} = (\cos t - t\sin t)\,\mathbf{i} + (\sin t + t\cos t)\,\mathbf{j} + (\sqrt{2}\,t^{1/2})\,\mathbf{k} \Rightarrow$
$|\mathbf{v}| = \sqrt{(\cos t - t\sin t)^2 + (\sin t + t\cos t)^2 + (\sqrt{2}\,t^{1/2})^2} = \sqrt{1 + t^2 + 2t} = \sqrt{(t+1)^2} = |t + 1| = t + 1$ since
$t \ge 0$. $\mathbf{T} = \frac{\mathbf{v}}{|\mathbf{v}|} = \left(\frac{\cos t - t\sin t}{t+1}\right)\mathbf{i} + \left(\frac{\sin t + t\cos t}{t+1}\right)\mathbf{j} + \left(\frac{\sqrt{2}\,t^{1/2}}{t+1}\right)\mathbf{k}$. Length $= \int_0^\pi (t + 1)\,dt =$
$\left[\frac{t^2}{2} + t\right]_0^\pi = \frac{\pi^2}{2} + \pi$

9. $\mathbf{r} = (4\cos t)\,\mathbf{i} + (4\sin t)\,\mathbf{j} + 3t\,\mathbf{k} \Rightarrow \mathbf{v} = (-4\sin t)\,\mathbf{i} + (4\cos t)\,\mathbf{j} + 3\,\mathbf{k} \Rightarrow |\mathbf{v}| = \sqrt{(-4\sin t)^2 + (4\cos t)^2 + 3^2}$
$= \sqrt{25} = 5$. $s(t) = \int_0^t 5\,d\tau = 5t$ Length $= s\left(\frac{\pi}{2}\right) = \frac{5\pi}{2}$

11. $\mathbf{r} = (e^t\cos t)\,\mathbf{i} + (e^t\sin t)\,\mathbf{j} + e^t\,\mathbf{k} \Rightarrow \mathbf{v} = (e^t\cos t - e^t\sin t)\,\mathbf{i} + (e^t\sin t + e^t\cos t)\,\mathbf{j} + e^t\,\mathbf{k} \Rightarrow$
$|\mathbf{v}| = \sqrt{(e^t\cos t - e^t\sin t)^2 + (e^t\sin t + e^t\cos t)^2 + (e^t)^2} = \sqrt{3e^{2t}} = \sqrt{3}\,e^t$
$s(t) = \int_0^t \sqrt{3}\,e^\tau\,d\tau = \sqrt{3}\,e^t - \sqrt{3}$. Length $= s(-\ln 4) = \sqrt{3}\,e^{-\ln 4} - \sqrt{3} = \frac{\sqrt{3}}{4} - \sqrt{3} = -\frac{3\sqrt{3}}{4}$

13. $\mathbf{r} = (\sqrt{2}\,t)\,\mathbf{i} + (\sqrt{2}\,t)\,\mathbf{j} + (1 - t^2)\,\mathbf{k} \Rightarrow \mathbf{v} = \sqrt{2}\,\mathbf{i} + \sqrt{2}\,\mathbf{j} - 2t\,\mathbf{k} \Rightarrow |\mathbf{v}| = \sqrt{(\sqrt{2})^2 + (\sqrt{2})^2 + (-2t)^2} = \sqrt{4 + 4t^2}$
$= 2\sqrt{1 + t^2}$ Length $= \int_0^1 2\sqrt{1 + t^2}\,dt = \left[2\left(\frac{t}{2}\sqrt{1 + t^2} + \frac{1}{2}\ln\left(t + \sqrt{1 + t^2}\right)\right)\right]_0^1 = \sqrt{2} + \ln\left(1 + \sqrt{2}\right)$

15. a) $\mathbf{r} = (\cos t)\,\mathbf{i} + (\sin t)\,\mathbf{j} + (1 - \cos t)\,\mathbf{k}$, $0 \le t \le 2\pi \Rightarrow x = \cos t$, $y = \sin t$, $z = 1 - \cos t$. Then $x^2 + y^2 = \cos^2 t + \sin^2 t = 1$, a right circular cylinder z–axis as the axis, radius = 1. ∴ $P(\cos t, \sin t, 1 - \cos t)$ lie on the cylinder $x^2 + y^2 = 1$. $t = 0 \Rightarrow P(1,0,0)$ is on the curve. $t = \frac{\pi}{2} \Rightarrow Q(0,1,1)$ is on the curve. $t = \pi \Rightarrow$ $R(-1,0,2)$ is on the curve. Then $\overrightarrow{PQ} - \mathbf{i} + \mathbf{j} + \mathbf{k}$, $\overrightarrow{PR} = -2\mathbf{i} + 2\mathbf{k}$. Thus, $\overrightarrow{PQ} \times \overrightarrow{PR} = \begin{vmatrix} \mathbf{i} & \mathbf{j} & \mathbf{k} \\ -1 & 1 & 1 \\ -2 & 0 & 2 \end{vmatrix} =$
$2\mathbf{i} + 2\mathbf{k}$ is a vector normal to the plane of P, Q, and R. Then the plane containing P, Q, and R has an equation $2x + 2z = 2(1) + 2(0)$ or $x + z = 1$. Any point on the curve will satisfy this equation since $x + z = \cos t + (1 - \cos t) = 1$. ∴ any point on the curve lies on the intersection of the cylinder $x^2 + y^2 = 1$

15. a) (Continued)

and the plane x + z = 1 ⇒ the curve is an ellipse.

b)

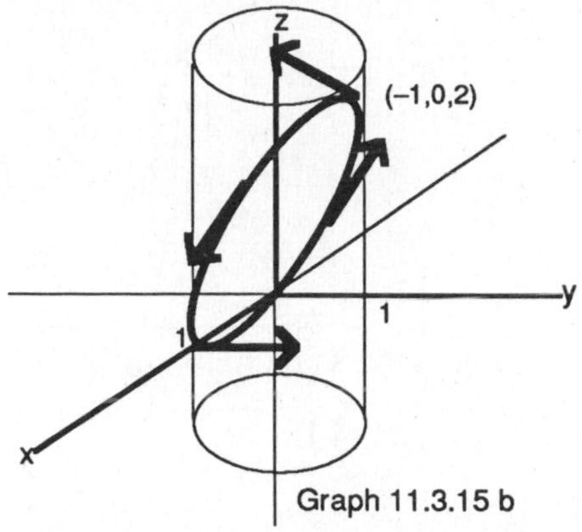

Graph 11.3.15 b

$\mathbf{v} = (-\sin t)\,\mathbf{i} + (\cos t)\,\mathbf{j} + (\sin t)\,\mathbf{k} \Rightarrow$

$|\mathbf{v}| = \sqrt{\sin^2 t + \cos^2 t + \sin^2 t} = \sqrt{1 + \sin^2 t} \Rightarrow$

$\mathbf{T} = \frac{\mathbf{v}}{|\mathbf{v}|} = \frac{(-\sin t)\,\mathbf{i} + (\cos t)\,\mathbf{j} + (\sin t)\,\mathbf{k}}{\sqrt{1 + \sin^2 t}} \Rightarrow \mathbf{T}_{t=0} = \mathbf{j},\ \mathbf{T}_{t=\pi/2} =$

$\frac{-\mathbf{i} + \mathbf{j}}{\sqrt{2}},\ \mathbf{T}_{t=\pi} = -\mathbf{j},\ \mathbf{T}_{t=3\pi/2} = \frac{\mathbf{i} - \mathbf{k}}{\sqrt{2}}$

c)

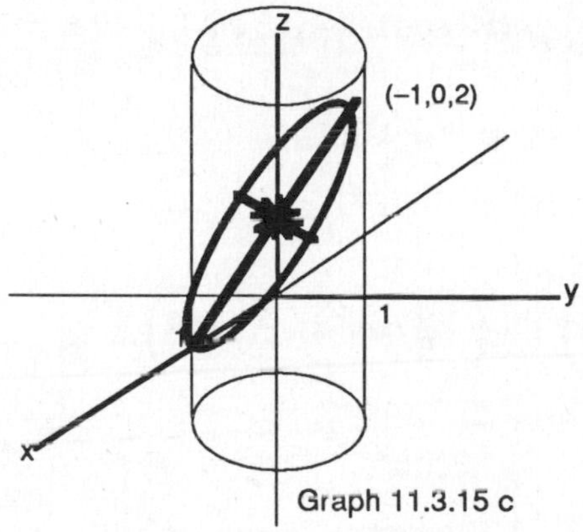

Graph 11.3.15 c

$\mathbf{a} = (-\cos t)\,\mathbf{i} - (\sin t)\,\mathbf{j} + (\cos t)\,\mathbf{k}$. $\mathbf{N} = \mathbf{i} + \mathbf{k}$ is normal to the plane x + z = 1. $\mathbf{N} \cdot \mathbf{a} = -\cos t + \cos t = 0 \Rightarrow \mathbf{a}$ is orthogonal to $\mathbf{N} \Rightarrow \mathbf{a}$ is parallel to the plane. $\mathbf{a}_{t=0} = -\mathbf{i} + \mathbf{k}$, $\mathbf{a}_{t=\pi/2} = -\mathbf{j}$, $\mathbf{a}_{t=\pi} = \mathbf{i} - \mathbf{k}$, $\mathbf{a}_{t=3\pi/2} = \mathbf{j}$

d) $|\mathbf{v}| = \sqrt{1 + \sin^2 t}$ (See part b). $\therefore$ $L = \int_0^{2\pi} \sqrt{1 + \sin^2 t}\, dt$

e) L ≈ 7.64 (by *Mathematica*)

SECTION 11.4 CURVATURE, TORSION, AND THE FRENET FRAME

1. $\mathbf{r} = t\,\mathbf{i} + \ln(\cos t)\,\mathbf{j} \Rightarrow \mathbf{v} = \mathbf{i} + \frac{-\sin t}{\cos t}\mathbf{j} = \mathbf{i} - \tan t\,\mathbf{j} \Rightarrow |\mathbf{v}| = \sqrt{1^2 + (-\tan t)^2} = \sqrt{\sec^2 t} = |\sec t| = \sec t$ since $-\frac{\pi}{2} < t < \frac{\pi}{2}$. $\mathbf{T} = \frac{\mathbf{v}}{|\mathbf{v}|} = \frac{1}{\sec t}\mathbf{i} - \frac{\tan t}{\sec t}\mathbf{j} = \cos t\,\mathbf{i} - \sin t\,\mathbf{j}$. $\frac{d\mathbf{T}}{dt} = -\sin t\,\mathbf{i} - \cos t\,\mathbf{j} \Rightarrow$ $\left|\frac{d\mathbf{T}}{dt}\right| = \sqrt{(-\sin t)^2 + (-\cos t)^2} = 1$. $\mathbf{N} = \frac{d\mathbf{T}/dt}{|d\mathbf{T}/dt|} = (-\sin t)\,\mathbf{i} - (\cos t)\,\mathbf{j}$. $\mathbf{a} = (-\sec^2 t)\,\mathbf{j} \Rightarrow$

$$\mathbf{v} \times \mathbf{a} = \begin{vmatrix} \mathbf{i} & \mathbf{j} & \mathbf{k} \\ 1 & -\tan t & 0 \\ 0 & -\sec^2 t & 0 \end{vmatrix} = (-\sec^2 t)\,\mathbf{k}.\ |\mathbf{v} \times \mathbf{a}| = \sqrt{(-\sec^2 t)^2} = \sec^2 t \Rightarrow \kappa = \frac{|\mathbf{v} \times \mathbf{a}|}{|\mathbf{v}|^3} = \frac{\sec^2 t}{\sec^3 t} = \cos t$$

3. $\mathbf{r} = (2t + 3)\,\mathbf{i} + \left(5 - t^2\right)\mathbf{j} \Rightarrow \mathbf{v} = 2\,\mathbf{i} - 2t\,\mathbf{j} \Rightarrow |\mathbf{v}| = \sqrt{2^2 + (-2t)^2} = 2\sqrt{1 + t^2}$. $\mathbf{T} = \frac{\mathbf{v}}{|\mathbf{v}|} = \frac{2}{2\sqrt{1+t^2}}\mathbf{i} + \frac{-2t}{2\sqrt{1+t^2}}\mathbf{j} = \frac{1}{\sqrt{1+t^2}}\mathbf{i} - \frac{t}{\sqrt{1+t^2}}\mathbf{j}$. $\frac{d\mathbf{T}}{dt} = = \frac{-t}{\left(\sqrt{1+t^2}\right)^3}\mathbf{i} - \frac{1}{\left(\sqrt{1+t^2}\right)^3}\mathbf{j} \Rightarrow$

3. (Continued)

$\left|\frac{dT}{dt}\right| = \sqrt{\left(\frac{-t}{\left(\sqrt{1+t^2}\right)^3}\right)^2 + \left(-\frac{1}{\left(\sqrt{1+t^2}\right)^3}\right)^2} = \sqrt{\frac{1}{(1+t^2)^2}} = \frac{1}{1+t^2}$. $\mathbf{N} = \frac{dT/dt}{|dT/dt|} = \frac{-t}{\sqrt{1+t^2}}\mathbf{i}$ $-\frac{1}{\sqrt{1+t^2}}\mathbf{j}$. $\mathbf{a} = -2\,\mathbf{j} \Rightarrow \mathbf{v} \times \mathbf{a} = \begin{vmatrix} \mathbf{i} & \mathbf{j} & \mathbf{k} \\ 2 & -2t & 0 \\ 0 & -2 & 0 \end{vmatrix} = -4\,\mathbf{k} \Rightarrow |\mathbf{v} \times \mathbf{a}| = \sqrt{(-4)^2} = 4$. $\kappa = \frac{|\mathbf{v} \times \mathbf{a}|}{|\mathbf{v}|^3} =$ $\frac{4}{\left(2\sqrt{1+t^2}\right)^3} = \frac{1}{2\left(\sqrt{1+t^2}\right)^3}$

5. $\mathbf{r} = (3\sin t)\,\mathbf{i} + (3\cos t)\,\mathbf{j} + 4t\,\mathbf{k} \Rightarrow \mathbf{v} = (3\cos t)\,\mathbf{i} + (-3\sin t)\,\mathbf{j} + 4\,\mathbf{k} \Rightarrow |\mathbf{v}| = \sqrt{(3\cos t)^2 + (-3\sin t)^2 + 4^2}$ $= \sqrt{25} = 5$. $\mathbf{T} = \frac{\mathbf{v}}{|\mathbf{v}|} = \frac{3\cos t}{5}\mathbf{i} - \frac{3\sin t}{5}\mathbf{j} + \frac{4}{5}\mathbf{k} \Rightarrow \frac{dT}{dt} = \left(-\frac{3}{5}\sin t\right)\mathbf{i} - \left(\frac{3}{5}\cos t\right)\mathbf{j}$

$\left|\frac{dT}{dt}\right| = \sqrt{\left(-\frac{3}{5}\sin t\right)^2 + \left(-\frac{3}{5}\cos t\right)^2} = \frac{3}{5}$. $\mathbf{N} = \frac{dT/dt}{|dT/dt|} = (-\sin t)\,\mathbf{i} - (\cos t)\,\mathbf{j}$

$\mathbf{a} = (-3\sin t)\,\mathbf{i} + (-3\cos t)\,\mathbf{j} \Rightarrow \mathbf{v} \times \mathbf{a} = \begin{vmatrix} \mathbf{i} & \mathbf{j} & \mathbf{k} \\ 3\cos t & -3\sin t & 4 \\ -3\sin t & -3\cos t & 0 \end{vmatrix} = (12\cos t)\,\mathbf{i} - (12\sin t)\,\mathbf{j} - 9\,\mathbf{k} \Rightarrow$

$|\mathbf{v} \times \mathbf{a}| = \sqrt{(12\cos t)^2 + (-12\sin t)^2 + (-9)^2} = \sqrt{225} = 15$. $\kappa = \frac{|\mathbf{v} \times \mathbf{a}|}{|\mathbf{v}|^3} = \frac{15}{5^3} = \frac{3}{25}$

$\mathbf{B} = \mathbf{T} \times \mathbf{N} = \begin{vmatrix} \mathbf{i} & \mathbf{j} & \mathbf{k} \\ \frac{3}{5}\cos t & -\frac{3}{5}\sin t & \frac{4}{5} \\ -\sin t & -\cos t & 0 \end{vmatrix} = \left(\frac{4}{5}\cos t\right)\mathbf{i} - \left(\frac{4}{5}\sin t\right)\mathbf{j} + \left(-\frac{3}{5}\cos^2 t - \frac{3}{5}\sin^2 t\right)\mathbf{k} =$

$\left(\frac{4}{5}\cos t\right)\mathbf{i} - \left(\frac{4}{5}\sin t\right)\mathbf{j} - \frac{3}{5}\mathbf{k}$. $\dot{\mathbf{a}} = (-3\cos t)\,\mathbf{i} + (3\sin t)\,\mathbf{j} \Rightarrow \tau = \frac{\begin{vmatrix} 3\cos t & -3\sin t & 4 \\ -3\sin t & -3\cos t & 0 \\ -3\cos t & 3\sin t & 0 \end{vmatrix}}{|\mathbf{v} \times \mathbf{a}|^2} =$ $\frac{-36\sin^2 t - 36\cos^2 t}{15^2} = -\frac{4}{25}$

7. $\mathbf{r} = (e^t\cos t)\,\mathbf{i} + (e^t\sin t)\,\mathbf{j} + 2\,\mathbf{k} \Rightarrow \mathbf{v} = (e^t\cos t - e^t\sin t)\,\mathbf{i} + (e^t\sin t + e^t\cos t)\,\mathbf{j} \Rightarrow$

$|\mathbf{v}| = \sqrt{(e^t\cos t - e^t\sin t)^2 + (e^t\sin t + e^t\cos t)^2} = \sqrt{2e^{2t}} = e^t\sqrt{2}$.

$\mathbf{a} = \left(e^t(\cos t - \sin t) + e^t(-\sin t - \cos t)\right)\mathbf{i} + \left(e^t(\sin t + \cos t) + e^t(\cos t - \sin t)\right)\mathbf{j} =$

$(-2e^t\sin t)\,\mathbf{i} + (2e^t\cos t)\,\mathbf{j} \Rightarrow \mathbf{v} \times \mathbf{a} = \begin{vmatrix} \mathbf{i} & \mathbf{j} & \mathbf{k} \\ e^t\cos t - e^t\sin t & e^t\sin t + e^t\cos t & 0 \\ -2e^t\sin t & 2e^t\cos t & 0 \end{vmatrix} = (2e^{2t})\,\mathbf{k} \Rightarrow$

$|\mathbf{v} \times \mathbf{a}| = \sqrt{(2e^{2t})^2} = 2e^{2t}$. $\kappa = \frac{|\mathbf{v} \times \mathbf{a}|}{|\mathbf{v}|^3} = \frac{2e^{2t}}{\left(e^t\sqrt{2}\right)^3} = \frac{1}{e^t\sqrt{2}}$.

$\mathbf{T} = \frac{\mathbf{v}}{|\mathbf{v}|} = \left(\frac{e^t\cos t - e^t\sin t}{e^t\sqrt{2}}\right)\mathbf{i} + \left(\frac{e^t\cos t + e^t\sin t}{e^t\sqrt{2}}\right)\mathbf{j} \Rightarrow \frac{dT}{dt} = \left(\frac{-\sin t - \cos t}{\sqrt{2}}\right)\mathbf{i} + \left(\frac{\cos t - \sin t}{\sqrt{2}}\right)\mathbf{j} \Rightarrow$

7. (Continued)

$$\left|\frac{dT}{dt}\right| = \sqrt{\left(\frac{-\sin t - \cos t}{\sqrt{2}}\right)^2 + \left(\frac{\cos t - \sin t}{\sqrt{2}}\right)^2} = 1.\ \mathbf{N} = \frac{dT/dt}{|dT/dt|} = \left(\frac{-\cos t - \sin t}{\sqrt{2}}\right)\mathbf{i} + \left(\frac{-\sin t + \cos t}{\sqrt{2}}\right)\mathbf{j}$$

$\dot{\mathbf{a}} = (-2e^t\sin t - 2e^t\cos t)\,\mathbf{i} + (2e^t\cos t - 2e^t\sin t)\,\mathbf{j} \Rightarrow$

$$\tau = \frac{\begin{vmatrix} e^t\cos t - e^t\sin t & e^t\sin t + e^t\cos t & 0 \\ -2e^t\sin t & 2e^t\cos t & 0 \\ -2e^t\sin t - 2e^t\cos t & 2e^t\cos t - 2e^t\sin t & 0 \end{vmatrix}}{|\mathbf{v} \times \mathbf{a}|^2} = 0.$$

$$\mathbf{B} = \mathbf{T} \times \mathbf{N} = \begin{vmatrix} \mathbf{i} & \mathbf{j} & \mathbf{k} \\ \frac{\cos t - \sin t}{\sqrt{2}} & \frac{\sin t + \cos t}{\sqrt{2}} & 0 \\ \frac{-\cos t - \sin t}{\sqrt{2}} & \frac{-\sin t + \cos t}{\sqrt{2}} & 0 \end{vmatrix} = \mathbf{k}$$

9. $\mathbf{r}(t) = (t^3/3)\,\mathbf{i} + (t^2/2)\,\mathbf{j},\ t > 0 \Rightarrow \mathbf{v} = t^2\,\mathbf{i} + t\,\mathbf{j} \Rightarrow |\mathbf{v}| = \sqrt{t^4 + t^2} = t\sqrt{t^2+1}$ since $t > 0$. $\mathbf{T} = \frac{\mathbf{v}}{|\mathbf{v}|} = \frac{t^2\,\mathbf{i} + t\,\mathbf{j}}{t\sqrt{t^2+1}} =$

$$\frac{t}{\sqrt{t^2+1}}\mathbf{i} + \frac{\mathbf{j}}{\sqrt{t^2+1}} \Rightarrow \frac{dT}{dt} = \frac{\mathbf{i}}{(t^2+1)^{3/2}} - \frac{t\,\mathbf{j}}{(t^2+1)^{3/2}} \Rightarrow \left|\frac{dT}{dt}\right| = \sqrt{\left(\frac{1}{(t^2+1)^{3/2}}\right)^2 + \left(\frac{-t}{(t^2+1)^{3/2}}\right)^2} =$$

$$\sqrt{\frac{1+t^2}{(t^2+1)^3}} = \frac{1}{t^2+1}.\ \mathbf{N} = \frac{dT/dt}{|dT/dt|} = \frac{\mathbf{i}}{\sqrt{t^2+1}} - \frac{t\,\mathbf{j}}{\sqrt{t^2+1}}$$

$$\mathbf{a} = 2t\,\mathbf{i} + \mathbf{j} \Rightarrow \mathbf{v} \times \mathbf{a} = \begin{vmatrix} \mathbf{i} & \mathbf{j} & \mathbf{k} \\ t^2 & t & 0 \\ 2t & 1 & 0 \end{vmatrix} = -t^2\,\mathbf{k} \Rightarrow |\mathbf{v} \times \mathbf{a}| = \sqrt{(-t^2)^2} = t^2.\ \therefore\ \kappa = \frac{|\mathbf{v} \times \mathbf{a}|}{|\mathbf{v}|^3} = \frac{t^2}{\left(t\sqrt{t^2+1}\right)^3} =$$

$$\frac{1}{t(t^2+1)^{3/2}}.\ \mathbf{B} = \mathbf{T} \times \mathbf{N} = \begin{vmatrix} \mathbf{i} & \mathbf{j} & \mathbf{k} \\ \frac{t}{\sqrt{t^2+1}} & \frac{1}{\sqrt{t^2+1}} & 0 \\ \frac{1}{\sqrt{t^2+1}} & \frac{-t}{\sqrt{t^2+1}} & 0 \end{vmatrix} = -\mathbf{k}\ \ \dot{\mathbf{a}} = 2\,\mathbf{i} \Rightarrow \tau = \frac{\begin{vmatrix} t^2 & t & 0 \\ 2t & 1 & 0 \\ 2 & 0 & 0 \end{vmatrix}}{|\mathbf{v} \times \mathbf{a}|^2} = 0$$

11. $\mathbf{r} = t\,\mathbf{i} + \left(a\cosh\frac{t}{a}\right)\mathbf{j},\ a > 0 \Rightarrow \mathbf{v} = \mathbf{i} + \left(\sinh\frac{t}{a}\right)\mathbf{j} \Rightarrow |\mathbf{v}| = \sqrt{1 + \sinh^2\left(\frac{t}{a}\right)} = \sqrt{\cosh^2\left(\frac{t}{a}\right)} = \cosh\frac{t}{a}$

$$\mathbf{T} = \frac{\mathbf{v}}{|\mathbf{v}|} = \frac{\mathbf{i} + \left(\sinh\frac{t}{a}\right)\mathbf{j}}{\cosh\left(\frac{t}{a}\right)} = \left(\operatorname{sech}\frac{t}{a}\right)\mathbf{i} + \left(\tanh\frac{t}{a}\right)\mathbf{j} \Rightarrow \frac{dT}{dt} = \left(-\frac{1}{a}\operatorname{sech}\frac{t}{a}\tanh\frac{t}{a}\right)\mathbf{i} + \left(\frac{1}{a}\operatorname{sech}^2\frac{t}{a}\right)\mathbf{j} \Rightarrow$$

$$\left|\frac{dT}{dt}\right| = \sqrt{\frac{1}{a^2}\operatorname{sech}^2\frac{t}{a}\tanh^2\frac{t}{a} + \frac{1}{a^2}\operatorname{sech}^4\frac{t}{a}} = \frac{1}{a}\operatorname{sech}\frac{t}{a} \Rightarrow \mathbf{N} = \frac{dT/dt}{|dT/dt|} =$$

11. (Continued)

$$\frac{\left(-\frac{1}{a}\operatorname{sech}\frac{t}{a}\tanh\frac{t}{a}\right)\mathbf{i}+\left(\frac{i}{a}\operatorname{sech}^2\frac{t}{a}\right)\mathbf{j}}{\frac{1}{a}\operatorname{sech}\frac{t}{a}}=\left(-\tanh\frac{t}{a}\right)\mathbf{i}+\left(\operatorname{sech}\frac{t}{a}\right)\mathbf{j}.. \ \mathbf{a}=\left(\frac{1}{a}\cosh\frac{t}{a}\right)\mathbf{j}\Rightarrow$$

$$\mathbf{v}\times\mathbf{a}=\begin{vmatrix}\mathbf{i}&\mathbf{j}&\mathbf{k}\\1&\sinh\frac{t}{a}&0\\0&\frac{1}{a}\cosh\frac{t}{a}&0\end{vmatrix}=\left(\frac{1}{a}\cosh\frac{t}{a}\right)\mathbf{k}\Rightarrow|\mathbf{v}\times\mathbf{a}|=\sqrt{\frac{1}{a^2}\cosh^2\frac{t}{a}}=\frac{1}{a}\cosh\frac{t}{a}\Rightarrow$$

$$\kappa=\frac{|\mathbf{v}\times\mathbf{a}|}{|\mathbf{v}|^3}=\frac{\frac{1}{a}\cosh\frac{t}{a}}{\cosh^3\frac{t}{a}}=\frac{1}{a\cosh^2\frac{t}{a}}.\ \mathbf{B}=\mathbf{T}\times\mathbf{N}=\begin{vmatrix}\mathbf{i}&\mathbf{j}&\mathbf{k}\\\operatorname{sech}\frac{t}{a}&\tanh\frac{t}{a}&0\\-\tanh\frac{t}{a}&\operatorname{sech}\frac{t}{a}&0\end{vmatrix}=\mathbf{k}$$

$$\dot{\mathbf{a}}=\frac{1}{a^2}\sinh\frac{t}{a}\mathbf{j}\Rightarrow\tau=\frac{\begin{vmatrix}1&\sinh\frac{t}{a}&0\\0&\frac{1}{a}\cosh\frac{t}{a}&0\\0&\frac{1}{a^2}\sinh\frac{t}{a}&0\end{vmatrix}}{|\mathbf{v}\times\mathbf{a}|^2}=0$$

13. $\mathbf{r}=(2t+3)\mathbf{i}+(t^2-1)\mathbf{j}\Rightarrow\mathbf{v}=2\mathbf{i}+2t\mathbf{j}\Rightarrow|\mathbf{v}|=\sqrt{2^2+(2t)^2}=2\sqrt{1+t^2}$

$a_T=2\left(\frac{1}{2}\right)(1+t^2)^{-1/2}(2t)=\frac{2t}{\sqrt{1+t^2}}$; $\mathbf{a}=2\mathbf{j}\Rightarrow|\mathbf{a}|=2\Rightarrow a_N=\sqrt{|\mathbf{a}|^2-a_T^2}=\sqrt{2^2-\left(\frac{2t}{\sqrt{1+t^2}}\right)^2}$

$=\frac{2}{\sqrt{1+t^2}}$. $\therefore\ \mathbf{a}=\frac{2t}{\sqrt{1+t^2}}\mathbf{T}+\frac{2}{\sqrt{1+t^2}}\mathbf{N}.$

15. $\mathbf{r}=(a\cos t)\mathbf{i}+(a\sin t)\mathbf{j}+bt\mathbf{k}\Rightarrow\mathbf{v}=(-a\sin t)\mathbf{i}+(a\cos t)\mathbf{j}+b\mathbf{k}\Rightarrow|\mathbf{v}|=\sqrt{(-a\sin t)^2+(a\cos t)^2+b^2}$

$=\sqrt{a^2+b^2}$. $a_T=0$. $\mathbf{a}=(-a\cos t)\mathbf{i}+(-a\sin t)\mathbf{j}\Rightarrow|\mathbf{a}|=\sqrt{(-a\cos t)^2+(-a\sin t)^2}=\sqrt{a^2}=|a|$.

$a_N=\sqrt{|\mathbf{a}|^2-a_T^2}=\sqrt{|a|^2-0^2}=|a|=|a|$. $\therefore\ \mathbf{a}=(0)\mathbf{T}+|a|\mathbf{N}=|a|\mathbf{N}$

17. $\mathbf{r}=(t+1)\mathbf{i}+2t\mathbf{j}+t^2\mathbf{k}\Rightarrow\ \mathbf{v}=\mathbf{i}+2\mathbf{j}+2t\mathbf{k}\Rightarrow\mathbf{v}(1)=\mathbf{i}+2\mathbf{j}+2\mathbf{k}\Rightarrow|\mathbf{v}|=\sqrt{1^2+2^2+(2t)^2}=$

$\sqrt{5+4t^2}$. $a_T=\frac{1}{2}\left(5+4t^2\right)^{-1/2}(8t)=4t\left(5+4t^2\right)^{-1/2}\Rightarrow a_T(1)=\frac{4}{\sqrt{9}}=\frac{4}{3}$. $\mathbf{a}=2\mathbf{k}\Rightarrow\mathbf{a}(1)=2\mathbf{k}\Rightarrow$

$|\mathbf{a}(1)|=2$. $a_N=\sqrt{|\mathbf{a}|^2-a_T^2}=\sqrt{2^2-\left(\frac{4}{3}\right)^2}=\frac{2\sqrt{5}}{3}$. $\therefore\ \mathbf{a}(1)=\frac{4}{3}\mathbf{T}+\frac{2\sqrt{5}}{3}\mathbf{N}$

19. $\mathbf{r}=t^2\mathbf{i}+(t+\frac{1}{3}t^3)\mathbf{j}+(t-\frac{1}{3}t^3)\mathbf{k}\Rightarrow\mathbf{v}=2t\mathbf{i}+(1+t^2)\mathbf{j}+(1-t^2)\mathbf{k}\Rightarrow|\mathbf{v}|=\sqrt{(2t)^2+(1+t^2)^2+(1-t^2)^2}$

$=\sqrt{2}\left(1+t^2\right)$. $a_T=2t\sqrt{2}\Rightarrow a_T(0)=0$. $\mathbf{a}=2\mathbf{i}+2t\mathbf{j}-2t\mathbf{k}\Rightarrow$

$\mathbf{a}(0)=2\mathbf{i}\Rightarrow|\mathbf{a}(0)|=2$. $a_N=\sqrt{|\mathbf{a}|^2-a_T^2}=\sqrt{2^2-0^2}=2$. $\therefore\ \mathbf{a}(0)=(0)\mathbf{T}+2\mathbf{N}=2\mathbf{N}$

21. $\mathbf{r} = (\cos t)\,\mathbf{i} + (\sin t)\,\mathbf{j} - \mathbf{k} \Rightarrow \mathbf{v} = (-\sin t)\,\mathbf{i} + (\cos t)\,\mathbf{j} \Rightarrow |\mathbf{v}| = \sqrt{(-\sin t)^2 + (\cos t)^2} = 1$. $\mathbf{T} = \frac{\mathbf{v}}{|\mathbf{v}|} =$ $(-\sin t)\,\mathbf{i} + (\cos t)\,\mathbf{j} \Rightarrow \mathbf{T}\left(\frac{\pi}{4}\right) = -\frac{\sqrt{2}}{2}\mathbf{i} + \frac{\sqrt{2}}{2}\mathbf{j}$. $\frac{d\mathbf{T}}{dt} = (-\cos t)\,\mathbf{i} - (\sin t)\,\mathbf{j} \Rightarrow \left|\frac{d\mathbf{T}}{dt}\right| = \sqrt{(-\cos t)^2 + (-\sin t)^2}$ $= 1$. $\mathbf{N} = \frac{d\mathbf{T}/dt}{|d\mathbf{T}/dt|} = (-\cos t)\,\mathbf{i} - (\sin t)\,\mathbf{j} \Rightarrow \mathbf{N}\left(\frac{\pi}{4}\right) = -\frac{\sqrt{2}}{2}\mathbf{i} - \frac{\sqrt{2}}{2}\mathbf{j}$. $\mathbf{r}\left(\frac{\pi}{4}\right) = \frac{\sqrt{2}}{2}\mathbf{i} + \frac{\sqrt{2}}{2}\mathbf{j} - \mathbf{k}$

$$\mathbf{B} = \mathbf{T} \times \mathbf{N} = \begin{vmatrix} \mathbf{i} & \mathbf{j} & \mathbf{k} \\ -\sin t & \cos t & 0 \\ -\cos t & -\sin t & 0 \end{vmatrix} = \mathbf{k} \Rightarrow \mathbf{B}\left(\frac{\pi}{4}\right) = \mathbf{k}.\ \ P = \left(\frac{\sqrt{2}}{2}, \frac{\sqrt{2}}{2}, -1\right)\left(\text{see } \mathbf{r}\left(\frac{\pi}{4}\right)\right) \text{ the}$$

osculating plane is $z = -1$ since $\mathbf{B}$ is the normal vector and $(0)x + (0)y + (1)z = (0)\left(\frac{\sqrt{2}}{2}\right) + (0)\left(\frac{\sqrt{2}}{2}\right) + (1)(-1)$.

The normal plane is $-x + y = 0$ since $\mathbf{T}$ is the normal vector and $-\frac{\sqrt{2}}{2}x + \frac{\sqrt{2}}{2}y + (0)z = \left(-\frac{\sqrt{2}}{2}\right)\left(\frac{\sqrt{2}}{2}\right) +$ $\left(\frac{\sqrt{2}}{2}\right)\left(\frac{\sqrt{2}}{2}\right) + (-1)(0) \Rightarrow -\frac{\sqrt{2}}{2}x + \frac{\sqrt{2}}{2}y = 0$. The rectifying plane is $x + y = \sqrt{2}$ since $\mathbf{N}$ is the normal vector and $-\frac{\sqrt{2}}{2}x - \frac{\sqrt{2}}{2}y + (0)z = \left(-\frac{\sqrt{2}}{2}\right)\left(\frac{\sqrt{2}}{2}\right) - \left(\frac{\sqrt{2}}{2}\right)\left(\frac{\sqrt{2}}{2}\right) + (-1)(0) \Rightarrow -\frac{\sqrt{2}}{2}x - \frac{\sqrt{2}}{2}y = -1$.

23. Yes, if the car is moving around a circle at a constant speed, the acceleration points along $\mathbf{N}$ toward the center of the circle and is of constant magnitude, but not $\mathbf{0}$.

25. If acceleration is perpendicular to the velocity, then $a_{\mathbf{T}} = 0 \Rightarrow |\mathbf{v}|$ is constant.

27. $a_{\mathbf{N}} = t$, $|\mathbf{v}| = t$ (from Example 7) $\Rightarrow t = \kappa t^2 \Rightarrow \kappa = \frac{1}{t} \Rightarrow \rho = \frac{1}{\kappa} = t$

29. $a_{\mathbf{N}} = 0 \Rightarrow 0 = \kappa|\mathbf{v}|^2 \Rightarrow \kappa = 0 \Rightarrow$ the curve is a straight line.

31. $\kappa = \frac{a}{a^2 + b^2} \Rightarrow \frac{d\kappa}{da} = \frac{-a^2 + b^2}{(a^2 + b^2)^2}$. If $\frac{d\kappa}{da} = 0$, then $-a^2 + b^2 = 0 \Rightarrow a = \pm b$. When $a = b$ $(b > 0)$, $\frac{d\kappa}{da} > 0$ if $a < b$ and $\frac{d\kappa}{da} < 0$ if $a > b$. $\therefore$ κ is at a maximum when $a = b \Rightarrow \kappa(b) = \frac{b}{b^2 + b^2} = \frac{1}{2b}$, the maximum value of κ.

33. $\mathbf{r} = f(t)\,\mathbf{i} + g(t)\,\mathbf{j} + h(t)\,\mathbf{k} \Rightarrow \mathbf{v} = f'(t)\,\mathbf{i} + g'(t)\,\mathbf{j} + h'(t)\,\mathbf{k}$. $\mathbf{v} \cdot \mathbf{k} = 0 \Rightarrow h'(t) = 0 \Rightarrow h(t) = C \Rightarrow$ $\mathbf{r} = f(t)\,\mathbf{i} + g(t)\,\mathbf{j} + C\,\mathbf{k}$. $\mathbf{r}_{t=a} = f(a)\,\mathbf{i} + g(a)\,\mathbf{j} + C\,\mathbf{k} = \mathbf{0} \Rightarrow f(a) = 0,\ g(a) = 0,\ C = 0$ $\therefore$ $\mathbf{r} = f(t)\,\mathbf{i} + g(t)\,\mathbf{j}$

35. $\mathbf{r} = t\,\mathbf{i} + (\sin t)\,\mathbf{j} \Rightarrow \mathbf{v} = \mathbf{i} + (\cos t)\,\mathbf{j} \Rightarrow |\mathbf{v}| = \sqrt{1^2 + (\cos t)^2} = \sqrt{1 + \cos^2 t} \Rightarrow \left|\mathbf{v}\left(\frac{\pi}{2}\right)\right| = \sqrt{1 + \cos^2\left(\frac{\pi}{2}\right)}$

$$= 1.\ \ \mathbf{a} = (-\sin t)\,\mathbf{j} \Rightarrow \mathbf{v} \times \mathbf{a} = \begin{vmatrix} \mathbf{i} & \mathbf{j} & \mathbf{k} \\ 1 & \cos t & 0 \\ 0 & -\sin t & 0 \end{vmatrix} = (-\sin t)\,\mathbf{k} \Rightarrow |\mathbf{v} \times \mathbf{a}| = \sqrt{(-\sin t)^2} = |\sin t| \Rightarrow$$

35. (Continued)

$|\mathbf{v} \times \mathbf{a}|\left(\frac{\pi}{2}\right) = \left|\sin\left(\frac{\pi}{2}\right)\right| = 1 \Rightarrow \kappa = \frac{|\mathbf{v} \times \mathbf{a}|}{|\mathbf{v}|^3} = \frac{1}{1^3} = 1.\ \therefore\ \rho = \frac{1}{1} = 1 \Rightarrow$ center is $\left(\frac{\pi}{2}, 0\right)$, $r = 1 \Rightarrow$

$\left(x - \frac{\pi}{2}\right)^2 + y^2 = 1$

37. a) $\mathbf{r} = t\,\mathbf{i} + (\ln(\sin t))\,\mathbf{j},\ 0 < t < \pi \Rightarrow x = t,\ y = \ln(\sin t) \Rightarrow \dot{x} = 1,\ \dot{y} = \frac{\cos t}{\sin t} = \cot t \Rightarrow \ddot{x} = 0,\ \ddot{y} = -\csc^2 t \Rightarrow$

$\kappa = \frac{|(1)(-\csc^2 t) - (\cot t)(0)|}{(1^2 + \cot^2 t)^{3/2}} = |\sin t| = \sin t$ since $0 < t < \pi$.

b) $\mathbf{r} = \left(\tan^{-1}(\sinh t)\right)\mathbf{i} + (\ln(\cosh t))\,\mathbf{j} \Rightarrow x = \tan^{-1}(\sinh t),\ y = \ln(\cosh t) \Rightarrow \dot{x} = \frac{\cosh t}{1 + \sinh^2 t} = \frac{1}{\cosh t} = \operatorname{sech} t,$

$\dot{y} = \frac{\sinh t}{\cosh t} = \tanh t \Rightarrow \ddot{x} = -\operatorname{sech} t \tanh t,\ \ddot{y} = \operatorname{sech}^2 t \Rightarrow \kappa = \frac{|\operatorname{sech} t \operatorname{sech}^2 t - \tanh t(-\operatorname{sech} t \tanh t)|}{\left(\operatorname{sech}^2 t + \tanh^2 t\right)^{3/2}} =$

$\frac{|\operatorname{sech} t(\operatorname{sech}^2 t + \tanh^2 t)|}{\left(\operatorname{sech}^2 t + \tanh^2 t\right)^{3/2}} = |\operatorname{sech} t| = \operatorname{sech} t$

39. a) $y = e^x \Rightarrow \mathbf{r} = x\,\mathbf{i} + e^x\,\mathbf{j} \Rightarrow f'(x) = e^x,\ f''(x) = e^x \Rightarrow \kappa = \frac{|e^x|}{\left(1 + e^{2x}\right)^{3/2}} \Rightarrow \kappa$ at $(0,1)$ is $\frac{e^0}{(1 + e^0)^{3/2}} = \frac{1}{2\sqrt{2}} \Rightarrow$

$\rho = \frac{1}{\kappa} = 2\sqrt{2}.\ \ \mathbf{v} = \mathbf{i} + e^x\,\mathbf{j} \Rightarrow |\mathbf{v}| = \sqrt{1 + e^{2x}} \Rightarrow \mathbf{T} = \frac{\mathbf{v}}{|\mathbf{v}|} = \frac{\mathbf{i} + e^x\,\mathbf{j}}{\sqrt{1 + e^{2x}}} \Rightarrow \mathbf{T}$ at $(0,1) = \frac{\mathbf{i} + \mathbf{j}}{\sqrt{2}} \Rightarrow$

$\mathbf{N}$ at $(0,1) = \frac{-\mathbf{i} + \mathbf{j}}{\sqrt{2}} \Rightarrow$ the center of the circle lies along the line $x = -\frac{1}{\sqrt{2}}t,\ y = \frac{1}{\sqrt{2}}t + 1$. Since the

radius is $2\sqrt{2}$, $\sqrt{\left(-\frac{1}{\sqrt{2}}t\right)^2 + \left(\frac{1}{\sqrt{2}}t + 1\right)^2} = 2\sqrt{2} \Rightarrow \sqrt{\frac{t^2}{2} + \frac{t^2}{2}} = 2\sqrt{2} \Rightarrow t^2 = 8 \Rightarrow t = \pm 2\sqrt{2}.$

$\therefore\ x = -\frac{1}{\sqrt{2}}(2\sqrt{2}) = -2,\ y = \frac{1}{\sqrt{2}}(2\sqrt{2}) + 1 = 3$ (Use $t = 2\sqrt{2}$ to be on the correct side of the curve.) Thus

the equation of the circle is $(x + 2)^2 + (y - 3)^2 = 8$.

b) $2(x + 2) + 2(y - 3)y' = 0 \Rightarrow y' = -\frac{x + 2}{y - 3} \Rightarrow y'' = \frac{(y - 3)(-1) + (x + 2)y'}{(y - 3)^2} = \frac{-8}{(y - 3)^3}.\ y'(0,1) = 1,\ f'(0) = 1,$

$y''(0,1) = 1,\ f''(0) = 1$

41. Components of **v**: –1.87001408, 0.708992989, 0.999999977

Components of **a**: –1.69606646, –2.03053933, 0

Speed: 2.23598383

Components of **T**: –0.836327193, 0.317083237, 0.447230417

Components of **N**: –0.641065166, –0.76748645, 0

Components of **B**: 0.343243285, –0.28670381, 0.845140807

Curvature: 0.505990677

43. Components of **v**: 1.99998356, $-796280802 \times 10^{-6}$, -0.162867596

Components of **a**: 0, -1.00000296, $-8.65198672 \times 10^{-3}$

Speed: 2.00660413

Components of **T**: 0.996700614, $-3.96830043 \times 10^{-6}$, -0.0811657837

Components of **N**: 0, -0.999962572, $-8.65163731 \times 10^{-3}$

Components of **B**: -0.0811627115, $8.62309222 \times 10^{-3}$, -0.99666331

Curvature: 0.248367078

SECTION 11.5 PLANETARY MOTION AND SATELLITES

1. $\frac{T^2}{a^3} = \frac{4\pi^2}{GM} \Rightarrow T^2 = \frac{4\pi^2}{GM} a^3 \Rightarrow T^2 = \frac{4\pi^2}{(6.6720 \times 10^{-11} \text{Nm}^2\text{kg}^{-2})(5.975 \times 10^{24} \text{ kg})} (6\,808\,000 \text{ m})^3 =$

$3.125 \times 10^7 \text{ sec}^2 \Rightarrow T = \sqrt{3125 \times 10^4 \text{ sec}^2} = 55.90 \times 10^2 \text{ sec} = 93.17$ minutes.

3. 92.25 minutes = 5535 seconds. $\frac{T^2}{a^3} = \frac{4\pi^2}{GM} \Rightarrow a^3 = \frac{GM}{4\pi^2} T^2 \Rightarrow a^3 =$

$\frac{(6.6720 \times 10^{-11} \text{Nm}^2\text{kg}^{-2})(5.975 \times 10^{24} \text{kg})}{4\pi^2} (5535 \text{ s})^2 = 3.094 \times 10^{20} \text{ m}^3 \Rightarrow a = \sqrt[3]{3.094 \times 10^{20} \text{ m}^3}$

$= 6.764 \times 10^6$ m = 6764 km. Mean distance from center of the Earth = $\frac{12758 \text{ km} + 183 \text{ km} + 589 \text{ km}}{2} =$

6765 km

5. $a = 22030 \text{ km} = 2.203 \times 10^7$ m. $T^2 = \frac{4\pi^2}{GM} a^3 \Rightarrow T^2 =$

$\frac{4\pi^2}{(6.670 \times 10^{-11} \text{Nm}^2\text{kg}^{-2})(6.418 \times 10^{23} \text{ kg})} (2.203 \times 10^7 \text{s})^3 = 9.857 \times 10^9 \text{ sec}^2 \Rightarrow T =$

$\sqrt{98.57 \times 10^8 \text{ sec}^2} = 9.928 \times 10^4 \text{ sec} = 1655$ minutes.

7. T = 1477.4 minutes = 88644 seconds. $a^3 = \frac{GMT^2}{4\pi^2} =$

$\frac{(6.6720 \times 10^{-11} \text{Nm}^2\text{kg}^{-2})(6.418 \times 10^{23} \text{kg})(88644 \text{ s})^2}{4\pi^2} = 8.523 \times 10^{21} \text{ m}^3 \Rightarrow$

$a = \sqrt[3]{8.523 \times 10^{21} \text{ m}^3} = 2.043 \times 10^7 \text{ m} = 20430$ km

9. $r = \frac{GM}{v^2} \Rightarrow v^2 = \frac{GM}{r} \Rightarrow |v| = \sqrt{\frac{GM}{r}} = \sqrt{\frac{(6.6720 \times 10^{-11} \text{ Nm}^2\text{kg}^{-2})(5.975 \times 10^{24} \text{ kg})}{r}} =$

$1.9966 \times 10^7 r^{-1/2}$ m/s

11. $e = \frac{r_0 v_0^2}{GM} - 1 \Rightarrow v_0^2 = \frac{GM(e+1)}{r_0} \Rightarrow v_0 = \sqrt{\frac{GM(e+1)}{r_0}}$

Circle: $e = 0 \Rightarrow v_0 = \sqrt{\frac{GM}{r_0}}$

Ellipse: $0 < e < 1 \Rightarrow \sqrt{\frac{GM}{r_0}} < v_0 < \sqrt{\frac{2GM}{r_0}}$

Parabola: $e = 1 \Rightarrow v_0 = \sqrt{\frac{2GM}{r_0}}$

Hyperbola: $e > 1 \Rightarrow v_0 > \sqrt{\frac{2GM}{r_0}}$

13. $\Delta \mathbf{A} = \frac{1}{2}\left|\mathbf{r}(t+\Delta t) \times \mathbf{r}(t)\right| \Rightarrow \frac{\Delta \mathbf{A}}{\Delta t} = \frac{1}{2}\left|\frac{\mathbf{r}(t+\Delta t)}{\Delta t} \times \mathbf{r}(t)\right| = \frac{1}{2}\left|\frac{\mathbf{r}(t+\Delta t) - \mathbf{r}(t) + \mathbf{r}(t)}{\Delta t} \times \mathbf{r}(t)\right| =$
$\frac{1}{2}\left|\frac{\mathbf{r}(t+\Delta t) - \mathbf{r}(t)}{\Delta t} \times \mathbf{r}(t)\right| + \frac{1}{2}\left|\mathbf{r}(t) \times \mathbf{r}(t)\right| = \frac{1}{2}\left|\frac{\mathbf{r}(t+\Delta t) - \mathbf{r}(t)}{\Delta t} \times \mathbf{r}(t)\right|$. $\therefore \frac{d\mathbf{A}}{dt} = \lim_{\Delta t \to 0} \frac{1}{2}\left|\frac{\mathbf{r}(t+\Delta t) - \mathbf{r}(t)}{\Delta t} \times \mathbf{r}(t)\right|$
$= \frac{1}{2}\left|\frac{d\mathbf{r}}{dt} \times \mathbf{r}(t)\right| = \frac{1}{2}\left|\mathbf{r}(t) \times \frac{d\mathbf{r}}{dt}\right| = \frac{1}{2}\left|\mathbf{r} \times \dot{\mathbf{r}}\right|$

15. $r = t,\ \theta = t \Rightarrow \frac{dr}{dt} = 1,\ \frac{d\theta}{dt} = 1.$ $\mathbf{v} = \frac{d\mathbf{r}}{dt} = \frac{d}{dt}\left[r\,\mathbf{u}_r\right] = \dot{r}\,\mathbf{u}_r + r\,\dot{\theta}\,\mathbf{u}_\theta = \mathbf{u}_r + r\,\mathbf{u}_\theta = (\cos\theta)\,\mathbf{i} + (\sin\theta)\,\mathbf{j} - r(\sin\theta)\,\mathbf{i} +$
$r(\cos\theta)\,\mathbf{j} = (\cos t - t\sin t)\,\mathbf{i} + (\sin t + t\cos t)\,\mathbf{j}$. $\mathbf{a} = (-2\sin t - t\cos t)\,\mathbf{i} + (2\cos t - t\sin t)\,\mathbf{j} \Rightarrow$

$$\mathbf{v} \times \mathbf{a} = \begin{vmatrix} \mathbf{i} & \mathbf{j} & \mathbf{k} \\ \cos t - t\sin t & \sin t + t\cos t & 0 \\ -2\sin t - t\cos t & 2\cos t - t\sin t & 0 \end{vmatrix} = (t^2+2)\,\mathbf{k} \Rightarrow |\mathbf{v} \times \mathbf{a}| = t^2 + 2.$$

$|\mathbf{v}| = \sqrt{(\cos t - t\sin t)^2 + (\sin t + t\cos t)^2} = \sqrt{t^2+1}$. Then $\kappa = \frac{|\mathbf{v} \times \mathbf{a}|}{|\mathbf{v}|^3} = \left(t^2+1\right)\sqrt{t^2+1}$

17. $\frac{d\theta}{dt} = 2,\ \theta = 2t,\ r = \cosh\theta = \cosh(2t).$ $\mathbf{a} = \left(\frac{d^2r}{dt^2} - r\left(\frac{d\theta}{dt}\right)^2\right)\mathbf{u}_r + \left(r\left(\frac{d^2\theta}{dt^2}\right) + 2\left(\frac{dr}{dt}\right)\left(\frac{d\theta}{dt}\right)\right)\mathbf{u}_\theta \Rightarrow$ the $\mathbf{u}_r$

component of acceleration $= \left(\frac{d^2r}{dt^2} - r\left(\frac{d\theta}{dt}\right)^2\right)\frac{dr}{dt} = 2\sinh(2t) \Rightarrow \frac{d^2r}{dt^2} = 4\cosh(2t).$

$\therefore \left(\frac{d^2r}{dt^2} - r\left(\frac{d\theta}{dt}\right)^2\right) = 4\cosh(2t) - \cosh(2t)(2^2) = 0$

SECTION 11.M MISCELLANEOUS EXERCISES

1. $\mathbf{r} = (4\cos t)\,\mathbf{i} + (\sqrt{2}\sin t)\,\mathbf{j} \Rightarrow x = 4\cos t \Rightarrow x^2 = 16\cos^2 t$

$y = \sqrt{2}\sin t \Rightarrow y^2 = 2\sin^2 t \Rightarrow 8y^2 = 16\sin^2 t \Rightarrow$

$x^2 + 8y^2 = 16 \Rightarrow \frac{x^2}{16} + \frac{y^2}{2} = 1.$ $t = 0 \Rightarrow x = 4,\ y = 0;$

$t = \frac{\pi}{4} \Rightarrow x = 2\sqrt{2},\ y = 1.$ $\mathbf{v} = (-4\sin t)\,\mathbf{i} + (\sqrt{2}\cos t)\,\mathbf{j}$

$\Rightarrow \mathbf{v}(0) = \sqrt{2}\,\mathbf{j},\ \mathbf{v}\left(\frac{\pi}{4}\right) = -2\sqrt{2}\,\mathbf{i} + \mathbf{j}.$

$\mathbf{a} = (-4\cos t)\,\mathbf{i} + (-\sqrt{2}\sin t)\,\mathbf{j} \Rightarrow \mathbf{a}(0) = -4\,\mathbf{i},\ \mathbf{a}\left(\frac{\pi}{4}\right) =$

$-2\sqrt{2}\,\mathbf{i} - \mathbf{j}.$

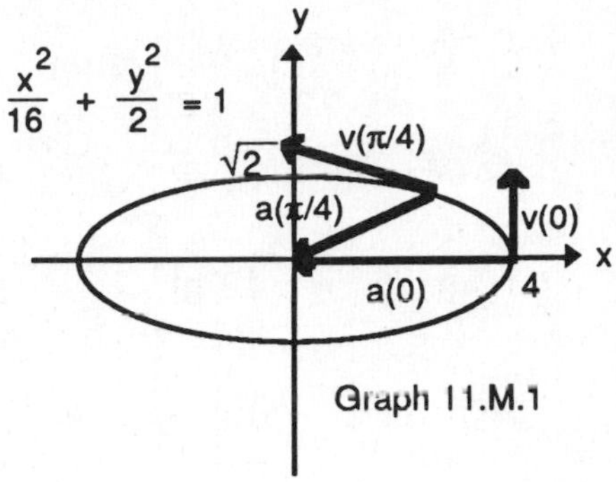

Graph 11.M.1

3. $\int_0^1 [(3+6t)\,\mathbf{i} + (4+8t)\,\mathbf{j} + (6\pi\cos \pi t)\,\mathbf{k}]\,dt = \left[3t + 3t^2\right]_0^1 \mathbf{i} + \left[4t + 4t^2\right]_0^1 \mathbf{j} + \left[-6\sin \pi t\right]_0^1 \mathbf{j} = 6\,\mathbf{i} + 8\,\mathbf{j}$

5. $\mathbf{r} = \int ((-\sin t)\,\mathbf{i} + (\cos t)\,\mathbf{j} + \mathbf{k})\,dt = (\cos t)\,\mathbf{i} + (\sin t)\,\mathbf{j} + t\,\mathbf{k} + \mathbf{C}.$ $\mathbf{r}(0) = \mathbf{j} \Rightarrow (\cos 0)\,\mathbf{i} + (\sin 0)\,\mathbf{j} + (0)\,\mathbf{k}$ $+ \mathbf{C} = \mathbf{j} \Rightarrow \mathbf{C} = \mathbf{j} - \mathbf{i} \Rightarrow \mathbf{r} = ((\cos t) - 1)\,\mathbf{i} + ((\sin t) + 1)\,\mathbf{j} + t\,\mathbf{k}$

7. $\frac{d\mathbf{r}}{dt} = \int 2\,\mathbf{j}\,dt = 2t\,\mathbf{j} + \mathbf{C}_1.$ $\frac{d\mathbf{r}}{dt}(0) = \mathbf{k} \Rightarrow 2(0)\,\mathbf{j} + \mathbf{C}_1 = \mathbf{k} \Rightarrow \mathbf{C}_1 = \mathbf{k}.$ $\therefore \frac{d\mathbf{r}}{dt} = 2t\,\mathbf{j} + \mathbf{k}.$

$\mathbf{r} = \int (2t\,\mathbf{j} + \mathbf{k})\,dt = t^2\,\mathbf{j} + t\,\mathbf{k} + \mathbf{C}_2.$ $\mathbf{r}(0) = \mathbf{i} \Rightarrow (0^2)\,\mathbf{j} + (0)\,\mathbf{k} + \mathbf{C}_2 = \mathbf{i} \Rightarrow \mathbf{C}_2 = \mathbf{i}$

$\therefore \mathbf{r} = \mathbf{i} + t^2\,\mathbf{j} + t\,\mathbf{k}$

9. $\mathbf{r} = (2\cos t)\,\mathbf{i} + (2\sin t)\,\mathbf{j} + t^2\,\mathbf{k} \Rightarrow \mathbf{v} = (-2\sin t)\,\mathbf{i} + (2\cos t)\,\mathbf{j} + 2t\,\mathbf{k} \Rightarrow$

$|\mathbf{v}| = \sqrt{(-2\sin t)^2 + (2\cos t)^2 + (2t)^2} = 2\sqrt{1+t^2}.$ Length $= \int_0^{\pi/4} 2\sqrt{1+t^2}\,dt =$

$\left[t\sqrt{1+t^2} + \ln\left|t + \sqrt{1+t^2}\right|\right]_0^{\pi/4} = \frac{\pi}{4}\sqrt{1 + \frac{\pi^2}{16}} + \ln\left(\frac{\pi}{4} + \sqrt{1 + \frac{\pi^2}{16}}\right)$

11. $\mathbf{r} = \frac{4}{9}(1+t)^{3/2}\,\mathbf{i} + \frac{4}{9}(1-t)^{3/2}\,\mathbf{j} + \frac{1}{3}t\,\mathbf{k} \Rightarrow \mathbf{v} = \frac{2}{3}(1+t)^{1/2}\,\mathbf{i} - \frac{2}{3}(1-t)^{1/2}\,\mathbf{j} + \frac{1}{3}\mathbf{k} \Rightarrow$

$|\mathbf{v}| = \sqrt{\left(\frac{2}{3}(1+t)^{1/2}\right)^2 + \left(-\frac{2}{3}(1-t)^{1/2}\right)^2 + \left(\frac{1}{3}\right)^2} = 1.$ $\mathbf{T} = \frac{2}{3}(1+t)^{1/2}\,\mathbf{i} - \frac{2}{3}(1-t)^{1/2} + \frac{1}{3}\mathbf{k} \Rightarrow$

$\mathbf{T}(0) = \frac{2}{3}\mathbf{i} - \frac{2}{3}\mathbf{j} + \frac{1}{3}\mathbf{k}.$ $\frac{d\mathbf{T}}{dt} = \frac{1}{3}(1+t)^{-1/2}\,\mathbf{i} + \frac{1}{3}(1-t)^{-1/2}\,\mathbf{j} \Rightarrow \left|\frac{d\mathbf{T}}{dt}\right| =$

$\sqrt{\left(\frac{1}{3}(1+t)^{-1/2}\right)^2 + \left(\frac{1}{3}(1-t)^{-1/2}\right)^2} = \frac{1}{3}\sqrt{\frac{2}{1-t^2}}.$ $\frac{d\mathbf{T}}{dt}(0) = \frac{1}{3}\mathbf{i} + \frac{1}{3}\mathbf{j} \Rightarrow \left|\frac{d\mathbf{T}}{dt}(0)\right| = \frac{\sqrt{2}}{3}.$

11. (Continued)

$$\therefore\ \mathbf{N}(0)=\frac{1}{\sqrt{2}}\mathbf{i}+\frac{1}{\sqrt{2}}\mathbf{j}.\ \mathbf{B}(0)=\mathbf{T}(0)\times\mathbf{N}(0)=\begin{vmatrix}\mathbf{i}&\mathbf{j}&\mathbf{k}\\ \frac{2}{3}&-\frac{2}{3}&\frac{1}{3}\\ \frac{1}{\sqrt{2}}&\frac{1}{\sqrt{2}}&0\end{vmatrix}=-\frac{1}{3\sqrt{2}}\mathbf{i}+\frac{1}{3\sqrt{2}}\mathbf{j}+\frac{4}{3\sqrt{2}}\mathbf{k}$$

$$\mathbf{a}=\frac{1}{3}(1+t)^{-1/2}\mathbf{i}+\frac{1}{3}(1-t)^{-1/2}\mathbf{j}\Rightarrow\mathbf{a}(0)=\frac{1}{3}\mathbf{i}+\frac{1}{3}\mathbf{j}.\ \mathbf{v}(0)=\frac{2}{3}\mathbf{i}-\frac{2}{3}\mathbf{j}+\frac{1}{3}\mathbf{k}.$$

$$\therefore\ \mathbf{v}(0)\times\mathbf{a}(0)=\begin{vmatrix}\mathbf{i}&\mathbf{j}&\mathbf{k}\\ \frac{2}{3}&-\frac{2}{3}&\frac{1}{3}\\ \frac{1}{3}&\frac{1}{3}&0\end{vmatrix}=-\frac{1}{9}\mathbf{i}+\frac{1}{9}\mathbf{j}+\frac{4}{9}\mathbf{k}\Rightarrow|\mathbf{v}\times\mathbf{a}|=\frac{\sqrt{2}}{3}\Rightarrow\kappa=\frac{|\mathbf{v}\times\mathbf{a}|}{|\mathbf{v}|^3}=\frac{\sqrt{2}/3}{1^3}=\frac{\sqrt{2}}{3}.$$

$$\dot{\mathbf{a}}=-\frac{1}{6}(1+t)^{-3/2}\mathbf{i}+\frac{1}{6}(1-t)^{-3/2}\mathbf{j}\Rightarrow\dot{\mathbf{a}}(0)=-\frac{1}{6}\mathbf{i}+\frac{1}{6}\mathbf{j}\Rightarrow\tau=\frac{\begin{vmatrix}\frac{2}{3}&-\frac{2}{3}&\frac{1}{3}\\ \frac{1}{3}&\frac{1}{3}&0\\ -\frac{1}{6}&\frac{1}{6}&0\end{vmatrix}}{|\mathbf{v}\times\mathbf{a}|^2}=\frac{1/27}{(\sqrt{2}/3)^2}=\frac{1}{6}$$

$t=0\Rightarrow\left(\frac{4}{9},\frac{4}{9},0\right)$ is the point on the curve. Osculating Plane: **B** is the normal vector $\Rightarrow$

$-\frac{1}{3\sqrt{2}}x+\frac{1}{3\sqrt{2}}y+\frac{4}{3\sqrt{2}}z=\frac{4}{9}\left(-\frac{1}{3\sqrt{2}}\right)+\frac{4}{9}\left(\frac{1}{3\sqrt{2}}\right)+0$ or $x-y-4z=0$ is the plane.

Rectifying Plane: **N** is the normal vector $\Rightarrow\frac{1}{\sqrt{2}}x+\frac{1}{\sqrt{2}}y=\frac{4}{9}\left(\frac{1}{\sqrt{2}}\right)+\frac{4}{9}\left(\frac{1}{\sqrt{2}}\right)$ or $x+y=\frac{8}{9}$ is the plane.

Normal Plane: **T** is the normal vector $\Rightarrow\frac{2}{3}x-\frac{2}{3}y+\frac{1}{3}z=\frac{4}{9}\left(\frac{2}{3}\right)-\frac{4}{9}\left(\frac{2}{3}\right)+0$ or $2x-2y+z=0$ is the plane.

13. $\mathbf{r}=t\,\mathbf{i}+\frac{1}{2}e^{2t}\mathbf{j},\ t=\ln 2\Rightarrow\mathbf{v}=\mathbf{i}+e^{2t}\mathbf{j}\Rightarrow|\mathbf{v}|=\sqrt{1+e^{4t}}\Rightarrow\mathbf{T}=\frac{\mathbf{v}}{|\mathbf{v}|}=\frac{\mathbf{i}+e^{2t}\mathbf{j}}{\sqrt{1+e^{4t}}}=\frac{1}{\sqrt{1+e^{4t}}}\mathbf{i}+\frac{e^{2t}}{\sqrt{1+e^{4t}}}\mathbf{j}$

$\Rightarrow\frac{d\mathbf{T}}{dt}=\frac{-2e^{4t}}{(1+e^{4t})^{3/2}}\mathbf{i}+\frac{2e^{2t}}{(1+e^{4t})^{3/2}}\mathbf{j}\Rightarrow\left|\frac{d\mathbf{T}}{dt}\right|=\sqrt{\frac{4e^{8t}+4e^{4t}}{(1+e^{4t})^3}}=\frac{2e^{2t}}{1+e^{4t}}\Rightarrow\mathbf{N}=\frac{d\mathbf{T}/dt}{|d\mathbf{T}/dt|}=\frac{-e^{2t}}{(1+e^{4t})^{1/2}}\mathbf{i}+\frac{1}{(1+e^{4t})^{1/2}}\mathbf{j}.\ \therefore\ \mathbf{T}(\ln 2)=\frac{1}{\sqrt{17}}\mathbf{i}+\frac{4}{\sqrt{17}}\mathbf{j},\ \mathbf{N}(\ln 2)=-\frac{4}{\sqrt{17}}\mathbf{i}+\frac{1}{\sqrt{17}}\mathbf{j}.\ \mathbf{B}(\ln 2)=\mathbf{T}(\ln 2)\times\mathbf{N}(\ln 2)=$

$$\begin{vmatrix}\mathbf{i}&\mathbf{j}&\mathbf{k}\\ \frac{1}{\sqrt{17}}&\frac{4}{\sqrt{17}}&0\\ -\frac{4}{\sqrt{17}}&\frac{1}{\sqrt{17}}&0\end{vmatrix}=\mathbf{k}.\ \mathbf{a}=2e^{2t}\mathbf{j}\Rightarrow\mathbf{a}(\ln 2)=8\mathbf{j}.\ \mathbf{v}(\ln 2)=\mathbf{i}+4\mathbf{j}\Rightarrow\mathbf{v}(\ln 2)\times\mathbf{a}(\ln 2)=\begin{vmatrix}\mathbf{i}&\mathbf{j}&\mathbf{k}\\ 1&4&0\\ 0&8&0\end{vmatrix}$$

$=8\mathbf{k}\Rightarrow|\mathbf{v}(\ln 2)\times\mathbf{a}(\ln 2)|=8.\ |\mathbf{v}(\ln 2)|=\sqrt{17}.\Rightarrow\kappa=\frac{8}{(\sqrt{17})^3}=\frac{8}{17\sqrt{17}}.\ \dot{\mathbf{a}}=4e^{2t}\mathbf{j}\Rightarrow\dot{\mathbf{a}}(\ln 2)=16\mathbf{j}\Rightarrow$

13. (Continued)

$$\tau = \frac{\begin{vmatrix} 1 & 4 & 0 \\ 0 & 8 & 0 \\ 0 & 16 & 0 \end{vmatrix}}{|\mathbf{v} \times \mathbf{a}|^2} = 0.$$ $t = \ln 2 \Rightarrow (\ln 2, 2, 0)$ is on the curve. Osculating Plane: **B** is the normal vector $\Rightarrow$ $z = 0$ is the plane. Rectifying Plane: **N** is the normal vector $\Rightarrow -\frac{4}{\sqrt{17}}x + \frac{1}{\sqrt{17}}y = -\frac{4}{\sqrt{17}}(\ln 2) + \frac{1}{\sqrt{17}}(2)$ or $4x - y = 4\ln 2 - 2$ is the plane. Normal Plane: **T** is the normal vector $\Rightarrow \frac{1}{\sqrt{17}}x + \frac{4}{\sqrt{17}}y = \frac{\ln 2}{\sqrt{17}} + \frac{8}{\sqrt{17}}$ or

$x + 4y = \ln 2 + 8$ is the plane.

15. $\mathbf{r} = (2 + 3t + 3t^2)\,\mathbf{i} + (4t + 4t^2)\,\mathbf{j} - (6\cos t)\,\mathbf{k} \Rightarrow \mathbf{v} = (3 + 6t)\,\mathbf{i} + (4 + 8t)\,\mathbf{j} + (6\sin t)\,\mathbf{k} \Rightarrow$

$|\mathbf{v}| = \sqrt{(3+6t)^2 + (4+8t)^2 + (6\sin t)^2} = \sqrt{25 + 100t + 100t^2 + 36\sin^2 t}$

$\frac{d|\mathbf{v}|}{dt} = \frac{1}{2}(25 + 100t + 100t^2 + 36\sin^2 t)^{-1/2}(100 + 200t + 72\sin t\cos t) \Rightarrow a_T(0) = \frac{d|\mathbf{v}|}{dt}(0) = 10.$

$\mathbf{a} = 6\,\mathbf{i} + 8\,\mathbf{j} + (6\cos t)\,\mathbf{k} \Rightarrow |\mathbf{a}| = \sqrt{6^2 + 8^2 + (6\cos t)^2} = \sqrt{100 + 36\cos^2 t} \Rightarrow |\mathbf{a}(0)| = \sqrt{136} = 2\sqrt{34}$

$a_N = \sqrt{|\mathbf{a}|^2 - a_T^2} = \sqrt{(2\sqrt{34})^2 - 10^2} = \sqrt{36} = 6. \therefore \mathbf{a}(0) = 10\,\mathbf{T} + 6\,\mathbf{N}$

17. $\mathbf{r} = (\sin t)\,\mathbf{i} + (\sqrt{2}\cos t)\,\mathbf{j} + (\sin t)\,\mathbf{k} \Rightarrow \mathbf{v} = (\cos t)\,\mathbf{i} - (\sqrt{2}\sin t)\,\mathbf{j} + (\cos t)\,\mathbf{k} \Rightarrow$

$|\mathbf{v}| = \sqrt{(\cos t)^2 + (-\sqrt{2}\sin t)^2 + (\cos t)^2} = \sqrt{2} \Rightarrow \mathbf{T} = \frac{\mathbf{v}}{|\mathbf{v}|} = \frac{(\cos t)\,\mathbf{i} - (\sqrt{2}\sin t)\,\mathbf{j} + (\cos t)\,\mathbf{k}}{\sqrt{2}} =$

$\left(\frac{1}{\sqrt{2}}\cos t\right)\mathbf{i} - (\sin t)\,\mathbf{j} + \left(\frac{1}{\sqrt{2}}\cos t\right)\mathbf{k}.$ $\frac{d\mathbf{T}}{dt} = \left(-\frac{1}{\sqrt{2}}\sin t\right)\mathbf{i} - (\cos t)\,\mathbf{j} - \left(\frac{1}{\sqrt{2}}\sin t\right)\mathbf{k} \Rightarrow$

$\left|\frac{d\mathbf{T}}{dt}\right| = \sqrt{\left(-\frac{1}{\sqrt{2}}\sin t\right)^2 + (-\cos t)^2 + \left(-\frac{1}{\sqrt{2}}\sin t\right)^2} = 1.$ $\mathbf{N} = \frac{d\mathbf{T}/dt}{|d\mathbf{T}/dt|} = \left(-\frac{1}{\sqrt{2}}\sin t\right)\mathbf{i} - (\cos t)\,\mathbf{j} -$

$$\left(\frac{1}{\sqrt{2}}\sin t\right)\mathbf{k}. \quad \mathbf{B} = \mathbf{T} \times \mathbf{N} = \begin{vmatrix} \mathbf{i} & \mathbf{j} & \mathbf{k} \\ \frac{1}{\sqrt{2}}\cos t & -\sin t & \frac{1}{\sqrt{2}}\cos t \\ -\frac{1}{\sqrt{2}}\sin t & -\cos t & -\frac{1}{\sqrt{2}}\sin t \end{vmatrix} = \frac{1}{\sqrt{2}}\mathbf{i} - \frac{1}{\sqrt{2}}\mathbf{k}$$

$$\mathbf{a} = (-\sin t)\,\mathbf{i} - (\sqrt{2}\cos t)\,\mathbf{j} - (\sin t)\,\mathbf{k} \Rightarrow \mathbf{v} \times \mathbf{a} = \begin{vmatrix} \mathbf{i} & \mathbf{j} & \mathbf{k} \\ \cos t & -\sqrt{2}\sin t & \cos t \\ -\sin t & -\sqrt{2}\cos t & -\sin t \end{vmatrix} = \sqrt{2}\,\mathbf{i} - \sqrt{2}\,\mathbf{k}$$

$|\mathbf{v} \times \mathbf{a}| = \sqrt{(\sqrt{2})^2 + (-\sqrt{2})^2} = \sqrt{4} = 2 \Rightarrow \kappa = \frac{|\mathbf{v} \times \mathbf{a}|}{|\mathbf{v}|^3} = \frac{2}{(\sqrt{2})^3} = \frac{1}{\sqrt{2}}.$

$$\dot{\mathbf{a}} = (-\cos t)\,\mathbf{i} + (\sqrt{2}\sin t)\,\mathbf{j} - (\cos t)\,\mathbf{k} \Rightarrow \tau = \frac{\begin{vmatrix} \cos t & -\sqrt{2}\sin t & \cos t \\ -\sin t & -\sqrt{2}\cos t & -\sin t \\ -\cos t & \sqrt{2}\sin t & -\cos t \end{vmatrix}}{|\mathbf{v} \times \mathbf{a}|^2} = \frac{0}{|\mathbf{v} \times \mathbf{a}|^2} = 0$$

19. $\mathbf{r} = (e^t \cos t)\,\mathbf{i} + (e^t \sin t)\,\mathbf{j} \Rightarrow \mathbf{v} = (e^t \cos t - e^t \sin t)\,\mathbf{i} + (e^t \sin t + e^t \cos t)\,\mathbf{j} \Rightarrow$
$\mathbf{a} = (e^t \cos t - e^t \sin t - e^t \sin t - e^t \cos t)\mathbf{i} + (e^t \sin t + e^t \cos t + e^t \cos t - e^t \sin t)\,\mathbf{j}$
$= (-2e^t \sin t)\,\mathbf{i} + (2e^t \cos t)\,\mathbf{j}$. Let θ be the angle between $\mathbf{r}$ and $\mathbf{a}$.
Then $\theta = \cos^{-1}\left(\frac{\mathbf{r}\cdot\mathbf{a}}{|\mathbf{r}||\mathbf{a}|}\right) = \cos^{-1}\left(\frac{-2e^{2t}\sin t \cos t + 2e^{2t} \sin t \cos t}{\sqrt{(e^t \cos t)^2 + (e^t \sin t)^2}\sqrt{(-2e^t \sin t)^2 + (2e^t \cos t)^2}}\right) =$
$\cos^{-1}\left(\frac{0}{e^{2t}}\right) = \cos^{-1} 0 = \frac{\pi}{2}$ for all t.

21. $\mathbf{r} = 2\,\mathbf{i} + \left(4 \sin \frac{t}{2}\right)\mathbf{j} + \left(3 - \frac{t}{\pi}\right)\mathbf{k} \Rightarrow \mathbf{r}\cdot(\mathbf{i} - \mathbf{j}) = 2(1) + \left(4 \sin \frac{t}{2}\right)(-1)$. $\mathbf{r}\cdot(\mathbf{i} - \mathbf{j}) = 0 \Rightarrow 2 - 4 \sin \frac{t}{2} = 0 \Rightarrow$
$\sin \frac{t}{2} = \frac{1}{2} \Rightarrow \frac{t}{2} = \frac{\pi}{6} \Rightarrow t = \frac{\pi}{3}$ (for the first time).

23. $9y = x^3 \Rightarrow 9\frac{dy}{dt} = 3x^2 \frac{dx}{dt} \Rightarrow \frac{dy}{dt} = \frac{1}{3}x^2 \frac{dx}{dt}$. If $\mathbf{r} = x\,\mathbf{i} + y\,\mathbf{j}$ where x and y are differentiable functions of t,
then $\mathbf{v} = \frac{dx}{dt}\mathbf{i} + \frac{dy}{dt}\mathbf{j}$. Then $\mathbf{v} \cdot \mathbf{i} = 4 \Rightarrow \frac{dx}{dt} = 4$. $\mathbf{v} \cdot \mathbf{j} = \frac{dy}{dt} = \frac{1}{3}x^2 \frac{dx}{dt} \Rightarrow$ at (3,3), $\mathbf{v} \cdot \mathbf{j} = \frac{1}{3}(3)^2(4) = 12$.
$\mathbf{a} = \frac{d^2x}{dt^2}\mathbf{i} + \frac{d^2y}{dt^2}\mathbf{j}$ and $\frac{d^2y}{dt^2} = \frac{2}{3}x\left(\frac{dx}{dt}\right)^2 + \frac{1}{3}x^2\frac{d^2x}{dt^2}$. $\mathbf{a} \cdot \mathbf{i} = -2 \Rightarrow \frac{d^2x}{dt^2} = -2 \Rightarrow \mathbf{a} \cdot \mathbf{j}$ (at (3,3)) $= \frac{d^2y}{dt^2} =$
$\frac{2}{3}(3)(4)^2 + \frac{1}{3}(3)^2(-2) = 26$.

25. a)

$\mathbf{r}(t) = (\pi t - \sin \pi t)\,\mathbf{i} + (1 - \cos \pi t)\,\mathbf{j}$

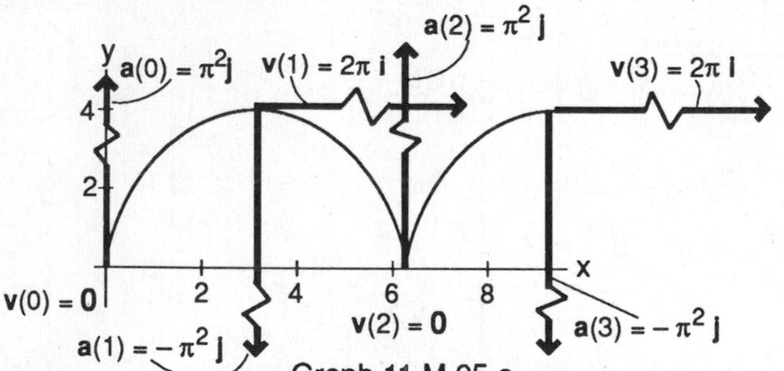

Graph 11.M.25 a

b) $\mathbf{v}(t) = (\pi - \pi \cos \pi t)\,\mathbf{i} + (\pi \sin \pi t)\,\mathbf{j} \Rightarrow \mathbf{v}(0) = \mathbf{0}$, $\mathbf{v}(1) = 2\pi\,\mathbf{i}$, $\mathbf{v}(2) = \mathbf{0}$, $\mathbf{v}(3) = 2\pi\,\mathbf{i}$.
$\mathbf{a}(t) = (\pi^2 \sin \pi t)\,\mathbf{i} + (\pi^2 \cos \pi t)\,\mathbf{j} \Rightarrow \mathbf{a}(0) = \pi^2\,\mathbf{j}$, $\mathbf{a}(1) = -\pi^2\,\mathbf{j}$, $\mathbf{a}(2) = \pi^2\,\mathbf{j}$, $\mathbf{a}(3) = -\pi^2\,\mathbf{j}$

c) The speed of the topmost point, Q, is the same as the speed of C. Since the circle makes $\frac{1}{2}$ revolution (π ft or 3.14 ft) per second, it moves 3.14 ft along the x–axis per second $\Rightarrow$ speed of C is 3.14 ft/sec.

27. Sphere: $\rho = a$. $y + z = 0$ is the plane $\Rightarrow z = -y \Rightarrow \rho \cos \phi = -\rho \sin \phi \sin \theta \Rightarrow \rho^2\cos^2\phi = \rho^2\sin^2\phi \sin^2\theta$
$\Rightarrow \rho^2\cos^2\phi + \rho^2\sin^2\phi = \rho^2\sin^2\phi \sin^2\theta + \rho^2\sin^2\phi \Rightarrow \rho^2 = \rho^2\sin^2\phi(\sin^2\theta + 1) \Rightarrow \frac{1}{\sin^2\theta + 1} = \sin^2\phi \Rightarrow$
$\sin \phi = \frac{1}{\sqrt{1 + \sin^2\theta}}$ Then $x = \rho \sin \phi \cos \theta = \frac{a \cos \theta}{\sqrt{1 + \sin^2\theta}}$, $y = \rho \sin \phi \sin \theta = \frac{a \sin \theta}{\sqrt{1 + \sin^2\theta}}$, $z = -y =$
$-\frac{a \sin \theta}{\sqrt{1 + \sin^2\theta}}$ $\therefore$ $\mathbf{r}(t) = \frac{a \cos \theta}{\sqrt{1 + \sin^2\theta}}\,\mathbf{i} + \frac{a \sin \theta}{\sqrt{1 + \sin^2\theta}}\,\mathbf{j} - \frac{a \sin \theta}{\sqrt{1 + \sin^2\theta}}\,\mathbf{k}$. Since the plane intersects the sphere in a circle of radius a, the length of the curve is $2\pi a$.

29. a) $x = v_0(\cos 40°)\,t$, $y = 7 + v_0(\sin 40°)\,t - \frac{1}{2}gt^2 = 7 + v_0(\sin 40°)\,t - 16t^2$. $x = 262\frac{5}{12}$ ft, $y = 0$ ft $\Rightarrow$

$262\frac{5}{12} = v_0(\cos 40°)\,t \Rightarrow v_0 = \frac{262.4167}{(\cos 40°)t} \Rightarrow 0 = 7 + \frac{262.4167}{(\cos 40°)t}(\sin 40°)\,t - 16t^2 \Rightarrow t^2 = 14.1996 \Rightarrow$

$t = 3.768$ sec. $\therefore\ v_0 = 90.91$ ft/sec.

b) Maximum height occurs at $t = \frac{3.768}{2}$ sec $= 1.884$ sec $\Rightarrow y = 7 + (90.91)(\sin 40°)(1.884) - 16(1.884)^2 =$

60.30 ft

c) The weight of the javelin is 1.32 lb $\Rightarrow$ the mass is $1.32\left(\frac{1}{32}\right) = 0.04125$ slugs. If $v_1 = 0$ ft/sec and $v_2 =$

90.91 ft/sec, the the work done is $w = \frac{1}{2}(0.04125)(90.91)^2 - \frac{1}{2}(0.04125)(0)^2 = 170.46$ ft–lbs.

31. $x^2 = \left(v_0^2\cos^2\alpha\right)t^2$, $\left(y + \frac{1}{2}gt^2\right)^2 = \left(v_0^2\sin^2\alpha\right)t^2 \Rightarrow x^2 + \left(y + \frac{1}{2}gt^2\right)^2 = v_0^2t^2$

33. $\kappa = \frac{|\dot{x}\ddot{y} - \dot{y}\ddot{x}|}{\left(\dot{x}^2 + \dot{y}^2\right)^{3/2}} \Rightarrow \rho = \frac{\left(\dot{x}^2 + \dot{y}^2\right)^{3/2}}{|\dot{x}\ddot{y} - \dot{y}\ddot{x}|} = \frac{\left(\dot{x}^2 + \dot{y}^2\right)^{3/2}}{|\dot{x}\ddot{y} - \dot{y}\ddot{x}|}\left(\frac{\left(\ddot{x}^2 + \ddot{y}^2 - \ddot{s}^2\right)^{1/2}}{\left(\ddot{x}^2 + \ddot{y}^2 - \ddot{s}^2\right)^{1/2}}\right) =$

$\frac{\left(\dot{x}^2 + \dot{y}^2\right)}{\left(\ddot{x}^2 + \ddot{y}^2 - \ddot{s}^2\right)^{1/2}}\left(\frac{\left(\dot{x}^2 + \dot{y}^2\right)^{1/2}\left(\ddot{x}^2 + \ddot{y}^2 - \ddot{s}^2\right)^{1/2}}{|\dot{x}\ddot{y} - \dot{y}\ddot{x}|}\right)$ Now $\left(\ddot{x}^2 + \ddot{y}^2 - \ddot{s}^2\right)^{1/2} =$

$\left[\left(\ddot{x}^2 + \ddot{y}^2\right)\frac{\dot{s}^2}{\dot{s}^2} - \left(\frac{\dot{x}\ddot{x} + \dot{y}\ddot{y}}{\dot{s}}\right)^2\right]^{1/2} = \left[\frac{\left(\ddot{x}^2 + \ddot{y}^2\right)\left(\dot{x}^2 + \dot{y}^2\right) - \left(\dot{x}\ddot{x} + \dot{y}\ddot{y}\right)^2}{\dot{s}^2}\right]^{1/2} = \frac{|\dot{x}\ddot{y} - \dot{y}\ddot{x}|}{\left(\dot{x}^2 + \dot{y}^2\right)^{1/2}}.$

$\therefore\ \rho = \frac{\left(\dot{x}^2 + \dot{y}^2\right)}{\left(\ddot{x}^2 + \ddot{y}^2 - \ddot{s}^2\right)^{1/2}}$

35. $\mathbf{r}(t) = \left(\int_0^t \cos\left(\frac{1}{2}\pi\theta^2\right)d\theta\right)\mathbf{i} + \left(\int_0^t \sin\left(\frac{1}{2}\pi\theta^2\right)d\theta\right)\mathbf{j} \Rightarrow \mathbf{v}(t) = \cos\left(\frac{\pi t^2}{2}\right)\mathbf{i} + \sin\left(\frac{\pi t^2}{2}\right)\mathbf{j} \Rightarrow |\mathbf{v}| = 1.$

$\mathbf{a}(t) = -\pi t\sin\left(\frac{\pi t^2}{2}\right)\mathbf{i} + \pi t\cos\left(\frac{\pi t^2}{2}\right)\mathbf{j}$. $\therefore\ \mathbf{v} \times \mathbf{a} = \begin{vmatrix} \mathbf{i} & \mathbf{j} & \mathbf{k} \\ \cos\left(\frac{\pi t^2}{2}\right) & \sin\left(\frac{\pi t^2}{2}\right) & 0 \\ -\pi t\sin\left(\frac{\pi t^2}{2}\right) & \pi t\cos\left(\frac{\pi t^2}{2}\right) & 0 \end{vmatrix} = \pi t\,\mathbf{k}$. Then $\kappa = \pi t$.

$\frac{ds}{dt} = |\mathbf{v}(t)| = 1 \Rightarrow ds = dt \Rightarrow s = t + C$. $\mathbf{r}(0) = \mathbf{0} \Rightarrow s = 0 \Rightarrow C = 0$. $\therefore\ \kappa = \pi s$.

37. $s = a\theta \Rightarrow \theta = \frac{s}{a} \Rightarrow \phi = \frac{s}{a} + \frac{\pi}{2} \Rightarrow \frac{d\phi}{ds} = \frac{1}{a}$. $\therefore\ \kappa = \left|\frac{1}{a}\right| = \frac{1}{a}$ if $a > 0$.

39. a) $\mathbf{r}(\theta) = (a\theta\cos\theta)\,\mathbf{i} + (a\theta\sin\theta)\,\mathbf{j} + b\theta\,\mathbf{k} \Rightarrow \mathbf{v}(\theta) = \left[(a\cos\theta - a\theta\sin\theta)\,\mathbf{i} + (a\sin\theta + a\theta\cos\theta)\,\mathbf{j} + b\,\mathbf{k}\right]$

$\Rightarrow |\mathbf{v}(\theta)| = (a^2 + a^2\theta^2 + b^2)^{1/2}\left(\frac{d\theta}{dt}\right) \Rightarrow \frac{d\theta}{dt} = \frac{\sqrt{2gb\theta}}{\sqrt{a^2 + a^2\theta^2 + b^2}}$

b) $s = \int_0^t |\mathbf{v}|\,dt = \int_0^t (a^2 + a^2\theta^2 + b^2)\frac{d\theta}{dt}\,dt = \int_0^t (a^2 + a^2\theta^2 + b^2)\,d\theta = \int_0^\theta (a^2 + a^2u^2 + b^2)^{1/2}\,du$

$= \int_0^\theta a\sqrt{\frac{a^2 + b^2}{a^2} + u^2}\,du = a\int_0^\theta \sqrt{c^2 + u^2}\,du$ where $c = \frac{a^2 + b^2}{a^2}$. Then $s =$

$a\left[\frac{u}{2}\sqrt{c^2 + u^2} + \frac{c^2}{2}\ln\left|u + \sqrt{c^2 + u^2}\right|\right]_0^\theta = \frac{a}{2}\left(\theta\sqrt{c^2 + \theta^2} + c^2\ln\left|\theta + \sqrt{c^2 + \theta^2}\right| - c^2\ln c\right)$

41. $r = \frac{(1+e)r_0}{1 + e\cos\theta} \Rightarrow \frac{dr}{d\theta} = \frac{(1+e)r_0(e\sin\theta)}{(1 + e\cos\theta)^2}.\ \frac{dr}{d\theta} = 0 \Rightarrow \frac{(1+e)r_0(e\sin\theta)}{(1 + e\cos\theta)^2} = 0 \Rightarrow (1+e)r_0(e\sin\theta) = 0 \Rightarrow$

$\sin\theta = 0 \Rightarrow \theta = 0$ or π. $\frac{d^2r}{d\theta^2} = \frac{er_0(1+e)(1 + e\cos\theta)^2(\cos\theta) - 2(1 + e\cos\theta)(-e\sin\theta)(\sin\theta)}{\left((1 + e\cos\theta)^2\right)^2} =$

$\frac{er_0(1+e)(\cos\theta + e\cos^2\theta + 2e\sin^2\theta)}{(1 + e\cos\theta)^3}.\ \frac{d^2r}{d\theta^2}(0) = \frac{er_0}{1+e} > 0$ since $e > 0$, $r_0 > 0 \Rightarrow$ minimum.

$\frac{d^2r}{d\theta^2}(\pi) = \frac{er_0(1+e)(-1+e)}{(1-e)^3} < 0$ since $0 < e < 1$ (for an ellipse) and $r_0 > 0 \Rightarrow$ maximum.

$\therefore$ r is a minimum when $\theta = 0$. Then $r(0) = \frac{(1+e)r_0}{1 + e\cos 0} = r_0$.

43. $f = f(\theta) \Rightarrow \frac{dr}{dt} = f'(\theta)\frac{d\theta}{dt} \Rightarrow \frac{d^2r}{dt^2} = f''(\theta)\left(\frac{d\theta}{dt}\right)^2 + f'(\theta)\frac{d^2\theta}{dt^2}.\ \mathbf{v} = \frac{dr}{dt}\mathbf{u}_r + r\frac{d\theta}{dt}\mathbf{u}_\theta = \left(\cos\theta\frac{dr}{dt} - r\sin\theta\frac{d\theta}{dt}\right)\mathbf{i} +$

$\left(\sin\theta\frac{dr}{dt} + r\cos\theta\frac{d\theta}{dt}\right)\mathbf{j} \Rightarrow |\mathbf{v}| = \left(\left(\frac{dr}{dt}\right)^2 + r^2\left(\frac{d\theta 2}{dt}\right)\right)^{1/2} = \left((f')^2 + f^2\right)^{1/2}\frac{d\theta}{dt}.\ |\mathbf{v} \times \mathbf{a}| = |\dot{x}\ddot{y} - \dot{y}\ddot{x}|$ where

$x = r\cos\theta$, $y = r\sin\theta$. Then $\frac{dx}{dt} = -r\sin\theta\frac{d\theta}{dt} + \cos\theta\frac{dr}{dt} \Rightarrow \frac{d^2x}{dt^2} = -2\sin\theta\frac{d\theta}{dt}\frac{dr}{dt} - r\cos\theta\left(\frac{d\theta}{dt}\right)^2 -$

$r\sin\theta\frac{d^2\theta}{dt^2} + \cos\theta\frac{d^2r}{dt^2}.\ \frac{dy}{dt} = r\cos\theta\frac{d\theta}{dt} + \sin\theta\frac{dr}{dt} \Rightarrow \frac{d^2y}{dt^2} = 2\cos\theta\frac{d\theta}{dt}\frac{dr}{dt} - r\sin\theta\left(\frac{d\theta}{dt}\right)^2 + r\cos\theta\frac{d^2\theta}{dt^2}$

$+ \sin\theta\frac{d^2r}{dt^2}$. Then $|\mathbf{v} \times \mathbf{a}|$ = (after much algebra) $r^2\left(\frac{d\theta}{dt}\right)^3 + r\frac{d^2\theta}{dt^2}\frac{dr}{dt} - r\frac{d\theta}{dt}\left(\frac{dr}{dt}\right)^2$.

$\therefore \kappa = \frac{r^2\left(\frac{d\theta}{dt}\right)^3 + r\frac{d^2\theta}{dt^2}\frac{dr}{dt} - r\frac{d\theta}{dt}\left(\frac{dr}{dt}\right)^2}{\left((f')^2 + f^2\right)^{3/2}} = \frac{r^2 + r\frac{d^2\theta}{dt^2}(f')\left(\frac{dt}{d\theta}\right)^2 - rf'' - rf'\frac{d^2\theta}{dt^2}\left(\frac{dt}{d\theta}\right)^2 + 2(f')^2}{\left((f')^2 + f^2\right)^{3/2}} =$

$\frac{f^2 - ff'' + 2(f')^2}{\left((f')^2 + f^2\right)^{3/2}}$

45. a) $\mathbf{u}_r = (\cos\theta)\,\mathbf{i} + (\sin\theta)\,\mathbf{j}$, $\mathbf{u}_\theta = (-\sin\theta)\,\mathbf{i} + (\cos\theta)\,\mathbf{j}$. $\frac{dx}{dt} = \mathbf{v}\cdot\mathbf{i} = \left(\frac{dr}{dt}\mathbf{u}_r + r\frac{d\theta}{dt}\mathbf{u}_\theta\right)\cdot\mathbf{i} = \frac{dr}{dt}\mathbf{y}_r\cdot\mathbf{i} + r\frac{d\theta}{dt}\mathbf{u}_\theta\cdot\mathbf{i} = \frac{dr}{dt}\cos\theta - r\frac{d\theta}{dt}\sin\theta = \dot{r}\cos\theta - r\dot{\theta}\sin\theta$. $\frac{dy}{dt} = \mathbf{v}\cdot\mathbf{j} = \left(\frac{dr}{dt}\mathbf{u}_r + r\frac{d\theta}{dt}\mathbf{u}_\theta\right)\cdot\mathbf{j} = \dot{r}\sin\theta + r\dot{\theta}\cos\theta$

b) $\frac{dr}{dt} = \mathbf{v}\cdot\mathbf{u}_r = \left(\frac{dx}{dt}\mathbf{i} + \frac{dy}{dt}\mathbf{j}\right)\cdot\left((\cos\theta)\,\mathbf{i} + (\sin\theta)\,\mathbf{j}\right) = \frac{dx}{dt}\cos\theta + \frac{dy}{dt}\sin\theta = \dot{x}\cos\theta + \dot{y}\sin\theta$. $r\frac{d\theta}{dt} = \mathbf{v}\cdot\mathbf{u}_\theta = \left(\frac{dx}{dt}\mathbf{i} + \frac{dy}{dt}\mathbf{j}\right)\left((-\sin\theta)\,\mathbf{i} + (\cos\theta)\,\mathbf{j}\right) = -\frac{dx}{dt}\sin\theta + \frac{dy}{dt}\cos\theta = -\dot{x}\sin\theta + \dot{y}\cos\theta$

47. a) $\mathbf{u}_\rho = \sin\phi\cos\theta\,\mathbf{i} + \sin\phi\sin\theta\,\mathbf{j} + \cos\phi\,\mathbf{k}$, $\mathbf{u}_\phi = \cos\phi\cos\theta\,\mathbf{i} + \cos\phi\sin\theta\,\mathbf{j} - \sin\phi\,\mathbf{k}$, $\mathbf{u}_\theta = \mathbf{u}_\rho \times \mathbf{u}_\phi = -\sin\theta\,\mathbf{i} + \cos\theta\,\mathbf{j}$

b) $\mathbf{u}_\rho\cdot\mathbf{u}_\phi = \sin\phi\cos\phi\cos^2\theta + \sin\phi\cos\phi\sin^2\theta - \sin\phi\cos\phi = 0$

c) $$\mathbf{u}_\rho \times \mathbf{u}_\phi = \begin{vmatrix} \mathbf{i} & \mathbf{j} & \mathbf{k} \\ \sin\phi\cos\theta & \sin\phi\sin\theta & \cos\phi \\ \cos\phi\cos\theta & \cos\phi\sin\theta & -\sin\phi \end{vmatrix} = (-\sin\theta)\,\mathbf{i} + z9\cos\theta)\,\mathbf{j} = \mathbf{u}_\theta$$

d) $\mathbf{u}_\rho \times \mathbf{u}_\phi = \mathbf{u}_\theta \Rightarrow$ Right handed frame.

49. a) $x = r\cos\theta \Rightarrow dx = \cos\theta\,dr - r\sin\theta\,d\theta$, $y = r\sin\theta \Rightarrow dy = \sin\theta\,dr + r\cos\theta\,d\theta$. Then $dx^2 = \cos^2\theta\,dr^2 - 2r\sin\theta\cos\theta\,dr\,d\theta + r^2\sin^2\theta\,d\theta^2$, $dy^2 = \sin^2\theta\,dr^2 + 2r\sin\theta\cos\theta\,dr\,d\theta + r^2\cos^2\theta\,d\theta^2$ $\therefore dx^2 + dy^2 + dz^2 = dr^2 + r^2d\theta^2 + dz^2$

b)

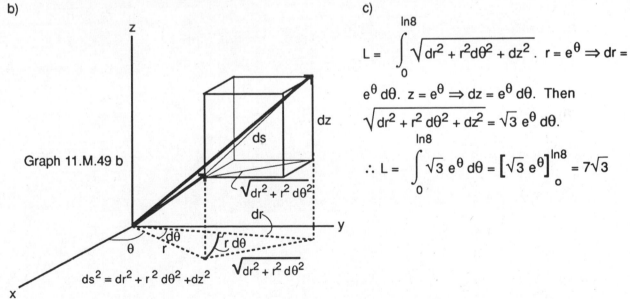

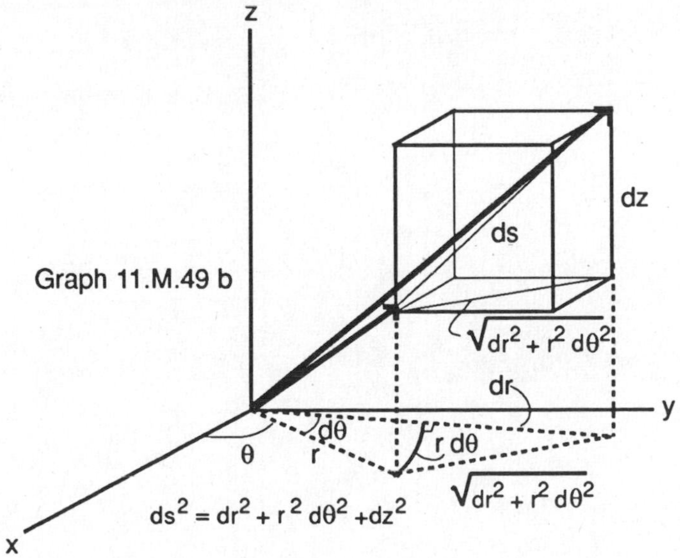

Graph 11.M.49 b

c) $L = \int_0^{\ln 8} \sqrt{dr^2 + r^2d\theta^2 + dz^2}$. $r = e^\theta \Rightarrow dr = e^\theta\,d\theta$. $z = e^\theta \Rightarrow dz = e^\theta\,d\theta$. Then $\sqrt{dr^2 + r^2\,d\theta^2 + dz^2} = \sqrt{3}\,e^\theta\,d\theta$.

$$\therefore L = \int_0^{\ln 8} \sqrt{3}\,e^\theta\,d\theta = \left[\sqrt{3}\,e^\theta\right]_0^{\ln 8} = 7\sqrt{3}$$

CHAPTER 12

FUNCTIONS OF TWO OR MORE VARIABLES AND THEIR DERIVATIVES

SECTION 12.1 FUNCTIONS OF TWO OR MORE INDEPENDENT VARIABLES

1. Domain: Set of all (x,y) so that $y - x \geq 0 \Rightarrow y \geq x$; Range: $z \geq 0$

   Level curves are straight lines of the form $y - x = c$ where $c \geq 0$.

3. Domain: All $(x,y) \neq (0,y)$; Range: All Real Numbers

   Level curves are parabolas with vertex (0,0) and the y–axis as axis.

5. Domain: All points in the xy–plane; Range: All positive Real Numbers

   Level curves are hyperbolas with the x and y axes as asympotes.

7. Domain: Set of all (x,y) so that $-1 \leq y - x \leq 1$; Range: $-\frac{\pi}{2} \leq z \leq \frac{\pi}{2}$

   Level curves are straight lines of the form $y - x = c$ where $-1 \leq c \leq 1$

9. Domain: Set of all (x,y) so that $x > 0$ and $y > 0$; Range: All Real Numbers

   Level curves are straight lines of the form $y = cx$ where $c > 0$, $x > 0$, and $y > 0$.

11. a)

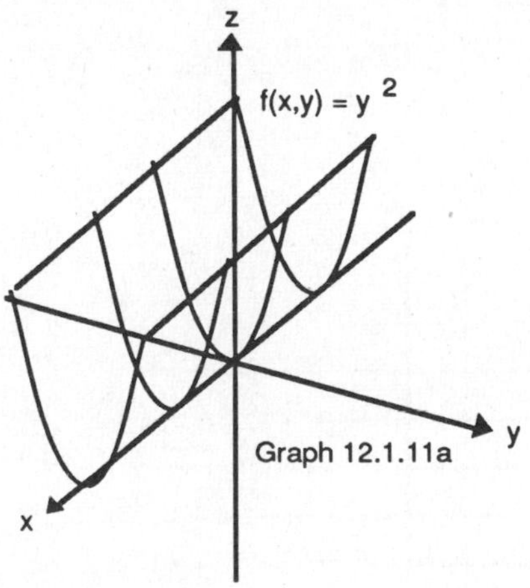

Graph 12.1.11a

b)

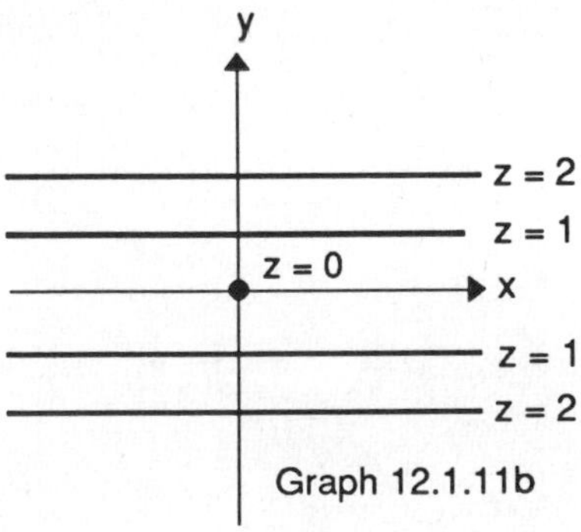

Graph 12.1.11b

13. a)

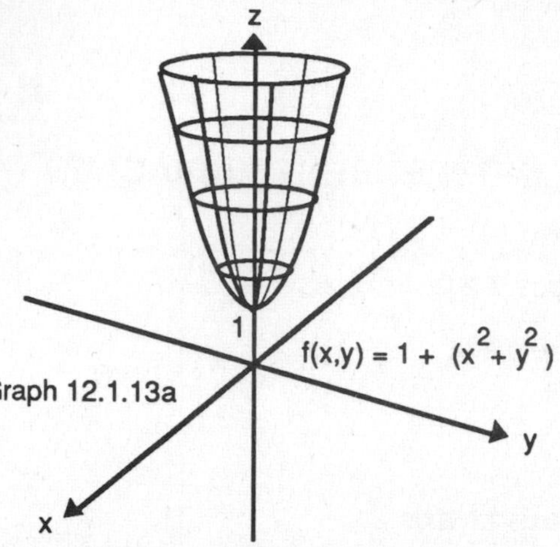

Graph 12.1.13a

b)

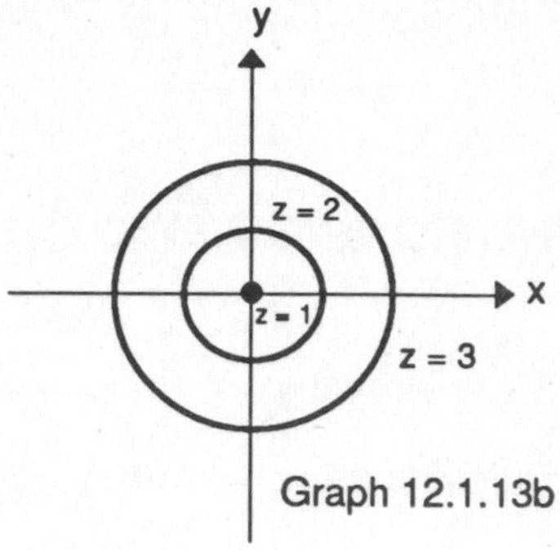

Graph 12.1.13b

15. a)

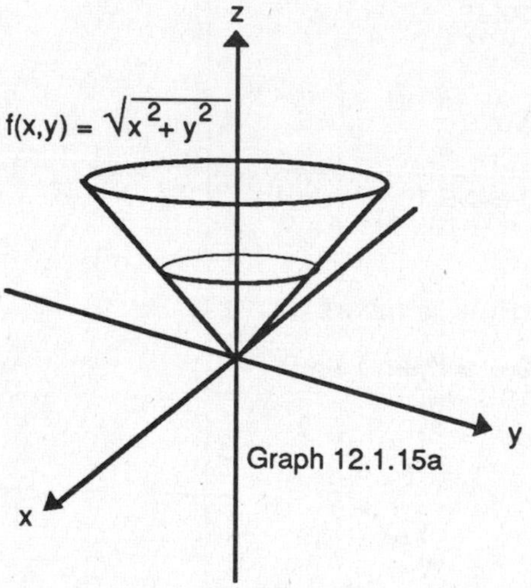

Graph 12.1.15a

b)

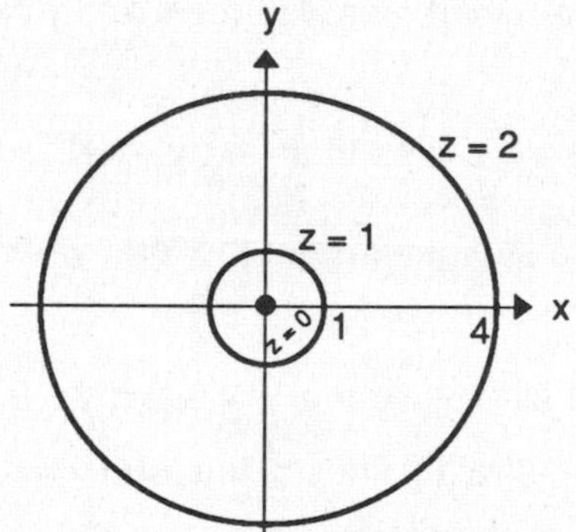

Graph 12.1.15b

17. a)

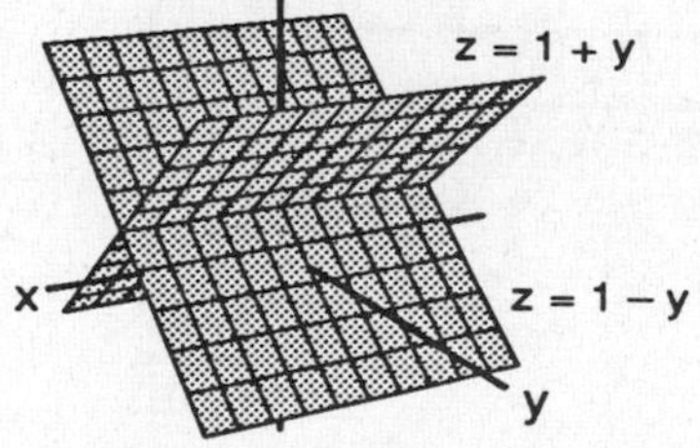

Graph 12.1.17 a

b)

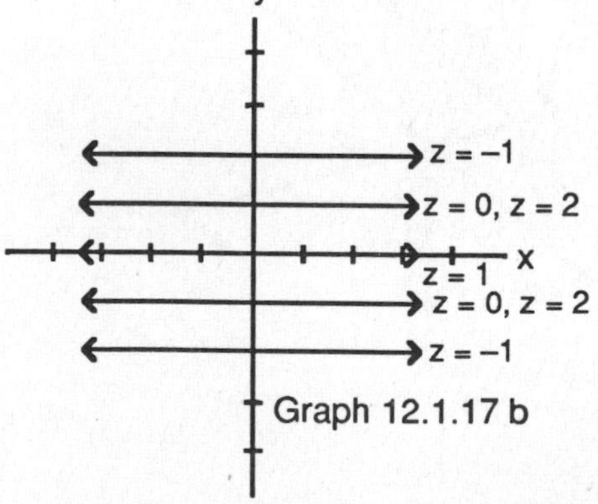

Graph 12.1.17 b

19. a)

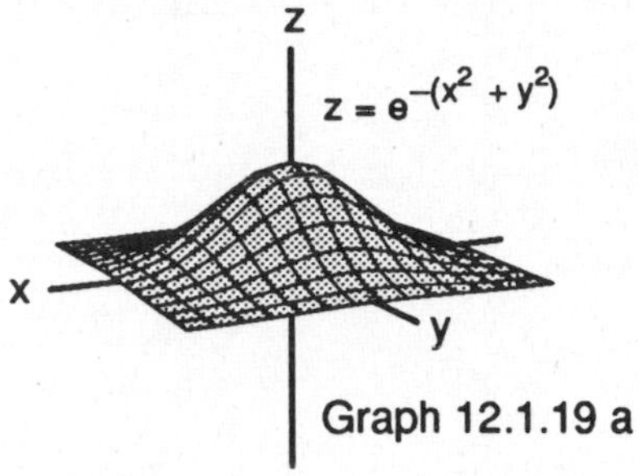

Graph 12.1.19 a

b)

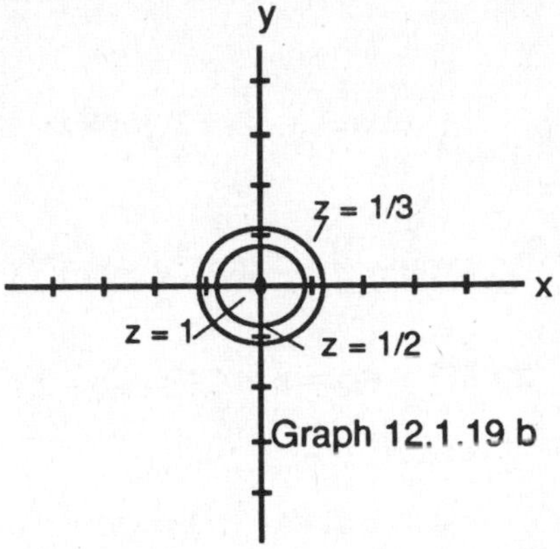

Graph 12.1.19 b

21. f

23. a

25. d

27. Domain: All (x,y,z);

Range: All Real Numbers.

Level surfaces are spheres with center (0,0,0).

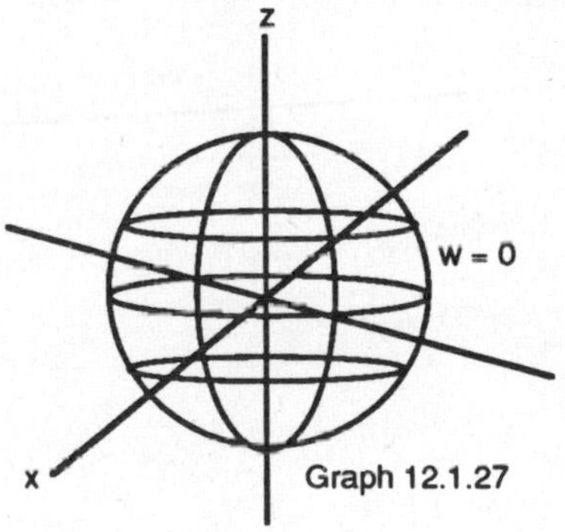

Graph 12.1.27

29. Domain: All (x,y,z);

Range: $-\frac{\pi}{2} < w < \frac{\pi}{2}$.

Level surfaces are paraboloids with the z–axis as axis.

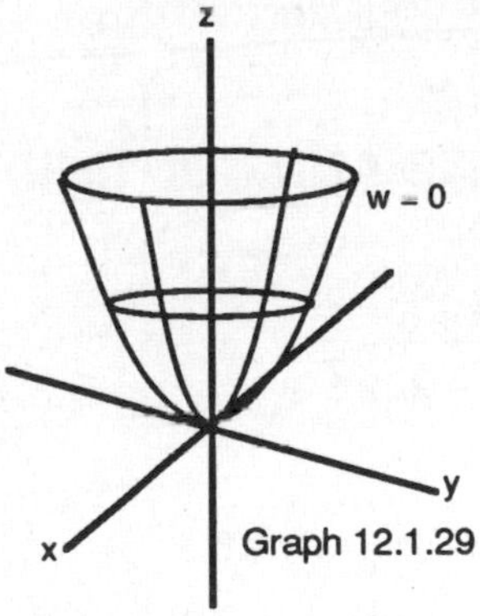

Graph 12.1.29

31. Domain: All (x,y,z)

Range: $0 < w \leq 1$

Level surfaces are pairs of parallel planes perpendicular to the z–axis, equidistant from (0,0,0).

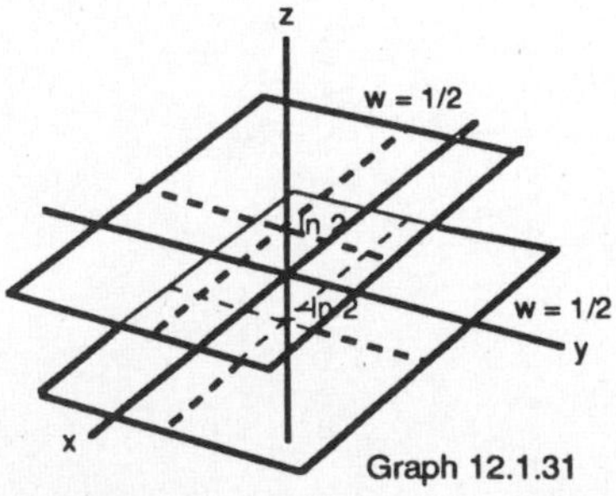

Graph 12.1.31

33. Domain: All (x,y,z) so that z > \r(

Range: All Real Numbers

Level surfaces are cones with the z–axis as axis.

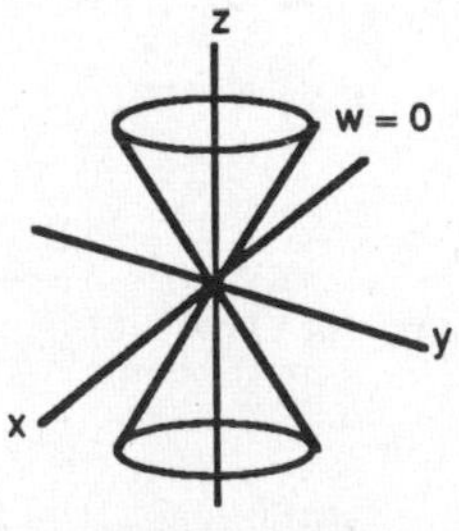

Graph 12.1.33

35. $f(x,y) = 16 - x^2 - y^2$ and $(2\sqrt{2},\sqrt{2}) \Rightarrow z = 16 - \left(2\sqrt{2}\right)^2 - \left(\sqrt{2}\right)^2 = 6 \Rightarrow 6 = 16 - x^2 - y^2 \Rightarrow x^2 + y^2 = 10$

37. $f(x,y) = \int_x^y \frac{dt}{1+t^2}\, dt$ at $\left(-\sqrt{2},\sqrt{2}\right) \Rightarrow z = \tan^{-1}y - \tan^{-1}x$. At $\left(-\sqrt{2},\sqrt{2}\right)$, $z = \tan^{-1}\sqrt{2} - \tan^{-1}\left(-\sqrt{2}\right) =$ $2\tan^{-1}\sqrt{2}$. $\therefore\ 2\tan^{-1}\sqrt{2} = \tan^{-1}y - \tan^{-1}x \Rightarrow \tan\left(\tan^{-1}\sqrt{2}\right) = \tan\left(\tan^{-1}y - \tan^{-1}x\right) \Rightarrow \frac{2\sqrt{2}}{1-\left(\sqrt{2}\right)^2} =$ $\frac{y-x}{1+xy} \Rightarrow -2\sqrt{2}(1+xy) = y - x \Rightarrow -2\sqrt{2} - 2\sqrt{2}\,xy = y - x$

39. $f(x,y,z) = \sqrt{x-y} - \ln z$ at $(3,-1,1) \Rightarrow w = \sqrt{x-y} - \ln z$. At $(3,-1,1)$, $w = \sqrt{3-(-1)} - \ln 1 = 2$. $\therefore\ 2 = \sqrt{x-y} -$ $\ln z \Rightarrow \ln z = \sqrt{x-y} - 2 \Rightarrow z = e^{\sqrt{x-y}-2}$

41. $f(x,y,z) = \sum_{n=0}^{\infty} \frac{(x+y)^n}{n!\,z^n}$ at $(\ln 2, \ln 4, 3) \Rightarrow w = \sum_{n=0}^{\infty} \frac{(x+y)^n}{n!\,z^n} = e^{(x+y)/z}$. At $(\ln 2, \ln 4, 3)$, $w = e^{(\ln 2 + \ln 4)/3} =$ $e^{(\ln 8)/3} = e^{\ln 2} = 2$. $\therefore\ 2 = e^{(x+y)/z} \Rightarrow \ln 2 = \frac{x+y}{z} \Rightarrow z\ln 2 = x + y$

43. $w = 4\left(\frac{Th}{d}\right)^{1/2} = 4\left(\frac{(290\text{ k})(16.8\text{ km})}{5\text{ k/km}}\right)^{1/2} = 124.86$ km. $\therefore$ must be 62.43 km south of Nantucket.

SECTION 12.2 LIMITS AND CONTINUITY

1. $\lim\limits_{(x,y)\to(0,0)} \frac{3x^2 - y^2 + 5}{x^2 + y^2 + 2} = \frac{5}{2}$

3. $\lim\limits_{(x,y)\to(0,\ln 2)} e^{x-y} = \frac{1}{2}$

5. $\lim\limits_{P\to(1,3,4)} \sqrt{x^2 + y^2 + z^2 - 1} = 5$

7. $\lim\limits_{(x,y)\to(0,\pi/4)} \sec x \tan y = 1$

9. $\lim\limits_{(x,y)\to(1,1)} \cos\left(\sqrt[3]{|xy| - 1}\right) = 1$

11. $\lim\limits_{(x,y)\to(0,0)} \frac{e^y \sin x}{x} = 1$

13. $\lim\limits_{\substack{(x,y)\to(1,1)\\ x\neq y}} \frac{x^2 - 2xy + y^2}{x - y} = \lim\limits_{(x,y)\to(1,1)} \frac{(x-y)^2}{x-y} = \lim\limits_{(x,y)\to(1,1)} (x - y) = 0$

15. $\lim\limits_{\substack{(x,y)\to(1,1)\\ x\neq 1}} \frac{xy - y - 2x + 2}{x - 1} = \lim\limits_{(x,y)\to(1,1)} \frac{(x-1)(y-2)}{x-1} = \lim\limits_{(x,y)\to(1,1)} (y - 2) = -1$

17. $\lim_{P\to(2,3,-6)} \sqrt{x^2+y^2+z^2} = 7$

19. $\lim_{P\to(3,3,0)} \left(\sin^2 x + \cos^2 y + \sec^2 z\right) = \sin^2 3 + \cos^2 3 + \sec^2 0 = 2$

21. $\lim_{P\to(-1/4,\pi/2,2)} \tan^{-1}(xyz) = \tan^{-1}\left(-\frac{\pi}{4}\right)$

23. a) Continuous at all (x,y)
    b) Continuous at all (x,y) except (0,0)

25. a) Continuous at all (x,y) except where $x = 0$ or $y = 0$
    b) Continuous at all (x,y)

27. a) Continuous at all (x,y,z)
    b) Continuous at all (x,y,z) except the interior of the cylinder $x^2 + y^2 = 1$

29. a) Continuous at all (x,y,z) so that $(x,y,z) \neq (x,y,0)$
    b) Continuous at all (x,y,z) except those on the sphere $x^2 + y^2 + z^2 = 1$

31. $\lim_{\substack{(x,y)\to(0,0) \\ \text{along } y = x, \\ x > 0}} \frac{x}{\sqrt{x^2+y^2}} = \lim_{(x,y)\to(0,0)} \frac{x}{\sqrt{x^2+x^2}} = \lim_{(x,y)\to(0,0)} \frac{x}{\sqrt{2}\,|x|} = \lim_{(x,y)\to(0,0)} \frac{x}{\sqrt{2}\,x} = \lim_{(x,y)\to(0,0)} \frac{1}{\sqrt{2}} = \frac{1}{\sqrt{2}}$

$\lim_{\substack{(x,y)\to(0,0) \\ \text{along } y = x, \\ x < 0}} \frac{x}{\sqrt{x^2+y^2}} = \lim_{(x,y)\to(0,0)} \frac{x}{\sqrt{2}\,|x|} = \lim_{(x,y)\to(0,0)} \frac{x}{\sqrt{2}(-x)} = \lim_{(x,y)\to(0,0)} -\frac{1}{\sqrt{2}} = -\frac{1}{\sqrt{2}}$

$\therefore$ consider paths along $y = x$ where $x > 0$ or $x < 0$.

33. $\lim_{\substack{(x,y)\to(0,0) \\ \text{along } y = kx^2}} \frac{x^4 - y^2}{x^4 + y^2} = \lim_{(x,y)\to(0,0)} \frac{x^4 - \left(kx^2\right)^2}{x^4 + \left(kx^2\right)^2} = \lim_{(x,y)\to(0,0)} \frac{x^4 - k^2x^4}{x^4 + k^2x^4} = \frac{1-k^2}{1+k^2} \Rightarrow$ different limits for different values of k. $\therefore$ consider paths along $y = kx^2$, k a constant.

35. $\lim_{\substack{(x,y)\to(0,0) \\ \text{along } y = kx,\ k \neq -1}} \frac{x - y}{x + y} = \lim_{(x,y)\to(0,0)} \frac{x - kx}{x + kx} = \frac{1-k}{1+k} \Rightarrow$ different limits for different values of k. $\therefore$ consider paths along $y = kx$, k a constant, $k \neq -1$.

37. $\lim_{\substack{(x,y)\to(0,0) \\ \text{along } y = kx^2,\ k \neq 0}} \frac{x^2 + y}{y} = \lim_{(x,y)\to(0,0)} \frac{x^2 + kx^2}{kx^2} = \frac{1+k}{k} \Rightarrow$ different limits for different values of k. $\therefore$ consider paths along $y = kx^2$, k a constant, $k \neq 0$.

39. a) $f(x,y)\big|_{y=mx} = \dfrac{2m}{1+m^2} = \dfrac{2\tan\theta}{1+\tan^2\theta} = \sin 2\theta$. The value of f(x,y) is $\sin 2\theta$ where $\tan\theta = m$ along $y = mx$.

b) Since $f(x,y)\big|_{y=mx} = \sin 2\theta$ and since $-1 \le \sin 2\theta \le 1$ for every θ, $\lim\limits_{(x,y)\to(0,0)} f(x,y)$ varies from -1 to 1 along $y = mx$.

41. $\lim\limits_{(x,y)\to(0,0)} 1 - \dfrac{x^2y^2}{3} = 1$, $\lim\limits_{(x,y)\to(0,0)} 1 = 1 \Rightarrow \lim\limits_{(x,y)\to(0,0)} \dfrac{\tan^{-1}xy}{xy} = 1$

43. $\lim\limits_{(x,y)\to(0,0)} \dfrac{x^3 - xy^2}{x^2+y^2} = \lim\limits_{r\to 0} \dfrac{r^3\cos^3\theta - r\cos\theta\, r^2\sin^2\theta}{r^2\cos^2\theta + r^2\sin^2\theta} = \lim\limits_{r\to 0} \dfrac{r\left(\cos^3\theta - \cos\theta\sin^2\theta\right)}{1} = 0$

45. $\lim\limits_{(x,y)\to(0,0)} \dfrac{y^2}{x^2+y^2} = \lim\limits_{r\to 0} \dfrac{r^2\sin^2\theta}{r^2} = \lim\limits_{r\to 0} (\sin^2\theta)$ does not exist since $\sin^2\theta$ is between 0 and 1 depending on the path.

47. $\lim\limits_{(x,y)\to(0,0)} \tan^{-1}\left[\dfrac{|x|+|y|}{x^2+y^2}\right] = \lim\limits_{r\to 0} \tan^{-1}\left[\dfrac{|r\cos\theta| + |r\sin\theta|}{r^2}\right] = \lim\limits_{r\to 0} \tan^{-1}\left[\dfrac{|r|\left(|\cos\theta| + |\sin\theta|\right)}{r^2}\right]$.

If $r \to 0^+$, then $\lim\limits_{r\to 0^+} \tan^{-1}\left[\dfrac{|r|\left(|\cos\theta|+|\sin\theta|\right)}{r^2}\right] = \lim\limits_{r\to 0^+} \tan^{-1}\left[\dfrac{r\left(|\cos\theta|+|\sin\theta|\right)}{r^2}\right] =$

$\lim\limits_{r\to 0^+} \tan^{-1}\left[\dfrac{\left(|\cos\theta|+|\sin\theta|\right)}{r}\right] = \dfrac{\pi}{2}$. If $r \to 0^-$, then $\lim\limits_{r\to 0^-} \tan^{-1}\left[\dfrac{|r|\left(|\cos\theta|+|\sin\theta|\right)}{r^2}\right] =$

$\lim\limits_{r\to 0^-} \tan^{-1}\left[\dfrac{-r\left(|\cos\theta|+|\sin\theta|\right)}{r^2}\right] = \lim\limits_{r\to 0^-} \tan^{-1}\left[\dfrac{-\left(|\cos\theta|+|\sin\theta|\right)}{r}\right] = \dfrac{\pi}{2}$. $\therefore$ the limit is $\dfrac{\pi}{2}$

49. Important Note: All points (x,y) so that x = 0 (the y–axis) are not in the domain of f.

Then $\lim\limits_{(x,y)\to(0,0)} \ln\left[\dfrac{y^2+3x}{3x^2+x}\right] = \lim\limits_{r\to 0} \ln\left[\dfrac{r^2\sin^2\theta + 3r\cos\theta}{3r^2\cos^2\theta + r\cos\theta}\right] = \lim\limits_{r\to 0} \ln\left[\dfrac{r\sin^2\theta + 3\cos\theta}{3r\cos^2\theta + \cos\theta}\right] =$

$\lim\limits_{r\to 0} \ln\left[\dfrac{\dfrac{r\sin^2\theta}{\cos\theta} + 3}{3r\cos\theta + 1}\right] = \ln 3$ except along $\theta = \dfrac{\pi}{2}$ or $\dfrac{3\pi}{2}$. But $\theta = \dfrac{\pi}{2}$ or $\dfrac{3\pi}{2} \Rightarrow x = 0$, such points are not in the domain. $\therefore \lim\limits_{(x,y)\to(0,0)} f(x,y) = \ln 3 \Rightarrow$ let $f(0,0) = \ln 3$.

51. Let $\delta = 0.1$. Then $\sqrt{x^2+y^2} < \delta \Rightarrow \sqrt{x^2+y^2} < 0.1 \Rightarrow x^2+y^2 < 0.01 \Rightarrow \left|x^2+y^2-0\right| < 0.01 \Rightarrow \left|f(x,y) - f(0,0)\right| < 0.01 = \varepsilon$.

53. Let $\delta = 0.005$. Then when $|x| < \delta$ and $|y| < \delta$, $\left|f(x,y) - f(0,0)\right| = \left|\frac{x+y}{x^2+1} - 0\right| = \left|\frac{x+y}{x^2+1}\right| \le |x+y| \le |x| + |y|$ $< 0.005 + 0.005 = 0.01 = \varepsilon$.

55. Let $\delta = \sqrt{0.015}$. Then when $\sqrt{x^2+y^2+z^2} < \delta = \sqrt{0.015}$, $\left|f(x,y,z) - f(0,0,0)\right| = \left|x^2+y^2+z^2-0\right| =$ $\left|x^2+y^2+z^2\right| = \left(\sqrt{x^2+y^2+z^2}\right)^2 < \left(\sqrt{0.015}\right)^2 = 0.015 = \varepsilon$.

57. Let $\delta = 0.005$. Then when $|x| < \delta$, $|y| < \delta$, and $|z| < \delta$, $\left|f(x,y,z) - f(0,0,0)\right| = \left|\frac{x+y+z}{x^2+y^2+z^2+1} - 0\right| =$ $\left|\frac{x+y+z}{x^2+y^2+z^2+1}\right| \le |x+y+z| \le |x| + |y| + |z| < 0.005 + 0.005 + 0.005 = 0.015 = \varepsilon$.

59. $\lim\limits_{(x,y,z)\to(x_0,y_0,z_0)} f(x,y,z) = \lim\limits_{(x,y,z)\to(x_0,y_0,z_0)} (x+y+z) = x_0 + y_0 + z_0 = f(x_0,y_0,z_0) \Rightarrow f$ is continuous at every (x_0,y_0,z_0).

SECTION 12.3 PARTIAL DERIVATIVES

1. $\frac{\partial f}{\partial x} = 4x, \frac{\partial f}{\partial y} = -3$

3. $\frac{\partial f}{\partial x} = 2x(y+2), \frac{\partial f}{\partial y} = x^2 - 1$

5. $\frac{\partial f}{\partial x} = 2y(xy-1), \frac{\partial f}{\partial y} = 2x(xy-1)$

7. $\frac{\partial f}{\partial x} = \frac{x}{\sqrt{x^2+y^2}}, \frac{\partial f}{\partial y} = \frac{y}{\sqrt{x^2+y^2}}$

9. $\frac{\partial f}{\partial x} = \frac{-1}{(x+y)^2}, \frac{\partial f}{\partial y} = \frac{-1}{(x+y)^2}$

11. $\frac{\partial f}{\partial x} = \frac{-y^2-1}{(xy-1)^2}, \frac{\partial f}{\partial y} = \frac{-x^2-1}{(xy-1)^2}$

13. $\frac{\partial f}{\partial x} = \frac{1}{x+y}, \frac{\partial f}{\partial y} = \frac{1}{x+y}$

15. $\frac{\partial f}{\partial x} = y\,e^{xy}\ln y, \frac{\partial f}{\partial y} = x\,e^{xy}\ln y + \frac{e^{xy}}{y}$

17. $\frac{\partial f}{\partial x} = -6\cos(3x-y^2)\sin(3x-y^2)$
$\frac{\partial f}{\partial y} = -4y\cos(3x-y^2)\sin(3x-y^2)$

19. $\frac{\partial f}{\partial x} = y\,x^{y-1}, \frac{\partial f}{\partial y} = x^y \ln x$

21. $\frac{\partial f}{\partial x} = -f(x), \frac{\partial f}{\partial y} = f(y)$

23. $f_x(x,y,z) = y^2$, $f_y(x,y,z) = 2xy$,
$f_z(x,y,z) = -4z$

25. $f_x(x,y,z) = 1$, $f_y(x,y,z) = -y(y^2+z^2)^{-1/2}$,
$f_z(x,y,z) = -z(y^2+z^2)^{-1/2}$

27. $f_x(x,y,z) = \dfrac{yz}{\sqrt{1-x^2y^2z^2}}$, $f_y(x,y,z) = \dfrac{xz}{\sqrt{1-x^2y^2z^2}}$, $f_z(x,y,z) = \dfrac{xy}{\sqrt{1-x^2y^2z^2}}$

29. $f_x(x,y,z) = \dfrac{1}{x+2y+3z}$, $f_y(x,y,z) = \dfrac{2}{x+2y+3z}$, $f_z(x,y,z) = \dfrac{3}{x+2y+3z}$

31. $f_x(x,y,z) = -2x\, e^{-(x^2+y^2+z^2)}$, $f_y(x,y,z) = -2y\, e^{-(x^2+y^2+z^2)}$, $f_z(x,y,z) = -2z\, e^{-(x^2+y^2+z^2)}$

33. $f_x(x,y,z) = \text{sech}^2(x + 2y + 3z$, $f_y(x,y,z) = 2\,\text{sech}^2(x + 2y + 3z)$, $f_z(x,y,z) = 3\,\text{sech}^2(x + 2y + 3z)$

35. $\dfrac{\partial f}{\partial t} = -2\pi \sin(2\pi t - \alpha)$, $\dfrac{\partial f}{\partial a} = \sin(2\pi t - \alpha)$

37. $\dfrac{\partial h}{\partial \rho} = \sin\phi\cos\theta$, $\dfrac{\partial h}{\partial \phi} = \rho\cos\phi\cos\theta$, $\dfrac{\partial h}{\partial \theta} = -\rho\sin\phi\sin\theta$

39. $W_P(P,V,p,v,g) = V$, $W_V(P,V,p,v,g) = P + \dfrac{pv^2}{2g}$, $W_p(P,V,p,v,g) = \dfrac{Vv^2}{2g}$, $W_v(P,V,p,v,g) = \dfrac{Vpv}{g}$, $W_g(P,V,p,v,g) =$
$-\dfrac{Vpv^2}{2g^2}$

41. $\dfrac{\partial f}{\partial x} = 1 + y$, $\dfrac{\partial f}{\partial y} = 1 + x$, $\dfrac{\partial^2 f}{\partial x^2} = 0$, $\dfrac{\partial^2 f}{\partial y^2} = 0$, $\dfrac{\partial^2 f}{\partial y \partial x} = \dfrac{\partial^2 f}{\partial x \partial y} = 1$

43. $\dfrac{\partial g}{\partial x} = 2xy + y\cos x$, $\dfrac{\partial g}{\partial y} = x^2 - \sin y + \sin x$, $\dfrac{\partial^2 g}{\partial x^2} = 2y - y\sin x$, $\dfrac{\partial^2 g}{\partial y^2} = -\cos y$, $\dfrac{\partial^2 g}{\partial y \partial x} = \dfrac{\partial^2 g}{\partial x \partial y} = 2x + \cos x$

45. $\dfrac{\partial r}{\partial x} = \dfrac{1}{x+y}$, $\dfrac{\partial r}{\partial y} = \dfrac{1}{x+y}$, $\dfrac{\partial^2 r}{\partial x^2} = \dfrac{-1}{(x+y)^2}$, $\dfrac{\partial^2 r}{\partial y^2} = \dfrac{-1}{(x+y)^2}$, $\dfrac{\partial^2 r}{\partial y \partial x} = \dfrac{\partial^2 r}{\partial x \partial y} = \dfrac{-1}{(x+y)^2}$

47. $\dfrac{\partial w}{\partial x} = \dfrac{2}{2x+3y}$, $\dfrac{\partial w}{\partial y} = \dfrac{3}{2x+3y}$, $\dfrac{\partial^2 w}{\partial y \partial x} = \dfrac{-6}{(2x+3y)^2}$ and $\dfrac{\partial^2 w}{\partial x \partial y} = \dfrac{-6}{(2x+3y)^2}$

49. $\dfrac{\partial w}{\partial x} = y^2 + 2xy^3 + 3x^2y^4$, $\dfrac{\partial w}{\partial y} = 2xy + 3x^2y^2 + 4x^3y^3$, $\dfrac{\partial^2 w}{\partial y \partial x} = 2y + 6xy^2 + 12x^2y^3$ and
$\dfrac{\partial^2 w}{\partial x \partial y} = 2y + 6xy^2 + 12x^2y^3$

51. a) x first b) y first c) x first d) x first e) y first f) y first

53. $y\dfrac{\partial x}{\partial x} + x\dfrac{\partial y}{\partial x} + 3z^2x\dfrac{\partial z}{\partial x} + z^3\dfrac{\partial x}{\partial x} - 2y\dfrac{\partial z}{\partial x} - 2z\dfrac{\partial y}{\partial x} = 0 \Rightarrow y + 3z^2x\dfrac{\partial z}{\partial x} + z^3 - 2y\dfrac{\partial z}{\partial x} = 0$. At (1,1,1),
$1 + 3\dfrac{\partial z}{\partial x} + 1 - 2\dfrac{\partial z}{\partial x} = 0 \Rightarrow \dfrac{\partial z}{\partial x} = -2$. (Note: since x and y are independent, $\dfrac{\partial y}{\partial x} = 0$.)

55. $a^2 = b^2 + c^2 - 2bc\cos A \Rightarrow 2a = 2bc\sin A\frac{\partial A}{\partial a} \Rightarrow \frac{\partial A}{\partial a} = \frac{a}{bc\sin A}$. Also $0 = 2b - 2c\cos A + 2bc\sin A\frac{\partial A}{\partial b}$
$\Rightarrow 2c\cos A - 2b = 2bc\sin A\frac{\partial A}{\partial b} \Rightarrow \frac{c\cos A - b}{bc\sin A} = \frac{\partial A}{\partial b}$

57. $1 = v_x \ln u + \frac{v}{u}u_x$ and $0 = u_x \ln v + \frac{u}{v}v_x$. $A = \begin{vmatrix} \ln u & \frac{v}{u} \\ \frac{v}{u} & \ln v \end{vmatrix} = \ln u \ln v - \frac{v^2}{u^2}$. $X = \begin{vmatrix} 1 & \frac{v}{u} \\ 0 & \ln v \end{vmatrix} = \ln v - \frac{v}{u}$

$$\therefore\ v_x = \frac{\ln v - \frac{v}{u}}{\ln u \ln v - \frac{v^2}{u^2}} = \frac{u^2\ln v - vu}{u^2\ln u\ln v - v^2}$$

59. $\frac{\partial f}{\partial x} = 2x, \frac{\partial f}{\partial y} = 2y, \frac{\partial f}{\partial z} = -4z \Rightarrow \frac{\partial^2 f}{\partial x^2} = 2, \frac{\partial^2 f}{\partial y^2} = 2, \frac{\partial^2 f}{\partial z^2} = -4 \Rightarrow \frac{\partial^2 f}{\partial x^2} + \frac{\partial^2 f}{\partial y^2} + \frac{\partial^2 f}{\partial z^2} = 2 + 2 + (-4) = 0$

61. $\frac{\partial f}{\partial x} = \frac{x}{x^2+y^2}, \frac{\partial f}{\partial y} = \frac{y}{x^2+y^2}, \frac{\partial^2 f}{\partial x^2} = \frac{y^2-x^2}{(x^2+y^2)^2}, \frac{\partial^2 f}{\partial y^2} = \frac{x^2-y^2}{(x^2+y^2)^2}$
$\therefore\ \frac{\partial^2 f}{\partial x^2} + \frac{\partial^2 f}{\partial y^2} = \frac{y^2-x^2}{(x^2+y^2)^2} + \frac{x^2-y^2}{(x^2+y^2)^2} = 0$

63. $\frac{\partial f}{\partial x} = 3e^{3x+4y}\cos 5z, \frac{\partial f}{\partial y} = 4e^{3x+4y}\cos 5z, \frac{\partial f}{\partial z} = -5e^{3x+4y}\sin 5z.\ \frac{\partial^2 f}{\partial x^2} = 9e^{3x+4y}\cos 5z,$
$\frac{\partial^2 f}{\partial y^2} = 16e^{3x+4y}\cos 5z, \frac{\partial^2 f}{\partial z^2} = -25e^{3x+4y}\cos 5z \Rightarrow \frac{\partial^2 f}{\partial x^2} + \frac{\partial^2 f}{\partial y^2} + \frac{\partial^2 f}{\partial z^2} = 9e^{3x+4y}\cos 5z + 16e^{3x+4y}\cos 5z -$
$25e^{3x+4y}\cos 5z = 0$

65. $f_x(x,y,z) = 2nx\left(x^2+y^2+z^2\right)^{n-1} \Rightarrow f_{xx}(x,y,z) = 2n\left(x^2+y^2+z^2\right)^{n-1} + 4n(n-1)x^2\left(x^2+y^2+z^2\right)^{n-2}$.
$f_y(x,y,z) = 2ny\left(x^2+y^2+z^2\right)^{n-1} \Rightarrow f_{yy}(x,y,z) = 2n\left(x^2+y^2+z^2\right)^{n-1} + 4n(n-1)y^2\left(x^2+y2+z^2\right)^{n-2}$.
$f_z(x,y,z) = 2nz\left(x^2+y^2+z^2\right)^{n-1} \Rightarrow f_{zz}(x,y,z) = 2n\left(x^2+y^2+z^2\right)^{n-1} + 4n(n-1)z^2\left(x^2+y^2+z^2\right)^{n-2}$.
$\therefore\ \frac{\partial^2 f}{\partial x^2} + \frac{\partial^2 f}{\partial y^2} + \frac{\partial^2 f}{\partial z^2} = 2n\left(x^2+y^2+z^2\right)^{n-1} + 4n(n-1)x^2\left(x^2+y^2+z^2\right)^{n-2} + 2n\left(x^2+y^2+z^2\right)^{n-1} +$
$4n(n-1)y^2\left(x^2+y2+z^2\right)^{n-2} + 2n\left(x^2+y^2+z^2\right)^{n-1} + 4n(n-1)z^2\left(x^2+y^2+z^2\right)^{n-2} = 6n\left(x^2+y^2+z^2\right)^{n-1}$
$+ 12n(n-1)\left(x^2+y^2+z^2\right)^{n-1} = (6n + 12n(n-1))\left(x^2+y^2+z^2\right)^{n-1} = 0 \Rightarrow 6n + 12n(n-1) = 0 \Rightarrow$
$12n^2 - 6n = 0 \Rightarrow 6n = 0$ or $2n - 1 = 0 \Rightarrow n = 0$ or $n = \frac{1}{2}$.

67. $\frac{\partial w}{\partial x} = \cos(x+ct), \frac{\partial w}{\partial t} = c\cos(x+ct).\ \frac{\partial^2 w}{\partial x^2} = -\sin(x+ct), \frac{\partial^2 w}{\partial t^2} = -c^2\sin(x+ct)$
$\therefore\ \frac{\partial^2 w}{\partial t^2} = c^2(-\sin(x+ct)) = c^2\frac{\partial^2 w}{\partial x^2}$

69. $\frac{\partial w}{\partial x} = \cos(x + ct) - 2\sin(2x + 2ct)$, $\frac{\partial w}{\partial t} = c\cos(x + ct) - 2c\sin(2x + 2ct)$. $\frac{\partial^2 w}{\partial x^2} = -\sin(x + ct) -$ $4\cos(2x + 2ct)$, $\frac{\partial^2 w}{\partial t^2} = -c^2\sin(x + ct) - 4c^2\cos(2x + 2ct)$ $\therefore$ $\frac{\partial^2 w}{\partial t^2} = c^2(-\sin(x + ct) - 4\cos(2x + 2ct)) =$ $c^2\frac{\partial^2 w}{\partial x^2}$

71. $\frac{\partial w}{\partial x} = 2\sec^2(2x - 2ct)$, $\frac{\partial w}{\partial t} = -2c\sec^2(2x - 2ct)$. $\frac{\partial^2 w}{\partial x^2} = 8\sec^2(2x - 2ct)\tan(2x - 2ct)$, $\frac{\partial^2 w}{\partial t^2} = 8c^2\sec^2(2x - 2ct)\tan(2x - 2ct)$ $\therefore$ $\frac{\partial^2 w}{\partial t^2} = c^2\left(8\sec^2(2x - 2ct)\tan(2x - 2ct)\right) = c^2\frac{\partial^2 w}{\partial x^2}$

73. $\frac{\partial w}{\partial t} = \frac{\partial f}{\partial u}\frac{\partial u}{\partial t} = \frac{\partial f}{\partial u}(ac) \Rightarrow \frac{\partial^2 w}{\partial t^2} = ac\frac{\partial^2 f}{\partial u^2}(ac) = a^2c^2\frac{\partial^2 f}{\partial u^2}$. $\frac{\partial w}{\partial x} = \frac{\partial f}{\partial u}\frac{\partial u}{\partial x} = \frac{\partial f}{\partial u}(a) \Rightarrow \frac{\partial^2 w}{\partial x^2} = a\frac{\partial^2 f}{\partial u^2}(a)$ $= a^2\frac{\partial^2 f}{\partial u^2}$.. $\therefore$ $\frac{\partial^2 w}{\partial t^2} = a^2c^2\frac{\partial^2 f}{\partial u^2}. = c^2\left(a^2\frac{\partial^2 f}{\partial u^2}.\right) = c^2\frac{\partial^2 w}{\partial x^2}$

SECTION 12.4 DIFFERENTIABILITY, LINEARIZATION, AND DIFFERENTIALS

1. a) $f(0,0) = 1$, $f_x(x,y) = 2x \Rightarrow f_x(0,0) = 0$, $f_y(x,y) = 2y \Rightarrow f_y(0,0) = 0 \Rightarrow L(x,y) = 1 + 0(x - 0) + 0(y - 0) = 1$

   b) $f(1,1) = 3$, $f_x(1,1) = 2$, $f_y(1,1) = 2 \Rightarrow L(x,y) = 3 + 2(x - 1) + 2(y - 1) = 2x + 2y - 1$

3. a) $f(0,0) = 1$, $f_x(x,y) = e^x\cos y \Rightarrow f_x(0,0) = 1$, $f_y(x,y) = -e^x\sin y \Rightarrow f_y(0,0) = 0 \Rightarrow L(x,y) = 1 + 1(x - 0) +$ $0(y - 0) = 1 + x$

   b) $f\left(0,\frac{\pi}{2}\right) = 0$, $f_x\left(0,\frac{\pi}{2}\right) = 0$, $f_y\left(0,\frac{\pi}{2}\right) = -1 \Rightarrow L(x,y) = 0 + 0(x - 0) - 1\left(y - \frac{\pi}{2}\right) = -y + \frac{\pi}{2}$

5. a) $f(0,0) = 5$, $f_x(x,y) = 3$ for all (x,y), $f_y(x,y) = -4$ for all $(x,y) \Rightarrow L(x,y) = 5 + 3(x - 0) - 4(y - 0) =$ $5 + 3x - 4y$

   b) $f(1,1) = 4$, $f_x(1,1) = 3$, $f_y(1,1) = -4 \Rightarrow L(x,y) = 4 + 3(x - 1) - 4(y - 1) = 3x - 4y + 5$

7. $f(2,1) = 3$, $f_x(x,y) = 2x - 3y \Rightarrow f_x(2,1) = 1$, $f_y(x,y) = -3x \Rightarrow f_y(2,1) = -6 \Rightarrow L(x,y) = 3 + 1(x - 2) - 6(y - 1)$ $= 7 + x - 6y$. $f_{xx}(x,y) = 2$, $f_{yy}(x,y) = 0$, $f_{xy}(x,y) = -3 \Rightarrow M = 3$. $\therefore$ $|E(x,y)| \le \frac{1}{2}(3)(|x - 2| + |y - 1|)^2 \le$ $\frac{3}{2}(0.1 + 0.1)^2 = 0.06$

9. $f(0,0) = 1$, $f_x(x,y) = \cos y \Rightarrow f_x(0,0) = 1$, $f_y(x,y) = 1 - x\sin y \Rightarrow f_y(0,0) = 1 \Rightarrow L(x,y) = 1 + 1(x - 0) +$ $1(y - 0) = x + y + 1$. $f_{xx}(x,y) = 0$, $f_{yy}(x,y) = 0$, $f_{xy}(x,y) = -\sin y \Rightarrow M = 1$. $\therefore$ $|E(x,y)| \le \frac{1}{2}(1)(|x| + |y|)^2 \le \frac{1}{2}(0.2 + 0.2)^2 = 0.08$

11. $f(0,0) = 1$, $f_x(x,y) = e^x \cos y \Rightarrow f_x(0,0) = 1$, $f_y(x,y) = -e^x \sin y \Rightarrow f_y(0,0) = 0 \Rightarrow L(x,y) = 1 + 1(x-0) + 0(y-0) = 1 + x$. $f_{xx}(x,y) = e^x \cos y$, $f_{yy}(x,y) = -e^x \cos y$, $f_{xy}(x,y) = -e^x \sin y$. $|x| \le 0.1 \Rightarrow -0.1 \le x \le 0.1$, $|y| \le 0.1 \Rightarrow -0.1 \le y \le 0.1. \Rightarrow$ max of $|f_{xx}(x,y)|$ on R is $e^{0.1}\cos(0.1) \le 1.11$, max of $|f_{yy}(x,y)|$ on R is $e^{0.1}\cos(0.1) \le 1.11$, max of $|f_{xy}(x,y)|$ on R is $e^{0.1}\sin(0.1) \le 0.002 \Rightarrow M = 1.11$.
$\therefore\ |E(x,y)| \le \frac{1}{2}(1.11)(|x| + |y|)^2 \le 0.555(0.1 + 0.1)^2 = 0.0222$

13. Let the width, w, be the long side. Then $A = lw \Rightarrow dA = A_l\, dl + A_w\, dw \Rightarrow dA = w\, dl + l\, dw$. Since $w > l$, dA is more sensitive to a change in w than l. $\therefore$ pay more attention to the width.

15. $T_x(x,y) = e^y + e^{-y}$, $T_y(x,y) = x(e^y - e^{-y}) \Rightarrow dT = T_x(x,y)\, dx + T_y(x,y)\, dy = (e^y + e^{-y})dx + x(e^y - e^{-y})dy \Rightarrow dT|_{(2,\ln 2)} = 2.5\, dx + 3.0\, dy$. If $|dx| \le 0.1$, $|dy| \le 0.02$, then the maximum possible error (estimate) $\le 2.5(0.1) + 3.0(0.02) = 0.31$ in magnitude.

17. $V_r = 2\pi rh$, $V_h = \pi r^2 \Rightarrow dV = V_r\, dr + V_h\, dh \Rightarrow dV = 2\pi rh\, dr + \pi r^2\, dh \Rightarrow dV|_{(5,12)} = 120\pi\, dr + 25\pi\, dh$. Since $|dr| \le 0.1$ cm, $|dh| \le 0.1$ cm, $dV \le 120\pi(0.1) + 25\pi(0.1) = 14.5\pi\ \text{cm}^3$. $V(5,12) = 300\pi\ \text{cm}^3 \Rightarrow$ Maximum percentage error $= \pm \frac{14.5\pi}{300\pi} \times 100 = \pm 4.83\%$

19. $df = f_x(x,y)\, dx + f_y(x,y)\, dy = 3x^2y^4dx + 4x^3y^3dy \Rightarrow df|_{(1,1)} = 3\, dx + 4\, dy$. Let $dx = dy \Rightarrow df = 7\, dx$. $|df| \le 0.1 \Rightarrow 7|dx| < 0.1 \Rightarrow |dx| \le \frac{0.1}{7} \approx 0.014$. $\therefore$ for the square, let $|x - 1| < 0.014$, $|y - 1| \le 0.014$

21. $dR = \left(\frac{R}{R_1}\right)^2 dR_1 + \left(\frac{R}{R_2}\right)^2 dR_2$ (See Exercise 20 above). R_1 changes from 20 to 20.1 ohms $\Rightarrow dR_1 =$ 0.1 ohms, R_2 changes from 25 to 24.9 ohms $\Rightarrow dR_2 = -0.1$ ohms. $\frac{1}{R} = \frac{1}{R_1} + \frac{1}{R_2} \Rightarrow R = \frac{100}{9}$ ohms.
$dR|_{(20,25)} = \frac{(100/9)^2}{(20)^2}(0.1) + \frac{(100/9)^2}{(25)^2}(-0.1) = 0.011$ ohms $\Rightarrow$ Percentage change $= \frac{dR}{R}\Big|_{(20,25)} \times 100$
$= \frac{0.011}{100/9} \times 100 \approx 0.099\%$

23. If the first partial derivatives are continuous throughout on open region R, Then $f(x,y) = f(x_0,y_0) + f_x(x_0,y_0)\Delta x + f_y(x_0,y_0)\Delta y + \varepsilon_1 \Delta x + \varepsilon_2 \Delta y$ (Equation 3, Section 12.4) where $\varepsilon_1, \varepsilon_2 \to 0$ as $\Delta x, \Delta y \to 0$. Then as $(x,y) \to (x_0,y_0)$, $\Delta x \to 0$ and $\Delta y \to 0 \Rightarrow \lim_{(x,y)\to(x_0,y_0)} f(x,y) = f(x_0,y_0) \Rightarrow$ f is continuous at every (x_0,y_0) in R.

25. a) $f(1,1,1) = 3$, $f_x(1,1,1) = y + z|_{(1,1,1)} = 2$, $f_y(1,1,1) = x + z|_{(1,1,1)} = 2$, $f_z(1,1,1) = y + x|_{(1,1,1)} = 2 \Rightarrow L(x,y,z) = 2x + 2y + 2z - 3$
b) $f(1,0,0) = 0$, $f_x(1,0,0) = 0$, $f_y(1,0,0) = 1$, $f_z(1,0,0) = 1 \Rightarrow L(x,y,z) = y + z$
c) $f(0,0,0) = 0$, $f_x(0,0,0) = 0$, $f_y(0,0,0) = 0$, $f_z(0,0,0) = 0 \Rightarrow L(x,y,z) = 0$

27. a) $f(1,0,0) = 1$, $f_x(1,0,0) = \frac{x}{\sqrt{x^2+y^2+z^2}}\Big|_{(1,0,0)} = 1$, $f_y(1,0,0) = \frac{y}{\sqrt{x^2+y^2+z^2}}\Big|_{(1,0,0)} = 0$,

$f_z(1,0,0) = \frac{z}{\sqrt{x^2+y^2+z^2}}\Big|_{(1,0,0)} = 0 \Rightarrow L(x,y,z) = x$

b) $f(1,1,0) = \sqrt{2}$, $f_x(1,1,0) = \frac{1}{\sqrt{2}}$, $f_y(1,1,0) = \frac{1}{\sqrt{2}}$, $f_z(1,1,0) = 0 \Rightarrow L(x,y,z) = \frac{1}{\sqrt{2}}x + \frac{1}{\sqrt{2}}y$

c) $f(1,2,2) = 3$, $f_x(1,2,2) = \frac{1}{3}$, $f_y(1,2,2) = \frac{2}{3}$, $f_z(1,2,2) = \frac{2}{3} \Rightarrow L(x,y,z) = \frac{1}{3}x + \frac{2}{3}y + \frac{2}{3}z$

29. a) $f(0,0,0) = 2$, $f_x(0,0,0) = e^x\Big|_{(0,0,0)} = 1$, $f_y(0,0,0) = -\sin(y+z)\Big|_{(0,0,0)} = 0$, $f_z(0,0,0) = -\sin(y+z)\Big|_{(0,0,0)}$

$= 0 \Rightarrow L(x,y,z) = 2 + x$

b) $f\left(0,\frac{\pi}{2},0\right) = 1$, $f_x\left(0,\frac{\pi}{2},0\right) = 1$, $f_y\left(0,\frac{\pi}{2},0\right) = -1$, $f_z\left(0,\frac{\pi}{2},0\right) = -1 \Rightarrow L(x,y,z) = x - y - z + \frac{\pi}{2} + 1$

c) $f\left(0,\frac{\pi}{4},\frac{\pi}{4}\right) = 1$, $f_x\left(0,\frac{\pi}{4},\frac{\pi}{4}\right) = 1$, $f_y\left(0,\frac{\pi}{4},\frac{\pi}{4}\right) = -1$, $f_z\left(0,\frac{\pi}{4},\frac{\pi}{4}\right) = -1 \Rightarrow L(x,y,z) = x - y - z + \frac{\pi}{2} + 1$

31. $f(x,y,z) = xz - 3yz + 2$ at $P_0(1,1,2) \Rightarrow f(1,1,2) = -2$. $f_x = z$, $f_y = -3z$, $f_z = x - 3y \Rightarrow L(x,y,z) = -2 + 2(x-1) - 6(y-1) - 2(z-2) = 2x - 6y - 2z + 6$. $f_{xx} = 0$, $f_{yy} = 0$, $f_{zz} = 0$, $f_{xy} = 0$, $f_{yz} = -3 \Rightarrow B = 3$.

$\therefore |E| \le \frac{1}{2}(3)(0.01 + 0.01 + 0.02)^2 = 0.0024$

33. $f(x,y,z) = xy + 2yz - 3xz$ at $P_0(1,1,0) \Rightarrow f(1,1,0) = 1$. $f_x = y - 3z$, $f_y = x + 2z$, $f_z = 2y - 3x \Rightarrow L(x,y,z) = 1 + (x-1) + (y-1) - z = x + y - z - 1$. $f_{xx} = 0$, $f_{yy} = 0$, $f_{zz} = 0$, $f_{xy} = 1$, $f_{xz} = -3$, $f_{yz} = 2 \Rightarrow B = 3$.

$\therefore |E| \le \frac{1}{2}(3)(0.01 + 0.01 + 0.01)^2 = 0.00135$

35. a) $dS = Szp)dp + S_x dx + S_w dw + S_h dh = C\left(\frac{x^4}{wh^3}dp + \frac{4px^3}{wh^3}dx - \frac{px^4}{w^2h^3}dw - \frac{3px^4}{wh^4}dh\right) =$

$C\left(\frac{px^4}{wh^3}\right)\left(\frac{1}{p}dp + \frac{4}{x}dx - \frac{1}{w}dw - \frac{3}{h}dh\right) = S_0\left(\frac{1}{p_0}dp + \frac{4}{x_0}dx - \frac{1}{w_0}dw - \frac{3}{h_0}dh\right) =$

$S_0\left(\frac{1}{100}dp + dx - 5\,dw - 30\,dh\right)$ where $p_0 = 100$ N/m, $x_0 = 4$ m, $w_0 = 0.2$ m, $h_0 = 0.1$ m.

b) More sensitive to a change in height.

37. $p(a,b,c) = abc \Rightarrow p_a = bc$, $p_b = ac$, $p_c = ab \Rightarrow dp = bc\,da + ac\,db + ab\,dc \Rightarrow \frac{dp}{p} = \frac{bc\,da + ac\,db + ab\,dc}{abc}$

$= \frac{da}{a} + \frac{db}{b} + \frac{dc}{c}$. Since $\left|\frac{da}{a} \times 100\right| = 2$, $\left|\frac{db}{b} \times 100\right| = 2$, $\left|\frac{dc}{c} \times 100\right| = 2$, $\left|\frac{dp}{p} \times 100\right| =$

$\left|\frac{da}{a} \times 100 + \frac{db}{b} \times 100 + \frac{dc}{c} \times 100\right| \le \left|\frac{da}{a} \times 100\right| + \left|\frac{db}{b} \times 100\right| + \left|\frac{dc}{c} \times 100\right| = 2 + 2 + 2 = 6$ or 6%

39. $A = \frac{1}{2}ab\sin C \Rightarrow A_a = \frac{1}{2}b\sin C$, $A_b = \frac{1}{2}a\sin C$, $A_C = \frac{1}{2}ab\cos C \Rightarrow dA = \frac{1}{2}b\sin C\,da + \frac{1}{2}a\sin C\,db + \frac{1}{2}ab\cos C\,dC$. $dC = \pm 2° = \pm 0.0349$ radians, $da = \pm 0.5$ ft, $db = \pm 0.5$ ft. At $a = 150$ ft, $b = 200$ ft, $C = 60°$, $dA = \frac{1}{2}(200)\sin 60°\,(\pm 0.5) + \frac{1}{2}(150)\sin 60°\,(\pm 0.5) + \frac{1}{2}(200)(150)\cos 60°\,(\pm 0.0349) = \pm 319.23\ \text{ft}^2$

41. $Q_k = \frac{1}{2}(2kM/h)^{-1/2}(2M/h)$, $Q_M = \frac{1}{2}(2kM/h)^{-1/2}(2k/h)$, $Q_h = \frac{1}{2}(2kM/h)^{-1/2}(-2kM/h^2) \Rightarrow$

$dQ = \frac{1}{2}(2kM/h)^{-1/2}(2M/h)\,dk + \frac{1}{2}(2kM/h)^{-1/2}(2k/h)\,dM + \frac{1}{2}(2kM/h)^{-1/2}(-2kM/h^2)\,dh =$

$\frac{1}{2}(2kM/h)^{-1/2}\left[\frac{2M}{h}\,dk + \frac{2k}{h}\,dM - \frac{2kM}{h^2}\,dh\right] \Rightarrow$

$dQ\Big|_{(2,20,0.05)} = \frac{1}{2}\left(2(2)(20)/0.05\right)^{-1/2}\left[\frac{2(20)}{0.05}\,dk + \frac{2(2)}{0.05}\,dM - \frac{2(2)(20)}{(0.05)^2}\,dh\right] =$

$(0.0125)(800\,dk + 80\,dM - 32000\,dh)$ $\therefore$ Q is most sensitive to changes in h.

SECTION 12.5 THE CHAIN RULE

1. $\frac{\partial w}{\partial x} = 2x$, $\frac{\partial w}{\partial y} = 2y$, $\frac{dx}{dt} = -\sin t$, $\frac{dy}{dt} = \cos t \Rightarrow \frac{dw}{dt} = -2x\sin t + 2y\cos t = -2\cos t\sin t + 2\sin t\cos t = 0$

$\Rightarrow \frac{dw}{dt}(\pi) = 0$

3. $\frac{\partial w}{\partial x} = \frac{1}{z}$, $\frac{\partial w}{\partial y} = \frac{1}{z}$, $\frac{\partial w}{\partial z} = \frac{-(x+y)}{z^2}$, $\frac{dx}{dt} = -2\cos t\sin t$, $\frac{dy}{dt} = 2\sin t\cos t$, $\frac{dz}{dt} = -\frac{1}{t^2} \Rightarrow$

$\frac{dw}{dt} = -\frac{2}{z}\cos t\sin t + \frac{2}{z}\sin t\cos t + \frac{x+y}{z^2t^2} = \frac{\cos^2 t + \sin^2 t}{\frac{1}{t^2}(t^2)} = 1 \Rightarrow \frac{dw}{dt}(3) = 1$

5. $\frac{\partial w}{\partial x} = 2ye^x$, $\frac{\partial w}{\partial y} = 2e^x$, $\frac{\partial w}{\partial z} = -\frac{1}{z}$, $\frac{dx}{dt} = \frac{2t}{t^2+1}$, $\frac{dy}{dt} = \frac{1}{t^2+1}$, $\frac{dz}{dt} = e^t \Rightarrow \frac{dw}{dt} = \frac{4yte^x}{t^2+1} + \frac{2e^x}{t^2+1} - \frac{e^t}{z} =$

$\frac{4t\tan^{-1} t\, e^{\ln(t^2+1)}}{t^2+1} + \frac{2(t^2+1)}{t^2+1} - \frac{e^t}{e^t} = 4t\tan^{-1} t + 1 \Rightarrow \frac{dw}{dt}(1) = \pi + 1$

7. a) $z = 4e^x \ln y$, $x = \ln(r\cos\theta)$, $y = r\sin\theta \Rightarrow \frac{\partial z}{\partial r} = 4e^x \ln y\left(\frac{1}{r\cos\theta}\right)\cos\theta + \frac{4e^x}{y}(\sin\theta) = \frac{4e^x \ln y}{r} +$

$\frac{4e^x\sin\theta}{y}$. $\frac{\partial z}{\partial \theta} = 4e^x \ln y\left(\frac{1}{r\cos\theta}\right)(-r\sin\theta) + \frac{4e^x}{y}(r\cos\theta) = -4e^x \ln y\tan\theta + \frac{4e^x r\cos\theta}{y}$.

As functions of r and θ only, $\frac{\partial z}{\partial r} = \frac{4e^{\ln(r\cos\theta)}\ln(r\sin\theta)}{r} + \frac{4e^{\ln(r\cos\theta)}\sin\theta}{r\sin\theta} = 4\cos\theta\ln(r\sin\theta) + 4\cos\theta$

$\frac{\partial z}{\partial \theta} = -4e^{\ln(r\cos\theta)}\ln(r\sin\theta)\tan\theta + \frac{4e^{\ln(r\cos\theta)}\,r\cos\theta}{r\sin\theta} = -4r\sin\theta\ln(r\sin\theta) + \frac{4r\cos^2\theta}{\sin\theta}$. To find the partial derivatives directly, let $z = 4e^{\ln(r\cos\theta)}\ln(r\sin\theta) = 4r\cos\theta\ln(r\sin\theta)$. Then find the partial with respect to r and the partial with respect to θ. The answers will be the same.

b) At $\left(2, \frac{\pi}{4}\right)$, $\frac{\partial z}{\partial r} = 4\cos\frac{\pi}{4}\ln(2\sin\frac{\pi}{4}) = 2\sqrt{2}\ln\sqrt{2} + 2\sqrt{2} = \sqrt{2}(\ln 2 + 2)$. $\frac{\partial z}{\partial \theta} = -4(2)\sin\frac{\pi}{4}\ln(2\sin\frac{\pi}{4}) +$

$\frac{4\cos^2\frac{\pi}{4}}{\sin\frac{\pi}{4}} = -4\sqrt{2}\ln\sqrt{2} + 2\sqrt{2} = -2\sqrt{2}\ln 2 + 2\sqrt{2}$

9. a) $w = xy + yz + xz,\ x = u + v,\ y = u - v,\ z = uv \Rightarrow \frac{\partial w}{\partial u} = (y + z)(1) + (x + z)(1) + (y + x)(v) = x + y + 2z + v(y + x)$. As a function of u an v only, $\frac{\partial w}{\partial u} = u + v + u - v + 2uv + v(u - v + u + v) = 2u + 4uv$.
$\frac{\partial w}{\partial v} = (y + z)(1) + (x + z)(-1) + (y + x)(u) = y - x + (y + x)u$. As a function of u and v only, $\frac{\partial w}{\partial v} = u - v - (u + v) + (u - v + u + v)u = -2v + 2u^2$. To find the partial derivatives directly, let $w = (u + v)(u - v) + (u - v)uv + (u + v)uv = u^2 - v^2 + 2u^2v$. Then find the partial with respect to u and the partial with respect to v. The answers will be the same as above.

b) At $\left(\frac{1}{2}, 1\right)$, $\frac{\partial w}{\partial u} = 2\left(\frac{1}{2}\right) + 4\left(\frac{1}{2}\right)(1) = 3$, $\frac{\partial w}{\partial v} = -2(1) + 2\left(\frac{1}{2}\right)^2 = -\frac{3}{2}$.

11. a) $u = \frac{p - q}{q - r},\ p = x + y + z,\ q = x - y - z,\ r = x + y - z \Rightarrow \frac{\partial u}{\partial x} = \frac{1}{q - r} + \frac{r - p}{(q - r)^2} + \frac{p - q}{(q - r)^2} = 0$. $\frac{\partial u}{\partial y} = \frac{1}{q - r} - \frac{r - p}{(q - r)^2} + \frac{p - q}{(q - r)^2} = \frac{2p - 2r}{(q - r)^2}$. As a function of x, y, and z only, $\frac{\partial u}{\partial y} = \frac{2(x + y + z) - 2(x + y - z)}{(2z - 2y)^2} = \frac{z}{(z - y)^2}$.
$\frac{\partial u}{\partial z} = \frac{1}{q - r} + \frac{r - p}{(q - r)^2} - \frac{p - q}{(q - r)^2} = \frac{-2p + 2q}{(q - r)^2}$. As a function of x, y, and z only, $\frac{\partial u}{\partial z} = \frac{-2(x + y + z) + 2(x + y - z)}{(2x - 2y)^2} = \frac{-y}{(z - y)^2}$. To find the partial derivatives directly, let $u = \frac{(x + y + z) - (x - y + z)}{(x - y + z) - (x + y - z)} = \frac{y}{z - y}$. Then find the partial with respect to x, the partial with respect to y, and the partial with respect to z. The answers will be the same as above.

b) At $\left(\sqrt{3}, 2, 1\right)$, $\frac{\partial u}{\partial x} = 0$, $\frac{\partial u}{\partial y} = \frac{1}{(1 - 2)^2} = 1$, $\frac{\partial u}{\partial z} = \frac{-2}{(1 - 2)^2} = -2$.

13. $\frac{dz}{dt} = \frac{\partial z}{\partial x}\frac{dx}{dt} + \frac{\partial z}{\partial y}\frac{dy}{dt}$

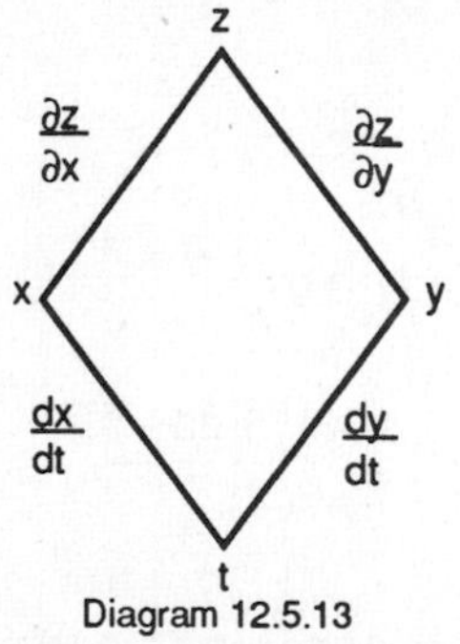

Diagram 12.5.13

15. a) $\frac{\partial w}{\partial u} = \frac{\partial w}{\partial x}\frac{\partial x}{\partial u} + \frac{\partial w}{\partial y}\frac{\partial y}{\partial u} + \frac{\partial w}{\partial z}\frac{\partial z}{\partial u}$

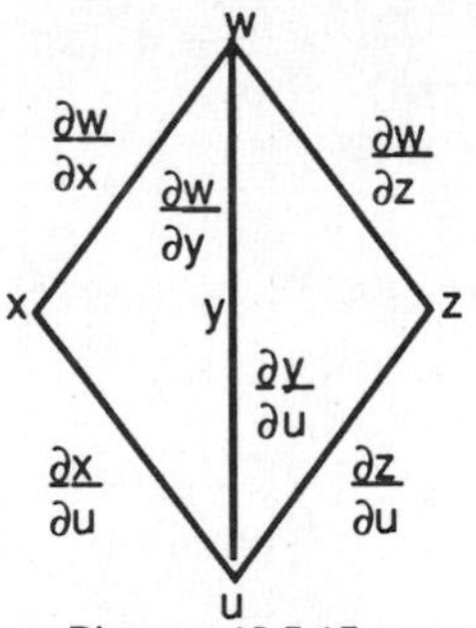

Diagram 12.5.15 a

15. b) $\frac{\partial w}{\partial v} = \frac{\partial w}{\partial x}\frac{\partial x}{\partial v} + \frac{\partial w}{\partial y}\frac{\partial y}{\partial v} + \frac{\partial w}{\partial z}\frac{\partial z}{\partial v}$

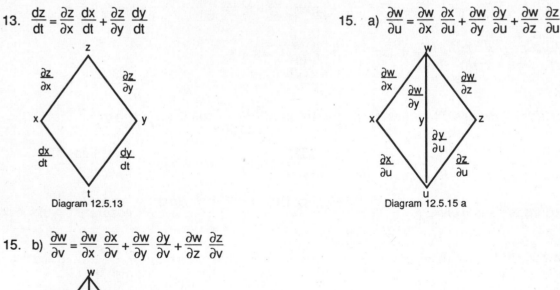

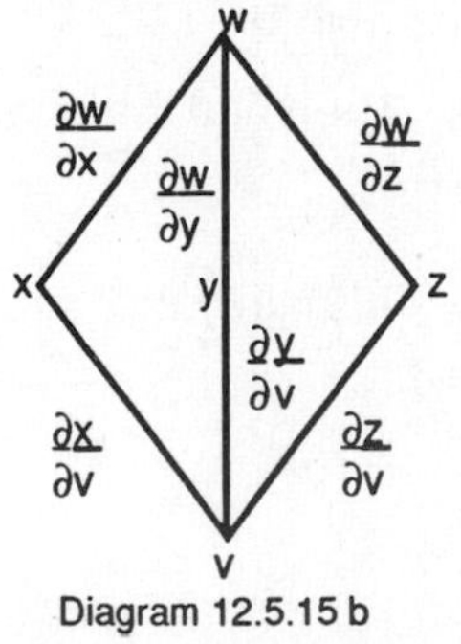

Diagram 12.5.15 b

17. $\dfrac{\partial w}{\partial u} = \dfrac{\partial w}{\partial x}\dfrac{\partial x}{\partial u} + \dfrac{\partial w}{\partial y}\dfrac{\partial y}{\partial u}$

$\dfrac{\partial w}{\partial v} = \dfrac{\partial w}{\partial x}\dfrac{\partial x}{\partial v} + \dfrac{\partial w}{\partial y}\dfrac{\partial y}{\partial v}$

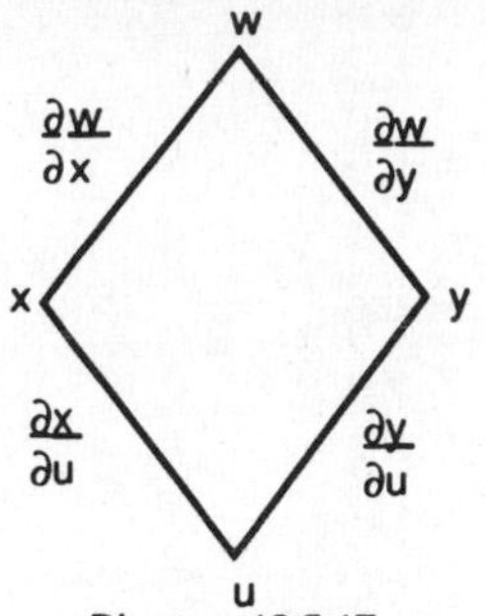

Diagram 12.5.17 a

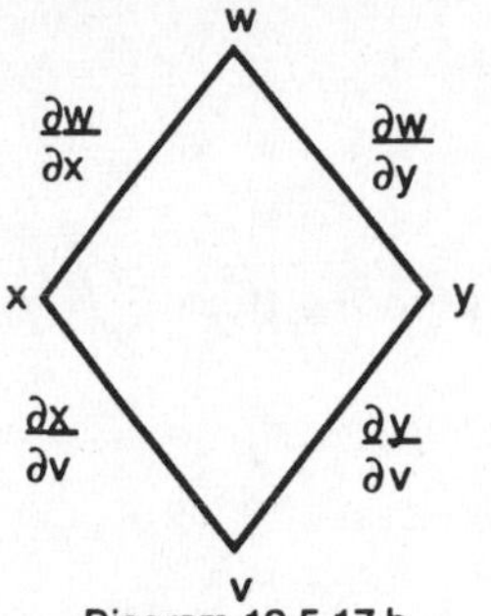

Diagram 12.5.17 b

19. $\dfrac{\partial z}{\partial t} = \dfrac{\partial z}{\partial x}\dfrac{\partial x}{\partial t} + \dfrac{\partial z}{\partial y}\dfrac{\partial y}{\partial t}$

$\dfrac{\partial z}{\partial s} = \dfrac{\partial z}{\partial x}\dfrac{\partial x}{\partial s} + \dfrac{\partial z}{\partial y}\dfrac{\partial y}{\partial s}$

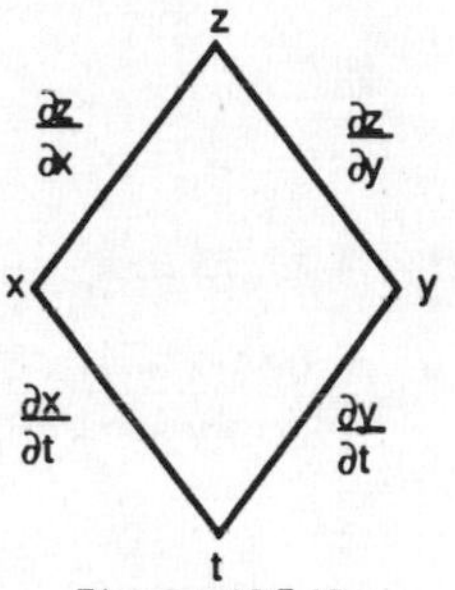

Diagram 12.5.19 a

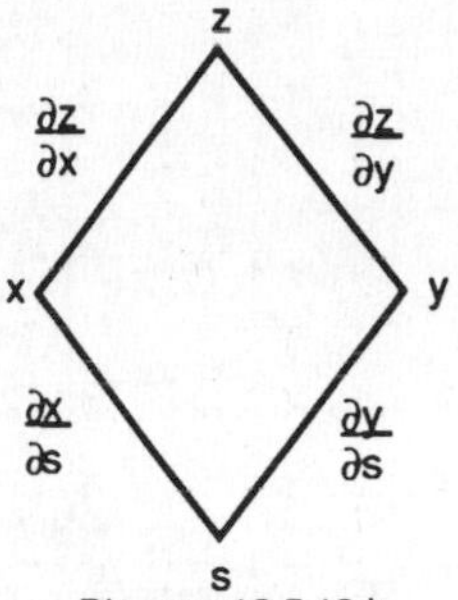

Diagram 12.5.19 b

21. $\dfrac{\partial w}{\partial s} = \dfrac{dw}{du}\dfrac{\partial u}{\partial s}$

$\dfrac{\partial w}{\partial t} = \dfrac{dw}{du}\dfrac{\partial u}{\partial t}$

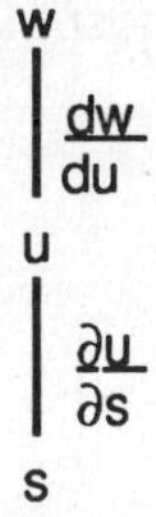

Diagram 12.5.21 a

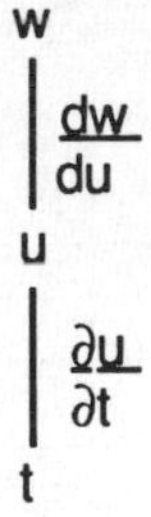

Diagram 12.5.21 b

23. $\dfrac{\partial w}{\partial r} = \dfrac{\partial w}{\partial x}\dfrac{dx}{dr} + \dfrac{\partial w}{\partial y}\dfrac{dy}{dr} = \dfrac{\partial w}{\partial x}\dfrac{dx}{dr}$

since $\dfrac{dy}{dr} = 0$

$\dfrac{\partial w}{\partial s} = \dfrac{\partial w}{\partial x}\dfrac{dx}{ds} + \dfrac{\partial w}{\partial y}\dfrac{dy}{ds} = \dfrac{\partial w}{\partial y}\dfrac{dy}{ds}$

since $\dfrac{dx}{ds} = 0$

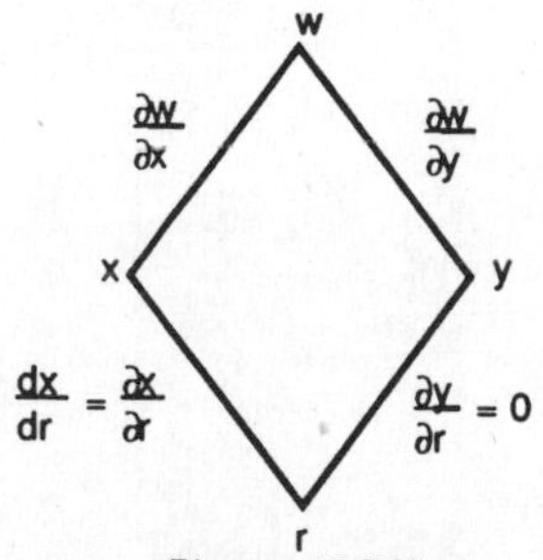

Diagram 12.5.23 a

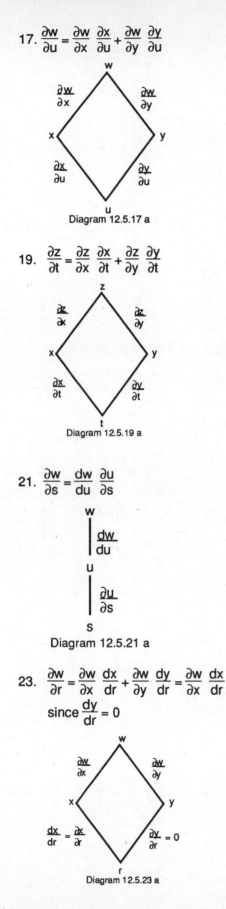

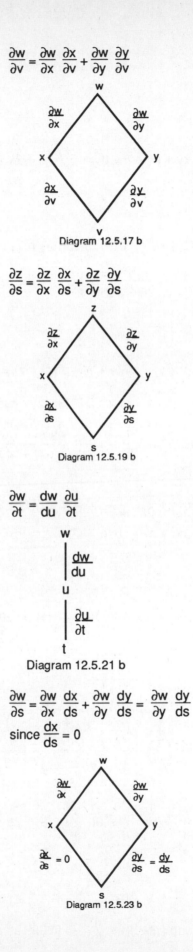

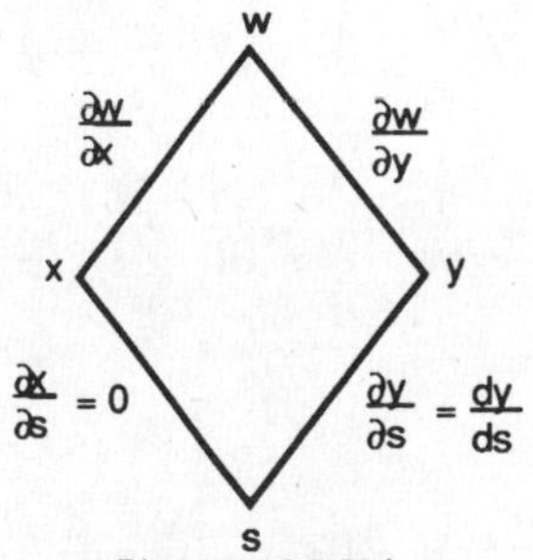

Diagram 12.5.23 b

25. Let $F(x,y) = x^3 - 2y^2 + xy = 0 \Rightarrow F_x(x,y) = 3x^2 + y,\ F_y(x,y) = -4y + x \Rightarrow \frac{dy}{dx} = -\frac{F_x}{F_y} = -\frac{3x^2+y}{-4y+x} \Rightarrow$

$\frac{dy}{dx}(1,1) = \frac{4}{3}$

27. Let $F(x,y) = x^2 + xy + y^2 - 7 = 0 \Rightarrow F_x(x,y) = 2x + y,\ F_y(x,y) = x + 2y \Rightarrow \frac{dy}{dx} = -\frac{F_x}{F_y} = -\frac{2x+y}{x+2y} \Rightarrow$

$\frac{dy}{dx}(1,2) = -\frac{4}{5}$

29. Let $F(x,y,z)) = z^3 - xy + yz + y^3 - 2 = 0 \Rightarrow F_x(x,y,z) = -y,\ F_y(x,y,z) = -x + z + 3y^2,\ F_z(x,y,z) = 3z^2 + y \Rightarrow$

$\frac{\partial z}{\partial x} = -\frac{F_x}{F_z} = -\frac{-y}{3z^2+y} = \frac{y}{3z^2+y} \Rightarrow \frac{\partial z}{\partial x}(1,1,1) = \frac{1}{4}.\ \frac{\partial z}{\partial y} = -\frac{F_y}{F_z} = -\frac{-x+z+3y^2}{3z^2+y} = \frac{x-z-3y^2}{3z^2+y} \Rightarrow$

$\frac{\partial z}{\partial y}(1,1,1) = -\frac{3}{4}$

31. Let $F(x,y,z) = \sin(x+y) + \sin(y+z) + \sin(x+z) = 0 \Rightarrow F_x(x,y,z) = \cos(x+y) + \cos(x+z)$,

$F_y(x,y,z) = \cos(x+y) + \cos(y+z),\ F_z(x,y,z) = \cos(y+z) + \cos(x+z) \Rightarrow \frac{\partial z}{\partial x} = -\frac{F_x}{F_z} = -\frac{\cos(x+y)+\cos(x+z)}{\cos(y+z)+\cos(x+z)}$

$\Rightarrow \frac{\partial z}{\partial x}(\pi,\pi,\pi) = -1.\ \frac{\partial z}{\partial y} = -\frac{F_y}{F_z} = -\frac{\cos(x+y)+\cos(y+z)}{\cos(y+z)+\cos(x+z)} \Rightarrow \frac{\partial z}{\partial y}(\pi,\pi,\pi) = -1$

33. $\frac{\partial w}{\partial r} = \frac{\partial w}{\partial x}\frac{\partial x}{\partial r} + \frac{\partial w}{\partial y}\frac{\partial y}{\partial r} + \frac{\partial w}{\partial z}\frac{\partial z}{\partial r} = 2(x+y+z)(1) + 2(x+y+z)(-\sin(r+s)) + 2(x+y+z)(\cos(r+s)) =$

$2(x+y+z)(1 - \sin(r+s) + \cos(r+s)) = 2(r - s + \cos(r+s) + \sin(r+s))(1 - \sin(r+s) + \cos(r+s)) \Rightarrow$

$\left.\frac{\partial w}{\partial r}\right|_{r=1,s=1} = 12$

35. $\frac{\partial w}{\partial v} = \frac{\partial w}{\partial x}\frac{\partial x}{\partial v} + \frac{\partial w}{\partial y}\frac{\partial y}{\partial v} = \left(2x - \frac{y}{x^2}\right)(-2) + \frac{1}{x}(1) = \left(2(u-2v+1) - \frac{2u+v-2}{(u-2v+1)^2}\right)(-2) + \frac{1}{u-2v+1} \Rightarrow$

$\left.\frac{\partial w}{\partial v}\right|_{u=0,v=0} = -7$

37. $\frac{\partial z}{\partial u} = \frac{dz}{dx}\frac{\partial x}{\partial u} = \frac{5}{1+x^2}\left(e^u + \ln v\right) = \frac{5}{1+\left(e^u+\ln v\right)^2}\left(e^u + \ln v\right) \Rightarrow \left.\frac{\partial z}{\partial u}\right|_{u=\ln 2,v=1} = 2$

$\frac{\partial z}{\partial v} = \frac{dz}{dx}\frac{\partial x}{\partial v} = \frac{5}{1+x^2}\left(\frac{1}{v}\right) = \frac{5}{1+\left(e^u+\ln v\right)^2}\left(\frac{1}{v}\right) \Rightarrow \left.\frac{\partial z}{\partial v}\right|_{u=\ln 2,v=1} = 1$

39. $\frac{\partial f}{\partial x} = \frac{\partial f}{\partial u}(1) + \frac{\partial f}{\partial v}(0) + \frac{\partial f}{\partial w}(-1) = \frac{\partial f}{\partial u} - \frac{\partial f}{\partial w}.\ \frac{\partial f}{\partial y} = \frac{\partial f}{\partial u}(-1) + \frac{\partial f}{\partial v}(1) + \frac{\partial f}{\partial w}(0) = -\frac{\partial f}{\partial u} + \frac{\partial f}{\partial v}.\ \frac{\partial f}{\partial z} = \frac{\partial f}{\partial u}(0) + \frac{\partial f}{\partial v}(-1) +$

$\frac{\partial f}{\partial w}(1) = -\frac{\partial f}{\partial v} + \frac{\partial f}{\partial w}.\ \therefore \frac{\partial f}{\partial x} + \frac{\partial f}{\partial y} + \frac{\partial f}{\partial z} = 0$

41. $\frac{dV}{dt} = \frac{\partial V}{\partial I}\frac{dI}{dt} + \frac{\partial V}{\partial R}\frac{dR}{dt}.\ V = IR \Rightarrow \frac{\partial V}{\partial I} = R,\ \frac{\partial V}{\partial R} = I \Rightarrow \frac{dV}{dt} = R\frac{dI}{dt} + I\frac{dR}{dt} \Rightarrow -0.01$ volts/sec = (600 ohms)$\frac{dI}{dt}$ +

(0.04 amps)(0.5 ohms/sec) $\Rightarrow \frac{dI}{dt} = -0.00005$ amps/sec.

43. $\frac{\partial w}{\partial x} = \frac{\partial w}{\partial u}(2x/2) + \frac{\partial w}{\partial v}(y) = x\frac{\partial w}{\partial u} + y\frac{\partial w}{\partial v} \Rightarrow w_{xx} = \frac{\partial w}{\partial u} + x\left(\frac{\partial^2 w}{\partial u^2}(x) + \frac{\partial^2 w}{\partial v \partial u}(y)\right) y\left(\frac{\partial^2 w}{\partial v^2}(y) + \frac{\partial^2 w}{\partial u \partial v}(x)\right) = \frac{\partial w}{\partial u} +$
$x^2\frac{\partial^2 w}{\partial u^2} + xy\frac{\partial^2 w}{\partial v \partial u} + xy\frac{\partial^2 w}{\partial u \partial v} + y^2\frac{\partial^2 w}{\partial v^2} \cdot \frac{\partial w}{\partial y} = \frac{\partial w}{\partial u}(-y) + \frac{\partial w}{\partial v}(x) = -y\frac{\partial w}{\partial u} + x\frac{\partial w}{\partial v} \Rightarrow w_{yy} = -\frac{\partial w}{\partial u} -$
$y\left(\frac{\partial^2 w}{\partial u^2}(-y) + \frac{\partial^2 w}{\partial v \partial u}(x)\right) + x\left(\frac{\partial^2 w}{\partial v^2}(x) + \frac{\partial^2 w}{\partial u \partial v}(-y)\right) = -\frac{\partial w}{\partial u} + y^2\frac{\partial^2 w}{\partial u^2} - xy\frac{\partial^2 w}{\partial v \partial u} - xy\frac{\partial^2 w}{\partial u \partial v} + y^2\frac{\partial^2 w}{\partial v^2}$. Then
$w_{xx} + w_{yy} = (x^2 + y^2)\frac{\partial^2 w}{\partial u^2} + (x^2 + y^2)\frac{\partial^2 w}{\partial v^2} = (x^2 + y^2)(w_{uu} + w_{vv}) = 0$ since $w_{uu} + w_{vv} = 0$

45. $f_x(x,y,z) = \cos t$, $f_y(x,y,z) = \sin t$, $f_z(x,y,z) = t^2 + t - 2$. $\frac{df}{dt} = \frac{\partial f}{\partial x}\frac{dx}{dt} + \frac{\partial f}{\partial y}\frac{dy}{dt} + \frac{\partial f}{\partial z}\frac{dz}{dt} = (\cos t)(-\sin t) +$
$(\sin t)(\cos t) + (t^2 + t - 2)(1) = t^2 + t - 2$. $\frac{df}{dt} = 0 \Rightarrow t^2 + t - 2 = 0 \Rightarrow t = -2$ or $t = 1$

$t = -2 \Rightarrow x = \cos(-2)$, $y = \sin(-2)$, $z = -2$; $t = 1 \Rightarrow x = \cos 1$, $y = \sin 1$, $z = 1$

47. a) $\frac{\partial T}{\partial x} = 8x - 4y$, $\frac{\partial T}{\partial y} = 8y - 4x$. $\frac{dT}{dt} = \frac{\partial T}{\partial x}\frac{dx}{dt} + \frac{\partial T}{\partial y}\frac{dy}{dt} = (8x - 4y)(-\sin t) + (8y - 4x)(\cos t) =$

$(8\cos t - 4\sin t)(-\sin t) + (8\sin t - 4\cos t)(\cos t) = 4\sin^2 t - 4\cos^2 t \Rightarrow \frac{d^2T}{dt^2} = 16\sin t\cos t$

$\frac{dT}{dt} = 0 \Rightarrow 4\sin^2 t - 4\cos^2 t = 0 \Rightarrow \sin^2 t = \cos^2 t \Rightarrow \sin t = \cos t$ or $\sin t = -\cos t \Rightarrow$

$t = \frac{\pi}{4}, \frac{5\pi}{4}$ or $\frac{3\pi}{4}, \frac{7\pi}{4}$ on the interval $0 \le t \le 2\pi$.

$\left.\frac{d^2T}{dt^2}\right|_{t=\pi/4} = 16\sin\frac{\pi}{4}\cos\frac{\pi}{4} > 0 \Rightarrow T$ has a minimum at $t = \frac{\pi}{4}$

$\left.\frac{d^2T}{dt^2}\right|_{t=3\pi/4} = 16\sin\frac{3\pi}{4}\cos\frac{3\pi}{4} < 0 \Rightarrow T$ has a maximum at $t = \frac{3\pi}{4}$

$\left.\frac{d^2T}{dt^2}\right|_{t=5\pi/4} = 16\sin\frac{5\pi}{4}\cos\frac{5\pi}{4} > 0 \Rightarrow T$ has a minimum at $t = \frac{5\pi}{4}$

$\left.\frac{d^2T}{dt^2}\right|_{t=7\pi/4} = 16\sin\frac{7\pi}{4}\cos\frac{7\pi}{4} < 0 \Rightarrow T$ has a maximum at $t = \frac{7\pi}{4}$

b) $T = 4x^2 - 4xy + 4y^2 \Rightarrow \frac{\partial T}{\partial x} = 8x - 4y$, $\frac{\partial T}{\partial y} = 8y - 4x$ (See part a above.)

$t = \frac{\pi}{4} \Rightarrow x = \cos\frac{\pi}{4} = \frac{\sqrt{2}}{2}$, $y = \sin\frac{\pi}{4} = \frac{\sqrt{2}}{2} \Rightarrow T\left(\frac{\pi}{4}\right) = 2$

$t = \frac{3\pi}{4} \Rightarrow x = \cos\frac{3\pi}{4} = -\frac{\sqrt{2}}{2}$, $y = \sin\frac{3\pi}{4} = \frac{\sqrt{2}}{2} \Rightarrow T\left(\frac{3\pi}{4}\right) = 6$

$t = \frac{5\pi}{4} \Rightarrow x = \cos\frac{5\pi}{4} = -\frac{\sqrt{2}}{2}$, $y = \sin\frac{5\pi}{4} = -\frac{\sqrt{2}}{2} \Rightarrow T\left(\frac{5\pi}{4}\right) = 2$

$t = \frac{7\pi}{4} \Rightarrow x = \cos\frac{7\pi}{4} = \frac{\sqrt{2}}{2}$, $y = \sin\frac{7\pi}{4} = -\frac{\sqrt{2}}{2} \Rightarrow T\left(\frac{7\pi}{4}\right) = 6$

$\therefore T_{max} = 6$ and $T_{min} = 2$.

SECTION 12.6 PARTIAL DERIVATIVES WITH CONSTRAINT VARIBLES

1. $w = x^2 + y^2 + z^2$ and $z = x^2 + y^2$

a) $\begin{pmatrix} y \\ z \end{pmatrix} \to \begin{pmatrix} x = x(y,z) \\ y = y \\ z = z \end{pmatrix} \to w \Rightarrow \left(\frac{\partial w}{\partial y}\right)_z = \frac{\partial w}{\partial x}\frac{\partial x}{\partial y} + \frac{\partial w}{\partial y}\frac{\partial y}{\partial y} + \frac{\partial w}{\partial z}\frac{\partial z}{\partial y}, \frac{\partial z}{\partial y} = 0$ and $\frac{\partial z}{\partial y} = 2x\frac{\partial x}{\partial y} + 2y\frac{\partial y}{\partial y} =$

$2x\frac{\partial x}{\partial y} + 2y \Rightarrow 0 = 2x\frac{\partial x}{\partial y} + 2y \Rightarrow \frac{\partial x}{\partial y} = -\frac{y}{x}$. $\therefore \left(\frac{\partial w}{\partial y}\right)_z = 2x\left(-\frac{y}{x}\right) + 2y(1) + 2z(0) = -2y + 2y = 0$

b) $\begin{pmatrix} x \\ z \end{pmatrix} \to \begin{pmatrix} x = x \\ y = y(x,z) \\ z = z \end{pmatrix} \to w \Rightarrow \left(\frac{\partial w}{\partial x}\right)_x = \frac{\partial w}{\partial x}\frac{\partial x}{\partial z} + \frac{\partial w}{\partial y}\frac{\partial y}{\partial z} + \frac{\partial w}{\partial z}\frac{\partial z}{\partial z}, \frac{\partial x}{\partial z} = 0$ and $\frac{\partial z}{\partial z} = 2x\frac{\partial x}{\partial z} + 2y\frac{\partial y}{\partial z} \Rightarrow$

$1 = 2y\frac{\partial y}{\partial z} \Rightarrow \frac{1}{2y} = \frac{\partial y}{\partial z}$. $\therefore \left(\frac{\partial w}{\partial z}\right)_x = 2x(0) + 2y\left(\frac{1}{2y}\right) + 2z(1) = 1 + 2z$

c) $\begin{pmatrix} y \\ z \end{pmatrix} \to \begin{pmatrix} x = x(y,z) \\ y = y \\ z = z \end{pmatrix} \to w \Rightarrow \left(\frac{\partial w}{\partial y}\right)_z = \frac{\partial w}{\partial x}\frac{\partial x}{\partial z} + \frac{\partial w}{\partial y}\frac{\partial y}{\partial z} + \frac{\partial w}{\partial z}\frac{\partial z}{\partial z}, \frac{\partial y}{\partial z} = 0$ and $\frac{\partial z}{\partial z} = 2x\frac{\partial x}{\partial z} + 2y\frac{\partial y}{\partial z} \Rightarrow$

$1 = 2x\frac{\partial x}{\partial z} \Rightarrow \frac{\partial x}{\partial z} = \frac{1}{2x}$. $\therefore \left(\frac{\partial w}{\partial z}\right)_y = 2x\left(\frac{1}{2x}\right) + 2y(0) + 2z(1) = 1 + 2z.$

3. $U = f(p,v,T)$ and $pv = nRT$

a) $\begin{pmatrix} p \\ v \end{pmatrix} \to \begin{pmatrix} p = p \\ v = v \\ T = \frac{pv}{nR} \end{pmatrix} \to U \Rightarrow \left(\frac{\partial U}{\partial p}\right)_v = \frac{\partial U}{\partial p}\frac{\partial p}{\partial p} + \frac{\partial U}{\partial v}\frac{\partial v}{\partial p} + \frac{\partial U}{\partial T}\frac{\partial T}{\partial p} = \frac{\partial U}{\partial p} + \frac{\partial U}{\partial v}(0) + \frac{\partial U}{\partial T}\left(\frac{v}{nR}\right) =$

$\frac{\partial U}{\partial p} + \frac{\partial U}{\partial T}\left(\frac{v}{nR}\right)$

b) $\begin{pmatrix} v \\ T \end{pmatrix} \to \begin{pmatrix} p = \frac{nRT}{v} \\ v = v \\ T = T \end{pmatrix} \to U \Rightarrow \left(\frac{\partial U}{\partial T}\right)_v = \frac{\partial U}{\partial p}\frac{\partial p}{\partial T} + \frac{\partial U}{\partial v}\frac{\partial v}{\partial T} + \frac{\partial U}{\partial T}\frac{\partial T}{\partial T} = \frac{\partial U}{\partial p}\left(\frac{nR}{v}\right) + \frac{\partial U}{\partial v}(0) + \frac{\partial U}{\partial T} =$

$\frac{\partial U}{\partial p}\left(\frac{nR}{v}\right) + \frac{\partial U}{\partial T}$

5. $w = x^2y^2 + yz - z^3$ and $x^2 + y^2 + z^2 = 6$

a) $\begin{pmatrix} x \\ y \end{pmatrix} \to \begin{pmatrix} x = x \\ y\ \ y \\ z = z(x,y) \end{pmatrix} \to w \Rightarrow \left(\frac{\partial w}{\partial y}\right)_x = \frac{\partial w}{\partial x}\frac{\partial x}{\partial y} + \frac{\partial w}{\partial y}\frac{\partial y}{\partial y} + \frac{\partial w}{\partial z}\frac{\partial z}{\partial y} = 2xy^2(0) + (2x^2y + z)(1) +$

$(y - 3z^2)\frac{\partial z}{\partial y} = 2x^2y + z + (y - 3z^2)\frac{\partial z}{\partial y}$. Now $2x\frac{\partial x}{\partial y} + 2y + 2z\frac{\partial z}{\partial y} = 0 \Rightarrow 2y + 2z\frac{\partial z}{\partial y} = 0 \Rightarrow \frac{\partial z}{\partial y} = -\frac{y}{z}$.

At $(4,2,1,-1)$, $\frac{\partial z}{\partial y} = -\frac{1}{-1} = 1 \Rightarrow \left(\frac{\partial w}{\partial y}\right)_x \Big|_{(4,2,1,-1)} = 2(2)^2(1) + (-1) + (1 - 3(-1)^2)(1) = 5$

5. b) $\begin{pmatrix} y \\ z \end{pmatrix} \to \begin{pmatrix} x = x(y,z) \\ y = y \\ z = z \end{pmatrix} \to w \Rightarrow \left(\frac{\partial w}{\partial y}\right)_z = \frac{\partial w}{\partial x}\frac{\partial x}{\partial y} + \frac{\partial w}{\partial y}\frac{\partial y}{\partial y} + \frac{\partial w}{\partial z}\frac{\partial z}{\partial y} = 2xy^2\frac{\partial x}{\partial y} + (2x^2y + z)(1) + (y - 3z^2)(0)$

$= 2xy^2\frac{\partial x}{\partial y} + 2x^2y + z$. Now $2x\frac{\partial x}{\partial y} + 2y + 2z\frac{\partial z}{\partial y} = 0 \Rightarrow 2x\frac{\partial x}{\partial y} + 2y = 0 \Rightarrow \frac{\partial x}{\partial y} = -\frac{y}{x}$. At (4,2,1,−1),

$\frac{\partial x}{\partial y} = -\frac{1}{2} \Rightarrow \left(\frac{\partial w}{\partial y}\right)\Big|_{(4,2,1,-1)} = 2(2)(1)^2\left(-\frac{1}{2}\right) + 2(2)^2(1) + (-1) = 5$

7. a) $x = r\cos\theta \Rightarrow \left(\frac{\partial x}{\partial r}\right)_\theta = \cos\theta$

b) $x^2 + y^2 = r^2 \Rightarrow 2x = 2r\left(\frac{\partial r}{\partial x}\right)_y \Rightarrow \left(\frac{\partial r}{\partial x}\right)_y = \frac{x}{r} = \frac{x}{\sqrt{x^2 + y^2}}$

9. $f(x,y,z) = 0$. Let x be a differentiable function of y and z. Then $\left(\frac{\partial x}{\partial y}\right)_z = -\frac{\partial f/\partial y}{\partial f/\partial z}$. Similarly, If y is a differentiable function of x and z, $\left(\frac{\partial y}{\partial z}\right)_x = -\frac{\partial f/\partial z}{\partial f/\partial x}$ and if z is a differentiable function of x and y, $\left(\frac{\partial z}{\partial x}\right)_y = -\frac{\partial f/\partial x}{\partial f/\partial y}$.. Then $\left(\frac{\partial x}{\partial y}\right)_z\left(\frac{\partial y}{\partial z}\right)_x\left(\frac{\partial z}{\partial x}\right)_y = \left(-\frac{\partial f/\partial y}{\partial f/\partial z}\right)\left(-\frac{\partial f/\partial z}{\partial f/\partial x}\right)\left(-\frac{\partial f/\partial x}{\partial f/\partial y}\right) = -1$

11. $w = x^2 - y^2 + 4z + t$ and $x + 2z + t = 25$. If x, y, and z are independent, then $1 + \frac{\partial t}{\partial x} = 0 \Rightarrow \frac{\partial t}{\partial x} = -1$.

$\therefore \left(\frac{\partial w}{\partial x}\right)_{y,z} = 2x(1) + (-2y)(0) + 4(0) + 1(-1) = 2x - 1$. If x, y, and t are independent, then $1 + 2\frac{\partial z}{\partial x} = 0 \Rightarrow$

$\frac{\partial z}{\partial x} = -\frac{1}{2}$. $\therefore \left(\frac{\partial w}{\partial x}\right)_{y,t} = 2x(1) + (-2y)(0) + 4\left(-\frac{1}{2}\right) + 1(0) = 2x - 2$.

SECTION 12.7 DIRECTIONAL DERIVATIVES

1. $\frac{\partial f}{\partial x} = 2x + \frac{z}{x} \Rightarrow \frac{\partial f}{\partial x}(1,1,1) = 3$. $\frac{\partial f}{\partial y} = 2y \Rightarrow \frac{\partial f}{\partial y}(1,1,1) = 2$. $\frac{\partial f}{\partial z} = -4z + \ln x \Rightarrow \frac{\partial f}{\partial z}(1,1,1) = -4$.

$\therefore \nabla f = 3\,\mathbf{i} + 2\,\mathbf{j} - 4\,\mathbf{k}$

3. $\frac{\partial f}{\partial x} = -\frac{x}{(x^2 + y^2 + z^2)^{3/2}} + \frac{1}{x} \Rightarrow \frac{\partial f}{\partial x}(1,2,-2) = -\frac{1}{27}$. $\frac{\partial f}{\partial y} = -\frac{y}{(x^2 + y^2 + z^2)^{3/2}} + \frac{1}{y} \Rightarrow \frac{\partial f}{\partial y}(1,2,-2) = \frac{23}{54}$. $\frac{\partial f}{\partial z} = -\frac{z}{(x^2 + y^2 + z^2)^{3/2}} + \frac{1}{z} \Rightarrow \frac{\partial f}{\partial z}(1,2,-2) = -\frac{23}{54}$. $\therefore \nabla f = -\frac{1}{27}\mathbf{i} + \frac{23}{54}\mathbf{j} - \frac{23}{54}\mathbf{k}$.

5. $x^2 + y^2 - z = 0 \Rightarrow \nabla f = 2x\,\mathbf{i} + 2y\,\mathbf{j} - \mathbf{k} \Rightarrow$

$\nabla f(1,1,2) = 2\,\mathbf{i} + 2\,\mathbf{j} - \mathbf{k}$

(Graph on the next page.)

7. $\frac{x^2}{2} + \frac{y^2}{2} - \frac{z^2}{2} = 0 \Rightarrow \nabla f = x\,\mathbf{i} + y\,\mathbf{j} - z\,\mathbf{k} \Rightarrow$

$\nabla f\left(1,1,-\sqrt{2}\right) = \mathbf{i} + \mathbf{j} + \sqrt{2}\,\mathbf{k}$

(Graph on the next page.)

5.

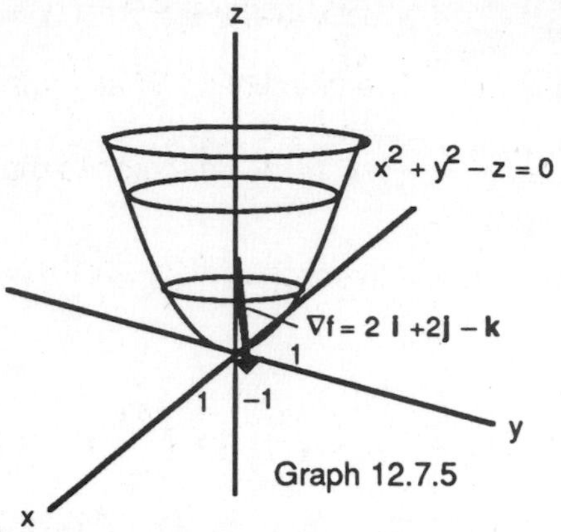

Graph 12.7.5

7.

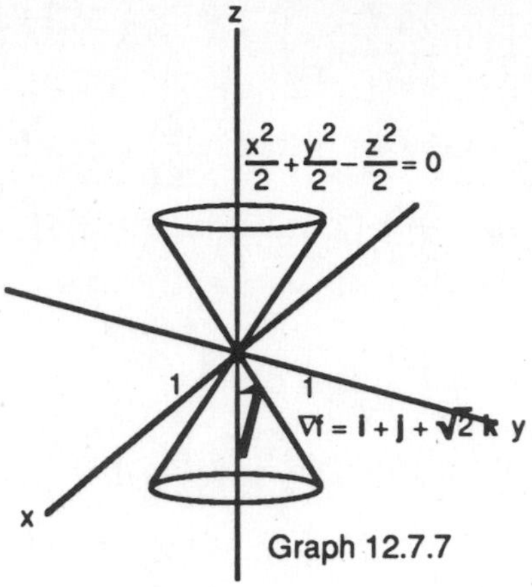

Graph 12.7.7

9. $\frac{\partial f}{\partial x} = -1, \frac{\partial f}{\partial y} = 1 \Rightarrow \nabla f = -\mathbf{i} + \mathbf{j}$

$-1 = y - x$ is the level curve.

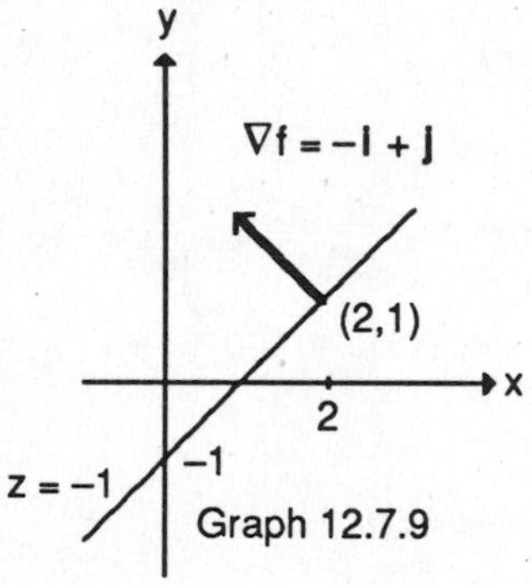

Graph 12.7.9

11. $\frac{\partial f}{\partial x} = -2x \Rightarrow \frac{\partial f}{\partial x}(-1,0) = 2.\ \frac{\partial f}{\partial y} = 1 \Rightarrow$

$\nabla f = 2\,\mathbf{i} + \mathbf{j}$. $-1 = y - x^2$ is the level curve.

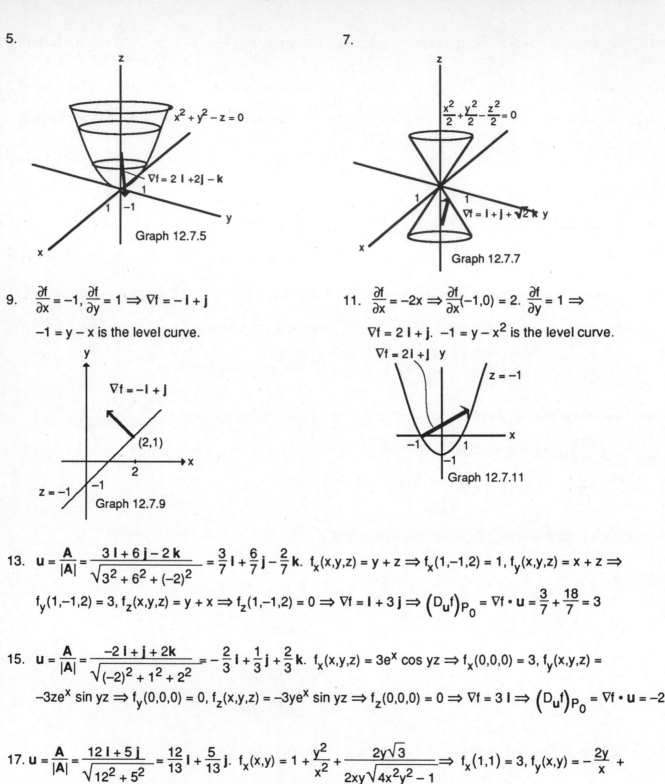

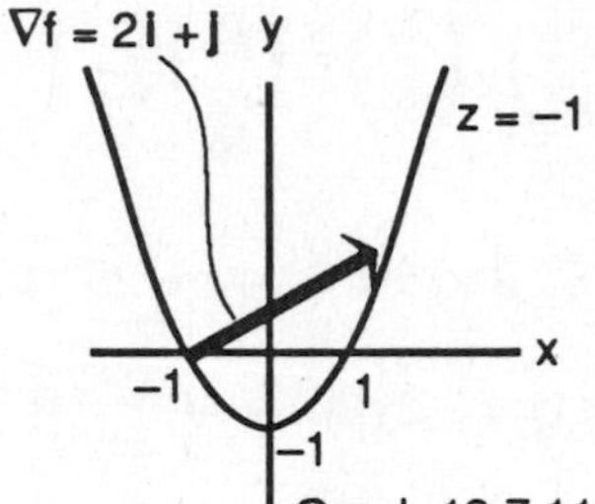

Graph 12.7.11

13. $\mathbf{u} = \frac{\mathbf{A}}{|\mathbf{A}|} = \frac{3\,\mathbf{i} + 6\,\mathbf{j} - 2\,\mathbf{k}}{\sqrt{3^2 + 6^2 + (-2)^2}} = \frac{3}{7}\mathbf{i} + \frac{6}{7}\mathbf{j} - \frac{2}{7}\mathbf{k}$. $f_x(x,y,z) = y + z \Rightarrow f_x(1,-1,2) = 1$, $f_y(x,y,z) = x + z \Rightarrow$
$f_y(1,-1,2) = 3$, $f_z(x,y,z) = y + x \Rightarrow f_z(1,-1,2) = 0 \Rightarrow \nabla f = \mathbf{i} + 3\,\mathbf{j} \Rightarrow \left(D_{\mathbf{u}}f\right)_{P_0} = \nabla f \cdot \mathbf{u} = \frac{3}{7} + \frac{18}{7} = 3$

15. $\mathbf{u} = \frac{\mathbf{A}}{|\mathbf{A}|} = \frac{-2\,\mathbf{i} + \mathbf{j} + 2\mathbf{k}}{\sqrt{(-2)^2 + 1^2 + 2^2}} = -\frac{2}{3}\mathbf{i} + \frac{1}{3}\mathbf{j} + \frac{2}{3}\mathbf{k}$. $f_x(x,y,z) = 3e^x \cos yz \Rightarrow f_x(0,0,0) = 3$, $f_y(x,y,z) =$
$-3ze^x \sin yz \Rightarrow f_y(0,0,0) = 0$, $f_z(x,y,z) = -3ye^x \sin yz \Rightarrow f_z(0,0,0) = 0 \Rightarrow \nabla f = 3\,\mathbf{i} \Rightarrow \left(D_{\mathbf{u}}f\right)_{P_0} = \nabla f \cdot \mathbf{u} = -2$

17. $\mathbf{u} = \frac{\mathbf{A}}{|\mathbf{A}|} = \frac{12\,\mathbf{i} + 5\,\mathbf{j}}{\sqrt{12^2 + 5^2}} = \frac{12}{13}\mathbf{i} + \frac{5}{13}\mathbf{j}$. $f_x(x,y) = 1 + \frac{y^2}{x^2} + \frac{2y\sqrt{3}}{2xy\sqrt{4x^2y^2 - 1}} \Rightarrow f_x(1,1) = 3$, $f_y(x,y) = -\frac{2y}{x} +$
$\frac{2x\sqrt{3}}{2xy\sqrt{4x^2y^2 - 1}} \Rightarrow f_y(1,1) = -1 \Rightarrow \nabla f = 3\,\mathbf{i} - \mathbf{j} \Rightarrow \left(D_{\mathbf{u}}f\right)_{P_0} = \nabla f \cdot \mathbf{u} = \frac{36}{13} - \frac{5}{13} = \frac{31}{13}$

19. The direction in which f increases most rapidly is in the direction of $\nabla f = (2xy + ye^{xy}\sin y)\,\mathbf{i} + (x^2 + xe^{xy}\sin y + e^{xy}\cos y)\,\mathbf{j}$. At (1,0), $\nabla f = 2\,\mathbf{j}$. The direction of ∇f at (1,0) is $\mathbf{u} = \mathbf{j}$. The derivative of f in this direction is $|\nabla f| = 2$. f decreases most rapidly in the direction $-\mathbf{u} = -\mathbf{j}$ and the derivative in this direction is $-|\nabla f| = -2$.

21. The direction in which f increases most rapidly is in the direction of $\nabla f = \left(e^y - \dfrac{4z^2}{x\sqrt{1-(\ln x)^2}}\right)\mathbf{i} + (x\,e^y)\,\mathbf{j} +$ $(8z\cos^{-1}(\ln x))\,\mathbf{k}$. At $\left(1, \ln 2, \frac{1}{2}\right)$, $\nabla f = \mathbf{i} + 2\,\mathbf{j} + 2\pi\,\mathbf{k} \Rightarrow |\nabla f| = \sqrt{5 + 4\pi^2}$. The direction of ∇f at $\left(1, \ln 2, \frac{1}{2}\right)$ is $\mathbf{u} = \dfrac{1}{\sqrt{5+4\pi^2}}(\mathbf{i} + \mathbf{j} + 2\pi\,\mathbf{k})$. The derivative of f in this direction is $|\nabla f| = \sqrt{5+4\pi^2}$. f decreases most rapidly in the direction $-\mathbf{u} = -\dfrac{1}{\sqrt{5+4\pi^2}}(\mathbf{i} + \mathbf{j} + 2\pi\,\mathbf{k})$. The derivative in this direction is $-|\nabla f| = -\sqrt{5+4\pi^2}$.

23. The direction in which f increases most rapidly is in the direction of $\nabla f = \left(\frac{2}{x}\right)\mathbf{i} + \left(\frac{2}{y}\right)\mathbf{j} + \left(\frac{2}{z}\right)\mathbf{k}$. At (1,1,1), $\nabla f = 2\,\mathbf{i} + 2\,\mathbf{j} + 2\,\mathbf{k} \Rightarrow |\nabla f| = \sqrt{4+4+4} = 2\sqrt{3}$. The direction of ∇f at (1,1,1) is $\mathbf{u} = \dfrac{1}{2\sqrt{3}}(2\,\mathbf{i} + 2\,\mathbf{j} + 2\,\mathbf{k}) = \dfrac{1}{\sqrt{3}}(\mathbf{i} + \mathbf{j} + \mathbf{k})$. The derivative in this direction is $|\nabla f| = 2\sqrt{3}$. f decreases most rapidly in the direction $-\mathbf{u} = -\dfrac{1}{\sqrt{3}}(\mathbf{i} + \mathbf{j} + \mathbf{k})$. The derivative in this direction is $-|\nabla f| = -2\sqrt{3}$.

25. $\nabla f = -2x\,\mathbf{i} + 2\,\mathbf{k} \Rightarrow \nabla f(2,0,2) = -4\,\mathbf{i} + 2\,\mathbf{k} \Rightarrow$ Tangent plane: $-4(x-2) + 2(z-2) = 0 \Rightarrow -4x + 2z + 4 = 0$; Normal line: $x = 2 - 4t,\ y = 0,\ z = 2 + 2t$

27. $\nabla f = (2x + 2y)\,\mathbf{i} + (2x - 2y)\,\mathbf{j} + 2z\,\mathbf{k} \Rightarrow \nabla f(1,-1,3) = 4\,\mathbf{j} + 6\,\mathbf{k} \Rightarrow$ Tangent plane: $4(y+1) + 6(z-3) = 0$ $\Rightarrow 2y + 3z = 7$; Normal line: $x = 1,\ y = -1 + 4t,\ z = 3 + 6t$

29. $\nabla f = \left(-\pi\sin\pi x - 2xy + ze^{xz}\right)\mathbf{i} + \left(-x^2 + z\right)\mathbf{j} + \left(xe^{xz} + y\right)\mathbf{k} \Rightarrow \nabla f(0,1,2) = 2\,\mathbf{i} + 2\,\mathbf{j} + \mathbf{k} \Rightarrow$ Tangent plane: $2(x-0) + 2(y-1) + 1(z-2) = 0 \Rightarrow 2x + 2y + z - 4 = 0$; Normal line: $x = 2t,\ y = 1 + 2t,$ $z = 2 + t$

31. $\nabla f = \mathbf{i} + 2y\,\mathbf{j} + 2\,\mathbf{k} \Rightarrow \nabla f(1,1,1) = \mathbf{i} + 2\,\mathbf{j} + 2\,\mathbf{k}$. $\nabla g = \mathbf{i}$ for all points P. $\mathbf{v} = \nabla f \times \nabla g \Rightarrow$

$$\mathbf{v} = \begin{vmatrix} \mathbf{i} & \mathbf{j} & \mathbf{k} \\ 1 & 2 & 2 \\ 1 & 0 & 0 \end{vmatrix} = 2\,\mathbf{j} - 2\,\mathbf{k} \Rightarrow \text{Tangent line: } x = 1,\ y = 1 + 2t,\ z = 1 - 2t$$

33. $\nabla f = \left(3x^2 + 6xy^2 + 4y\right)\mathbf{i} + \left(6x^2y + 3y^2 + 4x\right)\mathbf{j} - 2z\,\mathbf{k} \Rightarrow \nabla f(1,1,3) = 13\,\mathbf{i} + 13\,\mathbf{j} - 6\,\mathbf{k}$.

$$\nabla g = 2x\,\mathbf{i} + 2y\,\mathbf{j} + 2z\,\mathbf{k} \Rightarrow \nabla g(1,1,3) = 2\,\mathbf{i} + 2\,\mathbf{j} + 6\,\mathbf{k}.\ \mathbf{v} = \nabla f \times \nabla g \Rightarrow \mathbf{v} = \begin{vmatrix} \mathbf{i} & \mathbf{j} & \mathbf{k} \\ 13 & 13 & -6 \\ 2 & 2 & 6 \end{vmatrix} =$$

$90\,\mathbf{i} - 90\,\mathbf{j} \Rightarrow$ Tangent line: $x = 1 + 90t,\ y = 1 - 90t,\ z = 3$

35. $\nabla f = 2x\,\mathbf{i} + 2y\,\mathbf{j} \Rightarrow \nabla f(\sqrt{2},\sqrt{2}) = 2\sqrt{2}\,\mathbf{i} + 2\sqrt{2}\,\mathbf{j} \Rightarrow$ Tangent line: $2\sqrt{2}\left(x-\sqrt{2}\right) + 2\sqrt{2}\left(y-\sqrt{2}\right) = 0 \Rightarrow$
$\sqrt{2}\,x + \sqrt{2}\,y = 4 \Rightarrow x + y = 2\sqrt{2}$

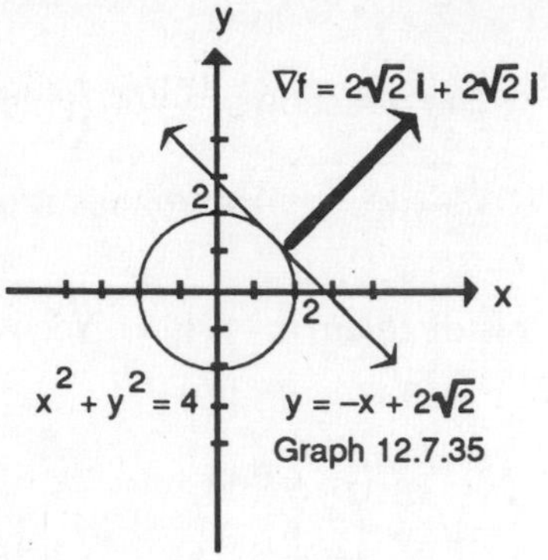

Graph 12.7.35

37. $\nabla f = y\,\mathbf{i} + x\,\mathbf{j} \Rightarrow \nabla f(2,-2) = -2\,\mathbf{i} + 2\,\mathbf{j} \Rightarrow$ Tangent line: $-2(x-2) + 2(y+2) = 0 \Rightarrow y = x - 4$

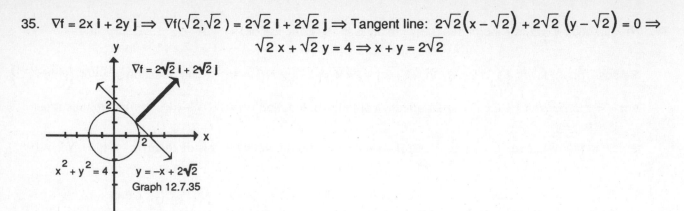

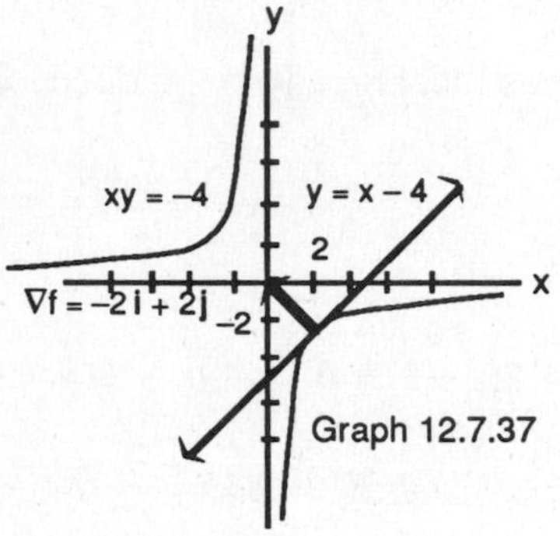

Graph 12.7.37

39. $\nabla f = (-\pi y \sin(\pi xy) + y^2)\,\mathbf{i} + (-\pi x \sin(\pi xy) + 2xy)\,\mathbf{j}$. At $(-1,1)$, $\nabla f = \mathbf{i} - 2\,\mathbf{j}$ The direction of $\mathbf{i} + \mathbf{j}$ is $\mathbf{u} = \frac{1}{\sqrt{2}}(\mathbf{i} + \mathbf{j})$. $\therefore\ df = (\nabla f \cdot \mathbf{u})\,ds = \frac{-1}{\sqrt{2}}(0.1) = -\frac{\sqrt{2}}{20}$

41. $\nabla f = \left(e^x \cos yz\right)\mathbf{i} - \left(ze^x \sin yz\right)\mathbf{j} - \left(ye^x \sin yz\right)\mathbf{k} \Rightarrow \nabla f(0,0,0) = \mathbf{i}$. $\mathbf{u} = \frac{\mathbf{A}}{|\mathbf{A}|} = \frac{2\,\mathbf{i} + \mathbf{j} - 2\,\mathbf{k}}{\sqrt{2^2 + 1^2 + (-2)^2}} =$
$\frac{2}{3}\mathbf{i} + \frac{1}{3}\mathbf{j} - \frac{2}{3}\mathbf{k} \Rightarrow \nabla f \cdot \mathbf{u} = \frac{2}{3}$. $\therefore\ df = (\nabla f \cdot \mathbf{u})ds = \frac{2}{3}(0.1) = \frac{1}{15}$ or 0.067

43. $\nabla f = y\,\mathbf{i} + (x + 2y)\,\mathbf{j} \Rightarrow \nabla f(3,2) = 2\,\mathbf{i} + 7\,\mathbf{j}$. $\mathbf{A}$, orthogonal to ∇f, is $\mathbf{A} = 7\,\mathbf{i} - 2\,\mathbf{j} \Rightarrow \mathbf{u} = \frac{\mathbf{A}}{|\mathbf{A}|} = \frac{7\,\mathbf{i} - 2\,\mathbf{j}}{\sqrt{7^2 + (-2)^2}} =$
$\frac{7}{\sqrt{53}}\mathbf{i} - \frac{2}{\sqrt{53}}\mathbf{j} \Rightarrow -\mathbf{u} = -\frac{7}{\sqrt{53}}\mathbf{i} + \frac{2}{\sqrt{53}}\mathbf{j}$

45. $\nabla f = 2x\,\mathbf{i} + 2y\,\mathbf{j} + 2z\,\mathbf{k} = (2\cos t)\,\mathbf{i} + (2\sin t)\,\mathbf{j} + 2t\,\mathbf{k}$. $\mathbf{v} = (-\sin t)\,\mathbf{i} + (\cos t)\,\mathbf{j} + \mathbf{k} \Rightarrow \mathbf{T} = \frac{\mathbf{v}}{|\mathbf{v}|} =$
$\frac{(-\sin t)\,\mathbf{i} + (\cos t)\,\mathbf{j} + \mathbf{k}}{\sqrt{(\sin t)^2 + (\cos t)^2 + 1^2}} = \left(\frac{-\sin t}{\sqrt{2}}\right)\mathbf{i} + \left(\frac{\cos t}{\sqrt{2}}\right)\mathbf{j} + \frac{1}{\sqrt{2}}\mathbf{k}.$
$\left(D_{\mathbf{T}}f\right)_{P_0} = \nabla f \cdot \mathbf{T} = (2\cos t)\left(\frac{-\sin t}{\sqrt{2}}\right) + (2\sin t)\left(\frac{\cos t}{\sqrt{2}}\right) + 2t\left(\frac{1}{\sqrt{2}}\right) = \frac{2t}{\sqrt{2}} \Rightarrow \left(D_{\mathbf{T}}f\right)\left(\frac{-\pi}{4}\right) = \frac{-\pi}{2\sqrt{2}}$,
$\left(D_{\mathbf{T}}f\right)(0) = 0$, $\left(D_{\mathbf{T}}f\right)\left(\frac{\pi}{4}\right) = \frac{\pi}{2\sqrt{2}}$

47. $\nabla f = f_x(1,2)\,\mathbf{i} + f_y(1,2)\,\mathbf{j}$. $\mathbf{u}_1 = \frac{\mathbf{i}+\mathbf{j}}{\sqrt{1^2+1^2}} = \frac{1}{\sqrt{2}}\mathbf{i} + \frac{1}{\sqrt{2}}\mathbf{j}$. $\left(D_{\mathbf{u}_1}f\right)(1,2) =$

$f_x(1,2)\left(\frac{1}{\sqrt{2}}\right) + f_y(1,2)\left(\frac{1}{\sqrt{2}}\right) = 2\sqrt{2} \Rightarrow f_x(1,2) + f_y(1,2) = 4$. $\mathbf{u}_2 = -\mathbf{j}$. $\left(D_{\mathbf{u}_2}f\right)(1,2) = f_x(1,2)(0) + f_y(1,2)(-1)$

$= -3 \Rightarrow -f_y(1,2) = -3 \Rightarrow f_y(1,2) = 3$. $\therefore$ $f_x(1,2) + 3 = 4 \Rightarrow f_x(1,2) = 1$. Then $\nabla f(1,2) = \mathbf{i} + 3\,\mathbf{j}$.

$\mathbf{u} = \frac{\mathbf{A}}{|\mathbf{A}|} = \frac{-\mathbf{i}-2\mathbf{j}}{\sqrt{(-1)^2+(-2)^2}} = -\frac{1}{\sqrt{5}}\mathbf{i} - \frac{2}{\sqrt{5}}\mathbf{j} \Rightarrow \left(D_{\mathbf{u}}f\right)_{P_0} = \nabla f \cdot \mathbf{u} = -\frac{1}{\sqrt{5}} - \frac{6}{\sqrt{5}} = -\frac{7}{\sqrt{5}}$

49. a) $V = lwh \Rightarrow V_l = wh,\ V_w = lh,\ V_h = lw \Rightarrow dV = wh\,dl + lh\,dw + lw\,dh \Rightarrow dV\big|_{(5,3,2)} = 6\,dl + 10\,dw + 15\,dh$

$dl = 1 \text{ in} = \frac{1}{12} \text{ ft},\ dw = 1 \text{ in} = \frac{1}{12} \text{ ft},\ dh = \frac{1}{2} \text{ in} = \frac{1}{24} \text{ ft} \Rightarrow dV = 6\left(\frac{1}{12}\right) + 10\left(\frac{1}{12}\right) + 15\left(\frac{1}{24}\right) = \frac{47}{24} \text{ ft}^3$

b) $A = \frac{1}{2}ab\sin C \Rightarrow A_a = \frac{1}{2}b\sin C,\ A_b = \frac{1}{2}a\sin C,\ A_C = \frac{1}{2}ab\cos C \Rightarrow dA = \frac{1}{2}b\sin C\,da + \frac{1}{2}a\sin C\,db +$

$\frac{1}{2}ab\cos C\,dC$. $dC = \pm 2° = \pm 0.0349$ radians, $da = \pm 0.5$ ft, $db = \pm 0.5$ ft. At $a = 150$ ft, $b = 200$ ft, $C =$

$60°$, $dA = \frac{1}{2}(200)\sin 60° (\pm 0.5) + \frac{1}{2}(150)\sin 60° (\pm 0.5) + \frac{1}{2}(200)(150)\cos 60° (\pm 0.0349) = \pm 319.23 \text{ ft}^2$

51. $x = g(t),\ y = h(t) \Rightarrow \mathbf{r} = g(t)\,\mathbf{i} + h(t)\,\mathbf{j} \Rightarrow \mathbf{v} = g'(t)\,\mathbf{i} + h'(t)\,\mathbf{j} \Rightarrow \mathbf{T} = \frac{\mathbf{v}}{|\mathbf{v}|} = \frac{g'(t)\,\mathbf{i} + h'(t)\,\mathbf{j}}{\sqrt{(g'(t))^2 + (h'(t))^2}}$

$z = f(x,y) \Rightarrow \frac{df}{dt} = \frac{\partial f}{\partial x}\frac{dx}{dt} + \frac{\partial f}{\partial y}\frac{dy}{dt} = \frac{\partial f}{\partial x}g'(t) + \frac{\partial f}{\partial y}h'(t)$. If $f(g(t),h(t)) = c$, then $\frac{df}{dt} = 0 \Rightarrow \frac{\partial f}{\partial x}g'(t) + \frac{\partial f}{\partial y}h'(t) = 0$.

Now $\nabla f = \frac{\partial f}{\partial x}\mathbf{i} + \frac{\partial f}{\partial y}\mathbf{j}$. $\therefore$ $\left(D_{\mathbf{T}}f\right) = \nabla f \cdot \mathbf{T} = \frac{\frac{\partial f}{\partial x}g'(t) + \frac{\partial f}{\partial y}h'(t)}{\sqrt{(g'(t))^2 + (h'(t))^2}} = 0 \Rightarrow \nabla f$ is normal to $\mathbf{T}$

53.

∇f:	0	4	0
Directional Derivative:	2.309401		
Plane:	4y = 0		
Normal Line:	x = 1, y = 4t, z = 1		

55.

∇f:	−1.818396	−1.818396	−0.4161477
Directional Derivative:	−1.85944		
Plane:	−1.818396 x − 1.818396 y − 0.416477 z = −4.05294		
Normal Line:	x = 1 − 1.818396 t, y = 1 − 1.818396 t, z = 1 − 0.4161477 t		

SECTION 12.8 MAXIMA, MINIMA, AND SADDLE POINTS

1. $f_x(x,y) = 2x + y + 3 = 0$ and $f_y(x,y) = x + 2y - 3 = 0 \Rightarrow x = -3,\ y = 3 \Rightarrow$ critical point is $(-3,3)$. $f_{xx}(-3,3) = 2$, $f_{yy}(-3,3) = 2$, $f_{xy}(-3,3) = 1 \Rightarrow f_{xx}f_{yy} - f_{xy}^2 = 3 > 0$ and $f_{xx} > 0 \Rightarrow$ minimum. $f(-3,3) = -5$, absolute minimum.

3. $f_x(x,y) = 5y - 14x + 3 = 0$ and $f_y(x,y) = 5x - 6 = 0 \Rightarrow x = \frac{6}{5},\ y = \frac{69}{25} \Rightarrow$ critical point is $\left(\frac{6}{5}, \frac{69}{25}\right)$. $f_{xx}\left(\frac{6}{5}, \frac{69}{25}\right) = -14$, $f_{yy}\left(\frac{6}{5}, \frac{69}{25}\right) = 0$, $f_{xy}\left(\frac{6}{5}, \frac{69}{25}\right) = 5 \Rightarrow f_{xx}f_{yy} - f_{xy}^2 = -25 < 0 \Rightarrow$ saddle point.

5. $f_x(x,y) = 2x + y + 3 = 0$ and $f_y(x,y) = x + 2 = 0 \Rightarrow x = -2,\ y = 1 \Rightarrow$ critical point is $(-2,1)$. $f_{xx}(-2,1) = 2$, $f_{yy}(-2,1) = 0$, $f_{xy}(-2,1) = 1 \Rightarrow f_{xx}f_{yy} - f_{xy}^2 = -1 \Rightarrow$ saddle point.

7 $f_x(x,y) = 2y - 10x + 4 = 0$ and $f_y(x,y) = 2x - 4y = 0 \Rightarrow x = \frac{4}{9},\ y = \frac{2}{9} \Rightarrow$ critical point is $\left(\frac{4}{9}, \frac{2}{9}\right)$. $f_{xx}\left(\frac{4}{9}, \frac{2}{9}\right) = -10$, $f_{yy}\left(\frac{4}{9}, \frac{2}{9}\right) = -4$, $f_{xy}\left(\frac{4}{9}, \frac{2}{9}\right) = 2 \Rightarrow f_{xx}f_{yy} - f_{xy}^2 = 36 > 0$ and $f_{xx} < 0 \Rightarrow$ maximum (absolute). $f\left(\frac{4}{9}, \frac{2}{9}\right) = -\frac{252}{81}$

9. $f_x(x,y) = 2x - 4y = 0$ and $f_y(x,y) = -4x + 2y + 6 = 0 \Rightarrow x = 2,\ y = 1 \Rightarrow$ critical point is $(2,1)$. $f_{xx}(2,1) = 2$, $f_{yy}(2,1) = 2$, $f_{xy}(2,1) = -4 \Rightarrow f_{xx}f_{yy} - f_{xy}^2 = -12 \Rightarrow$ saddle point.

11. $f_x(x,y) = 4x + 3y - 5 = 0$ and $f_y(x,y) = 3x + 8y + 2 = 0 \Rightarrow x = 2,\ y = -1 \Rightarrow$ critical point is $(2,-1)$. $f_{xx}(2,-1) = 4$, $f_{yy}(2,-1) = 8$, $f_{xy}(2,-1) = 3 \Rightarrow f_{xx}f_{yy} - f_{xy}^2 = 29 > 0$ and $f_{xx} > 0 \Rightarrow$ minimum (absolute). $f(2,-1) = -6$.

13. $f_x(x,y) = 2x - 4y + 5 = 0$ and $f_y(x,y) = -4x + 2y - 2 = 0 \Rightarrow x = \frac{1}{6},\ y = \frac{4}{3} \Rightarrow$ critical point is $\left(\frac{1}{6}, \frac{4}{3}\right)$. $f_{xx}\left(\frac{1}{6}, \frac{4}{3}\right) = 2$, $f_{yy}\left(\frac{1}{6}, \frac{4}{3}\right) = 2$, $f_{xy}\left(\frac{1}{6}, \frac{4}{3}\right) = -4 \Rightarrow f_{xx}f_{yy} - f_{xy}^2 = -12 \Rightarrow$ saddle point.

15. $f_x(x,y) = 2x - 2y - 2 = 0$ and $f_y(x,y) = -2x + 4y + 2 = 0 \Rightarrow x = 1,\ y = 0 \Rightarrow$ critical point is $(1,0)$. $f_{xx}(1,0) = 2$, $f_{yy}(1,0) = 4$, $f_{xy}(1,0) = -2 \Rightarrow f_{xx}f_{yy} - f_{xy}^2 = 4 > 0$ and $f_{xx} > 0 \Rightarrow$ minimum (absolute). $f(1,0) = 0$

17. $f_x(x,y) = 2 - 4x - 2y = 0$ and $f_y(x,y) = 2 - 2x - 2y = 0 \Rightarrow x = 0,\ y = 1 \Rightarrow$ critical point is $(0,1)$. $f_{xx}(0,1) = -4$, $f_{yy}(0,1) = -2$, $f_{xy}(0,1) = -2 \Rightarrow f_{xx}f_{yy} - f_{xy}^2 = 4 > 0$ and $f_{xx} < 0 \Rightarrow$ maximum (absolute). $f(0,1) = 4$

19. $f_x(x,y) = 3x^2 - 2y = 0$ and $f_y(x,y) = -3y^2 - 2x = 0 \Rightarrow x = 0, y = 0$ or $x = -\frac{2}{3}, y = \frac{2}{3} \Rightarrow$ critical points are $(0,0)$ and $\left(-\frac{2}{3},\frac{2}{3}\right)$. For (0,0): $f_{xx}(0,0) = 6x\big|_{(0,0)} = 0$, $f_{yy}(0,0) = -6y\big|_{(0,0)} = 0$, $f_{xy}(0,0) = -2 \Rightarrow$ $f_{xx}f_{yy} - f_{xy}^2 = -4 \Rightarrow$ saddle point. For $\left(-\frac{2}{3},\frac{2}{3}\right)$: $f_{xx}\left(-\frac{2}{3},\frac{2}{3}\right) = -4$, $f_{yy}\left(-\frac{2}{3},\frac{2}{3}\right) = -4$, $f_{xy}\left(-\frac{2}{3},\frac{2}{3}\right) = -2$ $\Rightarrow f_{xx}f_{yy} - f_{xy}^2 = 12 > 0$ and $f_{xx} < 0 \Rightarrow$ local maximum. $f\left(-\frac{2}{3},\frac{2}{3}\right) = \frac{170}{27}$

21. $f_x(x,y) = 12x - 6x^2 + 6y = 0$ and $f_y(x,y) = 6y + 6x = 0 \Rightarrow x = 0, y = 0$ or $x = 1, y = -1 \Rightarrow$ critical points are (0,0) and (1,–1). For (0,0): $f_{xx}(0,0) = 12 - 12x\big|_{(0,0)} = 12$, $f_{yy}(0,0)= 6$, $f_{xy}(0,0) = 6 \Rightarrow$ $f_{xx}f_{yy} - f_{xy}^2 = 36 > 0$ and $f_{xx} > 0 \Rightarrow$ local minimum. $f(0,0) = 0$. For (1,–1): $f_{xx}(1,-1) = 0$, $f_{yy}(1,-1) = 6$, $f_{xy}(1,-1) = 6 \Rightarrow f_{xx}f_{yy} - f_{xy}^2 = -36 \Rightarrow$ saddle point.

23. $f_x(x,y) = 3x^2 + 3y = 0$ and $f_y(x,y) = 3x + 3y^2 = 0 \Rightarrow x = 0, y = 0$ or $x = -1, y = -1 \Rightarrow$ critical points are (0,0) and (–1,–1). For (0,0): $f_{xx}(0,0) = 6x\big|_{(0,0)} = 0$, $f_{yy}(0,0) = 6y\big|_{(0,0)} = 0$, $f_{xy}(0,0) = 3 \Rightarrow$ $f_{xx}f_{yy} - f_{xy}^2 = -9 \Rightarrow$ saddle point. For (–1,–1): $f_{xx}(-1,-1) = -6$. $f_{yy}(-1,-1) = -6$, $f_{xy}(-1,-1) = 3 \Rightarrow$ $f_{xx}f_{yy} - f_{xy}^2 = 27 > 0$ and $f_{xx} < 0 \Rightarrow$ local maximum. $f(-1,-1) = 1$

25. $f_x(x,y) = \dfrac{-2x}{(x^2 + y^2 - 1)^2} = 0$ and $f_y(x,y) = \dfrac{-2y}{(x^2 + y^2 - 1)^2} = 0 \Rightarrow x = 0, y = 0 \Rightarrow$ critical point is (0,0). For (0,0): $f_{xx}(0,0) = -2$, $f_{yy}(0,0) = -2$, $f_{xy}(0,0) = 0 \Rightarrow$ $f_{xx}f_{yy} - f_{xy}^2 = 4 > 0$ and $f_{xx} =< 0 \Rightarrow$ maximum. $f(0,0) = -1$ not absolute since $f \to \infty$ as $x^2 + y^2 - 1 \to 0$.

27. $f_x(x,y) = \dfrac{8x}{(1 + x^2 + y^2)^2} = 0$ and $f_y(x,y) = \dfrac{8y}{(1 + x^2 + y^2)^2} = 0 \Rightarrow x = 0, y = 0 \Rightarrow$ critical point is (0,0). For (0,0): $f_{xx}(0,0) = 8$, $f_{yy}(0,0) = 8$, $f_{xy}(0,0) = 0 \Rightarrow$ $f_{xx}f_{yy} - f_{xy}^2 = 64 > 0$ and $f_{xx} > 0 \Rightarrow$ minimum. $f(0,0) = -4$ is absolute since $f(x,y) > -4$ for $(x,y) \neq (0,0)$.

29. $f_x(x,y) = y \cos x = 0$ and $f_y(x,y) = \sin x = 0 \Rightarrow x = n\pi$, n an integer, $y = 0 \Rightarrow$ critical points are $(n\pi,0)$ n an integer. (Note: cos x and sin x cannot both be 0 for the same x, so sin x must be $0 \Rightarrow \cos x \neq 0$ $\Rightarrow y = 0$.) For $(n\pi,0)$: $f_{xx}(n\pi,0) = 0$, $f_{yy}(n\pi,0) = 0$, $f_{xy}(n\pi,0) = 1$ if n is even and $f_{xy}(n\pi,0) = -1$ if n is odd $\Rightarrow$ $f_{xx}f_{yy} - f_{xy}^2 = -1 \Rightarrow$ saddle point. $f(n\pi,0) = 0$ for every n.

31.

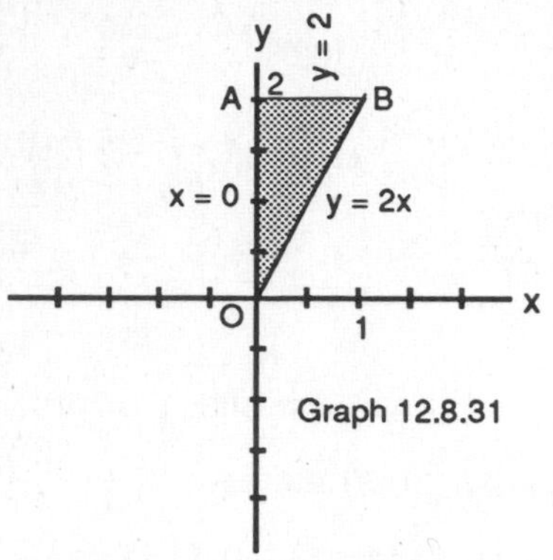

Graph 12.8.31

1. On OA, $f(x,y) = y^2 - 4y + 1 = f(0,y)$ on $0 \le y \le 2$. $y = 0 \Rightarrow f(0,0) = 1$. $y = 2 \Rightarrow f(0,2) = -3$. $f'(0,y) = 2y - 4 = 0 \Rightarrow y = 2 \Rightarrow f(0,2) = -3$.
2. On AB, $f(x,y) = 2x^2 - 4x - 3 = f(x,2)$ on $0 \le x \le 1$. $x = 0 \Rightarrow f(0,2) = -3$. $x = 1 \Rightarrow f(1,2) = -5$. $f'(x,2) = 4x - 4 = 0 \Rightarrow x = 1 \Rightarrow f(1,2) = -5$.
3. On OB, $f(x,y) = 6x^2 - 12x + 1 = f(x,2x)$ on $0 \le x \le 1$. Endpoint values have been found above. $f'(x,2x) = 12x - 12 = 0 \Rightarrow x = 1, y = 2 \Rightarrow (1,2)$, not an interior point of OB.
4. For interior points of the triangular region, $f_x(x,y) = 4x - 4 = 0$ and $f_y(x,y) = 2y - 4 = 0 \Rightarrow x = 1, y = 2 \Rightarrow (1,2)$, not an interior point of the region. $\therefore$ absolute maximum is 1 at (0,0); absolute minimum is –5 at (1,2)

33.

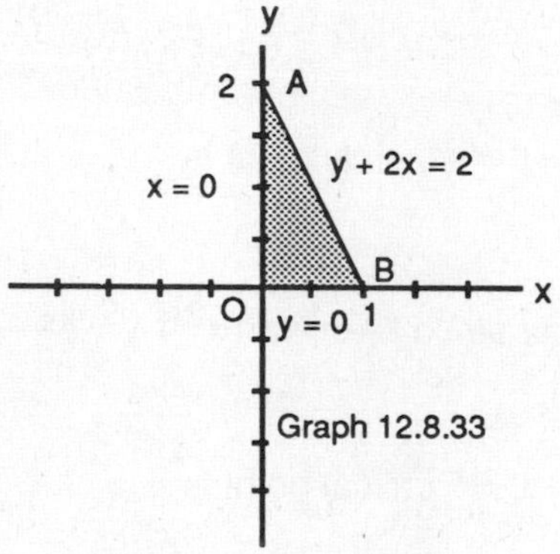

Graph 12.8.33

1. On OA, $f(x,y) = y^2 = f(0,y)$ on $0 \le y \le 2$. $f(0,0) = 0$. $f(0,2) = 4$. $f'(0,y) = 2y = 0 \Rightarrow y = 0, x = 0 \Rightarrow (0,0)$
2. On OB, $f(x,y) = x^2 = f(x,0)$ on $0 \le x \le 1$. $f(1,0) = 1$. $f'(x,0) = 2x = 0 \Rightarrow x = 0, y = 0 \Rightarrow (0,0)$
3. On AB, $f(x,y) = 5x^2 - 8x + 4 = f(x,-2x + 2)$ on $0 \le x \le 1$. $f(0,2) = 4$. $f'(x,-2x + 2) = 10x - 8 = 0 \Rightarrow x = \frac{4}{5}, y = \frac{2}{5}$. $f\left(\frac{4}{5}, \frac{2}{5}\right) = \frac{4}{5}$
4. For interior points of the triangular region, $f_x(x,y) = 2x = 0$ and $f_y(x,y) = 2y = 0 \Rightarrow x = 0, y = 0 \Rightarrow (0,0)$, not an interior point of the region. $\therefore$ absolute maximum is 4 at (0,2); absolute minimum is 0 at (0,0)

35.

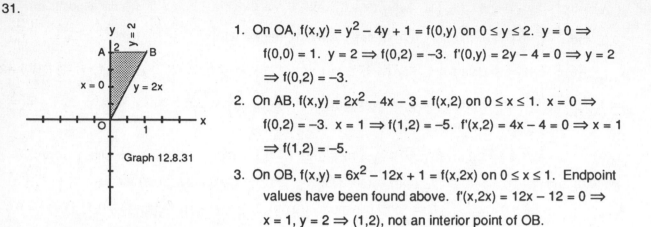

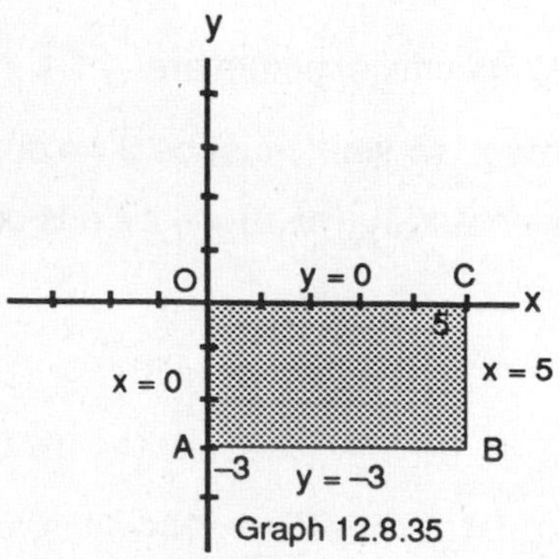

Graph 12.8.35

1. On OC, $T(x,y) = x^2 - 6x + 2 = T(x,0)$ on $0 \le x \le 5$. $T(0,0) = 2$. $T(5,0) = -3$. $T'(x,0) = 2x - 6 = 0 \Rightarrow x = 3, y = 0$. $T(3,0) = -7$
2. On CB, $T(x,y) = y^2 + 5y - 3 = T(5,y)$ on $-3 \le y \le 0$. $T(5,-3) = -9$ $T'(5,y) = 2y + 5 = 0 \Rightarrow y = -\frac{5}{2}, x = 5$. $T\left(5, -\frac{5}{2}\right) = -\frac{37}{4}$
3. On AB, $T(x,y) = x^2 - 9x + 11 = T(x,-3)$ on $0 \le x \le 5$. $T(0,-3) = 11$. $T'(x,-3) = 2x - 9 = 0 \Rightarrow x = \frac{9}{2}, y = -3$. $T\left(\frac{9}{2}, -3\right) = -\frac{37}{4}$
4. On AO, $T(x,y) = y^2 + 2 = T(0,y)$ on $-3 \le y \le 0$. $T'(0,y) = 2y = 0 \Rightarrow y = 0, x = 0$. (0,0) not an interior point of AO.
5. For interior points of the rectangular region, $T_x(x,y) = 2x + y - 6 = 0$ and $T_y(x,y) = x + 2y = 0 \Rightarrow x = 4, y = -2 \Rightarrow T(4,-2) = -10$. $\therefore$ absolute maximum is 11 at (0,–3); absolute minimum is –10 at (4,–2).

37.

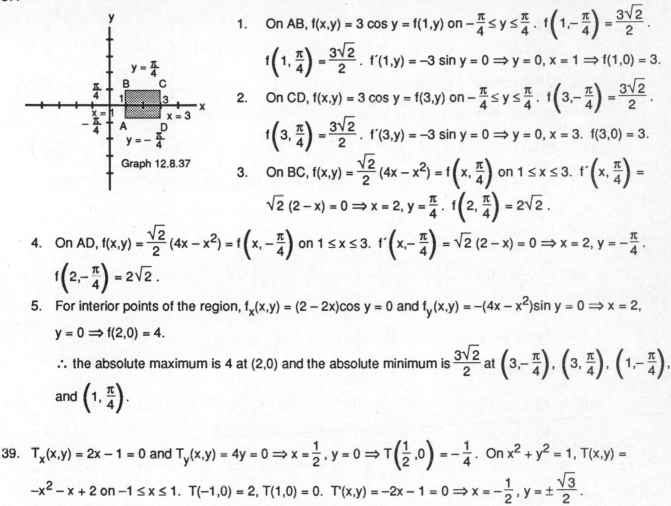

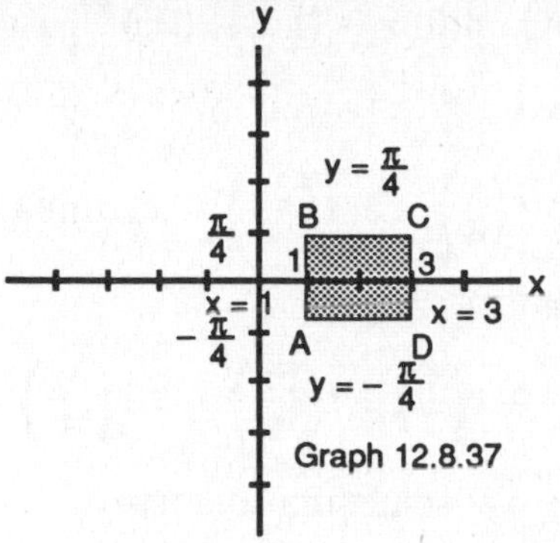

Graph 12.8.37

1. On AB, $f(x,y) = 3\cos y = f(1,y)$ on $-\frac{\pi}{4} \le y \le \frac{\pi}{4}$. $f\left(1,-\frac{\pi}{4}\right) = \frac{3\sqrt{2}}{2}$. $f\left(1,\frac{\pi}{4}\right) = \frac{3\sqrt{2}}{2}$. $f'(1,y) = -3\sin y = 0 \Rightarrow y = 0,\ x = 1 \Rightarrow f(1,0) = 3$.
2. On CD, $f(x,y) = 3\cos y = f(3,y)$ on $-\frac{\pi}{4} \le y \le \frac{\pi}{4}$. $f\left(3,-\frac{\pi}{4}\right) = \frac{3\sqrt{2}}{2}$. $f\left(3,\frac{\pi}{4}\right) = \frac{3\sqrt{2}}{2}$. $f'(3,y) = -3\sin y = 0 \Rightarrow y = 0,\ x = 3$. $f(3,0) = 3$.
3. On BC, $f(x,y) = \frac{\sqrt{2}}{2}(4x - x^2) = f\left(x,\frac{\pi}{4}\right)$ on $1 \le x \le 3$. $f'\left(x,\frac{\pi}{4}\right) = \sqrt{2}\,(2 - x) = 0 \Rightarrow x = 2,\ y = \frac{\pi}{4}$. $f\left(2,\frac{\pi}{4}\right) = 2\sqrt{2}$.
4. On AD, $f(x,y) = \frac{\sqrt{2}}{2}(4x - x^2) = f\left(x,-\frac{\pi}{4}\right)$ on $1 \le x \le 3$. $f'\left(x,-\frac{\pi}{4}\right) = \sqrt{2}\,(2 - x) = 0 \Rightarrow x = 2,\ y = -\frac{\pi}{4}$. $f\left(2,-\frac{\pi}{4}\right) = 2\sqrt{2}$.
5. For interior points of the region, $f_x(x,y) = (2 - 2x)\cos y = 0$ and $f_y(x,y) = -(4x - x^2)\sin y = 0 \Rightarrow x = 2$, $y = 0 \Rightarrow f(2,0) = 4$.

$\therefore$ the absolute maximum is 4 at (2,0) and the absolute minimum is $\frac{3\sqrt{2}}{2}$ at $\left(3,-\frac{\pi}{4}\right)$, $\left(3,\frac{\pi}{4}\right)$, $\left(1,-\frac{\pi}{4}\right)$, and $\left(1,\frac{\pi}{4}\right)$.

39. $T_x(x,y) = 2x - 1 = 0$ and $T_y(x,y) = 4y = 0 \Rightarrow x = \frac{1}{2},\ y = 0 \Rightarrow T\left(\frac{1}{2},0\right) = -\frac{1}{4}$. On $x^2 + y^2 = 1$, $T(x,y) = -x^2 - x + 2$ on $-1 \le x \le 1$. $T(-1,0) = 2$, $T(1,0) = 0$. $T'(x,y) = -2x - 1 = 0 \Rightarrow x = -\frac{1}{2},\ y = \pm\frac{\sqrt{3}}{2}$. $T\left(-\frac{1}{2},\frac{\sqrt{3}}{2}\right) = \frac{9}{4}$, $T\left(-\frac{1}{2},-\frac{\sqrt{3}}{2}\right) = \frac{9}{4}$. $\therefore$ hottest is $2\frac{1}{4}°$ at $\left(-\frac{1}{2},\frac{\sqrt{3}}{2}\right)$ and $\left(-\frac{1}{2},-\frac{\sqrt{3}}{2}\right)$; coldest is $-\frac{1}{4}°$ at $\left(\frac{1}{2},0\right)$.

41. a) $f_x(x,y) = 2x - 4y = 0$ and $f_y(x,y) = 2y - 4x = 0 \Rightarrow x = 0,\ y = 0$. $f_{xx}(0,0) = 2$, $f_{yy}(0,0) = 2$, $f_{xy}(0,0) = -4$ $\Rightarrow f_{xx}f_{yy} - f_{xy}^2 = -12 \Rightarrow$ saddle point.

b) $f_x(x,y) = 2x - 2 = 0$ and $f_y(x,y) = 2y - 4 = 0 \Rightarrow x = 1,\ y = 2$. $f_{xx}(1,2) = 2$, $f_{yy}(1,2) = 2$, $f_{xy}(1,2) = 0 \Rightarrow$ $f_{xx}f_{yy} - f_{xy}^2 = 4 > 0$ and $f_{xx} > 0 \Rightarrow$ local minimum at (1,2).

c) $f_x(x,y) = 9x^2 - 9 = 0$ and $f_y(x,y) = 2y + 4 = 0 \Rightarrow x = \pm 1,\ y = -2$. For (1,−2), $f_{xx}(1,-2) = 18x\big|_{(1,-2)} = 18$, $f_{yy}(1,-2) = 2$, $f_{xy}(1,-2) = 0 \Rightarrow f_{xx}f_{yy} - f_{xy}^2 = 36 > 0$ and $f_{xx} > 0 \Rightarrow$ local minimum at (1,−2). For (−1,−2), $f_{xx}(-1,-2) = -18$, $f_{yy}(-1,-2) = 2$, $f_{xy}(-1,-2) = 0 \Rightarrow f_{xx}f_{yy} - f_{xy}^2 = -36 \Rightarrow$ saddle point.

43. a) $x = 2\cos t,\ y = 2\sin t \Rightarrow f(t) = 4\cos t\sin t \Rightarrow \frac{df}{dt} = y(-2\sin t) + x(2\cos t) = -4\sin^2 t + 4\cos^2 t$.
$\frac{df}{dt} = 0 \Rightarrow 4\cos^2 t - 4\sin^2 t = 0 \Rightarrow \cos t = \sin t$ or $\cos t = -\sin t$

43. (Continued)

i) On the quarter circle $x^2 + y^2 = 4$ in the first quadrant, $\frac{df}{dt} = 0$ at $t = \frac{\pi}{4}$. $f(0) = 0$, $f\left(\frac{\pi}{2}\right) = 0$, $f\left(\frac{\pi}{4}\right) = 2$

$\therefore$ absolute minimum is 0 at $t = 0, \frac{\pi}{2}$; absolute maximum is 2 at $t = \frac{\pi}{4}$.

ii) On the half circle $x^2 + y^2 = 4$, $y \geq 0$, $\frac{df}{dt} = 0 \Rightarrow t = \frac{\pi}{4}, \frac{3\pi}{4}$. $f(0) = 0$, $f\left(\frac{\pi}{4}\right) = 2$, $f\left(\frac{3\pi}{4}\right) = -2$, $f(\pi) = 0$

$\therefore$ absolute minimum is -2 at $t = \frac{3\pi}{4}$; absolute maximum is 2 at $t = \frac{\pi}{4}$.

iii) On the full circle $x^2 + y^2 = 4$, $\frac{df}{dt} = 0 \Rightarrow t = \frac{\pi}{4}, \frac{3\pi}{4}, \frac{5\pi}{4}, \frac{7\pi}{4}$. $f(0) = 0$, $f\left(\frac{\pi}{4}\right) = f\left(\frac{5\pi}{4}\right) = 2$, $f\left(\frac{3\pi}{4}\right) = f\left(\frac{7\pi}{4}\right) = -2$, $f(2\pi) = 0$. $\therefore$ absolute minimum is -2 at $t = \frac{3\pi}{4}, \frac{7\pi}{4}$; absolute maximum is 2 at $t = \frac{\pi}{4}, \frac{5\pi}{4}$.

b) $x = 2\cos t$, $y = 2\sin t \Rightarrow f(t) = 2\cos t + 2\sin t \Rightarrow \frac{df}{dt} = -2\sin t + 2\cos t$. $\frac{df}{dt} = 0 \Rightarrow \cos t = \sin t$

i) On $0 \leq t \leq \frac{\pi}{2}$, $f(0) = 2$, $f\left(\frac{\pi}{2}\right) = 2$. $\frac{df}{dt} = 0 \Rightarrow t = \frac{\pi}{4} \Rightarrow f\left(\frac{\pi}{4}\right) = 2\sqrt{2}$. $\therefore$ absolute minimum is 2 at $t = 0, \frac{\pi}{2}$; absolute maximum is $2\sqrt{2}$ at $t = \frac{\pi}{4}$.

ii) On $0 \leq t \leq \pi$, $f(0) = 2$, $f(\pi) = -2$. $\frac{df}{dt} = 0 \Rightarrow t = \frac{\pi}{4}) \Rightarrow f\left(\frac{\pi}{4}\right) = 2\sqrt{2}$. $\therefore$ absolute minimum is -2 at $t = \pi$; absolute maximum is $2\sqrt{2}$ at $t = \frac{\pi}{4}$.

iii) On $0 \leq t \leq 2\pi$, $f(0) = 2$, $f(2\pi) = 2$. $\frac{df}{dt} = 0 \Rightarrow t = \frac{\pi}{4}, \frac{5\pi}{4}$. $f\left(\frac{\pi}{4}\right) = 2\sqrt{2}$, $f\left(\frac{5\pi}{4}\right) = -2\sqrt{2}$. $\therefore$ absolute minimum is $-2\sqrt{2}$ at $t = \frac{5\pi}{4}$; absolute maximum is $2\sqrt{2}$ at $t = \frac{\pi}{4}$.

c) $x = 2\cos t$, $y = 2\sin t \Rightarrow f(t) = 8\cos^2 t + 4\sin^2 t = 4\cos^2 t + 1 \Rightarrow \frac{df}{dt} = -8\cos t \sin t$. $\frac{df}{dt} = 0 \Rightarrow$ $\cos t = 0$ or $\sin t = 0$

i) On $0 \leq t \leq \frac{\pi}{2}$, $f(0) = 8$, $f\left(\frac{\pi}{2}\right) = 4$. $\frac{df}{dt} = 0 \Rightarrow t = 0, \frac{\pi}{2}$. $\therefore$ absolute minimum is 4 at $t = \frac{\pi}{2}$; absolute maximum is 8 at $t = 0$.

ii) On $0 \leq t \leq \pi$, $f(0) = 8$, $f(\pi) = 8$. $\frac{df}{dt} = 0 \Rightarrow t = 0, \pi, \frac{\pi}{2}$. $f\left(\frac{\pi}{2}\right) = 4$. $\therefore$ absolute minimum is 4 at $t = \frac{\pi}{2}$; absolute maximum is 8 at $t = 0, \pi$.

iii) On $0 \leq t \leq 2\pi$, $f(0) = 8$, $f(2\pi) = 8$. $\frac{df}{dt} = 0 \Rightarrow t = 0, \frac{\pi}{2}, \pi, \frac{3\pi}{2}, 2\pi \Rightarrow f\left(\frac{\pi}{2}\right) = 4$, $f(\pi) = 8$, $f\left(\frac{3\pi}{2}\right) = 4$.

$\therefore$ absolute minimum is 4 at $t = \frac{\pi}{2}, \frac{3\pi}{2}$; absolute maximum is 8 at $t = 0, \pi, 2\pi$.

45. a) $x = 3\cos t$, $y = 2\sin t \Rightarrow f(t) = 9 + 3\sin^2 t \Rightarrow \frac{df}{dt} = 6\sin t \cos t$. $\frac{df}{dt} = 0 \Rightarrow \sin t = 0$ or $\cos t = 0$

i) On the quarter ellipse, $\frac{x^2}{9} + \frac{y^2}{4} = 1$, in the first quadrant, $f(0) = 9$, $f\left(\frac{\pi}{2}\right) = 12$. $\frac{df}{dt} = 0 \Rightarrow t = 0, \frac{\pi}{2}$

$\therefore$ absolute maximum is 12 at $t = \frac{\pi}{2}$; absolute minimum is 9 at $t = 0$.

ii) On the half ellipse, $\frac{x^2}{9} + \frac{y^2}{4} = 1$, $y \geq 0$, $f(0) = 9$, $f(\pi) = 9$. $\frac{df}{dt} = 0 \Rightarrow t = 0, \frac{\pi}{2}, \pi$. $f\left(\frac{\pi}{2}\right) = 12$.

$\therefore$ absolute minimum is 9 at $t = 0, \pi$; absolute maximum is 12 at $t = \frac{\pi}{2}$.

45. (Continued)

iii) On the full ellipse, $\frac{x^2}{9}+\frac{y^2}{4}=1$, $f(0)=9$, $f(2\pi)=9$. $\frac{df}{dt}=0 \Rightarrow t=0, \frac{\pi}{2}, \pi, \frac{3\pi}{2}, 2\pi$. $f\left(\frac{\pi}{2}\right)=12$, $f(\pi)=9$, $f\left(\frac{3\pi}{2}\right)=12$. $\therefore$ absolute maximum is 12 at $t=\frac{\pi}{2}, \frac{3\pi}{2}$; absolute minimum is 9 at $t=0, \pi, 2\pi$.

b) $x=3\cos t$, $y=2\sin t \Rightarrow f(t)=6\cos t+6\sin t \Rightarrow \frac{df}{dt}=-6\sin t+6\cos t$. $\frac{df}{dt}=0 \Rightarrow \cos t=\sin t$.

i) On the quarter ellipse, $\frac{x^2}{9}+\frac{y^2}{4}=1$, $f(0)=6$, $f\left(\frac{\pi}{2}\right)=6$. $\frac{df}{dt}=0 \Rightarrow t=\frac{\pi}{4} \Rightarrow f\left(\frac{\pi}{4}\right)=6\sqrt{2} \Rightarrow$ absolute maximum is $6\sqrt{2}$ at $t=\frac{\pi}{4}$; absolute minimum is 6 at $t=0, \frac{\pi}{2}$.

ii) On the half ellipse, $\frac{x^2}{9}+\frac{y^2}{4}=1$, $y \ge 0$, $f(0)=6$, $f(\pi)=-6$. $\frac{df}{dt}=0 \Rightarrow t=\frac{\pi}{4} \Rightarrow f\left(\frac{\pi}{4}\right)=6\sqrt{2} \Rightarrow$ absolute maximum is $6\sqrt{2}$; absolute minimum is -6 at $t=\pi$.

iii) On the full ellipse, $\frac{x^2}{9}+\frac{y^2}{4}=1$, $f(0)=6$, $f(2\pi)=6$. $\frac{df}{dt}=0 \Rightarrow t=\frac{\pi}{4}, \frac{5\pi}{4} \Rightarrow f\left(\frac{\pi}{4}\right)=6\sqrt{2}$, $f\left(\frac{5\pi}{4}\right)=-6\sqrt{2} \Rightarrow$ absolute maximum is $6\sqrt{2}$ at $t=\frac{\pi}{4}$; absolute minimum is $-6\sqrt{2}$ at $t=\frac{5\pi}{4}$.

47.

k	x_k	y_k	x_k^2	$x_k y_k$
1	−1	2	1	−2
2	0	1	0	0
3	3	−4	9	−12
Σ	2	−1	10	−14

$m=\frac{2(-1)-3(-14)}{2^2-3(10)} \approx -1.5$, $b=\frac{1}{3}(-1-(-1.5)2) \approx 0.7$

$\therefore\ y=-1.5x+0.7$. $y\big|_{x=4}=-5.3$

49.

k	x_k	y_k	x_k^2	$x_k y_k$
1	0	0	0	0
2	1	2	1	2
3	2	3	4	6
Σ	3	5	5	8

$m=\frac{3(5)-3(8)}{3^2-3(5)}=1.5$, $b=\frac{1}{3}(5-1.5(3)) \approx 0.2$

$\therefore\ y=1.5x+0.2$. $y\big|_{x=4}=6.2$

51.

k	x_k	y_k	x_k^2	$x_k y_k$
1	12	5.27	144	63.24
2	18	5.68	324	102.24
3	24	6.25	576	150
4	30	7.21	900	216.3
5	36	8.20	1296	295.2
6	42	8.71	1764	365.82
Σ	162	41.32	5004	1192.8

$m=\frac{162(41.32)-6(1192.8)}{162^2-6(5004)} \approx 0.122$,

$b=\frac{1}{6}(41.32-(0.122)(162)) \approx 3.59$

$\therefore\ y=0.122x+3.59$

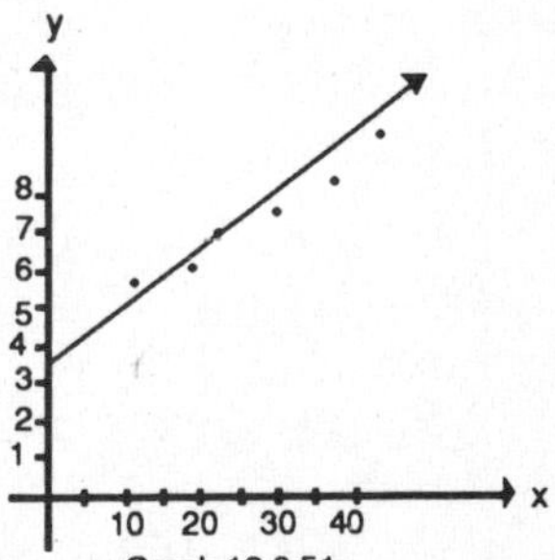

Graph 12.8.51

53. a)

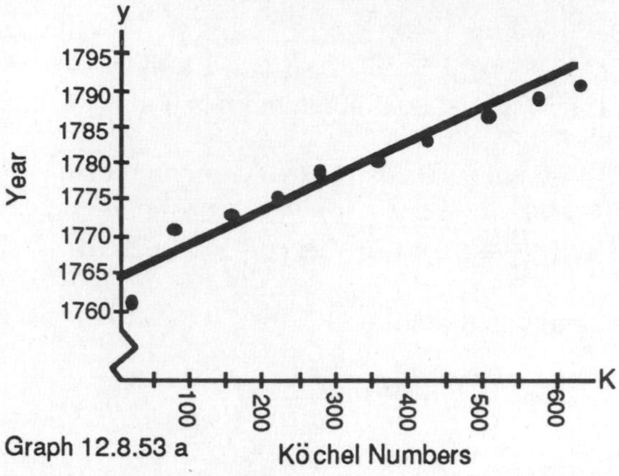

Graph 12.8.53 a

b) $m = \dfrac{(3201)(17785) - 10(5710292)}{(3201)^2 - 10(1430389)} \approx 0.0427$. $b = \dfrac{1}{10}(17785 - 0.0427(3201)) \approx 1764.8$.

$\therefore\ y = 0.0427\,K + 1764.8$

k	K_k	y_k	K_k^2	$K_k y_k$
1	1	1761	1	1761
2	75	1771	5625	132825
3	155	1772	24025	274660
4	219	1775	47961	388725
5	271	1777	73441	481567
6	351	1780	123201	624780
7	425	1783	180625	757775
8	503	1786	253009	898358
9	575	1789	330625	1028675
10	626	1791	391876	1121166
Σ	3201	17785	1430389	5710292

c) $K = 364 \Rightarrow y = .0427(364) + 1764.8 \approx 1780$

SECTION 12.9 LAGRANGE MULTIPLIERS

1. $\nabla f = y\,\mathbf{i} + x\,\mathbf{j}$, $\nabla g = 2x\,\mathbf{i} + 4y\,\mathbf{j}$. $\nabla f = \lambda \nabla g \Rightarrow y\,\mathbf{i} + x\,\mathbf{j} = \lambda(2x\,\mathbf{i} + 4y\,\mathbf{j}) \Rightarrow y = 2x\lambda$ and $x = 4y\lambda \Rightarrow$ $\lambda = \pm\frac{\sqrt{2}}{4}$ or $x = 0$. CASE 1: If $x = 0$, then $y = 0$ but $(0,0)$ not on the ellipse. $\therefore\ x \neq 0$.

   CASE 2: $x \neq 0 \Rightarrow \lambda = \pm\frac{\sqrt{2}}{4} \Rightarrow x = \pm\sqrt{2}\,y \Rightarrow (\pm\sqrt{2}\,y)^2 + 2y^2 = 1 \Rightarrow y = \pm\frac{1}{2}$. $\therefore$ f takes on its extreme values at $\left(\pm\frac{\sqrt{2}}{2}, \frac{1}{2}\right)$ and $\left(\pm\frac{\sqrt{2}}{2}, \frac{-1}{2}\right) \Rightarrow$ the extreme values of f are $\pm\frac{\sqrt{2}}{4}$..

3. $\nabla f = -2x\,\mathbf{i} - 2y\,\mathbf{j}$, $\nabla g = \mathbf{i} + 3\,\mathbf{j}$. $\nabla f = \lambda\,\nabla g \Rightarrow -2x\,\mathbf{i} - 2y\,\mathbf{j} = \lambda(\mathbf{i} + 3\,\mathbf{j}) \Rightarrow x = -\frac{\lambda}{2}$ and $y = -\frac{3\lambda}{2} \Rightarrow \lambda = -2$ $\Rightarrow x = 1$ and $y = 3 \Rightarrow$ f takes on its extreme value at $(1,3) \Rightarrow$ the extreme value of f is 39.

5. $\nabla f = 2xy\,\mathbf{i} + x^2\,\mathbf{j}$, $\nabla g = \mathbf{i} + \mathbf{j}$. $\nabla f = \lambda\,\nabla g \Rightarrow 2xy\,\mathbf{i} + x^2\,\mathbf{j} = \lambda(\mathbf{i} + \mathbf{j}) \Rightarrow 2xy = \lambda$ and $x^2 = \lambda \Rightarrow 2xy = x^2 \Rightarrow$ $x = 0, y = 3$ or $x = 2, y = 1$. $\therefore$ f takes on its extreme values at $(0,3)$ and $(2,1)$. $\therefore$ the extreme values of f are $f(0,3) = 0$ and $f(2,1) = 4$.

7. a) $\nabla f = \mathbf{i} + \mathbf{j}$, $\nabla g = y\,\mathbf{i} + x\,\mathbf{j}$. $\nabla f = \lambda\,\nabla g \Rightarrow \mathbf{i} + \mathbf{j} = \lambda(y\,\mathbf{i} + x\,\mathbf{j}) \Rightarrow 1 = \lambda y$ and $1 = \lambda x \Rightarrow y = \frac{1}{\lambda}$ and $x = \frac{1}{\lambda}$ $\Rightarrow \frac{1}{\lambda^2} = 16 \Rightarrow \lambda = \pm\frac{1}{4}$. Use $\lambda = \frac{1}{4}$ since $x > 0$, $y > 0$. Then $x = 4$, $y = 4 \Rightarrow$ the minimum value is 8 at $x = 4$, $y = 4$.

   b) $\nabla f = y\,\mathbf{i} + x\,\mathbf{j}$, $\nabla g = \mathbf{i} + \mathbf{j}$. $\nabla f = \lambda\,\nabla g \Rightarrow y\,\mathbf{i} + x\,\mathbf{j} = \lambda(\mathbf{i} + \mathbf{j}) \Rightarrow y = \lambda = x \Rightarrow y = x \Rightarrow y + y = 16 \Rightarrow y = 8$ $\Rightarrow x = 8 \Rightarrow f(8,8) = 64$ is the maximum value.

9. $V = \pi r^2 h \Rightarrow 16\pi = \pi r^2 h \Rightarrow 16 = r^2 h \Rightarrow g(r,h) = r^2 h - 16$. $\nabla S = (2\pi h + 4\pi r)\,\mathbf{i} + 2\pi r\,\mathbf{j}$, $\nabla g = 2rh\,\mathbf{i} + r^2\,\mathbf{j}$. $\nabla S = \lambda\,\nabla g \Rightarrow (2\pi rh + 4\pi r)\,\mathbf{i} + 2\pi r\,\mathbf{j} = \lambda(2rh\,\mathbf{i} + r^2\,\mathbf{j}) \Rightarrow 2\pi h + 4\pi r = 2rh\lambda$ and $2\pi r = \lambda r^2 \Rightarrow 0 = \lambda r^2 - 2\pi r$ $\Rightarrow r = 0$ or $\lambda = \frac{2\pi}{r}$. Now $r \neq 0 \Rightarrow \lambda = \frac{2\pi}{r} \Rightarrow 2\pi h + 4\pi r = 2rh\left(\frac{2\pi}{r}\right) \Rightarrow 2r = h \Rightarrow 16 = r^2(2r) \Rightarrow r = 2 \Rightarrow$ $h = 4$. $\therefore$ $r = 2$ cm, $h = 4$ cm give the smallest surface area.

11. $\nabla T = (8x - 4y)\,\mathbf{i} + (-4x + 2y)\,\mathbf{j}$, $\nabla g = 2x\,\mathbf{i} + 2y\,\mathbf{j}$. $\nabla T = \lambda\,\nabla g \Rightarrow (8x - 4y)\,\mathbf{i} + (-4x + 2y)\,\mathbf{j} = \lambda(2x\,\mathbf{i} + 2y\,\mathbf{j}) \Rightarrow$ $8x - 4y = 2\lambda x$ and $-4x + 2y = 2\lambda y \Rightarrow y = \frac{-2x}{\lambda - 1}$, $\lambda \neq 1 \Rightarrow 8x - 4\left(\frac{-2x}{\lambda - 1}\right) = 2\lambda x \Rightarrow x = 0$ or $\lambda = 0$ or $\lambda = 5$. $x = 0 \Rightarrow y = 0$. But $(0,0)$ not on $x^2 + y^2 = 25$. $\therefore\ x \neq 0 \Rightarrow \lambda = 0$ or $\lambda = 5$. $\lambda = 0 \Rightarrow y = 2x \Rightarrow$ $x^2 + (2x)^2 = 25 \Rightarrow x = \pm\sqrt{5} \Rightarrow y = \pm 2\sqrt{5}$. $\lambda = 5 \Rightarrow y = \frac{-2x}{4} = -\frac{1}{2}x \Rightarrow x^2 + \left(-\frac{1}{2}x\right)^2 = 25 \Rightarrow x = \pm 2\sqrt{5}$. $x = 2\sqrt{5} \Rightarrow y = -\sqrt{5}$, $x = -2\sqrt{5} \Rightarrow y = \sqrt{5}$. $T(\sqrt{5}, 2\sqrt{5}) = 0° = T(-\sqrt{5}, -2\sqrt{5})$, the minimum value; $T(2\sqrt{5}, -\sqrt{5}) = 125° = T(-2\sqrt{5}, \sqrt{5})$, the maximum value.

13. $\nabla f = 2x\,\mathbf{i} + 2y\,\mathbf{j}$, $\nabla g = (2x-2)\,\mathbf{i} + (2y-4)\,\mathbf{j}$. $\nabla f = \lambda\,\nabla g \Rightarrow 2x\,\mathbf{i} + 2y\,\mathbf{j} = \lambda((2x-2)\,\mathbf{i} + (2y-4)\,\mathbf{j}) \Rightarrow$ $2x = \lambda(2x-2)$ and $2y = \lambda(2y-4) \Rightarrow x = \dfrac{\lambda}{\lambda-1}$ and $y = \dfrac{2\lambda}{\lambda-1}$, $\lambda \neq 1 \Rightarrow y = 2x \Rightarrow x^2 - 2x + (2x)^2 - 4(2x)$ $= 0 \Rightarrow x = 0,\ y = 0$ or $x = 2,\ y = 4$. $\therefore$ $f(0,0) = 0$ is the minimum value, $f(2,4) = 20$ is the maximum value.

15. $\nabla f = \mathbf{i} - 2\,\mathbf{j} + 5\,\mathbf{k}$, $\nabla g = 2x\,\mathbf{i} + 2y\,\mathbf{j} + 2z\,\mathbf{k}$. $\nabla f = \lambda\,\nabla g \Rightarrow \mathbf{i} - 2\,\mathbf{j} + 5\,\mathbf{k} = \lambda(2x\,\mathbf{i} + 2y\,\mathbf{j} + 2z\,\mathbf{k}) \Rightarrow 1 = 2x\,\lambda$, $-2 = 2y\,\lambda$, and $5 = 2z\,\lambda \Rightarrow x = \dfrac{1}{2\lambda}$, $y = -\dfrac{1}{\lambda} = -2x$, $z = \dfrac{5}{2\lambda} = 5x \Rightarrow x^2 + (-2x)^2 + (5x)^2 = 30 \Rightarrow x = \pm 1$. $x = 1 \Rightarrow y = -2,\ z = 5$. $x = -1 \Rightarrow y = 2,\ z = -5$. $f(1,-2,5) = 30$, the maximum value; $f(-1,2,-5) = -30$, the minimum value.

17. Let $f(x,y,z) = x^2 + y^2 + z^2$ be the square of the distance to the origin. Then $\nabla f = 2x\,\mathbf{i} + 2y\,\mathbf{j} + 2z\,\mathbf{k}$, $\nabla g = y\,\mathbf{i} + x\,\mathbf{j} - \mathbf{k}$. $\nabla f = \lambda\,\nabla g \Rightarrow 2x\,\mathbf{i} + 2y\,\mathbf{j} + 2z\,\mathbf{k} = \lambda(y\,\mathbf{i} + x\,\mathbf{j} - \mathbf{k}) \Rightarrow 2x = \lambda y$, $2y = \lambda x$, and $2z = -\lambda$ $\Rightarrow x = \dfrac{\lambda y}{2} \Rightarrow 2y = \lambda\left(\dfrac{\lambda y}{2}\right) \Rightarrow y = 0$ or $\lambda = \pm 2$. $y = 0 \Rightarrow x = 0 \Rightarrow -z + 1 = 0 \Rightarrow z = 1$. $\lambda = 2 \Rightarrow x = y$, $z = -1 \Rightarrow x^2 - (-1) + 1 = 0 \Rightarrow x^2 + 2 = 0$, no solution. $\lambda = -2 \Rightarrow x = -y,\ z = 1 \Rightarrow (-y)y - 1 + 1 = 0 \Rightarrow$ $y = 0$, again. $\therefore$ $(0,0,1)$ is the point on the surface closest to the origin.

19. $\nabla f = \mathbf{i} + 2\,\mathbf{j} + 3\,\mathbf{k}$, $\nabla g = 2x\,\mathbf{i} + 2y\,\mathbf{j} + 2z\,\mathbf{k}$. $\nabla f = \lambda\,\nabla g \Rightarrow \mathbf{i} + 2\,\mathbf{j} + 3\,\mathbf{k} = \lambda(2x\,\mathbf{i} + 2y\,\mathbf{j} + 2z\,\mathbf{k}) \Rightarrow 1 = 2x\,\lambda$, $2 = 2y\lambda$, and $3 = 2z\lambda \Rightarrow x = \dfrac{1}{2\lambda}$, $y = \dfrac{1}{\lambda}$, and $z = \dfrac{3}{2\lambda} \Rightarrow y = 2x$ and $z = 3x \Rightarrow x^2 + (2x)^2 + (3x)^2 = 25 \Rightarrow$ $x = \pm\dfrac{5}{\sqrt{14}}$. $x = \dfrac{5}{\sqrt{14}} \Rightarrow y = \dfrac{10}{\sqrt{14}}$, $z = \dfrac{15}{\sqrt{14}}$. $x = -\dfrac{5}{\sqrt{14}} \Rightarrow y = -\dfrac{10}{\sqrt{14}}$, $z = -\dfrac{15}{\sqrt{14}}$. $f\left(\dfrac{5}{\sqrt{14}}, \dfrac{10}{\sqrt{14}}, \dfrac{15}{\sqrt{14}}\right) =$ $5\sqrt{14}$, the maximum value; $f\left(-\dfrac{5}{\sqrt{14}}, -\dfrac{10}{\sqrt{14}}, -\dfrac{15}{\sqrt{14}}\right) = -5\sqrt{14}$, the minimum value.

21. $\nabla f = yz\,\mathbf{i} + xz\,\mathbf{j} + xy\,\mathbf{k}$, $\nabla g = \mathbf{i} + \mathbf{j} + 2z\,\mathbf{k}$. $\nabla f = \lambda\,\nabla g \Rightarrow yz\,\mathbf{i} + xz\,\mathbf{j} + xy\,\mathbf{k} = \lambda(\mathbf{i} + \mathbf{j} + 2z\,\mathbf{k}) \Rightarrow yz = \lambda$, $xz =$ λ, and $xy = \lambda(2z) \Rightarrow yz = xz \Rightarrow z = 0$ or $y = x$. But $z > 0 \Rightarrow y = x \Rightarrow x^2 = 2z\lambda$ and $xz = \lambda$. Then $x^2 = 2z(xz)$ $\Rightarrow x = 0$ or $x = 2z^2$. But $x > 0 \Rightarrow x = 2z^2 \Rightarrow y = 2z^2 \Rightarrow 2z^2 + 2z^2 + z^2 = 16 \Rightarrow z = \pm\dfrac{4}{\sqrt{5}}$. Use $z = \dfrac{4}{\sqrt{5}}$ since $z > 0 \Rightarrow x = \dfrac{32}{5}$, $y = \dfrac{32}{5}$. $f\left(\dfrac{32}{5}, \dfrac{32}{5}, \dfrac{4}{\sqrt{5}}\right) = \dfrac{4096}{25\sqrt{5}}$

23. $\nabla U = (y+2)\,\mathbf{i} + x\,\mathbf{j}$, $\nabla g = 2\,\mathbf{i} + \mathbf{j}$. $\nabla U = \lambda\,\nabla g \Rightarrow (y+2)\,\mathbf{i} + x\,\mathbf{j} = \lambda(2\,\mathbf{i} + \mathbf{j}) \Rightarrow y + 2 = 2\lambda$ and $x = \lambda \Rightarrow$ $y + 2 = 2x \Rightarrow y = 2x - 2 \Rightarrow 2x + 2x - 2 = 30 \Rightarrow x = 8 \Rightarrow y = 14$. $\therefore$ $U(8,14) = \$128$, the maximum value of U under the constraint.

25. $\nabla f = \mathbf{i} + \mathbf{j}$, $\nabla g = y\,\mathbf{i} + x\,\mathbf{j}$. $\nabla f = \lambda\,\nabla g \Rightarrow \mathbf{i} + \mathbf{j} = \lambda(y\,\mathbf{i} + x\,\mathbf{j}) \Rightarrow 1 = y\lambda$ and $1 = x\lambda \Rightarrow y = x \Rightarrow y^2 = 16 \Rightarrow$ $y = \pm 4 \Rightarrow x = \pm 4$. But as $x \to \infty$, $y \to 0$ and $f(x,y) \to \infty$; as $x \to -\infty$, $y \to 0$ and $f(x,y) \to -\infty$.

27. Let $f(x,y,z) = x^2 + y^2 + z^2$. Maximize f subject to $g_1(x,y,z) = y + 2z - 12 = 0$ and $g_2(x,y,z) = x + y - 6 = 0$
$\nabla f = 2x\,\mathbf{i} + 2y\,\mathbf{j} + 2z\,\mathbf{k}$, $\nabla g_1 = \mathbf{j} + 2\,\mathbf{k}$, $\nabla g_2 = \mathbf{i} + \mathbf{j}$. Then $2x\,\mathbf{i} + 2y\,\mathbf{j} + 2z\,\mathbf{k} = \lambda(\mathbf{j} + 2\,\mathbf{k}) + \mu(\mathbf{i} + \mathbf{j}) \Rightarrow$
$2x\,\mathbf{i} + 2y\,\mathbf{j} + 2z\,\mathbf{k} = \mu\,\mathbf{i} + (\lambda + \mu)\,\mathbf{j} + 2\lambda\,\mathbf{k} \Rightarrow 2x = \mu,\ 2y = \lambda + \mu,\ 2z = 2\lambda \Rightarrow x = \frac{\mu}{2},\ z = \lambda \Rightarrow$
$2x = 2y - z \Rightarrow x = \frac{2y - z}{2}$. Then $y + 2z - 12 = 0$ and $\frac{2y - z}{2} + y - 6 = 0 \Rightarrow 4y - z - 12 = 0 \Rightarrow z = 4 \Rightarrow$
$y = 4 \Rightarrow x = 2 \Rightarrow (2,4,4)$ is the point closest.

29. Let $g_1(x,y,z) = z - 1 = 0$ and $g_2(x,y,z) = x^2 + y^2 + z^2 - 10 = 0 \Rightarrow \nabla g_1 = \mathbf{k}$, $\nabla g_2 = 2x\,\mathbf{i} + 2y\,\mathbf{j} + 2z\,\mathbf{k}$.
$\nabla f = 2xyz\,\mathbf{i} + x^2z\,\mathbf{j} + x^2y\,\mathbf{k} \Rightarrow 2xyz\,\mathbf{i} + x^2z\,\mathbf{j} + x^2y\,\mathbf{k} = \lambda(\mathbf{k}) + \mu(2x\,\mathbf{i} + 2y\,\mathbf{j} + 2z\,\mathbf{k}) \Rightarrow 2xyz = 2x\mu,\ x^2z =$
$2y\mu,\ x^2y = 2z + \lambda \Rightarrow xyz = x\mu \Rightarrow x = 0$ or $yz = \mu \Rightarrow \mu = y$ since $z = 1$. $x = 0 \Rightarrow 2y\mu = 0$ and $2z + \lambda = 0$
$\Rightarrow z = 1 \Rightarrow \lambda = -2 \Rightarrow y^2 - 9 = 0 \Rightarrow y = \pm 3 \Rightarrow (0,\pm 3,1)$. $\mu = y \Rightarrow x^2z = 2y^2 \Rightarrow x^2 = 2y^2$ since $z = 1$
$\Rightarrow 2y^2 + y^2 + 1 - 10 = 0 \Rightarrow 3y^2 - 9 = 0 \Rightarrow y = \pm\sqrt{3} \Rightarrow \mu = \pm\sqrt{3} \Rightarrow x^2 = 2(\pm\sqrt{3})(\pm\sqrt{3}) = 6 \Rightarrow x =$
$\pm\sqrt{6} \Rightarrow (\pm\sqrt{6}, \pm\sqrt{3}, 1)$. $f(0,\pm 3,1) = 1$. $f(\pm\sqrt{6},\pm\sqrt{3},1) = 6(\pm\sqrt{3}) + 1 = 1 \pm 6\sqrt{3} \Rightarrow$ Maximum of f
is $1 + 6\sqrt{3}$ at $(\pm\sqrt{6},\sqrt{3},1)$; minimum of f is $1 - 6\sqrt{3}$ at $(\pm\sqrt{6},-\sqrt{3},1)$

31. Let $g_1(x,y,z) = y - x = 0 \Rightarrow x = y$. Let $g_2(x,y,z) = x^2 + y^2 + z^2 - 4 = 0 \Rightarrow \nabla g_2 = 2x\,\mathbf{i} + 2y\,\mathbf{j} + 2z\,\mathbf{k}$. $\nabla g_1 =$
$-\mathbf{i} + \mathbf{j}$; $\nabla f = y\,\mathbf{i} + x\,\mathbf{j} + 2z\,\mathbf{k} \Rightarrow y\,\mathbf{i} + x\,\mathbf{j} + 2z\,\mathbf{k} = \lambda(-\mathbf{i} + \mathbf{j}) + \mu(2x\,\mathbf{i} + 2y\,\mathbf{j} + 2z\,\mathbf{k}) \Rightarrow y = -\lambda + 2x\mu,\ x =$
$\lambda + 2y\mu,\ 2z = 2z\mu \Rightarrow z = z\mu \Rightarrow z = 0$ or $\mu = 1$. $z = 0 \Rightarrow x^2 + y^2 - 4 = 0 \Rightarrow 2x^2 - 4 = 0$ (since $x = y$) $\Rightarrow$
$x^2 = 2 \Rightarrow x = \pm\sqrt{2} \Rightarrow y = \pm\sqrt{2} \Rightarrow (\pm\sqrt{2},\pm\sqrt{2},0)$. $\mu = 1 \Rightarrow y = -\lambda + 2x$ and $x = \lambda + 2y \Rightarrow x + y =$
$2(x + y) \Rightarrow 2x = 2(2x)$ since $x = y \Rightarrow x = 0 \Rightarrow y = 0 \Rightarrow z^2 - 4 = 0 \Rightarrow z = \pm 2 \Rightarrow (0,0,\pm 2)$.
$f(0,0,\pm 2) = 4$, $f(\pm\sqrt{2},\pm\sqrt{2},0) = 2 \Rightarrow$ Maximum value of f is 4 at $(0,0,\pm 2)$; minimum value of f is 2 at
$(\pm\sqrt{2},\pm\sqrt{2},0)$

SECTION 12.10 TAYLOR'S FORMULA

1. $f(x,y) = e^x\cos y \Rightarrow f(0,0) = 1$. $f_x(0,0) = e^x\cos y\big|_{(0,0)} = 1$, $f_y(0,0) = -e^x\sin y\big|_{(0,0)} = 0$, $f_{xx}(0,0) = e^x\cos y\big|_{(0,0)}$
$= 1$, $f_{yy}(0,0) = -e^x\cos y\big|_{(0,0)} = -1$, $f_{xy}(0,0) = -e^x\sin y\big|_{(0,0)} = 0$. $\therefore$ The quadratic approximation is
$e^x\cos y \approx 1 + x + \frac{1}{2}(x^2 - y^2)$.
$f_{xxx}(0,0) = e^x\cos y\big|_{(0,0)} = 1$, $f_{xxy}(0,0) = -e^x\sin y\big|_{(0,0)} = 0$, $f_{xyy}(0,0) = -e^x\cos y\big|_{(0,0)} = -1$, $f_{yyy}(0,0) =$
$e^x\sin y\big|_{(0,0)} = 0$. $\therefore$ the cubic approximation is $e^x\cos y \approx 1 + x + \frac{1}{2}(x^2 - y^2) + \frac{1}{6}(x^3 - 3xy^2)$

3. $f(x,y) = \sin(x^2 + y^2) \Rightarrow f(0,0) = 0$ $f_x(0,0) = 2x\cos(x^2 + y^2)\big|_{(0,0)} = 0$, $f_y(0,0) = 2y\cos(x^2 + y^2)\big|_{(0,0)} = 0$,
$f_{xx}(0,0) = 2\cos(x^2 + y^2) - 4x^2\sin(x^2 + y^2)\big|_{(0,0)} = 2$, $f_{yy}(0,0) = 2\cos(x^2 + y^2) - 4y^2\sin(x^2 + y^2)\big|_{(0,0)} = 2$,
$f_{xy}(0,0) = -4xy\sin(x^2 + y^2)\big|_{(0,0)} = 0$ $\therefore$ The quadratic approximation is $\sin(x^2 + y^2) \approx \frac{1}{2}(2x^2 + 2y^2) = x^2 + y^2$
$f_{xxx}(0,0) = -4x\sin(x^2 + y^2) - 8x\sin(x^2 + y^2) - 8x^3\cos(x^2 + y^2)\big|_{(0,0)} = 0$, $f_{xxy}(0,0) = -4y\sin(x^2 + y^2) -$
$8x^2y\cos(x^2 + y^2)\big|_{(0,0)} = 0$, $f_{xyy}(0,0) = -4x\sin(x^2 + y^2) - 8xy^2\cos(x^2 + y^2)\big|_{(0,0)} = 0$,

3. (Continued)

$f_{yyy}(0,0) = -4y\sin(x^2+y^2) - 8y\sin(x^2+y^2) - 8y^3\cos(x^2+y^2)\Big|_{(0,0)} = 0$

$\therefore$ the cubic approximation is $\sin(x^2+y^2) \approx x^2 + y^2$

5. $f(x,y) = \dfrac{1}{1-x-y} \Rightarrow f(0,0) = 1$. $f_x(0,0) = \dfrac{1}{(1-x-y)^2}\Big|_{(0,0)} = 1$, $f_y(0,0) = \dfrac{1}{(1-x-y)^2}\Big|_{(0,0)} = 1$,

$f_{xx}(0,0) = \dfrac{2}{(1-x-y)^3}\Big|_{(0,0)} = 2$, $f_{yy}(0,0) = \dfrac{2}{(1-x-y)^3}\Big|_{(0,0)} = 2$, $f_{xy}(0,0) = \dfrac{2}{(1-x-y)^3}\Big|_{(0,0)} = 2$.

$\therefore$ the quadratic approximation is $\dfrac{1}{1-x-y} \approx 1 + x + y + \dfrac{1}{2}(2x^2 + 4xy + 2y^2) = 1 + x + y + x^2 + 2xy + y^2$.

$f_{xxx}(0,0) = \dfrac{6}{(1-x-y)^4}\Big|_{(0,0)} = 6$, $f_{xxy}(0,0) = \dfrac{6}{(1-x-y)^4}\Big|_{(0,0)} = 6$, $f_{xyy}(0,0) = \dfrac{6}{(1-x-y)^4}\Big|_{(0,0)} = 6$,

$f_{yyy}(0,0) = m\dfrac{6}{(1-x-y)^4}\Big|_{(0,0)} = 6$. $\therefore$ the cubic approximation is $\dfrac{1}{1-x-y} \approx 1 + x + y + x^2 + 2xy + y^2 +$

$\dfrac{1}{6}(6x^3 + 18x^2y + 18xy^2 + 6y^3) = 1 + x + y + x^2 + 2xy + y^2 + x^3 + 3x^2y + 3xy^2 + y^3$.

7. $f(x,y) = \cos x\cos y \Rightarrow f(0,0) = 1$. $f_x(0,0) = -\sin x\cos y\Big|_{(0,0)} = 0$, $f_y(0,0) = -\cos x\sin y\Big|_{(0,0)} = 0$, $f_{xx}(0,0) =$ $-\cos x\cos y\Big|_{(0,0)} = -1$, $f_{yy}(0,0) = -\cos x\cos y\Big|_{(0,0)} = -1$, $f_{xy}(0,0) = \sin x\sin y\Big|_{(0,0)} = 0$. $\therefore$ the quadratic approximation is $\cos x\cos y \approx 1 + \dfrac{1}{2}(-x^2 - y^2) = 1 - \dfrac{1}{2}x^2 - \dfrac{1}{2}y^2$. Since all partial derivatives of f are products of sines and cosines, the absolute value of these derivatives is less than or equal to 1 $\Rightarrow$ $E(x,y) \le \dfrac{1}{6}[(0.1)^3 + 3(0.1)^3 + 3(0.1)^3 + 0.1)^3] \le 0.0013$.

9. a) If $F(t) = f(x_0 + th, y_0 + tk, z_0 + tm)$, then $F'(t) = f_x\dfrac{dx}{dt} + f_y\dfrac{dy}{dt} + f_z\dfrac{dz}{dt} = hf_x + kf_y + mf_z$. $F''(t) = \dfrac{\partial F'(t)}{\partial x}\dfrac{dx}{dt} +$ $\dfrac{\partial F'(t)}{\partial y}\dfrac{dy}{dt} + \dfrac{\partial F'(t)}{\partial z}\dfrac{dz}{dt} = h\dfrac{\partial F'(t)}{\partial x} + k\dfrac{\partial F'(t)}{\partial y} + m\dfrac{\partial F'(t)}{\partial z} = h[hf_{xx} + kf_{yx} + mf_{zx}] + k[hf_{xy} + kf_{yy} + mf_{zy}] +$ $m[hf_{xz} + kf_{yz} + mf_{zz}] = h^2f_{xx} + 2khf_{xy} + 2mhf_{xz} + k^2f_{yy} + 2mkf_{yz} + m^2f_{zz} = h^2f_{xx} + k^2f_{yy} + m^2f_{zz} +$ $2khf_{xy} + 2hmf_{xz} + 2kmf_{yz}$. By Taylor's polynomial, $F(t) = F(t_0) + F'(t_0)t + \dfrac{1}{2}F''(c)t^2$, $0 \le c \le t_0$. Then $f(x_0 + h, y_0 + k, z_0 + m) \Rightarrow t = 1 \Rightarrow f(x_0 + h, y_0 + k, z_0 + m) = f(x_0,y_0,z_0) + hf_x(x_0,y_0,z_0) + kf_y(x_0,y_0,z_0) +$ $mf_z(x_0,y_0,z_0) + \left[\dfrac{1}{2}[h^2f_{xx} + k^2f_{yy} + m^2f_{zz} + 2hkf_{xy} + 2hmf_{xz} + 2kmf_{yz}]\Big|_{(x_0+ch,y_0+ck,z_0+cm)}\right]$ for some c so that $0 \le c \le 1$.

b) Note: $x = x_0 + h \Rightarrow x - x_0 = h$. Similarly $y - y_0 = k$, $z - z_0 = m$. If all the second ordered partials are bounded by M, then $|E| \le \dfrac{1}{2}M\left[|h|^2 + |k|^2 + |m|^2 + 2|h||k| + 2|h||m| + 2|k||m|\right]$

$= \dfrac{1}{2}M\left((|h| + |k|)^2 + 2(|h| + |k|)|m| + |m|^2\right) = \dfrac{1}{2}M(|h| + |k| + |m|)^2 = \dfrac{1}{2}M\left(|x - x_0| + |y - y_0| + |z - z_0|\right)^2$

SECTION 12.M MISCELLANEOUS EXERCISES

1.

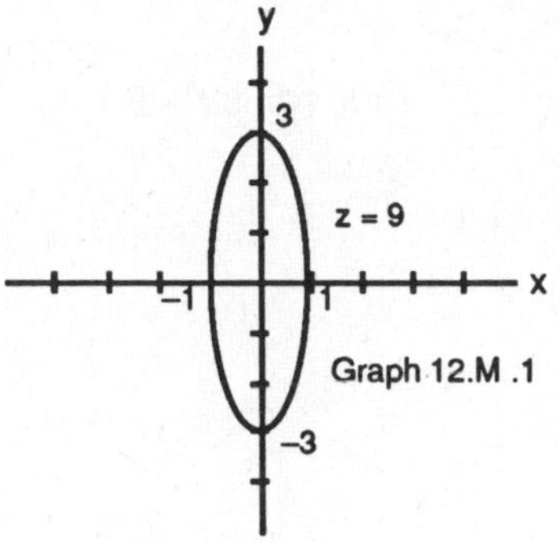

Graph 12.M .1

Domain: All points in the xy–plane

Range: $f(x,y) \geq 0$

Level curves are ellipses with major axis along the y–axis and minor axis along the x–axis.

3.

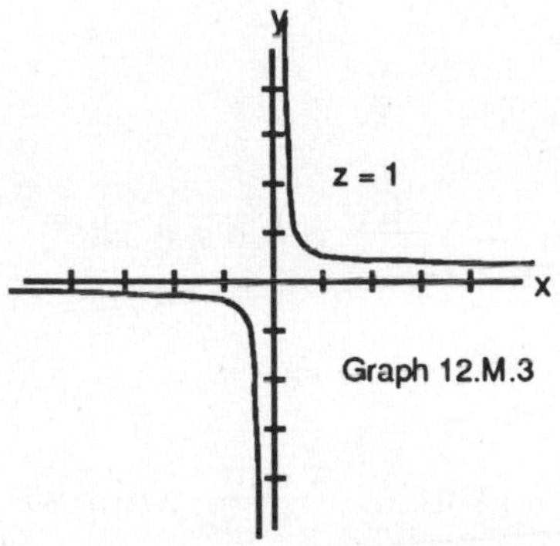

Graph 12.M.3

Domain: All (x,y) such that $x \neq 0$ or $y \neq 0$

Range: $f(x,y) \neq 0$

Level curves are hyperbolas rotated 45° or 135°.

5.

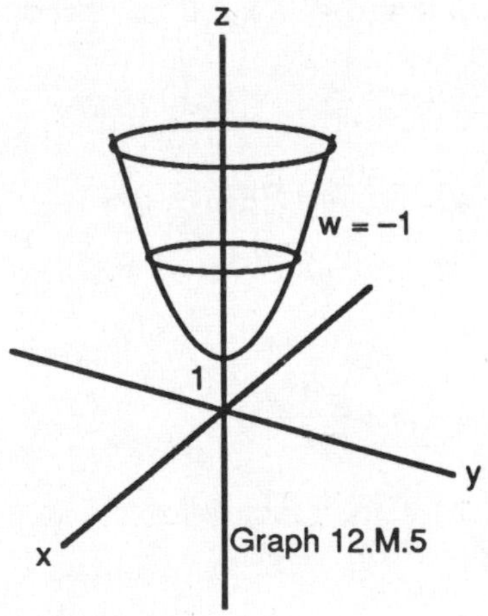

Graph 12.M.5

Domain: All (x,y,z)

Range: All Real Numbers

Level surfaces are paraboloids of revolution with the z–axis as axis.

7.

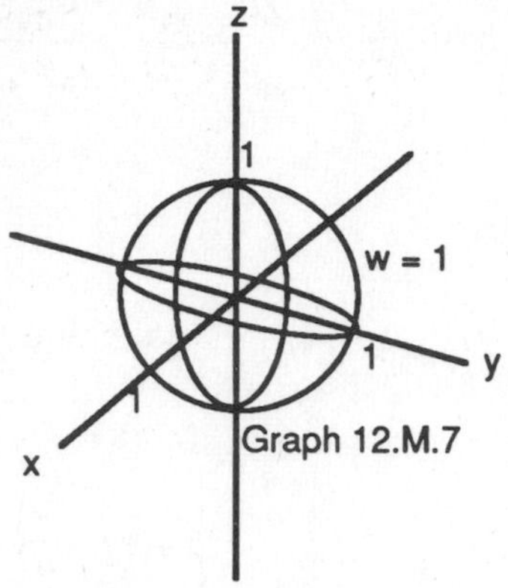

Graph 12.M.7

Domain: All (x,y,z) such that $(x,y,z) \neq (0,0,0)$

Range: $f(x,y,z) > 0$

Level surfaces are spheres with center (0,0,0) and radius $r > 0$.

9. $\displaystyle\lim_{(x,y)\to(\pi,\ln 2)} e^y \cos x = e^{\ln 2}\cos\pi = -2$

11. $\displaystyle\lim_{(x,y)\to(1,1)} \frac{x^2-y^2}{x-y} = \lim_{(x,y)\to(1,1)} (x+y) = 2$

13. $\displaystyle\lim_{(x,y)\to(0,0)} y\sin\frac{1}{x} = 0$ since $-1 \le \sin\frac{1}{x} < 1$

15. Let $y = kx^2$. Then $\displaystyle\lim_{(x,y)\to(0,0)} \frac{y}{x^2-y} = \lim_{(x,kx^2)\to(0,0)} \frac{kx^2}{x^2-kx^2} = \lim_{(x,kx^2)\to(0,0)} \frac{k}{1-k}$ which will differ, depending on the value of k.

17. a) Let $y = mx$. Then $\displaystyle\lim_{(x,y)\to(0,0)} \frac{x^2-y^2}{x^2+y^2} = \lim_{(x,mx)\to(0,0)} \frac{x^2-m^2x^2}{x^2+m^2x^2} = \lim_{(x,mx)\to(0,0)} \frac{1-m^2}{1+m^2}$ which will differ, depending on m $\Rightarrow$ limit does not exist no matter what f(0,0) is.

b) Let $y = -x$, Then $\displaystyle\lim_{(x,y)\to(0,0)} \frac{\sin(x-y)}{|x|+|y|} = \lim_{(x,-x)\to(0,0)} \frac{\sin 2x}{2|x|} = 1$ if $x > 0$ and -1 if $x < 0 \Rightarrow$ limit does not exist $\Rightarrow$ f is not continuous at (0,0).

19. $\dfrac{\partial g}{\partial r} = \cos\theta + \sin\theta,\ \dfrac{\partial g}{\partial\theta} = -r\sin\theta + r\cos\theta$

21. $\dfrac{\partial f}{\partial R_1} = -\dfrac{1}{{R_1}^2},\ \dfrac{\partial f}{\partial R_2} = -\dfrac{1}{{R_2}^2},\ \dfrac{\partial f}{\partial R_3} = -\dfrac{1}{{R_3}^2}$

23. $\dfrac{\partial P}{\partial n} = \dfrac{RT}{V},\ \dfrac{\partial P}{\partial R} = \dfrac{nT}{V},\ \dfrac{\partial P}{\partial T} = \dfrac{nR}{V},\ \dfrac{\partial P}{\partial V} = -\dfrac{nRT}{V^2}$

25. $\dfrac{\partial f}{\partial x} = \dfrac{1}{y},\ \dfrac{\partial f}{\partial y} = 1 - \dfrac{x}{y^2} \Rightarrow \dfrac{\partial^2 f}{\partial x^2} = 0,$

$\dfrac{\partial^2 f}{\partial y^2} = \dfrac{2x}{y^3},\ \dfrac{\partial^2 f}{\partial y\partial x} = \dfrac{\partial^2 f}{\partial x\partial y} = -\dfrac{1}{y^2}$

27. $\dfrac{\partial f}{\partial x} = 1 + y - 15x^2 + \dfrac{2x}{x^2+1},\ \dfrac{\partial f}{\partial y} = x \Rightarrow \dfrac{\partial^2 f}{\partial x^2} = -30x + \dfrac{2-2x^2}{(x^2+1)^2},\ \dfrac{\partial^2 f}{\partial y^2} = 0,\ \dfrac{\partial^2 f}{\partial y\partial x} = \dfrac{\partial^2 f}{\partial x\partial y} = 1$

29. $f_x(0,0) = \displaystyle\lim_{h\to 0} \frac{f(h,0)-f(0,0)}{h} = \lim_{h\to 0}\frac{0-0}{h} = 0$ since $f(h,0) = 0,\ h \neq 0$. $f_{xy} = \displaystyle\lim_{h\to 0}\frac{f_x(0,h)-f_x(0,0)}{h}$.

$f_x(x,y) = \dfrac{x^2y-y^3}{x^2+y^2} + \dfrac{4x^2y^3}{(x^2+y^2)^2} \Rightarrow f_x(0,h) = -\dfrac{h^3}{h^2} = -h$. Then $f_{xy}(0,0) = \displaystyle\lim_{h\to 0}\frac{-h-0}{h} = -1$. $f_y(0,0) =$

29. (Continued)

$\lim\limits_{h \to 0} \frac{f(0,h) - f(0,0)}{h} = \lim\limits_{h \to 0} \frac{0 - 0}{h} = 0$. $f_{yx}(0,0) = \lim\limits_{h \to 0} \frac{f_y(h,0) - f_y(0,0)}{h}$. $f_y(x,y) = \frac{x^3 - xy^2}{x^2 + y^2} - \frac{4x^3y^2}{(x^2 + y^2)^2}$ $\Rightarrow f_y(h,0) = \frac{h^3}{h^2} = h$. Then $f_{yx}(0,0) = \lim\limits_{h \to 0} \frac{h - 0}{h} = 1$.

31. $f\left(\frac{\pi}{4}, \frac{\pi}{4}\right) = \frac{1}{2}$, $f_x\left(\frac{\pi}{4}, \frac{\pi}{4}\right) = \cos x \cos y\Big|_{(\pi/4,\pi/4)} = \frac{1}{2}$, $f_y\left(\frac{\pi}{4}, \frac{\pi}{4}\right) = -\sin x \sin y\Big|_{(\pi/4,\pi/4)} = -\frac{1}{2} \Rightarrow L(x,y) = \frac{1}{2} + \frac{1}{2}\left(x - \frac{\pi}{4}\right) - \frac{1}{2}\left(y - \frac{\pi}{4}\right) = \frac{1}{2} + \frac{1}{2}x - \frac{1}{2}y$. $f_{xx}(x,y) = -\sin x \cos y$, $f_{yy}(x,y) = -\sin x \cos y$, $f_{xy}(x,y) = -\cos x \sin y$. $\therefore$ maximum of $|f_{xx}|$, $|f_{yy}|$, and $|f_{xy}|$ is $1 \Rightarrow M = 1 \Rightarrow |E(x,y)| \le \frac{1}{2}(1)\left(\left|x - \frac{\pi}{4}\right| + \left|y - \frac{\pi}{4}\right|\right)^2$ ≤ 0.02.

33. a) $f(1,0,0) = 0$, $f_x(1,0,0) = y - 3z\Big|_{(1,0,0)} = 0$, $f_y(1,0,0) = x + 2z\Big|_{(1,0,0)} = 1$, $f_z(1,0,0) = 2y - 3x\Big|_{(1,0,0)} = -3 \Rightarrow L(x,y,z) = 0(x - 1) + (y - 0) - 3(z - 0) = y - 3z$.

b) $f(1,1,0) = 1$, $f_x(1,1,0) = 1$, $f_y(1,1,0) = 1$, $f_z(1,1,0) = -1 \Rightarrow L(x,y,z) = 1 + (x - 1) + (y - 1) - 1(z - 0) = x + y - z - 1$

35. $dV = 2\pi rh\, dr + \pi r^2\, dh \Rightarrow dV\Big|_{(1.5,5280)} = 2\pi(1.5)(5280)\, dr + \pi(1.5)^2\, dh = 15840\pi\, dr + 2.25\pi\, dh$. Be more careful with the diameter since it has a greater effect on dV.

37. $dI = \frac{1}{R}dV - \frac{V}{R^2}dR \Rightarrow dI\Big|_{(24,100)} = \frac{1}{100}dV - \frac{24}{100^2}dR \Rightarrow dI\Big|_{dV=-1,dR=-20} = 0.038$. % change in V = $-\frac{1}{24} = -4.17\%$; % change in R = $-\frac{20}{100} = -20\%$. $I = \frac{24}{100} = 0.24 \Rightarrow$ Estimated % change in I = $\frac{dI}{I} \times 100 = \frac{0.038}{0.24} \times 100 = 15.83\%$

39. a) $|u| \le 0.02$, $|v| \le 0.03$. $y = uv \Rightarrow y_u = v$, $y_v = u \Rightarrow dy = v\,du + u\,dv$. Percentage change in $u \le 2\% \Rightarrow |du| \le 0.02$, percentage change in $v \le 3\% \Rightarrow |dv| \le 0.03$. $\frac{dy}{y} = \frac{v\,du + u\,dv}{uv} = \frac{du}{u} + \frac{dv}{v}$. Then $\left|\frac{dy}{y} \times 100\right| = \left|\frac{du}{u} \times 100 + \frac{dv}{v} \times 100\right| \le \left|\frac{du}{u} \times 100\right| + \left|\frac{dv}{v} \times 100\right| \le 2\% + 3\% = 5\%$

b) $z = u + v \Rightarrow z_u = du$, $z_v = dv \Rightarrow dz,z) = \frac{du + dv}{u + v} = \frac{du}{u + v} + \frac{dv}{u + v} \le \frac{du}{u} + \frac{dv}{v}$ (since $u > 0$, $v > 0$). Then $\left|\frac{dz}{z} \times 100\right| = \left|\frac{du}{u + v} \times 100 + \frac{dv}{u + v} \times 100\right| \le \left|\frac{du}{u} \times 100 + \frac{dv}{v} \times 100\right| \le \left|\frac{du}{u} \times 100\right| + \left|\frac{dv}{v} \times 100\right| = \left|\frac{dy}{y} \times 100\right|$

41. $\frac{\partial w}{\partial x} = y\cos(xy + \pi)$, $\frac{\partial w}{\partial y} = x\cos(xy + \pi)$, $\frac{dx}{dt} = e^t$, $\frac{dy}{dt} = \frac{1}{t + 1} \Rightarrow \frac{dw}{dt} = y\cos(xy + \pi)\, e^t + x\cos(xy + \pi)\left(\frac{1}{t + 1}\right)$ $= e^t \ln(t + 1)\cos(e^t \ln(t + 1) + \pi) + \frac{e^t}{t + 1}\cos(e^t \ln(t + 1) + \pi) \Rightarrow \frac{dw}{dt}\Big|_{t=0} = -1$

43. $\frac{\partial w}{\partial x} = 2\cos(2x - y)$, $\frac{\partial w}{\partial y} = -\cos(2x - y)$, $\frac{\partial x}{\partial r} = 1$, $\frac{\partial x}{\partial s} = \cos s$, $\frac{\partial y}{\partial r} = s$, $\frac{\partial y}{\partial s} = r \Rightarrow \frac{\partial w}{\partial r} = 2\cos(2x - y)(1) +$ $(-\cos(2x - y)(s)) = 2\cos(2r + 2\sin s - rs) - s\cos(2r + 2\sin s - rs) \Rightarrow \left.\frac{\partial w}{\partial r}\right|_{r=\pi, s=0} = 2.$

$\frac{\partial w}{\partial s} = 2\cos(2x - y)(\cos s) + (-\cos(2x - y)(r) = 2\cos(2r + 2\sin s - rs)(\cos s) - r\cos(2r + 2\sin s - rs) \Rightarrow$

$\left.\frac{\partial w}{\partial s}\right|_{r=\pi, s=0} = 2 - \pi$

45. $F_x = -1 - y\cos xy$, $F_y = -2y - x\cos xy$. $\frac{dy}{dx} = -\frac{F_x}{F_y} = -\frac{-1 - y\cos xy}{-2y - x\cos xy} = \frac{1 + y\cos xy}{-2y - x\cos xy} \Rightarrow$

$\left.\frac{dy}{dx}\right|_{(x,y)=(0,1)} = -1$

47. $\frac{\partial f}{\partial x} = y + z$, $\frac{\partial f}{\partial y} = x + z$, $\frac{\partial f}{\partial z} = y + x$, $\frac{dx}{dt} = -\sin t$, $\frac{dy}{dt} = \cos t$, $\frac{dz}{dt} = -2\sin 2t \Rightarrow \frac{df}{dt} = -(y + z)\sin t + (x + z)\cos t$ $- 2(y + x)\sin 2t = -(\sin t + \cos 2t)\sin t + (\cos t + \cos 2t)\cos t - 2(\sin t + \cos t)\sin 2t \Rightarrow \left.\frac{df}{dt}\right|_{t=1} =$

$-(\sin 1 + \cos 2)\sin 1 + (\cos 1 + \cos 2)\cos 1 - 2(\sin 1 + \cos 1)\sin 2$

49. $\frac{\partial w}{\partial x} = \frac{\partial w}{\partial r}\frac{\partial r}{\partial x} + \frac{\partial w}{\partial \theta}\frac{\partial \theta}{\partial x} = \frac{\partial w}{\partial r}\left(\frac{x}{\sqrt{x^2 + y^2}}\right) + \frac{\partial w}{\partial \theta}\left(\frac{-y}{x^2 + y^2}\right) = \cos\theta\,\frac{\partial w}{\partial r} - \frac{\sin\theta}{r}\frac{\partial w}{\partial \theta}$. $\frac{\partial w}{\partial y} = \frac{\partial w}{\partial r}\frac{\partial r}{\partial y} + \frac{\partial w}{\partial \theta}\frac{\partial \theta}{\partial y} =$ $\frac{\partial w}{\partial r}\left(\frac{y}{\sqrt{x^2 + y^2}}\right) + \frac{\partial w}{\partial \theta}\left(\frac{x}{x^2 + y^2}\right) = \sin\theta\,\frac{\partial w}{\partial r} + \frac{\cos\theta}{r}\frac{\partial w}{\partial \theta}$

51. $\frac{\partial u}{\partial y} = b$, $\frac{\partial u}{\partial x} = a$. $\frac{\partial w}{\partial x} = \frac{dw}{du}\frac{\partial u}{\partial x} = a\frac{dw}{du}$, $\frac{\partial w}{\partial y} = \frac{dw}{du}\frac{\partial u}{\partial y} = b\frac{dw}{du}$. Then $\frac{1}{a}\frac{\partial w}{\partial x} = \frac{dw}{du}$, $\frac{1}{b}\frac{\partial w}{\partial y} = \frac{dw}{du} \Rightarrow \frac{1}{a}\frac{\partial w}{\partial x} = \frac{1}{b}\frac{\partial w}{\partial y} \Rightarrow$ $b\frac{\partial w}{\partial x} = a\frac{\partial w}{\partial y}$

53. $e^u\cos v - x = 0 \Rightarrow e^u\cos v\frac{\partial u}{\partial x} - e^u\sin v\frac{\partial v}{\partial x} = 1$. $e^u\sin v - y = 0 \Rightarrow e^u\sin v\frac{\partial u}{\partial x} + e^u\cos v\frac{\partial v}{\partial x} = 0$. Solving this system yields $\frac{\partial u}{\partial x} = e^{-u}\cos v$, $\frac{\partial v}{\partial x} = -e^{-u}\sin v$. Similarly, $e^u\cos v - x = 0 \Rightarrow e^u\cos v\frac{\partial u}{\partial y} - e^u\sin v\frac{\partial v}{\partial y} = 0$ and $e^u\sin v - y = 0 \Rightarrow e^u\sin v\frac{\partial u}{\partial y} + e^u\cos v\frac{\partial v}{\partial y} = 1$. Solving this system yields $\frac{\partial u}{\partial y} = e^{-u}\sin v$, $\frac{\partial v}{\partial y} = e^{-u}\cos v$.

$\therefore \left[\left(\frac{\partial u}{\partial x}\mathbf{i} + \frac{\partial u}{\partial y}\mathbf{j}\right)\cdot\left(\frac{\partial v}{\partial x}\mathbf{i} + \frac{\partial v}{\partial y}\mathbf{j}\right)\right] = \left[\left(e^{-u}\cos v\right)\mathbf{i} + \left(e^{-u}\sin v\right)\mathbf{j}\right]\cdot\left[\left(-e^{-u}\sin v\right)\mathbf{i} + \left(e^{-u}\cos v\right)\mathbf{j}\right] = 0 \Rightarrow$

the vectors are orthogonal.

55. $g(u,v) = \int_u^v f(t)\,dt \Rightarrow \frac{dy}{dx} = \frac{\partial g}{\partial u}\frac{du}{dx} + \frac{\partial g}{\partial v}\frac{dv}{dx} = \left(\frac{\partial}{\partial u}\int_u^v f(t)\,dt\right)\frac{du}{dx} + \left(\frac{\partial}{\partial v}\int_u^v f(t)\,dt\right)\frac{dv}{dx} = \left(-\frac{\partial}{\partial u}\int_v^u f(t)\,dt\right)\frac{du}{dx} +$ $\left(\frac{\partial}{\partial v}\int_u^v f(t)\,dt\right)\frac{dv}{dx} = -f(u(x))\frac{du}{dx} + f(v(x))\frac{dv}{dx} = f(v(x))\frac{dv}{dx} - f(u(x))\frac{du}{dx}$

57. a) Let u = tx, v = ty. Then $F = f(u,v) = t^n f(x,y)$. With t, x, and y as independent, $\frac{\partial F}{\partial t} = nt^{n-1}f(x,y)$.

$\therefore x\frac{\partial f}{\partial u} + y\frac{\partial f}{\partial v} = nt^{n-1}f(x,y)$. Now $\frac{\partial f}{\partial x} = \frac{\partial f}{\partial u}\frac{\partial u}{\partial x} = \frac{\partial f}{\partial u}t \Rightarrow \frac{1}{t}\frac{\partial f}{\partial x} = \frac{\partial f}{\partial u}$. $\frac{\partial f}{\partial y} = \frac{\partial f}{\partial v}\frac{\partial v}{\partial y} = \frac{\partial f}{\partial v}t \Rightarrow \frac{1}{t}\frac{\partial f}{\partial y} = \frac{\partial f}{\partial v}$.

Thus $\frac{x}{t}\frac{\partial f}{\partial x} + \frac{y}{t}\frac{\partial f}{\partial y} = nt^{n-1}f(x,y)$, true for all $t \neq 0$. Let $t = 1 \Rightarrow x\frac{\partial f}{\partial x} + y\frac{\partial f}{\partial y} = n\, f(x,y)$

b) From $x\frac{\partial f}{\partial u} + y\frac{\partial f}{\partial v} = nt^{n-1}f(x,y)$, differentiate with respect to t again. $x\frac{\partial^2 f}{\partial u^2}\frac{\partial u}{\partial t} + x\frac{\partial^2 f}{\partial v\partial u}\frac{\partial x}{\partial t} + y\frac{\partial^2 f}{\partial v^2}\frac{\partial v}{\partial t} +$ $y\frac{\partial^2 f}{\partial u\partial v}\frac{\partial u}{\partial t} = n(n-1)t^{n-2}f(x,y) \Rightarrow x^2\frac{\partial^2 f}{\partial u^2} + 2xy\frac{\partial^2 f}{\partial u\partial v} + y^2\frac{\partial^2 f}{\partial v^2} = n(n-1)t^{n-2}f(x,y)$. Now $\frac{\partial^2 f}{\partial x^2} =$ $\frac{\partial^2 f}{\partial u^2}\frac{\partial u}{\partial x}t = t^2\frac{\partial^2 f}{\partial u^2}$ and $\frac{\partial^2 f}{\partial y^2} = \frac{\partial^2 f}{\partial v^2}\frac{\partial v}{\partial y}t = t^2\frac{\partial^2 f}{\partial v^2}$ and $\frac{\partial^2 f}{\partial y\partial x} = \frac{\partial^2 f}{\partial v\partial u}\frac{\partial v}{\partial y}t = t^2\frac{\partial^2 f}{\partial v\partial u} \Rightarrow \frac{1}{t^2}\frac{\partial^2 f}{\partial x^2} = \frac{\partial^2 f}{\partial u^2}, \frac{1}{t^2}\frac{\partial^2 f}{\partial y^2} = \frac{\partial^2 f}{\partial v^2}$, and $\frac{1}{t^2}\frac{\partial^2 f}{\partial y\partial x} = \frac{\partial^2 f}{\partial v\partial u}$. Thus $\frac{x^2}{t^2}\frac{\partial^2 f}{\partial x^2} + \frac{2xy}{t^2}\frac{\partial^2 f}{\partial y\partial x} + \frac{y^2}{t^2}\frac{\partial^2 f}{\partial y^2} = n(n-1)t^{n-2}f(x,y)$ for $t \neq 0$. Again, let $t = 1 \Rightarrow$

$$x^2\frac{\partial^2 f}{\partial x^2} + 2xy\frac{\partial^2 f}{\partial y\partial x} + y^2\frac{\partial^2 f}{\partial y^2} = n(n-1)f(x,y)$$

59. $\nabla f = (-\sin x\cos y)\,\mathbf{i} - (\cos x\sin y)\,\mathbf{j} \Rightarrow \nabla f\big|_{(\pi/4,\pi/4)} = -\frac{1}{2}\mathbf{i} - \frac{1}{2}\mathbf{j} \Rightarrow |\nabla f| = \sqrt{\left(-\frac{1}{2}\right)^2 + \left(-\frac{1}{2}\right)^2} = \frac{1}{\sqrt{2}}$

$\mathbf{u} = \frac{\nabla f}{|\nabla f|} = \frac{-\frac{1}{2}\mathbf{i} - \frac{1}{2}\mathbf{j}}{\frac{1}{\sqrt{2}}} = -\frac{\sqrt{2}}{2}\mathbf{i} - \frac{\sqrt{2}}{2}\mathbf{j}$. f increases most rapidly in the direction $\mathbf{u} = -\frac{\sqrt{2}}{2}\mathbf{i} - \frac{\sqrt{2}}{2}\mathbf{j}$;

decreases most rapidly in the direction $-\mathbf{u} = \frac{\sqrt{2}}{2}\mathbf{i} + \frac{\sqrt{2}}{2}\mathbf{j}$. $\left(D_{\mathbf{u}}f\right)_{P_0} = \frac{\sqrt{2}}{2}$, $\left(D_{-\mathbf{u}}f\right)_{P_0} = -\frac{\sqrt{2}}{2}$.

$\mathbf{u}_1 = \frac{\mathbf{A}}{|\mathbf{A}|} = \frac{3\,\mathbf{i} + 4\,\mathbf{j}}{\sqrt{3^2 + 4^2}} = \frac{3}{5}\mathbf{i} + \frac{4}{5}\mathbf{j}$. $\left(D_{\mathbf{u}_1}f\right)_{P_0} = \nabla f \cdot \mathbf{u}_1 = -\frac{7}{10}$.

61. $\nabla f = \left(\frac{2}{2x+3y+6z}\right)\mathbf{i} + \left(\frac{3}{2x+3y+6z}\right)\mathbf{j} + \left(\frac{6}{2x+3y+6z}\right)\mathbf{k} \Rightarrow \nabla f\big|_{(-1,-1,1)} = 2\,\mathbf{i} + 3\,\mathbf{j} + 6\,\mathbf{k}$.

$\mathbf{u} = \frac{\nabla f}{|\nabla f|} = \frac{2\,\mathbf{i} + 3\,\mathbf{j} + 6\,\mathbf{k}}{\sqrt{2^2 + 3^2 + 6^2}} = \frac{2}{7}\mathbf{i} + \frac{3}{7}\mathbf{j} + \frac{6}{7}\mathbf{k}$. f increases most rapidly in the direction $\mathbf{u} = \frac{2}{7}\mathbf{i} + \frac{3}{7}\mathbf{j} + \frac{6}{7}\mathbf{k}$;

decreases most rapidly in the direction $-\mathbf{u} = -\frac{2}{7}\mathbf{i} - \frac{3}{7}\mathbf{j} - \frac{6}{7}\mathbf{k}$. $\left(D_{\mathbf{u}}f\right)_{P_0} = \nabla f \cdot \mathbf{u} = 7$,

$\left(D_{-\mathbf{u}}f\right)_{P_0} = -7$. $\mathbf{u}_1 = \frac{\mathbf{A}}{|\mathbf{A}|} = \frac{2}{7}\mathbf{i} + \frac{3}{7}\mathbf{j} + \frac{6}{7}\mathbf{k}$ since $\mathbf{A} = \nabla f$. $\Rightarrow \left(D_{\mathbf{u}_1}f\right)_{P_0} = 7$.

63.

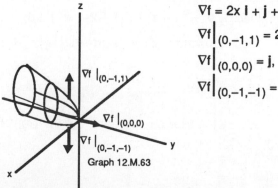

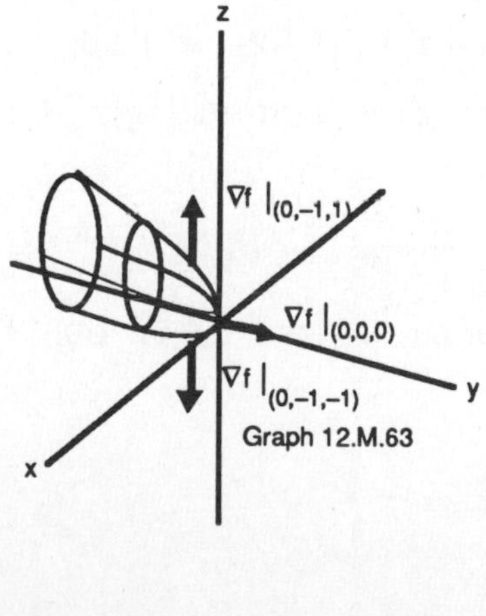

Graph 12.M.63

$\nabla f = 2x\,\mathbf{i} + \mathbf{j} + 2z\,\mathbf{k} \Rightarrow$

$\nabla f\big|_{(0,-1,1)} = 2\,\mathbf{k}$,

$\nabla f\big|_{(0,0,0)} = \mathbf{j}$,

$\nabla f\big|_{(0,-1,-1)} = -2\,\mathbf{k}$

65. $\nabla f = 2x\,\mathbf{i} - \mathbf{j} - 5\,\mathbf{k} \Rightarrow \nabla f\Big|_{(2,-1,1)} = 4\,\mathbf{i} - \mathbf{j} - 5\,\mathbf{k} \Rightarrow$ Tangent Plane: $4(x-2) - (y+1) - 5(z-1) = 0 \Rightarrow$ $4x - y - 5z = 4$; Normal Line: $x = 2 + 4t,\ y = -1 - t,\ z = 1 - 5t$

67.

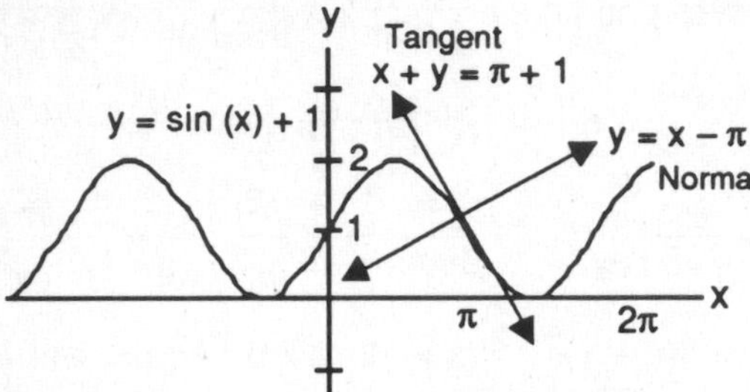

Graph 12.M.67

$\nabla f = (-\cos x)\,\mathbf{i} + \mathbf{j} \Rightarrow \nabla f\Big|_{(\pi,1)} = \mathbf{i} + \mathbf{j} \Rightarrow$ Tangent Line: $(x-\pi) + (y-1) = 0 \Rightarrow x + y = \pi + 1$; Normal Line: $y - 1 = 1(x - \pi) \Rightarrow y = x - \pi + 1$

69. Let $f(x,y,z) = x^2 + 2y + 2z - 4$ and $g(x,y,z) = y - 1$ at $P_0\left(1,1,\frac{1}{2}\right)$. $\nabla f = 2x\,\mathbf{i} + 2\,\mathbf{j} + 2\,\mathbf{k}\Big|_{\left(1,1,\frac{1}{2}\right)} = 2\,\mathbf{i} + 2\,\mathbf{j} + 2\,\mathbf{k}$. $\nabla g = \mathbf{j} \Rightarrow \nabla f \times \nabla g = \begin{vmatrix} \mathbf{i} & \mathbf{j} & \mathbf{k} \\ 2 & 2 & 2 \\ 0 & 1 & 0 \end{vmatrix} = -2\,\mathbf{i} + 2\,\mathbf{k} \Rightarrow$ the line is $x = 1 - 2t,\ y = 1,\ z = \frac{1}{2} + 2t$.

71. $(y+z)^2 + (z-x)^2 = 16 \Rightarrow \nabla f = -2(z-x)\,\mathbf{i} + 2(y+z)\,\mathbf{j} + 2(y + 2z - x)\,\mathbf{k}$. We want x constant $\Rightarrow \frac{\partial f}{\partial x} = 0 \Rightarrow$ $-2(z-x) = 0 \Rightarrow z = x \Rightarrow (y+z)^2 + (z-z)^2 = 16 \Rightarrow y + z = \pm 4$. Let $x = t \Rightarrow z = t \Rightarrow y = -t \pm 4$. $\therefore$ the points are $(t, -t \pm 4, t)$, t a real number.

73. $\mathbf{r} = (\cos 3t)\,\mathbf{i} + (\sin 3t)\,\mathbf{j} + et\,\mathbf{k} \Rightarrow \mathbf{v}(t) = (-3\sin 3t)\,\mathbf{i} + (3\cos 3t)\,\mathbf{j} + 3\,\mathbf{k} \Rightarrow \mathbf{v}\left(\frac{\pi}{3}\right) = -3\,\mathbf{j} + 3\,\mathbf{k} \Rightarrow \mathbf{u} = -\frac{1}{\sqrt{2}}\,\mathbf{j} + \frac{1}{\sqrt{2}}\,\mathbf{k}$. $f(x,y,z) = xyz \Rightarrow \nabla f = yz\,\mathbf{i} + xz\,\mathbf{j} + xy\,\mathbf{k}$. $t = \frac{\pi}{3}$ yields the point on the helix $(-1,0,\pi) \Rightarrow$ $\nabla f\Big|_{(1,0,\pi)} = -\pi\,\mathbf{j} \Rightarrow \nabla f \cdot \mathbf{u} = (-\pi\,\mathbf{j}) \cdot \left(-\frac{1}{\sqrt{2}}\,\mathbf{j} + \frac{1}{\sqrt{2}}\,\mathbf{k}\right) = \frac{\pi}{\sqrt{2}}$.

75. Let $\nabla f = a\,\mathbf{i} + b\,\mathbf{j}$ at (1,2). The direction toward (2,2) is determined by $\mathbf{v}_1 = (2-1)\,\mathbf{i} + (2-2)\,\mathbf{j} = \mathbf{i} = \mathbf{u}$. $\nabla f \cdot \mathbf{u} = 2 \Rightarrow a = 2$. The direction toward (1,1) is determined by $\mathbf{v}_2 = (1-1)\,\mathbf{i} + (1-2)\,\mathbf{j} = -\mathbf{j} = \mathbf{u}$. $\nabla f \cdot \mathbf{u} = -2 \Rightarrow -b = -2 \Rightarrow b = 2$. $\therefore\ \nabla f = 2\,\mathbf{i} + 2\,\mathbf{j}$. The direction toward (4,6) is determined by $\mathbf{v}_3 = 3\,\mathbf{i} + 4\,\mathbf{j}$ $\Rightarrow \mathbf{u} = \frac{3}{5}\,\mathbf{i} + \frac{4}{5}\,\mathbf{j} \Rightarrow \nabla f \cdot \mathbf{u} = \frac{14}{5}$.

77. $D_{\mathbf{u}}f = |\nabla f| = \left|\frac{x}{\sqrt{x^2+y^2+z^2}}\,\mathbf{i} + \frac{y}{\sqrt{x^2+y^2+z^2}}\,\mathbf{j} + \frac{z}{\sqrt{x^2+y^2+z^2}}\,\mathbf{k}\right| = 1$ for any $(x,y,z) \neq (0,0,0)$. But $\nabla f = \frac{x\,\mathbf{i} + y\,\mathbf{j} + z\,\mathbf{k}}{\sqrt{x^2+y^2+z^2}}$ does not exist at (0,0,0).

79. Let $f(x,y,z) = xy + z - 2 \Rightarrow \nabla f = y\,\mathbf{i} + z\,\mathbf{j} + \mathbf{k}$. At (1,1,1), $\nabla f = \mathbf{i} + \mathbf{j} + \mathbf{k} \Rightarrow$ the normal line is $x = 1 + t$, $y = 1 + t$, $z = 1 + t$. $\therefore t = -1 \Rightarrow x = 0, y = 0, z = 0$.

81. $f(x,y,z) = xz^2 - yz + \cos xy - 1 \Rightarrow \nabla f = (z^2 - y \sin xy)\,\mathbf{i} + (-z - x \sin xy)\,\mathbf{j} + (2xz - y)\,\mathbf{k}$. At (0,0,1), $\nabla f = \mathbf{i} - \mathbf{j}$ $\Rightarrow$ The tangent plane is $x - y = 0$. $\mathbf{r} = (\ln t)\,\mathbf{i} + (t \ln t)\,\mathbf{j} + t\,\mathbf{k} \Rightarrow \mathbf{r}' = \left(\frac{1}{t}\right)\mathbf{i} + (\ln t + 1)\,\mathbf{j} + \mathbf{k} \Rightarrow \mathbf{r}'(1) = \mathbf{i} + \mathbf{j} + \mathbf{k}$. Since $(\mathbf{i} + \mathbf{j} + \mathbf{k}) \cdot (\mathbf{i} - \mathbf{j}) = 0$, $\mathbf{r}$ is parallel to the plane. $\mathbf{r}(1) = 0\,\mathbf{i} + 0\,\mathbf{j} + \mathbf{k} \Rightarrow \mathbf{r}$ is contained in the plane.

83. a) True, since $\left(f_x(x_0,y_0)\,\mathbf{i} + f_y(x_0,y_0)\,\mathbf{j}\right) \cdot \mathbf{u} = \nabla f \cdot \mathbf{u} = D_{\mathbf{u}}f$

b) False, $D_{\mathbf{u}}f$ is a scalar.

c) True, see Properties of the Directional Derivative, Section 12.7.

d) True, see Section 12.7.

85. $\frac{\partial w}{\partial r} = \frac{\partial w}{\partial x}\frac{\partial x}{\partial r} + \frac{\partial w}{\partial y}\frac{\partial y}{\partial r} = \frac{\partial w}{\partial x}(\cos\theta) + \frac{\partial w}{\partial y}(\sin\theta)$. $\frac{\partial w}{\partial \theta} = \frac{\partial w}{\partial x}\frac{\partial x}{\partial \theta} + \frac{\partial w}{\partial y}\frac{\partial y}{\partial \theta} = \frac{\partial w}{\partial x}(-r\sin\theta) + \frac{\partial w}{\partial y}(r\cos\theta)$. $\frac{\partial w}{\partial r}\mathbf{u}_r = \left[\frac{\partial w}{\partial x}(\cos\theta) + \frac{\partial w}{\partial y}(\sin\theta)\right]\left((\cos\theta)\,\mathbf{i} + (\sin\theta)\,\mathbf{j}\right)$ and $\frac{1}{r}\frac{\partial w}{\partial \theta}\mathbf{u}_\theta =$ $\left[\frac{\partial w}{\partial x}(-\sin\theta) + \frac{\partial w}{\partial y}(\cos\theta)\right]\left((-\sin\theta)\,\mathbf{i} + (\cos\theta)\,\mathbf{j}\right) \Rightarrow \frac{\partial w}{\partial r}\mathbf{u}_r + \frac{1}{r}\frac{\partial w}{\partial \theta}\mathbf{u}_\theta + \frac{\partial w}{\partial z}\mathbf{k} = \frac{\partial w}{\partial x}\left(\cos^2\theta + \sin^2\theta\right)\mathbf{i} +$ $\frac{\partial w}{\partial y}\left(\cos^2\theta + \sin^2\theta\right)\mathbf{j} + \frac{\partial w}{\partial z}\mathbf{k} = \frac{\partial w}{\partial x}\mathbf{i} + \frac{\partial w}{\partial y}\mathbf{j} + \frac{\partial w}{\partial z}\mathbf{k}$.

87. $f_x(x,y) = 2x - y + 2 = 0$ and $f_y(x,y) = -x + 2y + 2 = 0 \Rightarrow x = -2, y = -2 \Rightarrow (-2,-2)$ is the critical point. $f_{xx}(-2,-2) = 2$, $f_{yy}(-2,-2) = 2$, $f_{xy}(-2,-2) = -1 \Rightarrow f_{xx}f_{yy} - f_{xy}^2 = 3 > 0$ and $f_{xx} > 0 \Rightarrow$ Minimum (absolute). $f(-2,-2) = -8$

89. $f_x(x,y) = 6y - 3x^2 = 0$ and $f_y(x,y) = 6x - 2y = 0 \Rightarrow x = 0, y = 0$ or $x = 6, y = 18 \Rightarrow$ critical points are (0,0) and (6,18). For (0,0): $f_{xx}(0,0) = -6x\big|_{(0,0)} = 0$, $f_{yy}(0,0) = -2$, $f_{xy}(0,0) = 6 \Rightarrow$ $f_{xx}f_{yy} - f_{xy}^2 = -36 \Rightarrow$ Saddle Point. $f(0,0) = 0$. For (6,18): $f_{xx}(6,18) = -36$, $f_{yy}(6,18) = -2$, $f_{xy}(6,18) = 6$ $\Rightarrow f_{xx}f_{yy} - f_{xy}^2 = 36 > 0$ and $f_{xx} < 0 \Rightarrow$ maximum (local since $y = 0$ and $x < 0 \Rightarrow f(x,y)$ increases without bound). $f(6,18) = 108$.

91. $f_x(x,y) = 3x^2 - 3y = 0$ and $f_y(x,y) = 3y^2 - 3x = 0 \Rightarrow x = 0, y = 0$ or $x = 1, y = 1 \Rightarrow$ critical points are (0,0) and (1,1). For (0,0): $f_{xx}(0,0) = 6x\big|_{(0,0)} = 0$, $f_y(0,0) = 6y\big|_{(0,0)} = 0$, $f_{xy}(,0) = -3 \Rightarrow f_{xx}f_{yy} - f_{xy}^2 = -9 \Rightarrow$ Saddle Point. $f(0,0) = 15$. For (1,1): $f_{xx}(1,1) = 6$, $f_{yy}(1,1) = 6$, $f_{xy}(1,1) = -3 \Rightarrow f_{xx}f_{yy} - f_{xy}^2 = 27 > 0$ and $f_{xx} > 0 \Rightarrow$ Minimum (local since $y = 0$, $x < 0 \Rightarrow f(x,y)$ decreases without bound). $f(1,1) = 14$.

93.

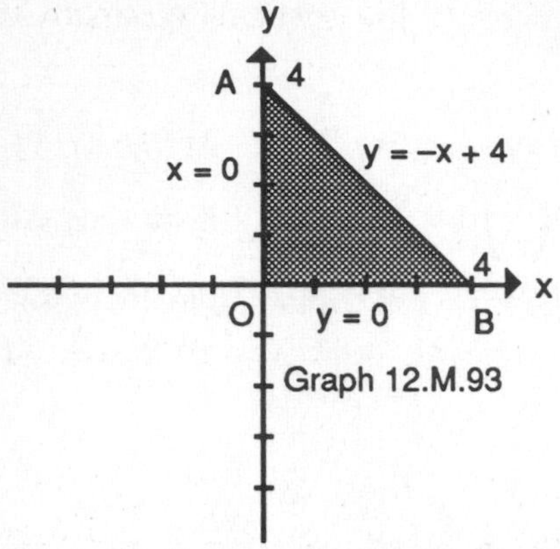

1. On OA, $f(x,y) = y^2 + 3y = f(0,y)$ for $0 \le y \le 4$. $f(0,0) = 0$, $f(0,4) = 28$. $f'(0,y) = 2y + 3 = 0 \Rightarrow y = -\frac{3}{2}$. But $\left(0,-\frac{3}{2}\right)$ is not in the region.
2. On AB, $f(x,y) = x^2 - 10x + 28 = f(x,-x+4)$ for $0 \le x \le 4$. $f(4,0) = 4$. $f'(x,-x+4) = 2x - 10 = 0 \Rightarrow x = 5, y = -1$. But (5,−1) not in the region.
3. On OB, $f(x,y) = x^2 - 3x = f(x,0)$ for $0 \le x \le 4$. $f'(x,0) = 2x - 3 \Rightarrow x = \frac{3}{2}, y = 0 \Rightarrow \left(\frac{3}{2},0\right)$ is a critical point. $f\left(\frac{3}{2},0\right) = -\frac{9}{4}$
4. For the interior of the triangular region, $f_x(x,y) = 2x + y - 3 = 0$ and $f_y(x,y) = x + 2y + 3 = 0 \Rightarrow x = 3, y = -3$. But (3,−3) is not in the region. ∴ the absolute maximum is 28 at (0,4); the absolute minimum is $-\frac{9}{4}$ at (3/2,0).

95.

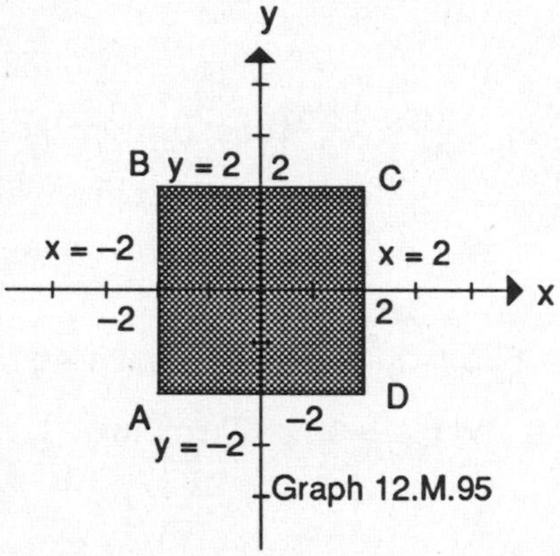

1. On AB, $f(x,y) = y^2 - y - 4 = f(-2,y)$ for $-2 \le y \le 2$. $f(-2,-2) = 2$, $f(-2,2) = -2$. $f'(-2,y) = 2y - 1 \Rightarrow y = \frac{1}{2}, x = -2 \Rightarrow \left(-2,\frac{1}{2}\right)$ is a critical point. $f\left(-2,\frac{1}{2}\right) = -\frac{17}{4}$.
2. On BC, $f(x,y) = -2 = f(x,2)$ for $-2 \le x \le 2$. $f(2,2) = -2$. $f'(x,2) = 0 \Rightarrow$ no critical points in the interior of BC.
3. On CD, $f(x,y) = y^2 - 5y + 4 = f(2,y)$ for $-2 \le y \le 2$. $f(2,-2) = 18$. $f'(2,y) = 2y - 5 = 0 \Rightarrow y = \frac{5}{2}, x = 2 \Rightarrow \left(2,\frac{5}{2}\right)$ which is not in the region.
4. On AD, $f(x,y) = 4x + 10 = f(x,-2)$ for $-2 \le x \le 2$. $f'(x,-2) = 4 \Rightarrow$ no critical points in the interior of AD.
5. For the interior of the square, $f_x(x,y) = -y - 2 = 0$ and $f_y(x,y) = 2y - x - 3 = 0 \Rightarrow x = 1, y = 2 \Rightarrow (1,2)$ is a critical point. $f(1,2) = -2$ ∴ the absolute maximum is 18 at (2,−2); the absolute minimum is $-\frac{17}{4}$ at $\left(-2,\frac{1}{2}\right)$.

97.

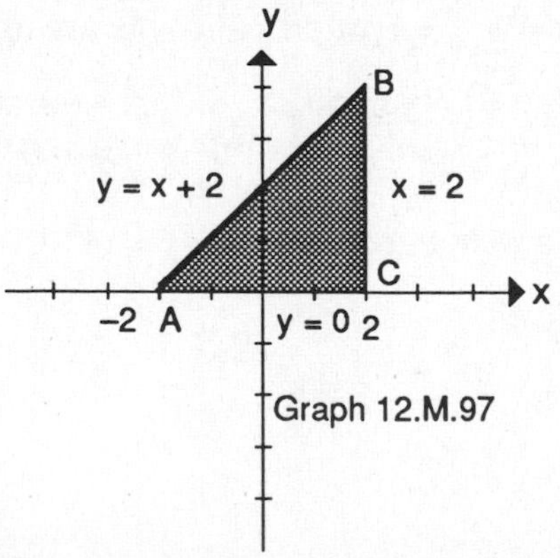

1. On AB, $f(x,y) = -2x + 4 = f(x,x+2)$ for $-2 \le x \le 2$. $f(-2,0) = 8$, $f(2,4) = 0$. $f'(x,x+2) = -2 \Rightarrow$ no critical points in the interior of AB.
2. On BC, $f(x,y) = -y^2 + 4y = f(2,y)$ for $0 \le y \le 4$. $f(2,0) = 0$. $f'(2,y) = -2y + 4 = 0 \Rightarrow y = 2, x = 2 \Rightarrow (2,2)$ is a critical point. $f(2,2) = 4$.
3. On AC, $f(x,y) = x^2 - 2x = f(x,0)$ for $-2 \le x \le 2$. $f'(x,0) = 2x - 2 = 0 \Rightarrow x = 1, y = 0 \Rightarrow (1,0)$ is a critical point. $f(1,0) = -1$.
4. For the interior of the triangular region, $f_x(x,y) = 2x - 2 = 0$ and

97. (Continued)

$f_y(x,y) = -2y + 4 = 0 \Rightarrow x = 1,\ y = 2 \Rightarrow (1,2)$ is a critical point. $f(1,2) = 3$. $\therefore$ the absolute maximum is 8 at $(-2,0)$; the absolute minimum is -1 at $(1,0)$.

99.

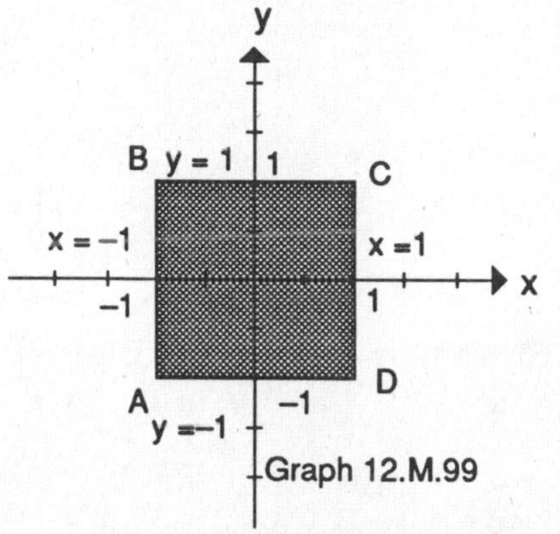

Graph 12.M.99

1. On $x = 1$, $-1 \le y \le 1$, $f(1,y) = y^3 - 3y^2 + 4 \Rightarrow f'(1,y) = 3y^2 - 6y = 0$ if $y = 0$ or 2 (not in the interval) and $x = 1$. $f(1,0) = 4$. At the corners, $f(1,1) = 2$, $f(1,-1) = 0$.
2. On $y = 1$, $-1 \le x \le 1$, $f(x,1) = x^3 + 3x^2 - 2 \Rightarrow f'(x,1) = 3x^2 + 6x = 0 \Rightarrow x = 0,\ x = -2$ (not in the interval) and $y = 1$. $f(0,1)\ -2$. At the corner $(-1,1)$, $f(-1,1) = 0$.
3. On $x = -1$, $-1 \le y \le 1$, $f(-1,y) = y^3 - 3y^2 + 2 \Rightarrow f'(-1,y) = 3y^2 - 6y - 0 \Rightarrow y = 0,\ y = 2$ (not in the interval) and $x = -1 \Rightarrow f(-1,0) = 2$. At the corner $(-1,-1)$, $f(-1,-1) = -2$.
4. On $y = -1$, $-1 \le x \le 1$, $f(x,-1) = x^3 + 3x^2 - 4 \Rightarrow f'(x,-1) = 3x^2 + 6x = 0 \Rightarrow x = 0$, $x = -2$ (not in the interval) and $y = -1 \Rightarrow f(0,-1) = -4$
5. In the interior: $f_x = 3x^2 + 6x$ and $f_y = 3y^2 - 6y$. Setting equal to 0 and solving, we get $x = 0, -2$ and $y = 0,2$. $f(0,0) = 0$ (other points are not in the region).

$\therefore$ the absolute maximum is 4 at $(1,0)$ and the absolute minimum is -4 at $(0,-1)$.

101. $\frac{\partial z}{\partial x} = 3x^2 - 9y,\ \frac{\partial z}{\partial y} = 3y^2 - 9x$. Setting these equal to 0 and solving, we get $y = \frac{1}{3}x^2 \Rightarrow 3\left(\frac{1}{3}x^2\right)^2 - 9x = 0$ $\Rightarrow \frac{1}{3}x^4 - 9x = 0 \Rightarrow x(x^3 - 27) = 0 \Rightarrow x = 0$ or $x = 3$. $x = 0 \Rightarrow y = 0$ or $(0,0)$. $x = 3 \Rightarrow y = 3$ or $(3,3)$.
$\frac{\partial^2 z}{\partial x^2} = 6x,\ \frac{\partial^2 z}{\partial y^2} = 6y$, and $\frac{\partial^2 z}{\partial x \partial y} = -9$. For $(0,0)$, $\frac{\partial^2 z}{\partial x^2}\frac{\partial^2 z}{\partial y^2} - \left(\frac{\partial^2 z}{\partial x \partial y}\right)^2 = -81 \Rightarrow$ no extremum. For $(3,3)$, $\frac{\partial^2 z}{\partial x^2}\frac{\partial^2 z}{\partial y^2} - \left(\frac{\partial^2 z}{\partial x \partial y}\right)^2 = 243 > 0$ and $\frac{\partial^2 z}{\partial x^2} = 18 > 0 \Rightarrow$ minimum.

103. Let $f(x,y) = x^2 + y^2$ be the square of the distance to the origin. $\nabla f = 2x\,\mathbf{i} + 2y\,\mathbf{j}$, $\nabla g = y^2\,\mathbf{i} + 2xy\,\mathbf{j}$.
$\nabla f = \lambda \nabla g \Rightarrow 2x\,\mathbf{i} + 2y\,\mathbf{j} = \lambda(y^2\,\mathbf{i} + 2xy\,\mathbf{j}) \Rightarrow 2x = \lambda y^2$ and $2y = 2xy\lambda \Rightarrow 2y = \lambda y^2(y\lambda) \Rightarrow y = 0$ (not on $xy^2 = 54$) or $\lambda^2 y^2 - 2 = 0 \Rightarrow y^2 = \frac{2}{\lambda^2}$. $2y = 2xy\lambda \Rightarrow 1 = x\lambda$ since $y \ne 0 \Rightarrow x = \frac{1}{\lambda}$.
$\therefore \frac{1}{\lambda}\left(\frac{2}{\lambda^2}\right) = 54 \Rightarrow \lambda^3 = \frac{1}{27} \Rightarrow \lambda = \frac{1}{3} \Rightarrow x = 3,\ y^2 = 18 \Rightarrow y = \pm 3\sqrt{2} \Rightarrow$ the points nearest to the origin are $(3, \pm 3\sqrt{2})$.

105. $\nabla T = 400yz^2\,\mathbf{i} + 400xz^2\,\mathbf{j} + 800xyz\,\mathbf{k}$, $\nabla g = 2x\,\mathbf{i} + 2y\,\mathbf{j} + 2z\,\mathbf{k}$. $\nabla T = \lambda\,\nabla g \Rightarrow 400yz^2\,\mathbf{i} + 400xz^2\,\mathbf{j} + 800xyz\,\mathbf{k} = \lambda(2x\,\mathbf{i} + 2y\,\mathbf{j} + 2z\,\mathbf{k}) \Rightarrow 400yz^2 = 2x\lambda$, $400xz^2 = 2y\lambda$, and $800xyz = 2z\lambda$. Solving this system yields the following points: $(0,\pm1,0)$, $(\pm1,0,0)$, $\left(\pm\frac{1}{2},\pm\frac{1}{2},\pm\frac{\sqrt{2}}{2}\right)$ $T(0,\pm1,0) = 0$, $T(\pm1,0,0) = 0$, $T\left(\pm\frac{1}{2},\pm\frac{1}{2},\pm\frac{\sqrt{2}}{2}\right) = \pm50$. $\therefore$ 50 is the maximum at $\left(\frac{1}{2},\frac{1}{2},\pm\frac{\sqrt{2}}{2}\right)$ and $\left(-\frac{1}{2},-\frac{1}{2},\pm\frac{\sqrt{2}}{2}\right)$; -50 is the minimum at $\left(\frac{1}{2},-\frac{1}{2},\pm\frac{\sqrt{2}}{2}\right)$ and $\left(-\frac{1}{2},\frac{1}{2},\pm\frac{\sqrt{2}}{2}\right)$

107. Let $f(x,y,z) = x^2 + y^2 + z^2$ (the square of the distance to the origin) and $g(x,y,z) = xyz - 1$. Then $\nabla f = \lambda\nabla g$

$$\Rightarrow \begin{cases} 2x = \lambda yz \\ 2y = \lambda xz \\ 2z = \lambda xy \end{cases} \Rightarrow \begin{cases} 2x^2 = \lambda xyz \\ 2y^2 = \lambda xyz \end{cases} \Rightarrow 2x^2 = 2y^2 \Rightarrow y = \pm x. \text{ Similarly, } z = \pm x. \ \therefore\ x(\pm x)(\pm x) = 1 \Rightarrow$$

$x^3 = \pm1 \Rightarrow x = \pm1 \Rightarrow$ the points are $(1,1,1)$, $(1,-1,-1)$, $(-1,-1,1)$, and $(-1,1,-1)$.

109. Let $f(x,y,z) = \frac{x^2}{a^2} + \frac{y^2}{b^2} + \frac{z^2}{c^2} - 1 \Rightarrow \nabla f = \frac{2x}{a^2}\,\mathbf{i} + \frac{2y}{b^2}\,\mathbf{j} + \frac{2z}{c^2}\,\mathbf{k} \Rightarrow$ the equation of a plane tangent at a point $P_0(x_0,y_0,y_0)$ is $\left(\frac{2x_0}{a^2}\right)x + \left(\frac{2y_0}{b^2}\right)y + \left(\frac{2z_0}{c^2}\right)z = \frac{2x_0}{a^2} + \frac{2y_0}{b^2} + \frac{2z_0}{c^2} = 2$ or $\frac{x_0}{a^2}x + \frac{y_0}{b^2}y + \frac{z_0}{c^2}z = \frac{x_0}{a^2} + \frac{y_0}{b^2} + \frac{z_0}{c^2} = 1$. The intercepts of the plane are $\left(\frac{a^2}{x_0},0,0\right)\left(0,\frac{b^2}{y_0},0\right)$ and $\left(0,0,\frac{c^2}{z_0}\right)$ The volume of the tetrahedron formed by the plane and the coordinate planes is $V = \frac{1}{3}\left(\frac{1}{2}\right)\left(\frac{a^2}{x_0}\right)\left(\frac{b^2}{y_0}\right)\left(\frac{c^2}{z_0}\right)$ Therefore we need to maximize $V(x,y,z) = \frac{(abc)^2}{6}(xyz)^{-1}$ subject to the constraint $f(x,y,z) = \frac{x^2}{a^2} + \frac{y^2}{b^2} + \frac{z^2}{c^2} - 1$. Then $-\frac{(abc)^2}{6}\left(\frac{1}{x^2yz}\right) = \frac{2x}{a^2}\lambda$, $-\frac{(abc)^2}{6}\left(\frac{1}{xy^2z}\right) = \frac{2y}{b^2}\lambda$, and $-\frac{(abc)^2}{6}\left(\frac{1}{xyz^2}\right) = \frac{2z}{c^2}\lambda$. Multiply the first equation by yz, the second by xz, and the third by xy. Equate the first and second $\Rightarrow a^2y^2 = b^2x^2 \Rightarrow y = \frac{b}{a}x$, $x > 0$. Equate the first and third $\Rightarrow a^2z^2 = c^2x^2 \Rightarrow z = \frac{c}{a}x$, $x > 0$. Substitute into $f(x,y,z) = 0 \Rightarrow x = \sqrt{\frac{a}{3}}$, $y = \sqrt{\frac{b}{3}}$, $c = \sqrt{\frac{c}{3}}$, and $V = \frac{\sqrt{3}\,abc}{2}$

111. $w = e^{rt}\sin\pi x \Rightarrow w_t = re^{rt}\sin\pi x$ and $w_x = \pi e^{rt}\cos\pi x \Rightarrow w_{xx} = -\pi^2e^{rt}\sin\pi x$. $w_{xx} = \frac{1}{c^2}w_t$ where c^2 is the positive constant determined by the material of the rod $\Rightarrow -\pi^2e^{rt}\sin\pi x = \frac{1}{c^2}\left(r\,e^{rt}\sin\pi x\right)$

$\Rightarrow (r + c^2\pi^2)e^{rt}\sin\pi x = 0 \Rightarrow r = -c^2\pi^2$. $\therefore$ $w = e^{-c^2\pi^2t}\sin\pi x$.

113.

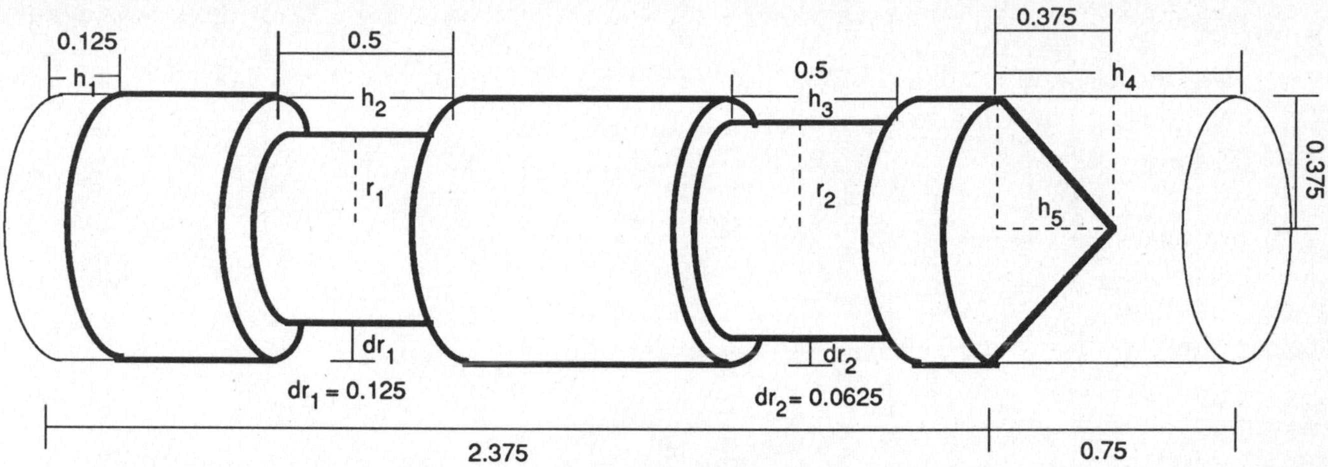

Figure 12.M.113

The mass is $m = \frac{F}{a} = \frac{V\delta}{a} = \frac{\delta}{a}V$ where δ is the density, a is the acceleration due to gravity. Thus the mass is directly proportional to the volume. The measured volume is: $V = \pi(0.375)^2(3.25) - \pi(0.125)(0.375)^2 - \pi(0.5)(0.375^2 - 0.250^2) - \pi(0.5)(0.375^2 - 0.3125^2) - \pi\left((0.375)^2(0.75) - \frac{1}{3}(0.375)^3\right) - 0.9143 \text{ in}^3$.

In general, $V = \pi(0.375)^2(3.25) - \pi(0.375)^2h_1 - \pi h_2(0.375^2 - r_1^2) - \pi h_3(0.375^2 - r_2^2) - \pi\left(0.375^2h_4 - \frac{1}{3}(0.375)^2h_5\right) \Rightarrow$ V is a function of h_1, h_2, h_3, h_4, h_5, r_1, and $r_2 \Rightarrow dV = -\pi(0.375)^2dh_1 - \pi(0.375^2 - r_1^2)dh_2 + 2\pi h_2r_1dr_1 - \pi(0.375^2 - r_2^2)dh_3 + 2\pi h_3r_2dr_2 - \pi(0.375)^2dh_4 + \frac{1}{3}\pi(0.375)^2dh_5$ where the absolute value of each differential is less than or equal to 0.0005 in. $\Rightarrow dV \le 0.00195 \text{ in}^3$. The percentage error in the mass is $\frac{dm}{m} \times 100 = \frac{\frac{\delta}{a}dV}{\frac{\delta}{a}V} \times 100 = \frac{0.00195}{0.9143} \times 100 = 0.213\ \%$

CHAPTER 13

MULTIPLE INTEGRALS

13.1 DOUBLE INTEGRALS

1. $$\int_0^3 \int_0^2 \left(4 - y^2\right) dy\, dx = \int_0^3 \left[4y - \frac{y^3}{3}\right]_0^2 dx = \frac{16}{3}\int_0^3 dx = 16$$

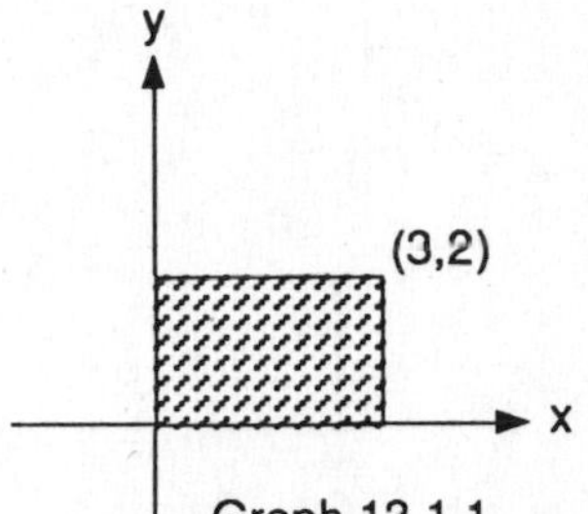

Graph 13.1.1

3. $$\int_{-1}^0 \int_{-1}^1 (x + y + 1)\, dx\, dy = \int_{-1}^0 \left[\frac{x^2}{2} + yx + x\right]_{-1}^1 dy =$$
$$\int_{-1}^0 2y + 2\, dy = 1$$

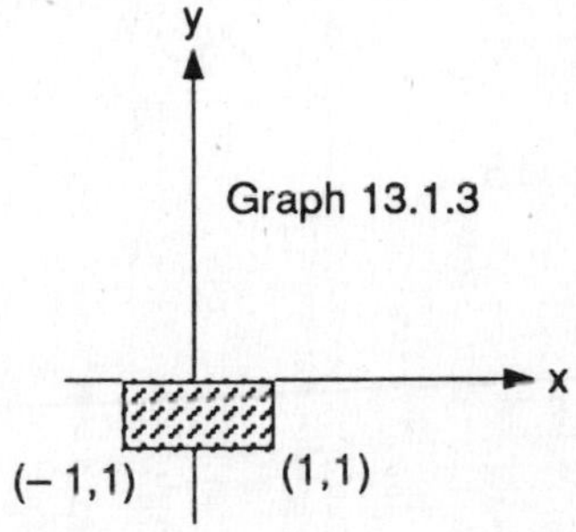

Graph 13.1.3

5. $$\int_0^\pi \int_0^x (x \sin y)\, dy\, dx = \int_0^\pi [-x \cos y]_0^x\, dx =$$
$$\int_0^\pi (x - x \cos x)\, dx = \frac{\pi^2}{2} + 2$$

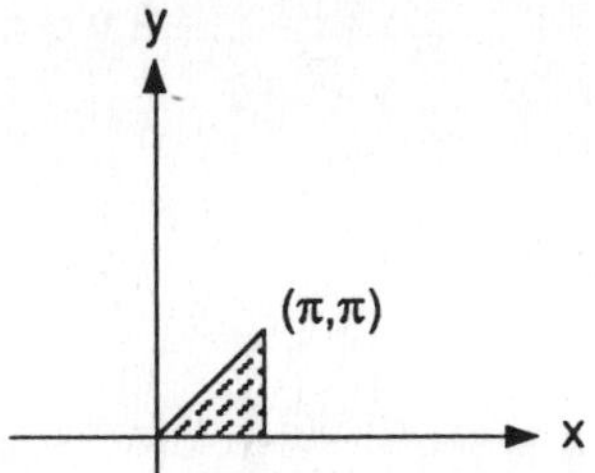

Graph 13.1.5

7. $$\int_1^{\ln 8} \int_0^{\ln y} e^{x+y}\, dx\, dy = \int_1^{\ln 8} \left[e^{x+y}\right]_0^{\ln y} dy =$$
$$\int_1^{\ln 8} y\, e^y - e^y\, dy = \left[(y-1)e^y - e^y\right]_1^{\ln 8} = 8 \ln(8) - 16 + e$$

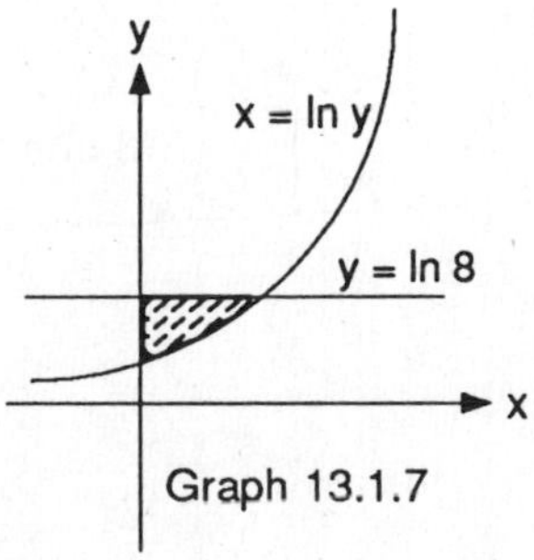

Graph 13.1.7

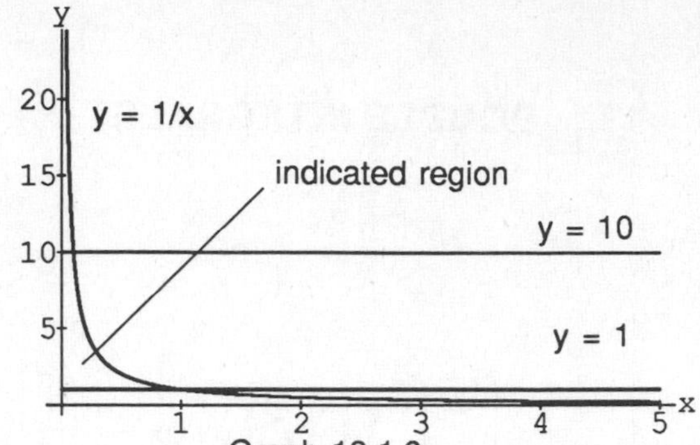

Graph 13.1.9

9. $\displaystyle\int_{10}^{1}\int_{0}^{1/y} y\, e^{xy}\, dx\, dy = \int_{10}^{1} \left[e^{xy}\right]_0^{1/y} dy =$

$\displaystyle\int_{10}^{1} e - 1\, dy = [ey - y]_{10}^{1} = 9 - 9e$

11. $\displaystyle\int_1^2\int_x^{2x} \frac{x}{y}\, dy\, dx = \int_1^2 [x \ln y]_x^{2x}\, dx = \ln(2)\int_1^2 x\, dx = \frac{\ln 8}{2}$

13. $\displaystyle\int_1^2\int_1^2 \frac{1}{xy}\, dy\, dx = \int_1^2 \frac{1}{x}(\ln 2 - \ln 1)\, dx = \ln 2\int_1^2 \frac{1}{x}\, dx = (\ln 2)^2$

15. $\displaystyle\int_0^1\int_0^{1-u} v - \sqrt{u}\, dv\, du = \int_0^1 \left[\frac{v^2}{2} - v\sqrt{u}\right]_0^{1-u} du = \int_0^1 \frac{1 - 2u + u^2}{2} - \sqrt{u}(1-u)\, du = -\frac{1}{10}$

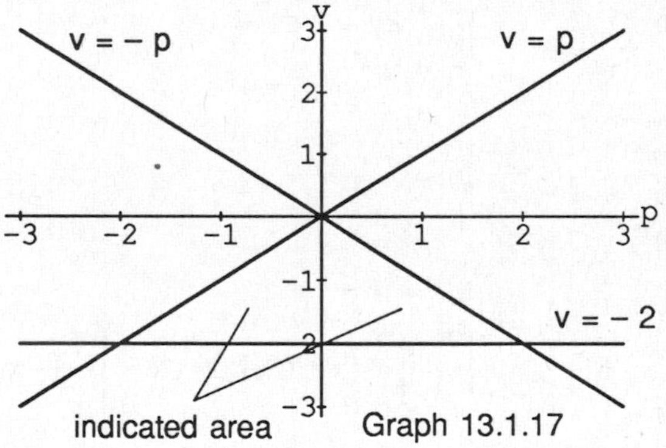

Graph 13.1.17

17. $\displaystyle\int_{-2}^{0}\int_{v}^{-v} 2\, dp\, dv = \int_{-2}^{0} [p]_v^{-v}\, dv =$

$\displaystyle 2\int_{-2}^{0} -2v\, dv = -2\left[2v^2\right]_{-2}^{0} = 8$

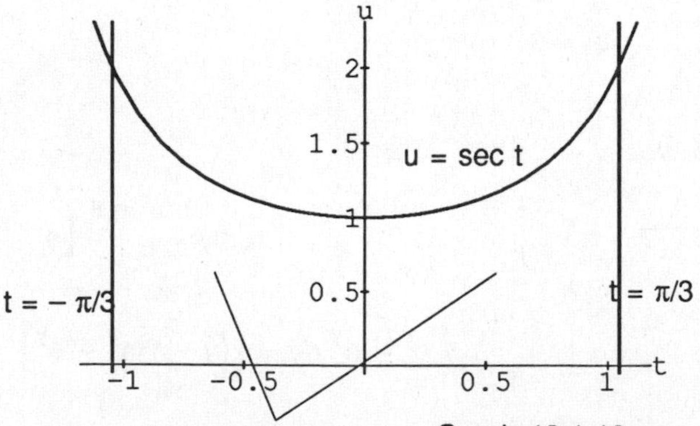

Graph 13.1.19

19. $\displaystyle\int_{-\pi/3}^{\pi/3}\int_{0}^{\sec t} 3\cos t\, du\, dt =$

$\displaystyle\int_{-\pi/3}^{\pi/3} 3\cos t[u]_0^{\sec t}\, dt = 3\int_{-\pi/3}^{\pi/3} dt = 2\pi$

21. $\int_2^4 \int_0^{(4-y)/2} dx\, dy = \int_2^4 \frac{4-y}{2} dy = 1$

$\int_0^1 \int_2^{4-2x} dy\, dx = \int_0^1 2 - 2x\, dx = 1$

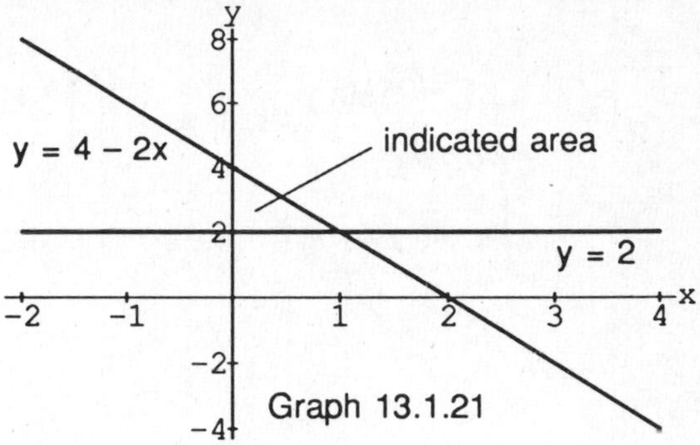

Graph 13.1.21

23. $\int_0^1 \int_{\sqrt{y}}^1 dx\, dy = \int_0^1 1 - y^{1/2} dy = \left[y - \frac{2}{3}y^{3/2}\right]_0^1 = \frac{1}{3}$

$\int_0^1 \int_0^{x^2} dy\, dx = \int_0^1 x^2 dx = \left[\frac{x^3}{3}\right]_0^1 = \frac{1}{3}$

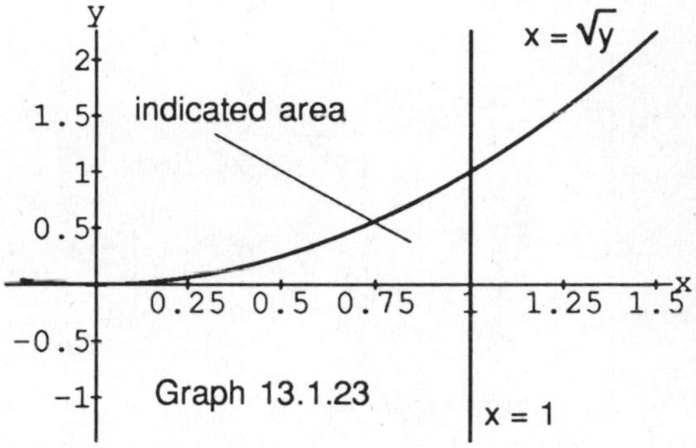

Graph 13.1.23

25. $\int_0^{3/2} \int_0^{9-4x^2} 16x\, dy\, dx =$

$\int_0^{3/2} 16x\left(9 - 4x^2\right)dx = \left[72x^2 - 16x^4\right]_0^{3/2} = 81$

$\int_0^9 \int_0^{\left(\sqrt{9-y}\right)/2} 16x\, dx\, dy = \int_0^9 18 - 2y\, dy =$

$\left[18y - y^2\right]_0^9 = 81$

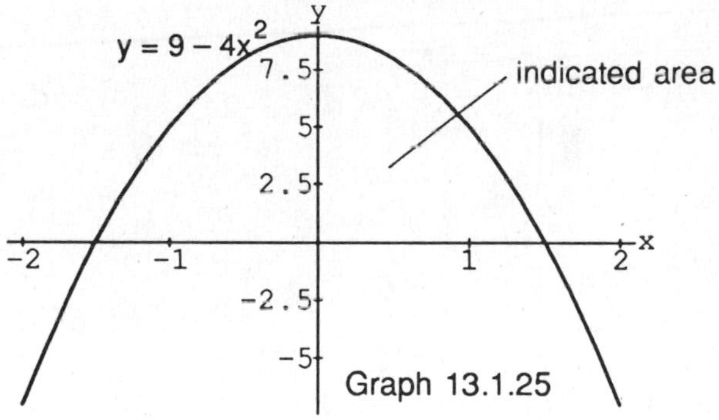

Graph 13.1.25

27. $\int_1^\infty \int_{e^{-x^2}}^1 \frac{1}{x^3 y} dy\, dx = \int_1^\infty \frac{1}{x^2} dx = -\lim_{b \to \infty} \left(\frac{1}{b} - 1\right) = 1$

29. $\int_{-\infty}^\infty \int_{-\infty}^\infty \frac{1}{\left(x^2+1\right)\left(y^2+1\right)} dx\, dy = 4\int_0^\infty \frac{2}{y^2+1} \lim_{b \to \infty} (\arctan b - \arctan 0) dy =$

$2\pi \lim_{b \to \infty} \int_0^b \frac{1}{y^2+1} dy = \pi^2$

31. $\int_0^\pi \int_x^\pi \frac{\sin y}{y}\, dy\, dx = \int_0^\pi \int_0^y \frac{\sin y}{y}\, dx\, dy =$

$\int_0^\pi \sin y\, dy = 2$

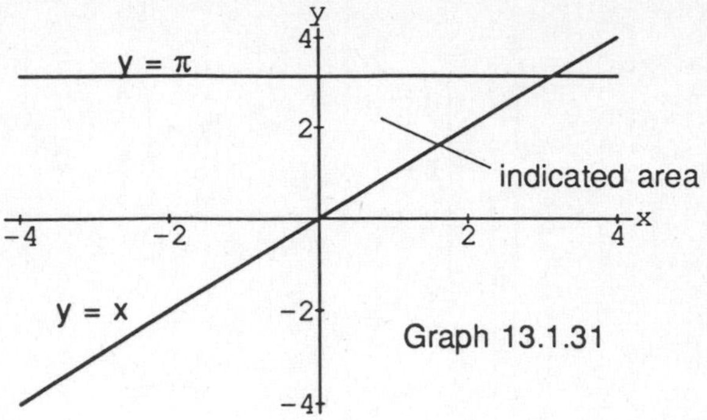

Graph 13.1.31

33. $\int_0^2 \int_x^2 2y^2 \sin xy\, dy\, dx = \int_0^2 \int_0^y 2y^2 \sin xy\, dx\, dy =$

$\int_0^2 [-2y \cos xy]_0^y\, dy = \int_0^2 -\cos y^2(2y) + 2y\, dy =$

$4 - \sin 4$

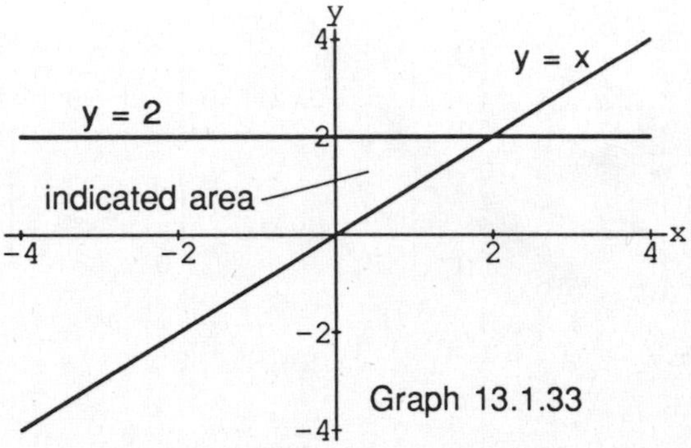

Graph 13.1.33

35. $\int_0^{2\sqrt{\ln 3}} \int_{y/2}^{\sqrt{\ln 3}} e^{x^2}\, dx\, dy =$

$\int_0^{\sqrt{\ln 3}} \int_0^{2x} e^{x^2}\, dy\, dx = \int_0^{\sqrt{\ln 3}} e^{x^2}\, 2x\, dx =$

$\left[e^{x^2}\right]_0^{\sqrt{\ln 3}} = 2$

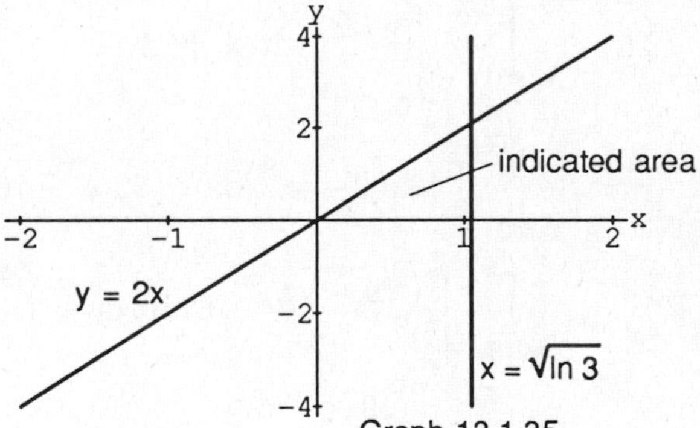

Graph 13.1.35

37. $\int_0^3 \int_{\sqrt{x/3}}^1 e^{y^3}\, dy\, dx = \int_0^1 \int_0^{3y^2} e^{y^3}\, dx\, dy =$

$\int_0^1 e^{y^3}\left(3y^2\right) dy = \left[e^{y^3}\right]_0^1 = e - 1$

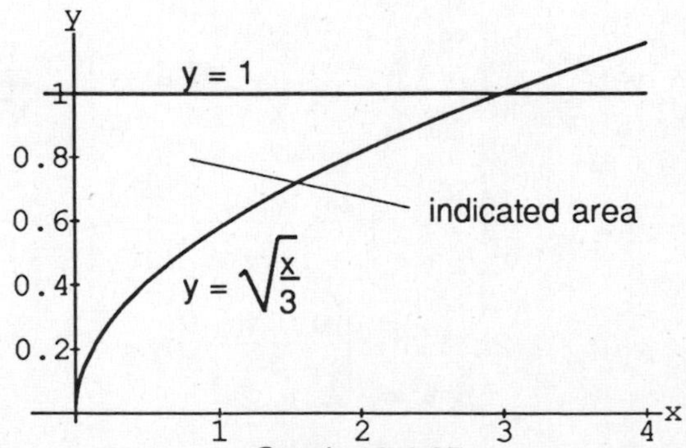

Graph 13.1.37

39. $V = \int_0^1 \int_x^{2-x} x^2 + y^2 \, dy \, dx = \int_0^1 \left[x^2 y + \frac{y^3}{3} \right]_x^{2-x} dx = \int_0^1 2x^2 - \frac{7x^3}{3} + \frac{(2-x)^3}{3} \, dx =$

$\left[\frac{2x^3}{3} - \frac{7x^4}{12} - \frac{(2-x)^4}{12} \right]_0^1 = \frac{4}{3}$

41. $V = \int_{-4}^{1} \int_{3x}^{4-x^2} (x+4) \, dy \, dx = \int_{-4}^{1} \left[(xy + 4y) \right]_{3x}^{4-x^2} dx = \int_{-4}^{1} \left(-x^3 - 7x^2 - 8x + 16 \right) dx = \frac{625}{12}$

43. $V = \int_0^2 \int_0^3 4 - y^2 \, dx \, dy = \int_0^2 \left[4x - y^2 x \right]_0^3 dy = \int_0^2 \left(12 - 3y^2 \right) dy = 16$

45. $V = \int_0^2 \int_0^{2-x} 12 - 3y^2 \, dy \, dx = \int_0^2 \left[12y - y^3 \right]_0^{2-x} dx = \int_0^2 24 - 12x - (2-x)^3 \, dx =$

$\left[42x - 6x^2 + \frac{(2-x)^4}{4} \right]_0^2 = 20$

47. $V = 2 \int_1^2 \int_0^{1/x} x + 1 \, dy \, dx = 2 \int_1^2 \left[xy + y \right]_0^{1/x} dx = 2 \int_1^2 1 + \frac{1}{x} dx = 2 \left[x + \ln x \right]_1^2 = 2 + \ln 4$

49. $\int_0^{\sqrt{3}} \int_{x/\sqrt{3}}^{\sqrt{4-x^2}} \sqrt{4-x^2} \, dy \, dx = \int_0^{\sqrt{3}} 4 - x^2 + \frac{(4-x^2)^{1/2}(-2x)}{2\sqrt{3}} \, dx =$

$\left[4x - \frac{x^3}{3} + \frac{(4-x^2)^{3/2}}{3\sqrt{3}} \right]_0^{\sqrt{3}} = \frac{20\sqrt{3}}{9}$

51. $V = \int_0^1 \int_x^{2-x} x^2 + y^2 \, dy \, dx =$

$\int_0^1 \left[x^2 y + \frac{y^3}{3} \right]_x^{2-x} dx = \int_0^1 2x^2 - \frac{7x^3}{3} + \frac{(2-x)^3}{3} \, dx =$

$\left[\frac{2x^3}{3} - \frac{7x^4}{12} - \frac{(2-x)^4}{12} \right]_0^1 = \frac{4}{3}$

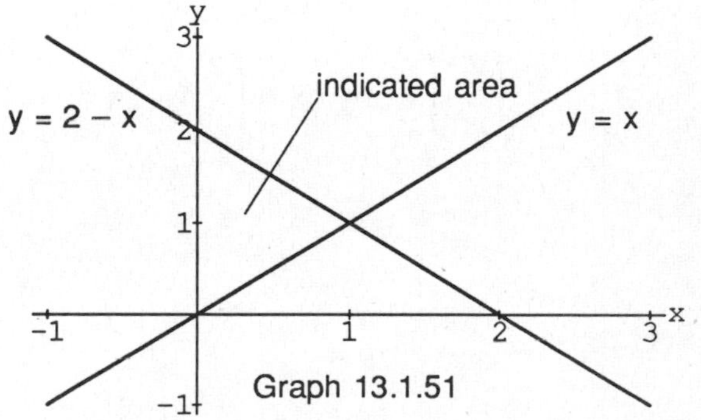

Graph 13.1.51

53. Outside the ellipse, $\frac{x^2}{4} + \frac{y^2}{2} = 1$, the integrand is negative, while on the ellipse the integrand is zero.

$\therefore$ the maximum must occur inside the ellipse, $\{ (x,y) \mid 1 > \frac{x^2}{4} + \frac{y^2}{2} \}$.

55. $\int_1^3 \int_1^x \frac{1}{xy}\, dy\, dx = 0.603$

57. $\int_0^1 \int_0^1 \tan^{-1} xy\, dy\, dx = 0.233$

13.2 AREAS, MOMENTS, AND CENTERS OF MASS

1. $\int_0^2 \int_0^{2-x} dy\, dx = \int_0^2 2 - x\, dx = 2$

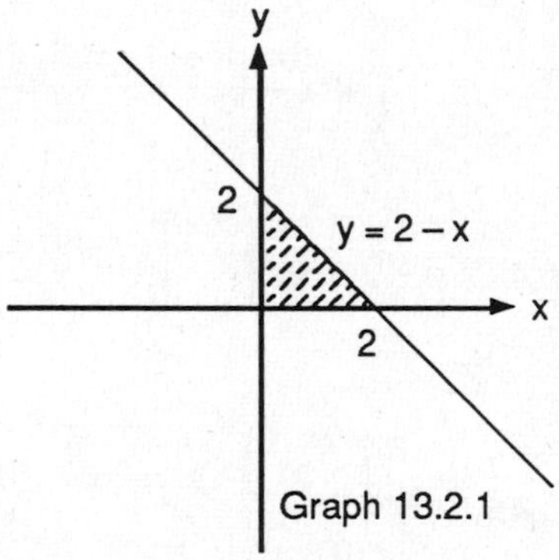

Graph 13.2.1

3. $\int_0^2 \int_{2x}^4 dy\, dx = \int_0^2 4 - 2x\, dx = 4$

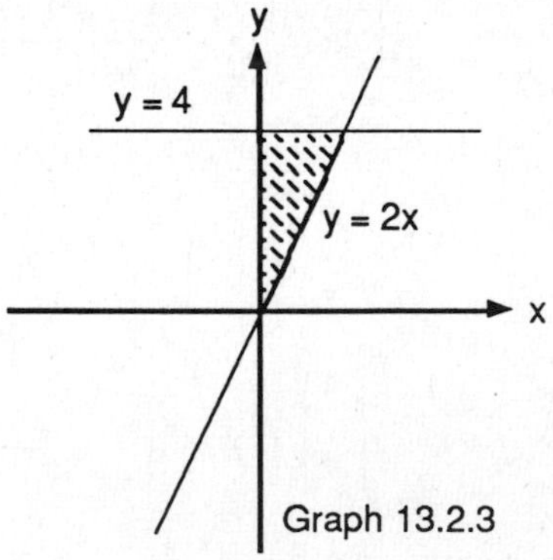

Graph 13.2.3

5. $$\int_0^1 \int_{y^2}^{2y-y^2} dx\,dy = \int_0^1 2y - 2y^2\,dy = \frac{1}{3}$$

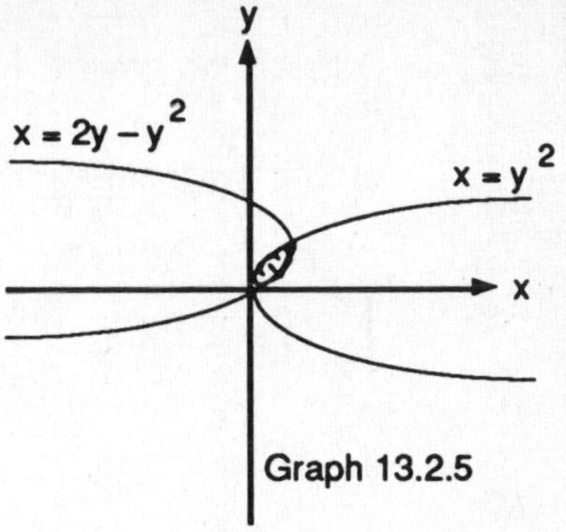

Graph 13.2.5

7. $$2\int_0^1 \int_{-1}^{2\sqrt{1-x^2}} dy\,dx = 2\int_0^1 2\sqrt{1-x^2} + 1\,dx = \pi + 2$$

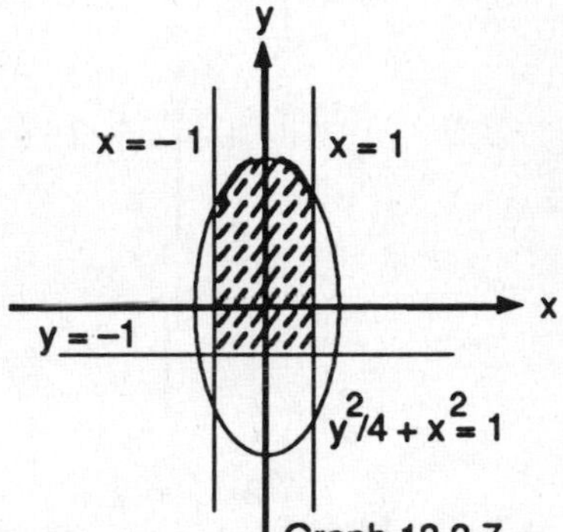

Graph 13.2.7

9. $$\int_0^6 \int_{y^2/3}^{2y} dx\,dy = \int_0^6 \left(2y - y^2/3\right) dy = 12$$

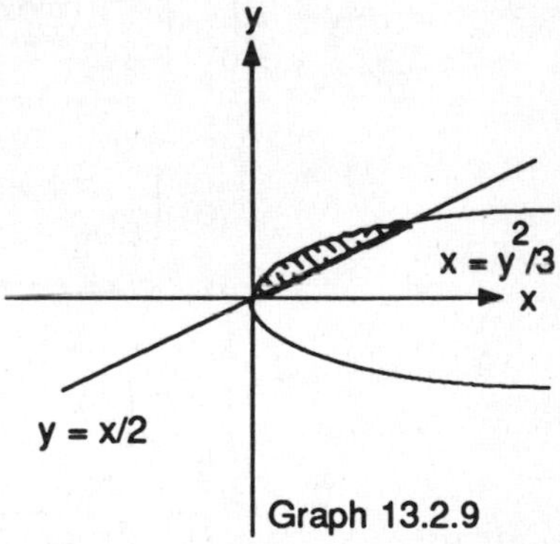

Graph 13.2.9

11. $$\int_0^{\pi/4} \int_{\sin x}^{\cos x} dy\,dx = \int_0^{\pi/4} (\cos x - \sin x)\,dx = \sqrt{2} - 1$$

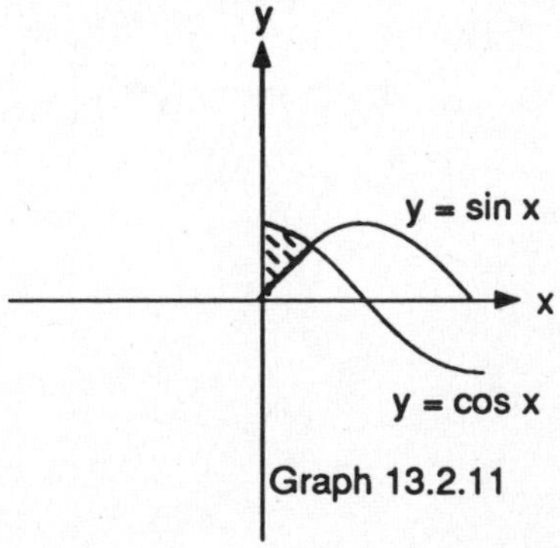

Graph 13.2.11

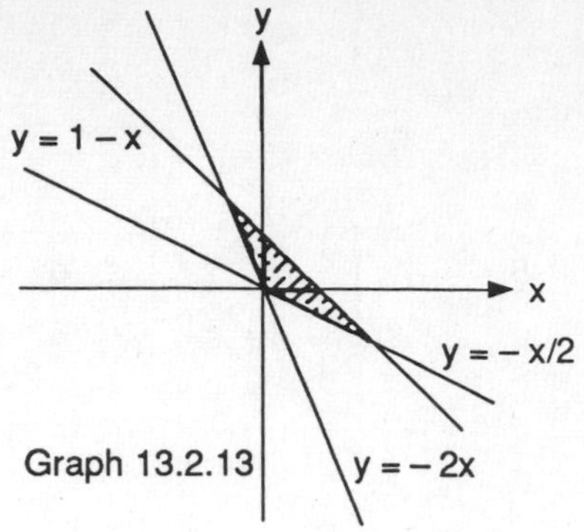

Graph 13.2.13

13. $\int_{-1}^{0}\int_{-2x}^{1-x} dy\,dx + \int_{0}^{2}\int_{-x/2}^{1-x} dy\,dx = \int_{-1}^{0} (1 + x)\,dx + \int_{0}^{2} (1 - x/2)\,dx = \frac{3}{2}$

15. $M = \int_0^1 \int_x^{2-x^2} 3\,dy\,dx = 3\int_0^1 2 - x^2 - x\,dx = \frac{7}{2}$

$M_y = \int_0^1 \int_x^{2-x^2} 3x\,dy\,dx = 3\int_0^1 [xy]_x^{2-x^2}\,dx = 3\int_0^1 \left(2x - x^3 - x^2\right)dx = \frac{5}{4}$

$M_x = \int_0^1 \int_x^{2-x^2} 3y\,dy\,dx = \frac{3}{2}\int_0^1 \left[y^2\right]_x^{2-x^2}\,dx = \frac{3}{2}\int_0^1 4 - 5x^2 + x^4\,dx = \frac{19}{5} \Rightarrow \overline{x} = \frac{5}{14},\ \overline{y} = \frac{38}{35}$

17. $M = \int_0^2 \int_{y^2/2}^{4-y} dx\,dy = \int_0^2 4 - y - \frac{y^2}{2}\,dy = \frac{14}{3}$

$M_y = \int_0^2 \int_{y^2/2}^{4-y} x\,dx\,dy = \frac{1}{2}\int_0^2 \left[x^2\right]_{y^2/2}^{4-y}\,dx = \frac{1}{2}\int_0^2 16 - 8y + y^2 - \frac{y^4}{4}\,dy = \frac{128}{15}$

$M_x = \int_0^2 \int_{y^2/2}^{4-y} y\,dx\,dy = \int_0^2 y\left(4 - y - \frac{y^2}{2}\right)dy = \frac{10}{3} \Rightarrow \overline{x} = \frac{64}{35},\ \overline{y} = \frac{5}{7}$

19. $M = 2\int_0^1 \int_0^{\sqrt{1-x^2}} dy\,dx = 2\int_0^1 \sqrt{1 - x^2}\,dx = \frac{\pi}{2}$

$M_x = 2\int_0^1 \int_0^{\sqrt{1-x^2}} y\,dy\,dx = \int_0^1 \left[y^2\right]_0^{\sqrt{1-x^2}}\,dx = \int_0^1 1 - x^2\,dx = \frac{2}{3} \Rightarrow \overline{x} = 0,$

by symmetry; $\overline{y} = \frac{4}{3\pi}$

21. $M = \int_0^a \int_0^{\sqrt{a^2-x^2}} dy\,dx = \frac{\pi a^2}{4}$; $M_y = \int_0^a \int_0^{\sqrt{a^2-x^2}} x\,dy\,dx =$

$\int_0^a [xy]_0^{\sqrt{a^2-x^2}} dx = -\frac{1}{2}\int_0^a \sqrt{a^2-y^2}(-2x)\,dx = \frac{a^3}{3} \Rightarrow \overline{x} = \overline{y} = \frac{4a}{3\pi}$, by symmetry

23. $M = \int_0^\pi \int_0^{\sin x} dy\,dx = \int_0^\pi \sin x\,dx = 2$; $M_x = \int_0^\pi \int_0^{\sin x} y\,dy\,dx = \frac{1}{2}\int_0^\pi \left[y^2\right]_0^{\sin x} dx =$

$\frac{1}{4}\int_0^\pi 1 - \cos 2x\,dx = \frac{\pi}{4} \Rightarrow \overline{x} = \frac{\pi}{2}, \overline{y} = \frac{\pi}{8}$

25. $I_x = \int_{-a}^a \int_{-b}^b y^2\,dy\,dx = \int_{-a}^a \frac{2b^3}{3}dx = \frac{4ab^3}{3}$, $I_y = \int_{-b}^b \int_{-a}^a x^2\,dx\,dy =$

$\int_{-b}^b \frac{2a^3}{3}dy = \frac{4a^3b}{3} \Rightarrow I_o = I_x + I_y = \frac{4ab^3}{3} + \frac{4a^3b}{3} = \frac{4ab\left(b^2+a^2\right)}{3}$

27. $M = \int_{-\infty}^0 \int_0^{\exp(x)} dy\,dx = \int_{-\infty}^0 e^x\,dx = \lim_{t \to -\infty} \int_t^0 e^x\,dx = 1$

$M_y = \int_{-\infty}^0 \int_0^{\exp(x)} x\,dy\,dx = \int_{-\infty}^0 x\,e^x\,dx = \lim_{t \to -\infty} \int_t^0 x\,e^x\,dx = -1$

$M_x = \int_{-\infty}^0 \int_0^{\exp(x)} y\,dy\,dx = \frac{1}{2}\int_{-\infty}^0 e^{2x}\,dx = \frac{1}{2}\lim_{t \to -\infty} \int_t^0 e^{2x}\,dx = \frac{1}{4} \Rightarrow \overline{x} = -1, \overline{y} = \frac{1}{4}$

29. $M = \int_0^2 \int_{-y}^{y-y^2} (x+y)\,dx\,dy = \int_0^2 \left[\frac{x^2}{2} + xy\right]_{-y}^{y-y^2} dy =$

$\int_0^2 \left(\frac{y^4}{2} - 2y^3 + 2y^2\right)dy = \frac{8}{15}$; $I_x = \int_0^2 \int_{-y}^{y-y^2} y^2(x+y)\,dx\,dy = \int_0^2 \left[\frac{x^2y^2}{2} + xy^3\right]_{-y}^{y-y^2} dy =$

$\int_0^2 \left(\frac{y^6}{2} - 2y^5 + 2y^4\right)dy = \frac{64}{105}$; $R_x = \sqrt{\frac{I_x}{M}} = \sqrt{\frac{8}{7}}$

31. $M = \int_0^1 \int_x^{2-x} (6x + 3y + 3)\,dy\,dx = \int_0^1 -12x^2 + 12\,dx = 8$; $M_y = \int_0^1 \int_x^{2-x} x(6x + 3y + 3)\,dy\,dx =$

$\int_0^1 12x - 12x^3\,dx = 3$; $M_x = \int_0^1 \int_x^{2-x} y(6x + 3y + 3)\,dy\,dx =$

$\int_0^1 14 - 6x - 6x^2 - 2x^3\,dx = \frac{17}{2} \Rightarrow \overline{x} = \frac{3}{8}, \overline{y} = \frac{17}{16}$

33. $M = \int_0^1 \int_0^6 (x + y + 1)\,dx\,dy = \int_0^1 (6y + 24)\,dy = 27$

$M_x = \int_0^1 \int_0^6 y(x + y + 1)\,dx\,dy = \int_0^1 y(6y + 24)\,dy = 14$

$M_y = \int_0^1 \int_0^6 x(x + y + 1)\,dx\,dy = \int_0^1 (18y + 90)\,dy = 99 \Rightarrow \overline{x} = \frac{11}{3},\ \overline{y} = \frac{14}{27}$

$I_y = \int_0^1 \int_0^6 x^2(x + y + 1)\,dx\,dy = 216\int_0^1 \left(\frac{y}{3} + \frac{11}{6}\right) dy = 432 \Rightarrow R_y = \sqrt{\frac{I_y}{M}} = 4$

35. $M = \int_{-1}^1 \int_0^{x^2} (7y + 1)\,dy\,dx = \int_{-1}^1 \left(\frac{7x^4}{2} + x^2\right) dx = \frac{31}{15}$

$M_x = \int_{-1}^1 \int_0^{x^2} y(7y + 1)\,dy\,dx = \int_{-1}^1 \left(\frac{7x^6}{3} + \frac{x^4}{2}\right) dx = \frac{13}{15}$

$M_y = \int_{-1}^1 \int_0^{x^2} x(7y + 1)\,dy\,dx = \int_{-1}^1 \left(\frac{7x^5}{2} + x^3\right) dx = 0 \Rightarrow \overline{x} = 0,\ \overline{y} = \frac{13}{31}$

$I_y = \int_{-1}^1 \int_0^{x^2} x^2(7y + 1)\,dy\,dx = \int_{-1}^1 x^2\left(\frac{7x^4}{2} + x^2\right) dx = \frac{7}{5} \Rightarrow R_y = \sqrt{\frac{I_y}{M}} = \sqrt{\frac{21}{31}}$

37. $M = \int_0^1 \int_{-y}^{y} (1 + y)\,dx\,dy = \int_0^1 \left(2y^2 + 2y\right) dy = \frac{5}{3}$

$M_x = \int_0^1 \int_{-y}^{y} y(1 + y)\,dx\,dy = 2\int_0^1 \left(y^3 + y^2\right) dy = \frac{7}{6}$

$M_y = \int_0^1 \int_{-y}^{y} x(1 + y)\,dx\,dy = \int_0^1 0\,dy = 0 \Rightarrow \overline{x} = 0, \overline{y} = \frac{7}{10}$

$I_x = \int_0^1 \int_{-y}^{y} y^2(1 + y)\,dx\,dy = \int_0^1 y^2\left(2y^2 + 2y\right) dy = \frac{9}{10} \Rightarrow R_x = \sqrt{\frac{I_x}{M}} = \frac{3\sqrt{6}}{10}$

$I_y = \int_0^1 \int_{-y}^{y} x^2(1 + y)\,dx\,dy = \frac{1}{3}\int_0^1 y^2\left(2y^2 + 2y\right) dy = \frac{3}{10} \Rightarrow R_y = \sqrt{\frac{I_y}{M}} = \frac{3\sqrt{2}}{10}$

$I_o = I_x + I_y = \frac{6}{5}$ and $R_o = \sqrt{\frac{I_o}{M}} = \frac{3\sqrt{2}}{5}$

39. a) Average $= \frac{1}{\pi^2}\int_0^{\pi}\int_0^{\pi} \sin(x+y)\,dy\,dx = \frac{1}{\pi^2}\int_0^{\pi}\left[-\cos(x+y)\right]_0^{\pi} dx =$

$\frac{1}{\pi^2}\int_0^{\pi} -\cos(x+\pi) + \cos x\,dx = \frac{1}{\pi^2}\left[-\sin(x+\pi) + \sin x\right]_0^{\pi} = 0$

b) Average $= \frac{1}{\pi^2/2}\int_0^{\pi}\int_0^{\pi/2} \sin(x+y)\,dy\,dx = \frac{2}{\pi^2}\int_0^{\pi}\left[-\cos(x+y)\right]_0^{\pi/2} dx =$

$\frac{2}{\pi^2}\int_0^{\pi} -\cos(x+\pi/2) + \cos x\,dx = \frac{2}{\pi^2}\left[-\sin(x+\pi/2) + \sin x\right]_0^{\pi} = \frac{4}{\pi^2}$

41. Average height $= \frac{1}{4}\int_0^2\int_0^2 x^2 + y^2\,dy\,dx = \frac{1}{4}\int_0^2\left[x^2y + \frac{y^3}{3}\right]_0^2 dx = \frac{1}{4}\int_0^2 2x^2 + \frac{8}{3}dx =$

$\frac{1}{2}\left[\frac{x^3}{3} + \frac{4x}{3}\right]_0^2 = \frac{8}{3}$

43. $M = \int_{-1}^{1}\int_0^{a(1-x^2)} dy\,dx = 2a\int_0^1 1 - x^2\,dx = 2a\left[x - \frac{x^3}{3}\right]_0^1 = \frac{4a}{3}$, $M_x =$

$\int_{-1}^{1}\int_0^{a(1-x^2)} y\,dy\,dx = \frac{2a^2}{2}\int_0^1 1 - 2x^2 + x^4\,dx = a^2\left[x - \frac{2x^3}{3} + \frac{x^5}{5}\right]_0^1 = \frac{8a^2}{15} \therefore \bar{y} = \frac{M_x}{M} =$

$\frac{8a^2/15}{4a/3} = \frac{2a}{5}$; The angle formed by the c. m., O and fulcrum plus 45° must remain less than 90°, i. e. $\tan^{-1}\left(\frac{2a}{5}\right) + \frac{\pi}{4} < \frac{\pi}{2} \Rightarrow a < \frac{5}{2}$.

45. $M = \int_0^1\int_{-1/\sqrt{1-x^2}}^{1/\sqrt{1-x^2}} dy\,dx = \int_0^1 \frac{2}{\sqrt{1-x^2}}dx = \pi$

$M_y = \int_0^1\int_{-1/\sqrt{1-x^2}}^{1/\sqrt{1-x^2}} x\,dy\,dx = \int_0^1 \frac{2x}{\sqrt{1-x^2}}dx = 2 \Rightarrow \bar{x} = \frac{2}{\pi}$, $\bar{y} = 0$ by symmetry

47. a) $\bar{x} = \frac{M_y}{M} = 0 \Rightarrow M_y = \int_R\int x\,\delta(x,y)\,dy\,dx = 0$

b) $I_L = \int_R\int (x-h)^2\,\delta(x,y)\,dA = \int_R\int x^2\,\delta(x,y)\,dA - \int_R\int 2hx\,\delta(x,y)\,dA + \int_R\int h^2\,\delta(x,y)\,dA =$

$I_y - 0 + h^2\int_R\int \delta(x,y)\,dA = I_{c.m.} + mh^2$

49. $M_{x_{p_1 \cup p_2}} = \int_{R_1}\int y\, dA_1 + \int_{R_2}\int y\, dA_2 = M_{x_1} + M_{x_2} \Rightarrow \overline{x} = \frac{M_{x_1} + M_{x_2}}{m_1 + m_2}$ and likewise

$\overline{y} = \frac{M_{y_1} + M_{y_2}}{m_1 + m_2}$. $\therefore$ $\mathbf{C} = \overline{x}\,\mathbf{i} + \overline{y}\,\mathbf{j} = \frac{1}{m_1 + m_2}\left[\left(M_{x_1} + M_{x_2}\right)\mathbf{i} + \left(M_{y_1} + M_{y_2}\right)\mathbf{j}\right] =$

$\frac{1}{m_1 + m_2}\left[\left(m_1\overline{x}_1 + m_2\overline{x}_2\right)\mathbf{i} + \left(m_1\overline{y}_1 + m_2\overline{y}_2\right)\mathbf{j}\right] =$

$\frac{1}{m_1 + m_2}\left[m_1\left(\overline{x}_1\mathbf{i} + \overline{y}_1\mathbf{j}\right) + m_2\left(\overline{x}_2\mathbf{i} + \overline{y}_2\mathbf{j}\right)\right] = \frac{m_1\mathbf{C}_1 + m_2\mathbf{C}_2}{m_1 + m_2}$.

51. a) $\mathbf{C} = \frac{8(\mathbf{i} + 3\mathbf{j}) + 2(3\mathbf{i} + 3.5\mathbf{j})}{8 + 2} = \frac{14\mathbf{j} + 31\mathbf{k}}{10} \Rightarrow \overline{x} = \frac{7}{5},\ \overline{y} = \frac{31}{10}$

b) $\mathbf{C} = \frac{8(\mathbf{i} + 3\mathbf{j}) + 6(5\mathbf{i} + 2\mathbf{j})}{14} = \frac{38\mathbf{i} + 36\mathbf{j}}{14} = \overline{x} = \frac{19}{7},\ \overline{y} = \frac{18}{7}$

c) $\mathbf{C} = \frac{2(3\mathbf{i} + 3.5\mathbf{j}) + 6(5\mathbf{i} + 2\mathbf{j})}{8} = \frac{36\mathbf{i} + 19\mathbf{j}}{8} \Rightarrow \overline{x} = \frac{9}{2},\ \overline{y} = \frac{19}{8}$

d) $\mathbf{C} = \frac{8(\mathbf{i} + 3\mathbf{j}) + 2(3\mathbf{i} + 3.5\mathbf{j}) + 6(5\mathbf{i} + 2\mathbf{j})}{16} = \frac{44\mathbf{i} + 43\mathbf{j}}{16} \Rightarrow \overline{x} = \frac{11}{4},\ \overline{y} = \frac{43}{16}$

53. Place the midpoint of the triangle's base at the origin and above the semicircle. From Pappus's formula we have $\mathbf{C} = \frac{ah(0\mathbf{i} + \frac{h}{3}\mathbf{j}) + \frac{\pi a}{2}(0\mathbf{i} - \frac{4a}{3\pi}\mathbf{j})}{ah + \frac{\pi a}{2}} = \frac{\left(\frac{ah^2 - 2a^2}{3}\right)\mathbf{j}}{ah + \frac{\pi a}{2}}$. If $\frac{ah^2 - 2a^2}{3} = 0 \Rightarrow h^2 = 2a$, then the centroid is on the boundary. If $\frac{a}{3}\left(h^2 - 2a\right) > 0 \Rightarrow h^2 > 2a$, then the centroid is inside T.

13.3 DOUBLE INTEGRALS IN POLAR FORM

1. $\int_{-1}^{1}\int_{0}^{\sqrt{1-x^2}} dy\, dx = \int_{0}^{\pi}\int_{0}^{1} r\, dr\, d\theta = \frac{1}{2}\int_{0}^{\pi} d\theta = \frac{\pi}{2}$

3. $\int_{0}^{1}\int_{0}^{\sqrt{1-y^2}}\left(x^2 + y^2\right) dx\, dy = \int_{0}^{\pi/2}\int_{0}^{1} r^3\, dr\, d\theta = \frac{1}{4}\int_{0}^{\pi/2} d\theta = \frac{\pi}{8}$

5. $\int_{-a}^{a}\int_{-\sqrt{a^2-x^2}}^{\sqrt{a^2-x^2}} dy\, dx = \int_{0}^{2\pi}\int_{0}^{a} r\, dr\, d\theta = \frac{a^2}{2}\int_{0}^{2\pi} d\theta = \pi a^2$

7. $\int_{0}^{\pi/4}\int_{0}^{\sqrt{2}} \cos\theta\, r^2\, dr\, d\theta = \frac{2\sqrt{2}}{3}\int_{0}^{\pi/4} \cos\theta\, d\theta = \frac{2}{3}$

9. $\displaystyle\int_0^3 \int_0^{\sqrt{3}x} \frac{1}{\sqrt{x^2+y^2}}\, dy\, dx = \int_0^{\pi/3} \int_0^{3\sec\theta} dr\, d\theta = \int_9^{\pi/3} 3\sec\theta\, d\theta = 3\ln\left(2+\sqrt{3}\right)$

11. $\displaystyle\int_0^1 \int_0^{\sqrt{1-x^2}} 5\sqrt{x^2+y^2}\, dy\, dx = \int_0^{\pi/2} \int_0^1 5r^2\, dr\, d\theta = \frac{5}{3}\int_0^{\pi/2} d\theta = \frac{5\pi}{6}$

13. $\displaystyle\int_0^2 \int_0^{\sqrt{1-(x-1)^2}} \frac{x+y}{x^2+y^2}\, dy\, dx = \int_0^{\pi/2} \int_0^{2\cos\theta} \frac{r(\cos\theta+\sin\theta)}{r^2}\, r\, dr\, d\theta =$

$\displaystyle\int_0^{\pi/2} 2\cos^2\theta + 2\sin\theta\cos\theta\, d\theta = \left[\theta + \frac{\sin 2\theta}{2} + \sin^2\theta\right]_0^{\pi/2} = \frac{\pi+2}{2}$

15. $\displaystyle\int_{-1}^1 \int_{-\sqrt{1-x^2}}^{\sqrt{1-x^2}} \frac{2}{\left(1+x^2+y^2\right)^2}\, dy\, dx = 4\int_0^{\pi/2} \int_0^1 \frac{2r}{(1+r^2)^2}\, dr\, d\theta = 2\int_0^{\pi/2} d\theta = \pi$

17. $\displaystyle\int_0^{\pi/2} \int_0^{2\sqrt{2-\sin 2\theta}} r\, dr\, d\theta = 2\int_0^{\pi/2} \left(2-\sin 2\theta\right) d\theta = 2(\pi-1)$

19. $\displaystyle 2\int_0^{\pi/6} \int_0^{12\cos 3\theta} r\, dr\, d\theta = 144\int_0^{\pi/6} \cos^2 3\theta\, d\theta = 12\pi$

21. $\displaystyle A = \int_0^{\pi/2} \int_0^{1+\sin\theta} r\, dr\, d\theta = \frac{1}{2}\int_0^{\pi/2} \frac{3}{2} + 2\sin\theta - \frac{\cos 2\theta}{2}\, d\theta = \frac{3\pi+8}{8}$

23. $\displaystyle A = 2\int_0^{\pi/2} \int_0^{\sec^2(\theta/2)} r\, dr\, d\theta = 2\int_0^{\pi/2} \sec^4\left(\frac{\theta}{2}\right) d\theta = \frac{2}{3}\left[\sec^2\left(\frac{\theta}{2}\right)\tan\left(\frac{\theta}{2}\right)\right]_0^{\pi/2} +$

$\displaystyle\frac{2}{3}\int_0^{\pi/2} \sec^2\left(\frac{\theta}{2}\right) d\theta = \left[\frac{\sec^2\left(\frac{\theta}{2}\right)\tan\left(\frac{\theta}{2}\right)}{3/2} + \frac{4}{3}\tan\left(\frac{\theta}{2}\right)\right]_0^{\pi/2} = \frac{8}{3}$

25. $\displaystyle\int_0^{2\pi} \int_0^{\sqrt{3}/2} \frac{1}{1-r^2}\, r\, dr\, d\theta = \ln(2)\int_0^{2\pi} d\theta = \pi\ln 4$

27. $\displaystyle M_x = \int_0^{\pi} \int_0^{1-\cos\theta} 3r^2\sin\theta\, dr\, d\theta = \int_0^{\pi} \left(1-\cos\theta\right)^3 \sin\theta\, d\theta = 4$

29. $\displaystyle M = 2\int_0^{\pi} \int_0^{1+\cos\theta} r\, dr\, d\theta = \int_0^{\pi} \left(1+\cos\theta\right)^2 d\theta = \frac{3\pi}{2}$

$\displaystyle M_y = \int_0^{2\pi} \int_0^{1+\cos\theta} \cos\theta\, r^2\, dr\, d\theta = \int_0^{2\pi} \frac{4\cos\theta}{3} + \frac{15}{24} + \cos 2\theta - \sin^2\theta\cos\theta + \frac{\cos 4\theta}{4}\, d\theta =$

$\displaystyle\frac{5\pi}{4} \Rightarrow \overline{x} = \frac{5}{6},\ \overline{y} = 0$ by symmetry

31. $M = 2\int_{\pi/6}^{\pi/2}\int_{a}^{2a\sin\theta}\frac{1}{r}r\,dr\,d\theta = 2\int_{\pi/6}^{\pi/2} 2a\sin\theta - a\,d\theta = 2[-2a\cos\theta - a\theta]_{\pi/6}^{\pi/2} = \frac{2\pi(3\sqrt{3}-\pi)}{3}$

33. $V = 2\int_{0}^{\pi/2}\int_{1}^{1+\cos\theta} r\cos\theta\, r\,dr\,d\theta = \frac{2}{3}\int_{0}^{\pi/2} 3\cos^2\theta + 3\cos^3\theta + \cos^4\theta\,d\theta =$

$\frac{2}{3}\left[\frac{15\theta}{8} + \sin(2\theta) + 3\sin(\theta) - \sin^3\theta + \frac{\sin(4\theta)}{32}\right]_0^{\pi/2} = \frac{5\pi}{8} + \frac{4}{3}$

35. $\text{Average} = \frac{4}{\pi}\int_{0}^{\pi/2}\int_{0}^{1}\sqrt{1-r^2}\,r\,dr\,d\theta = -\frac{4}{3\pi}\int_{0}^{\pi/2} d\theta = \frac{2}{3}$

37. $\text{Average} = \frac{1}{a^2\pi}\int_{0}^{2\pi}\int_{0}^{a}\left((r\cos\theta - h)^2 + r^2\sin^2\theta\right) r\,dr\,d\theta = \frac{1}{\pi}\int_{0}^{2\pi}\frac{a^2}{4} - \frac{2ah\cos\theta}{3} + \frac{h^2}{2}\,d\theta =$

$\frac{1}{\pi}\left[\frac{a^2\theta}{4} - \frac{2ah\sin\theta}{3} + \frac{h^2\theta}{2}\right]_0^{2\pi} = \frac{a^2 + 2h^2}{2}$

13.4 TRIPLE INTEGRALS IN RECTANGULAR COORDINATES

1. $\int_{0}^{1}\int_{0}^{1-z}\int_{0}^{2} dx\,dy\,dz = 2\int_{0}^{1}\int_{0}^{1-z} dy\,dz = 2\int_{0}^{1} 1 - z\,dz = 1$

3. $\int_{0}^{1}\int_{0}^{2-2x}\int_{0}^{3-3x-3y/2} dz\,dy\,dx = \int_{0}^{1}\int_{0}^{2-2x} 3 - 3x - \frac{3}{2}y\,dy\,dx = \int_{0}^{1} 3 - 6x + 3x^2\,dx = 1,$

$\int_{0}^{2}\int_{0}^{1-y/2}\int_{0}^{3-3x-3y/2} dz\,dx\,dy, \quad \int_{0}^{1}\int_{0}^{3-3x}\int_{0}^{2-2x-2z/3} dy\,dz\,dx,$

$\int_{0}^{3}\int_{0}^{1-z/3}\int_{0}^{2-2x-2z/3} dy\,dx\,dz, \quad \int_{0}^{2}\int_{0}^{3-3y/2}\int_{0}^{1-y/2-z/3} dx\,dz\,dy,$

$\int_{0}^{3}\int_{0}^{2-2z/3}\int_{0}^{1-y/2-z/3} dx\,dy\,dz$

5. $\int_{0}^{1}\int_{0}^{1}\int_{0}^{1} x^2 + y^2 + z^2\,dz\,dy\,dx = \int_{0}^{1}\int_{0}^{1}\left(x^2 + y^2 + \frac{1}{3}\right) dy\,dx = \int_{0}^{1} x^2 + \frac{2}{3}\,dx = 1$

7. $\int_{1}^{e}\int_{1}^{e}\int_{1}^{e}\frac{1}{xyz}\,dx\,dy\,dz = \int_{1}^{e}\int_{1}^{e}\frac{\ln x}{yz}\,dy\,dz = \int_{1}^{e}\frac{1}{z}\,dz = 1$

9. $\int_{0}^{1}\int_{0}^{\pi}\int_{0}^{\pi} y\sin z\,dx\,dy\,dz = \int_{0}^{1}\int_{0}^{\pi}\pi y\sin z\,dy\,dz = \frac{\pi^3}{2}\int_{0}^{1}\sin z\,dz = \frac{\pi^3}{2}(1 - \cos 1)$

11. $\int_0^3 \int_0^{\sqrt{9-x^2}} \int_0^{\sqrt{9-x^2}} dz\, dy\, dx = \int_0^3 \int_0^{\sqrt{9-x^2}} \sqrt{9-x^2}\, dy\, dx = \int_0^3 \left(9-x^2\right) dx = 18$

13. $\int_0^1 \int_0^{2-x} \int_0^{2-x-y} dz\, dy\, dx = \int_0^1 \int_0^{2-x} 2-x-y\, dy\, dx = \int_0^1 \frac{x^2}{2} - 2x + 2\, dx = \frac{7}{6}$

15. $\int_0^{\pi} \int_0^{\pi} \int_0^{\pi} \cos(u+v+w)\, du\, dv\, dw = \int_0^{\pi} \int_0^{\pi} \sin(w+v+\pi) - \sin(w+v)\, dv\, dw =$

$\int_0^{\pi} \left(-\cos(w+2\pi) + \cos(w+\pi) + \cos(w) - \cos(w+\pi)\right) dw = 0$

17. $\int_0^{\pi/4} \int_0^{\ln \sec v} \int_{-\infty}^{2t} e^x\, dx\, dt\, dv = \int_0^{\pi/4} \int_0^{\ln \sec v} e^{2t}\, dt\, dv = \int_0^{\pi/4} \frac{\sec^2 v}{2}\, dv =$

$\left[\frac{\tan v}{2}\right]_0^{\pi/4} = \frac{1}{2} - \frac{\pi}{8} = \frac{4-\pi}{8}$

19. a) $\int_{-1}^{1} \int_0^{1-x^2} \int_{x^2}^{1-z} dy\, dz\, dx$ b) $\int_0^1 \int_{-\sqrt{1-z}}^{\sqrt{1-z}} \int_{x^2}^{1-z} dy\, dx\, dz$

c) $\int_0^1 \int_0^{1-z} \int_{-\sqrt{y}}^{\sqrt{y}} dx\, dy\, dz$ d) $\int_0^1 \int_0^{1-y} \int_{-\sqrt{y}}^{\sqrt{y}} dx\, dz\, dy$

e) $\int_0^1 \int_{-\sqrt{y}}^{\sqrt{y}} \int_0^{1-y} dz\, dx\, dy$

21. $\int_0^4 \int_0^1 \int_{2y}^2 \frac{4\cos x^2}{2\sqrt{z}}\, dx\, dy\, dz = \int_0^4 \int_0^2 \int_0^{x/2} \frac{4\cos x^2}{2\sqrt{z}}\, dy\, dx\, dz = \frac{1}{2}\int_0^4 \int_0^2 \frac{(\cos x^2)(x)}{\sqrt{z}}\, dx\, dz =$

$\frac{1}{2}\int_0^4 (\sin 4)\, z^{-1/2}\, dz = \left[(\sin 4) z^{1/2}\right]_0^4 = 2\sin 4$

23. $\int_0^1 \int_{\sqrt[3]{z}}^1 \int_0^{\ln 3} \frac{\pi e^{2x} \sin(\pi y^2)}{y^2}\, dx\, dy\, dz = \int_0^1 \int_{\sqrt[3]{z}}^1 \frac{4\pi \sin\left(\pi y^2\right)}{y^2}\, dy\, dz =$

$\int_0^1 \int_0^{y^3} \frac{4\pi \sin\left(\pi y^2\right)}{y^2}\, dz\, dy = 2\int_0^1 \sin(\pi y^2)(2\pi y)\, dy = 2\left[-\cos(\pi y^2)\right]_0^1 = 4$

25. average $= \frac{1}{8}\int_0^2 \int_0^2 \int_0^2 x^2 + 9\, dz\, dy\, dx = \frac{1}{8}\int_0^2 \int_0^2 \left(2x^2 + 18\right) dy\, dx = \frac{1}{8}\int_0^2 \left(4x^2 + 36\right) dx = \frac{31}{3}$

27. average $= \int_0^1 \int_0^1 \int_0^1 x^2 + y^2 + z^2\, dz\, dy\, dx = \int_0^1 \int_0^1 \left(x^2 + y^2 + \frac{1}{3}\right) dy\, dx = \int_0^1 \left(x^2 + \frac{2}{3}\right) dx = 1$

29. $V = \int_0^1 \int_{-1}^1 \int_0^{y^2} dz\, dy\, dx = \int_0^1 \int_{-1}^1 y^2\, dy\, dx = \frac{2}{3}\int_0^1 dx = \frac{2}{3}$

31. $V = \int_0^4 \int_0^{\sqrt{4-x}} \int_0^{2-y} dz\,dy\,dx = \int_0^4 \int_0^{\sqrt{4-x}} (2-y)\,dy\,dx = \int_0^4 2\sqrt{4-x} - \left(\frac{4-x}{2}\right) dx = \frac{20}{3}$

33. $V = \int_0^2 \int_0^{4-x^2} \int_0^{4-x^2-y} dz\,dy\,dx = \int_0^2 \int_0^{4-x^2} 4 - x^2 - y\,dy\,dx = \int_0^2 8 - 4x^2 + \frac{x^4}{2}\,dx = \frac{128}{15}$

35. $V = 8 \int_0^1 \int_0^{\sqrt{1-x^2}} \int_0^{\sqrt{1-x^2}} dz\,dy\,dx = 8 \int_0^1 \int_0^{\sqrt{1-x^2}} \sqrt{1-x^2}\,dy\,dx = 8\int_0^1 1 - x^2\,dx = \frac{16}{3}$

37. $\int_0^2 \int_0^{2-x} \int_{(2-x-y)/2}^{4-2x-2y} dz\,dy\,dx = \int_0^2 \int_0^{2-x} 3 - \frac{3x}{2} - \frac{3y}{2}\,dy\,dz =$

$\int_0^2 6 - 6x + \frac{3x^2}{2} - \frac{3(2-x)^2}{4}\,dz = \left[6x - 3x^2 + \frac{x^3}{3} + \frac{(2-x)^3}{4}\right]_0^2 = 2$

39. $V = \int_{-2}^2 \int_{-\sqrt{4-x^2}}^{\sqrt{4-x^2}} \int_0^{3-x} dz\,dy\,dx = \int_{-2}^2 \int_{-\sqrt{4-x^2}}^{\sqrt{4-x^2}} (3-x)\,dy\,dx =$

$\int_{-2}^2 (3-x)\left(2\sqrt{4-x^2}\right) dx = 12\pi$

41. $V = 2\int_0^1 \int_0^{1-y^2} \int_0^{x^2+y^2} dz\,dx\,dy = 2\int_0^1 \int_0^{1-y^2} x^2 + y^2\,dx\,dy = \frac{2}{3}\int_0^1 1 - y^6\,dy = \frac{4}{7}$

43. $\int_0^1 \int_0^{4-a-x^2} \int_a^{4-x^2-y} dz\,dy\,dx = \frac{4}{15} \Rightarrow \int_0^1 \int_0^{4-a-x^2} 4 - x^2 - y - a\,dy\,dx = \frac{4}{15} \Rightarrow$

$\int_0^1 4(4-a-x^2) - a(4-a-x^2) - x^2(4-a-x^2) - \frac{(4-a-x^2)^2}{2}\,dx = \frac{4}{15} \Rightarrow$

$\frac{15a^2 - 2110a + 203}{30} = \frac{4}{15} \Rightarrow a = 3 \text{ or } \frac{13}{3}$

45. The integral is maximized when D is $\frac{x^2}{18} + \frac{y^2}{18} + \frac{z^2}{9} < 1$ since, $18 - x^2 - y^2 - 2z^2 > 0 \Rightarrow$

$x^2 + y^2 + 2z^2 < 18 \Rightarrow \frac{x^2}{18} + \frac{y^2}{18} + \frac{z^2}{9} < 1.$

13.5 MASSES AND MOMENTS IN THREE DIMENSIONS

1. $I_x = \int_{-c/2}^{c/2}\int_{-b/2}^{b/2}\int_{-a/2}^{a/2}\left(y^2+z^2\right)dx\,dy\,dz = 4a\int_0^{c/2}\int_0^{b/2}\left(y^2+z^2\right)dy\,dz =$
$4a\int_0^{c/2}\left(\frac{b^3}{24}+\frac{z^2b}{2}\right)dx = \frac{abc}{12}\left(b^2+c^2\right) \Rightarrow I_x = \frac{M}{12}\left(b^2+c^2\right);\ R_x = \sqrt{\frac{b^2+c^2}{12}},$
$R_y = \sqrt{\frac{a^2+c^2}{12}},\ R_z = \sqrt{\frac{a^2+b^2}{12}}$

3. $I_x = \int_0^a\int_0^b\int_0^c\left(y^2+z^2\right)dz\,dy\,dx = \int_0^a\int_0^b cy^2+\frac{c^3}{3}dy\,dx = \int_0^a\frac{cb^3}{3}+\frac{c^3b}{3}dx = \frac{abc\left(b^2+c^2\right)}{3} =$
$\frac{M}{3}\left(b^2+c^2\right);\ I_y = \frac{M}{3}\left(a^2+c^2\right)$ and $I_z = \frac{M}{3}\left(a^2+b^2\right)$ by symmetry, where $M = abc$

5. $M = 4\int_0^1\int_0^1\int_{4y^2}^4 dz\,dy\,dx = 4\int_0^1\int_0^1 4-4y^2\,dy\,dx = 16\int_0^1\frac{2}{3}dx = \frac{32}{3}$
$M_{xy} = 4\int_0^1\int_0^1\int_{4y^2}^4 z\,dz\,dy\,dx = 2\int_0^1\int_0^1\left(16-16y^4\right)dy\,dx = \frac{128}{5}\int_0^1 dx = \frac{128}{5} \Rightarrow$
$\overline{z} = \frac{12}{5},\ \overline{x} = \overline{y} = 0$ by symmetry; $I_z = 4\int_0^1\int_0^1\int_{4y^2}^4\left(x^2+y^2\right)dz\,dy\,dx =$
$16\int_0^1\int_0^1 x^2-x^2y^2+y^2-y^4\,dy\,dx = 16\int_0^1\frac{2x^2}{3}+\frac{2}{15}dx = \frac{256}{45};$
$I_x = 4\int_0^1\int_0^1\int_{4y^2}^4\left(y^2+z^2\right)dz\,dy\,dx = 4\int_0^1\int_0^1\left(4y^2+\frac{64}{3}\right)-\left(4y^4+\frac{64y^6}{3}\right)dy\,dx =$
$4\int_0^1\frac{1976}{105}dx = \frac{7904}{105};\ I_y = 4\int_0^1\int_0^1\int_{4y^2}^4\left(x^2+z^2\right)dz\,dy\,dx =$
$4\int_0^1\int_0^1\left(4x^2+\frac{64}{3}\right)-\left(4x^2y^2+\frac{64y^6}{3}\right)dy\,dx = 4\int_0^1\frac{8}{3}x^2+\frac{128}{7}dx = \frac{4832}{63}$

7. a) $M = 4\int_0^2\int_0^{\sqrt{4-x^2}}\int_{x^2+y^2}^4 dz\,dy\,dx = 4\int_0^{\pi/2}\int_0^2\int_{r^2}^4 r\,dz\,dr\,d\theta =$
$4\int_0^{\pi/2}\int_0^2 4r-r^3\,dr\,d\theta = 4\int_0^{\pi/2}4\,d\theta = 8\pi;\ M_{xy} = \int_0^{2\pi}\int_0^2\int_{r^2}^4 r\,dz\,dr\,d\theta =$
$\int_0^{2\pi}\int_0^2\frac{r}{2}\left(16-r^4\right)dr\,d\theta = \frac{32\pi}{3}\int_0^{2\pi}d\theta = \frac{64\pi}{3} \Rightarrow \overline{z} = \frac{8}{3},\ \overline{x} = \overline{y} = 0$ by symmetry

b) $M = 8\pi;\ 4\pi = \int_0^{2\pi}\int_0^{\sqrt{c}}\int_{r^2}^{c} r\,dz\,dr\,d\theta = \int_0^{2\pi}\int_0^{\sqrt{c}} cr - r^3\,dr\,d\theta = \int_0^{2\pi}\left(\frac{c^2}{4}\right)d\theta = \frac{c^2\pi}{2} \Rightarrow$

$c^2 = 8 \Rightarrow c = 2\sqrt{2}$, since $c > 0$

9. $I_L = \int_{-2}^{2}\int_{-2}^{4}\int_{-1}^{1-y/2}\left((y-6)^2 + z^2\right)dz\,dy\,dx = \int_{-2}^{2}\int_{-2}^{4}\frac{(y-6)^2(4-y)}{2} + \frac{(2-y)^3}{24} + \frac{1}{3}\,dy\,dx =$

$4\int_{-2}^{4}\frac{13t^3}{24} + 5t^2 + 16t + \frac{49}{3}\,dt = 1386$, where $t = 2 - y$; $M = 36$, $R_L = \sqrt{\frac{I_L}{M}} = \sqrt{\frac{77}{2}}$

11. $M = 8,\ I_L = \int_0^4\int_0^2\int_0^1\left(z^2 + (y-2)^2\right)dz\,dy\,dx = \int_0^4\int_0^2 y^2 - 4y + \frac{13}{3}\,dy\,dx = \frac{10}{3}\int_0^4 dx = \frac{40}{3} \Rightarrow$

$R_L = \sqrt{\frac{I_L}{M}} = \sqrt{\frac{5}{3}}$

13. $M = \int_0^2\int_0^{2-x}\int_0^{2-x-y} 2x\,dz\,dy\,dx = \int_0^2\int_0^{2-x} 4x - 2x^2 - 2xy\,dy\,dx = \int_0^2 x^3 - 4x^2 + 4x\,dx = \frac{4}{3}$

$M_{xy} = \int_0^2\int_0^{2-x}\int_0^{2-x-y} 2xz\,dz\,dy\,dx = \int_0^2\int_0^{2-x} x(2-x-y)^2\,dy\,dx = \int_0^2 \frac{x(2-x)^3}{3}\,dx = \frac{8}{15};$

$M_{xz} = \frac{8}{15}$ by symmetry; $M_{yz} = \int_0^2\int_0^{2-x}\int_0^{2-x-y} 2x^2\,dz\,dy\,dx = \int_0^2\int_0^{2-x} 2x^2(2-x-y)\,dy\,dx =$

$\int_0^2\left(2x - x^2\right)^2 dx = \frac{16}{15} \Rightarrow \overline{x} = \frac{4}{5},\ \overline{y} = \overline{z} = \frac{2}{5}$

15. $M = \int_0^1\int_0^1\int_0^1 (x + y + z + 1)\,dz\,dy\,dx = \int_0^1\int_0^1\left(x + y + \frac{3}{2}\right)dy\,dx = \int_0^1 (x + 2)\,dx = \frac{5}{2}$

$M_{xy} = \int_0^1\int_0^1\int_0^1 (x + y + z + 1)z\,dz\,dy\,dx = \frac{1}{2}\int_0^1\int_0^1\left(x + y + \frac{5}{3}\right)dy\,dx =$

$\frac{1}{2}\int_0^1\left(x + \frac{13}{6}\right)dx = \frac{4}{3} \Rightarrow M_{xy} = M_{yz} = M_{xz} = \frac{4}{3}$ by symmetry $\therefore\ \overline{x} = \overline{y} = \overline{z} = \frac{8}{15}$

$I_z = \int_0^1\int_0^1\int_0^1 (x + y + z + 1)\left(x^2 + y^2\right)dz\,dy\,dx = \int_0^1\int_0^1\left(x + y + \frac{3}{2}\right)\left(x^2 + y^2\right)dy\,dx =$

$\int_0^1 x^3 + 2x^2 + \frac{1}{3}x + \frac{3}{4}\,dx = \frac{11}{6} \Rightarrow I_x = I_y = I_z = \frac{11}{6}$ by symmetry $\Rightarrow R_x = R_y = R_z = \sqrt{\frac{I_z}{M}} = \sqrt{\frac{11}{15}}$

17. a) $\overline{x} = \frac{M_{yz}}{M} = 0 \Rightarrow \iiint_R x\,\delta(x,y,z)\,dx\,dy\,dz = 0 \Rightarrow M_{yz} = 0$

b) $I_L = \iiint_R |\mathbf{v} - h\mathbf{i}|^2\,dm = \iiint_R |(x-h)\mathbf{i} + y\mathbf{j}|^2\,dm = \iiint_R x^2 - 2xh + h^2 + y^2\,dm =$

$\iiint_R x^2 + y^2\,dm - 2h\iiint_R x\,dm + h^2\iiint_R dm = I_x - 0 + h^2 m = I_{c..m.} + h^2 m$

19. a) The center of mass is at (a/2,b/2,c/2). $I_z = I_{c.m.} + abc\left(\frac{\sqrt{a^2+b^2}}{2}\right)^2 \Rightarrow I_{c.m.} = I_z - \frac{abc(a^2+b^2)}{4} =$

$\frac{abc(a^2+b^2)}{3} - \frac{abc(a^2+b^2)}{4} = \frac{abc(a^2+b^2)}{12}$, $R_{c.m.} = \sqrt{\frac{I_{c.m.}}{M}} = \frac{\sqrt{a^2+b^2}}{2\sqrt{3}}$

b) $I_L = I_{c.m.} + abc\left(\frac{a^2+9b^2}{4}\right) = \frac{abc(a^2+b^2)}{12} + abc\left(\frac{a^2+b^2}{4}\right) = \frac{abc(a^2+7b^2)}{3}$,

$R_{2b} = \sqrt{\frac{I_L}{M}} = \frac{\sqrt{a^2+7b^2}}{\sqrt{3}}$

21. $M_{xz_{B_1 \cup B_2}} = \iiint y\,dV_1 + \iiint y\,dV_2 = M_{(xz)_1} + M_{(xz)_2} \Rightarrow \overline{x} = \frac{M_{(xz)_1} + M_{(xz)_2}}{m_1 + m_2}$ and

likewise $\overline{y} = \frac{M_{(xz)_1} + M_{(xz)_2}}{m_1 + m_2}$ and $\overline{z} = \frac{M_{(xy)_1} + M_{(xy)_2}}{m_1 + m_2}$. $\therefore \mathbf{C} = \overline{x}\,\mathbf{i} + \overline{y}\,\mathbf{j} + \overline{z}\,\mathbf{k} =$

$\frac{1}{m_1 + m_2}\left[\left(M_{(xz)_1} + M_{(xz)_2}\right)\mathbf{i} + \left(M_{(yz)_1} + M_{(yz)_2}\right)\mathbf{j} + \left(M_{(xy)_1} + M_{(xy)_2}\right)\mathbf{k}\right] =$

$\frac{1}{m_1 + m_2}\left[\left(m_1\overline{x}_1 + m_2\overline{x}_2\right)\mathbf{i} + \left(m_1\overline{y}_1 + m_2\overline{y}_2\right)\mathbf{j} + \left(m_1\overline{z}_1 + m_2\overline{z}_2\right)\mathbf{k}\right] =$

$\frac{1}{m_1 + m_2}\left[m_1\left(\overline{x}_1\mathbf{i} + \overline{y}_1\mathbf{j} + \overline{z}_1\mathbf{k}\right) + m_2\left(\overline{x}_2\mathbf{i} + \overline{y}_2\mathbf{j} + \overline{z}_2\mathbf{k}\right)\right] = \frac{m_1\mathbf{C}_1 + m_2\mathbf{C}_2}{m_1 + m_2}$

23. $\mathbf{C} = \frac{\frac{\pi a^2 h}{3}\left(\frac{h}{4}\mathbf{k}\right) + \frac{2\pi a^3}{3}\left(-\frac{3a}{8}\mathbf{k}\right)}{m_1 + m_2} = \frac{\frac{a^2\pi}{3}\left(\frac{h^2 - 3a^2}{4}\mathbf{k}\right)}{m_1 + m_2}$, where $m_1 = \frac{\pi a^2 h}{3}$ and $m_2 = \frac{2\pi a^3}{3}$.

If $\frac{h^2 - 3a^2}{4} = 0 \Rightarrow h = a\sqrt{3}$, then the centroid is on the common base.

13.5 TRIPLE INTEGRALS IN CYLINDRICAL AND SPHERICAL COORDINATES

1. $\int_0^{2\pi}\int_0^1\int_r^{\sqrt{2-r^2}} r\,dz\,dr\,d\theta = \int_0^{2\pi}\int_0^1 \left(\left(2-r^2\right)^{1/2}r - r^2\right) dr\,d\theta =$

$\int_0^{2\pi}\left(\frac{2^{3/2}}{3}-\frac{2}{3}\right)d\theta = \frac{4\pi\left(\sqrt{2}-1\right)}{3}$

3. $\int_0^{2\pi}\int_0^{\theta/2\pi}\int_0^{3+24r^2} r\,dz\,dr\,d\theta = \int_0^{2\pi}\int_0^{\theta/2\pi}\left(3+24r^2\right) r\,dr\,d\theta = \frac{3}{2}\int_0^{2\pi}\frac{\theta^2}{4\pi^2}+\frac{4\theta^4}{16\pi^4}\,d\theta = \frac{17\pi}{5}$

5. $\int_0^{2\pi}\int_0^1\int_r^{\left(2-r^2\right)^{-1/2}} 3\,r\,dz\,dr\,d\theta = 3\int_0^{2\pi}\int_0^1 \left(2-r^2\right)^{-1/2}r - r^2\,dr\,d\theta =$

$3\int_0^{2\pi}\left(\sqrt{2}-\frac{4}{3}\right)d\theta = \pi\left(6\sqrt{2}-8\right)$

7. $\int_0^{\pi}\int_0^{\pi}\int_0^{2\sin\phi} \rho^2\sin\phi\,d\rho\,d\phi\,d\theta = \frac{8}{3}\int_0^{\pi}\int_0^{\pi}\sin^4\phi\;d\phi\,d\theta = \frac{2}{3}\int_0^{\pi}\frac{3\pi}{2}d\theta = \pi^2$

9. $\int_0^{2\pi}\int_0^{\pi}\int_0^{(1-\cos\phi)/2} \rho^2\sin\phi\,d\rho\,d\phi\,d\theta = \frac{1}{24}\int_0^{2\pi}\int_0^{\pi}\left(1-\cos\phi\right)^3\sin\phi\;d\phi\,d\theta =$

$\frac{1}{6}\int_0^{2\pi}d\theta = \frac{\pi}{3}$

11. $\int_0^{2\pi}\int_0^{\pi/3}\int_{\sec\phi}^{2} 3\rho^2\sin\phi\,d\rho\,d\phi\,d\theta = \int_0^{2\pi}\int_0^{\pi/3}\left(8-\sec^3\phi\right)\sin\phi\;d\phi\,d\theta = \frac{5}{2}\int_0^{2\pi}d\theta = 5\pi$

13. $\int_0^{2\pi}\int_0^{3}\int_0^{z/3} r^3\,dr\,dz\,d\theta = \int_0^{2\pi}\int_0^{3}\frac{z^4}{324}dz\,d\theta = \int_0^{2\pi}\frac{3}{20}d\theta = \frac{3\pi}{10}$

15. $\int_0^1\int_0^{\sqrt{z}}\int_0^{2\pi}(r^2\cos^2\theta + z^2)\,r\,d\theta\,dr\,dz = \int_0^1\int_0^{2\pi}\pi r^3 + 2\pi r z^2\,dr\,dz =$

$\int_0^1\frac{\pi z^2}{4}+\pi z^3\,dz = \left[\frac{\pi z^3}{12}+\frac{\pi z^4}{4}\right]_0^1 = \frac{\pi}{3}$

17. $\int_0^2\int_{-\pi}^{0}\int_{\pi/4}^{\pi/2}\rho^3\sin 2\phi\,d\phi\,d\theta\,d\rho = \int_0^2\int_{-\pi}^{0}\frac{\rho^3}{2}d\theta\,d\rho = \int_0^2\frac{\rho^3\pi}{2}d\rho = \left[\frac{\pi\rho^4}{8}\right]_0^2 = 2\pi$

19. $\int_0^1\int_0^{\pi}\int_0^{\pi/4}12\rho\sin^3\phi\,d\phi\,d\theta\,d\rho = \int_0^1\int_0^{\pi}8\rho-\frac{10\rho}{\sqrt{2}}\,d\theta\,d\rho = \pi\int_0^1 8\rho-\frac{10\rho}{\sqrt{2}}\,d\rho =$

$\pi\left[4\rho^2-\frac{5\rho^2}{\sqrt{2}}\right]_0^1 = \frac{(4\sqrt{2}-5)\pi}{\sqrt{2}}$

21. a) $8\int_0^{\pi/2}\int_0^{\pi/2}\int_0^2 \rho^2 \sin\phi \, d\rho \, d\phi \, d\theta$

b) $8\int_0^{\pi/2}\int_0^2\int_0^{\sqrt{4-r^2}} r \, dz \, dr \, d\theta$

c) $8\int_0^2\int_0^{\sqrt{4-x^2}}\int_0^{\sqrt{4-x^2-y^2}} dz \, dy \, dx$

23. $\int_{-\pi/2}^{\pi/2}\int_0^{\cos\theta}\int_0^{3r^2} f(r,\theta,z) \, r \, dz \, dr \, d\theta$

25. a) $V = \int_0^{2\pi}\int_0^{\pi/3}\int_{\sec\phi}^2 \rho^2 \sin\phi \, d\rho \, d\phi \, d\theta$

b) $V = \int_0^{2\pi}\int_0^{\sqrt{3}}\int_1^{\sqrt{4-r^2}} r \, dz \, dr \, d\theta$

c) $V = \int_{-\sqrt{3}}^{\sqrt{3}}\int_{-\sqrt{3-x^2}}^{\sqrt{3-x^2}}\int_1^{\sqrt{4-x^2-y^2}} dz \, dy \, dx$

d) $V = \int_0^{2\pi}\int_1^{\sqrt{3}} (4-r^2)^{1/2} r - r \, dr \, d\theta = \int_0^{2\pi}\left[(4-r^2)^{3/2} - \frac{r^2}{2}\right]_1^{\sqrt{3}} d\theta = \frac{5}{6}\int_0^{2\pi} d\theta = \frac{5\pi}{3}$

27. $V = 4\int_0^{\pi/2}\int_0^1\int_0^{r^2} r \, dz \, dr \, d\theta = 4\int_0^{\pi/2}\int_0^1 r^3 \, dr \, d\theta = \int_0^{\pi/4} d\theta = \frac{\pi}{2}$

29. $V = 4\int_0^{\pi/2}\int_0^2\int_0^{4-r^2} r \, dz \, dr \, d\theta = 4\int_0^{\pi/2}\int_0^2 \left(4r - r^3\right) dr \, d\theta = 16\int_0^{\pi/2} d\theta = 8\pi$

31. $V = 4\int_0^{\pi/2}\int_0^1\int_{4r^2}^{5-r^2} r \, dz \, dr \, d\theta = 4\int_0^{\pi/2}\int_0^1 \left(5 - 5r^2\right) r \, dr \, d\theta = 5\int_0^{\pi/2} d\theta = \frac{5\pi}{2}$

33. $V = 8\int_0^{\pi/2}\int_0^1\int_0^{\sqrt{4-r^2}} r \, dz \, dr \, d\theta = 8\int_0^{\pi/2}\int_0^1 \left(4-r^2\right)^{1/2} r \, dr \, d\theta =$

$-\frac{8}{3}\int_0^{\pi/2} \left(3^{3/2} - 8\right) d\theta = \frac{4\pi\left(8 - 3\sqrt{3}\right)}{3}$

35. $\text{average} = \frac{1}{2\pi}\int_0^{2\pi}\int_0^1\int_{-1}^1 r^2 \, dz \, dr \, d\theta = \frac{1}{2\pi}\int_0^{2\pi}\int_0^1 2r^2 \, dr \, d\theta = \frac{1}{3\pi}\int_0^{2\pi} d\theta = \frac{2}{3}$

37. $M = 4\int_0^{\pi/2}\int_0^1\int_0^r r\,dz\,dr\,d\theta = 4\int_0^{\pi/2}\int_0^1 r^2\,dr\,d\theta = \frac{4}{3}\int_0^{\pi/2} d\theta = \frac{2\pi}{3}$

$M_{xy} = \int_0^{2\pi}\int_0^1\int_0^r r\,z\,dz\,dr\,d\theta = \frac{1}{2}\int_0^{2\pi}\int_0^1 r^3\,dr\,d\theta = \frac{1}{8}\int_0^{2\pi} d\theta = \frac{\pi}{4} \Rightarrow$

$\overline{z} = \frac{3}{8}$, $\overline{x} = \overline{y} = 0$ by symmetry

39. $M = 12\pi$, $I_z = \int_0^{2\pi}\int_1^2\int_0^4 r^3\,dz\,dr\,d\theta = 4\int_0^{2\pi}\int_1^2 r^3\,dr\,d\theta = 15\int_0^{2\pi} d\theta = 30\pi \Rightarrow$

$R_z = \sqrt{\frac{I_z}{M}} = \sqrt{\frac{5}{2}}$

41. a) $I_z = \int_0^{2\pi}\int_0^1\int_{-1}^1 r^3\,dz\,dr\,d\theta = 2\int_0^{2\pi}\int_0^1 r^3\,dr\,d\theta = \frac{1}{2}\int_0^{2\pi} d\theta = \pi$

b) $I_x = \int_0^{2\pi}\int_0^1\int_{-1}^1 \left(r^2\sin^2\theta + z^2\right) r\,dz\,dr\,d\theta = \int_0^{2\pi}\int_0^1 \left(2r^2\sin^2\theta + \frac{2}{3}\right) r\,dr\,d\theta =$

$\int_0^{2\pi}\left(\frac{\sin^2\theta}{2} + \frac{1}{3}\right)d\theta = \frac{7\pi}{6}$

43. $I_z = \int_0^{2\pi}\int_0^a\int_{-\sqrt{a^2-r^2}}^{\sqrt{a^2-r^2}} r^3\,dz\,dr\,d\theta = -\int_0^{2\pi}\int_0^a r^2\left(a^2 - r^2\right)^{1/2}(-2r)\,dr\,d\theta = \int_0^{2\pi}\frac{4a^5}{15}\,d\theta = \frac{8\pi a^5}{15}$

45. $M = \frac{2a^2h\pi}{3}$, $M_{xy} = \int_0^{2\pi}\int_0^a\int_0^{(h\sqrt{a^2-r^2})/a} z\,r\,dz\,dr\,d\theta = \frac{h^2}{2a^2}\int_0^{2\pi}\int_0^a a^2r - r^3\,dr\,d\theta =$

$\frac{h^2}{2a^2}\int_0^{2\pi}\frac{a^4}{2} - \frac{a^4}{4}\,d\theta = \frac{a^2h^2\pi}{4} \Rightarrow \overline{z} = \frac{\pi a^2h^2}{4}\cdot\frac{3}{2a^2h\pi} = \frac{3}{8}h \therefore \overline{x} = \overline{y} = 0$, by symmetry,

and $\overline{z} = \frac{3}{8}h$

47. Place the cone's axis of symmetry along the z–axis with the vertex at the origin. $M = \frac{\pi r^2 h}{3}$,

$M_{xy} = \int_0^{2\pi}\int_0^{ah/c}\int_{cr/a}^h z\,r\,dz\,dr\,d\theta = \frac{1}{2}\int_0^{2\pi}\int_0^{ah/c} h^2r - \frac{c^2r^3}{a^2}\,dr\,d\theta = \frac{1}{2}\int_0^{2\pi}\frac{h^4a^2}{2c^2} - \frac{a^2h^4}{4c^2}\,d\theta =$

$\frac{a^2h^4\pi}{4c^2} = \frac{r^2h^2\pi}{4}$, since $r = \frac{ah}{c} \Rightarrow \overline{z} = \frac{r^2h^2\pi}{4}\cdot\frac{3}{\pi r^2 h} = \frac{3}{4}h$ and $\overline{x} = \overline{y} = 0$ by symmetry $\therefore$ the centroid is one fourth of the way from the base to the vertex.

49. a) $M = \int_0^{2\pi}\int_0^1\int_r^1 z\,r\,dz\,dr\,d\theta = \frac{1}{2}\int_0^{2\pi}\int_0^1 r - r^3\,dr\,d\theta = \frac{1}{8}\int_0^{2\pi} d\theta = \frac{\pi}{4}$

$M_{xy} = \int_0^{2\pi}\int_0^1\int_r^1 z^2\,r\,dz\,dr\,d\theta = \frac{1}{3}\int_0^{2\pi}\int_0^1 r - r^4\,dr\,d\theta = \frac{1}{10}\int_0^{2\pi} d\theta = \frac{\pi}{5} \Rightarrow \overline{z} = \frac{4}{5}$,

$\overline{x} = \overline{y} = 0$ by symmetry; $I_z = \int_0^{2\pi}\int_0^1\int_r^1 z\,r^3\,dz\,dr\,d\theta = \frac{1}{2}\int_0^{2\pi}\int_0^1 r^3 - r^5\,dr\,d\theta =$

$\frac{1}{24}\int_0^{2\pi} d\theta = \frac{\pi}{12} \Rightarrow R_z = \sqrt{\frac{I_z}{M}} = \sqrt{\frac{1}{3}}$

b) $M = \int_0^{2\pi}\int_0^1\int_r^1 z^2 r\,dz\,dr\,d\theta = \frac{\pi}{5}$, see part (a) for details

$M_{xy} = \int_0^{2\pi}\int_0^1\int_r^1 z^3\,r\,dz\,dr\,d\theta = \frac{1}{4}\int_0^{2\pi}\int_0^1 r - r^5\,dr\,d\theta = \frac{1}{12}\int_0^{2\pi} d\theta = \frac{\pi}{6} \Rightarrow \overline{z} = \frac{5}{6}$,

$\overline{x} = \overline{y} = 0$ by symmetry; $I_z = \int_0^{2\pi}\int_0^1\int_r^1 z^2\,r^3\,dz\,dr\,d\theta = \frac{1}{3}\int_0^{2\pi}\int_0^1 r^3 - r^6\,dr\,d\theta =$

$\frac{1}{28}\int_0^{2\pi} d\theta = \frac{\pi}{14} \Rightarrow R_z = \sqrt{\frac{I_z}{M}} = \sqrt{\frac{5}{14}}$

51. $\int_0^{2\pi}\int_{\pi/3}^{2\pi/3} \frac{a^3}{3}\sin\phi\,d\phi\,d\theta = \frac{a^3}{3}\int_0^{2\pi} d\theta = \frac{2\pi a^3}{3} \Rightarrow V = \frac{4}{3}\pi a^3 - \frac{2\pi a^3}{3} = \frac{2\pi a^3}{3}$

53. $V = \int_0^{2\pi}\int_0^{\pi/3}\int_{\sec\phi}^2 \rho^2\sin\phi\,d\rho\,d\phi\,d\theta = \frac{1}{3}\int_0^{2\pi}\int_0^{\pi/3} 8\sin\phi - \tan\phi\sec^2\phi\,d\phi\,d\theta =$

$\frac{5}{6}\int_0^{2\pi} d\theta = \frac{5\pi}{3}$

55. $V = 8\int_0^{\pi/2}\int_1^{\sqrt{2}}\int_0^r r\,dz\,dr\,d\theta = 8\int_0^{\pi/2}\int_1^{\sqrt{2}}\int_0^r dz\,r\,dr\,d\theta = 8\int_0^{\pi/2}\int_1^{\sqrt{2}} r^2\,dr\,d\theta =$

$\frac{2\sqrt{2}-1}{3}\int_0^{\pi/2} d\theta = \frac{4\pi(2\sqrt{2}-1)}{3}$

57. average $= \frac{3}{4\pi}\int_0^{2\pi}\int_0^{\pi}\int_0^1 \rho^3\sin\phi\,d\rho\,d\phi\,d\theta = \frac{3}{16\pi}\int_0^{2\pi}\int_0^{\pi}\sin\phi\,d\phi\,d\theta = \frac{3}{8\pi}\int_0^{2\pi} d\theta = \frac{3}{4}$

59. $M = \int_0^{2\pi}\int_0^{\pi/4}\int_0^a \rho^2\sin\phi\,d\rho\,d\phi\,d\theta = \frac{a^3}{3}\int_0^{2\pi}\int_0^{\pi/4}\sin\phi\,d\phi\,d\theta =$

$\frac{a^3}{3}\int_0^{2\pi}\frac{\sqrt{2}-1}{\sqrt{2}}\,d\theta = \frac{\pi a^3\left(2-\sqrt{2}\right)}{3}$

$M_{xy} = \int_0^{2\pi}\int_0^{\pi/4}\int_0^a \rho^3\sin\phi\cos\phi\,d\rho\,d\phi\,d\theta = \frac{a^3}{4}\int_0^{2\pi}\int_0^{\pi/4}\sin\phi\cos\phi\,d\phi\,d\theta =$

$\frac{a^4}{16}\int_0^{2\pi} d\theta = \frac{\pi a^4}{8} \Rightarrow \overline{z} = \frac{3\left(2+\sqrt{2}\right)a}{16}$, $\overline{x} = \overline{y} = 0$ by symmetry

13.7 SUBSTITUTIONS IN MULTIPLE INTEGRALS

1. $\int_0^4 \int_{y/2}^{1+y/2} \frac{2x-y}{2}\, dx\, dy = \int_0^4 \left[\frac{x^2}{2} - \frac{xy}{2}\right]_{y/2}^{1+y/2} dy = \frac{1}{2}\int_0^4 dy = 2$

3. a) $x = \frac{u+v}{3}$, $y = \frac{v-2u}{3}$, $J(u,v) = \begin{vmatrix} 1/3 & 1/3 \\ -2/3 & 1/3 \end{vmatrix} = \frac{1}{9} + \frac{2}{9} = \frac{1}{3}$

b) $\int_4^7 \int_{-1}^2 \frac{vu}{3}\, dv\, du = \frac{1}{2}\int_4^7 u\, du = \frac{33}{4}$

5. $J(u,v) = \begin{vmatrix} v^{-1} & -uv^{-2} \\ v & u \end{vmatrix} = v^{-1}u + v^{-1}u = \frac{2u}{v}$; $\int_1^3 \int_1^2 (v+u)\left(\frac{2u}{v}\right) dv\, du =$

$\int_1^3 2u + (\ln 4)\, u^2\, du = 8 + \frac{26 \ln 4}{3}$

7. $J(r,\theta) = \begin{vmatrix} a\cos\theta & -ar\sin\theta \\ b\sin\theta & br\cos\theta \end{vmatrix} = abr\cos^2\theta + abr\sin^2\theta = abr$

$I_0 = \int_{-a}^{a} \int_{(-b/a)\sqrt{a^2-x^2}}^{(b/a)\sqrt{a^2-x^2}} \left(x^2+y^2\right) dy\, dx = \int_0^{2\pi}\int_0^1 r^3\left(a^2\cos^2\theta + b^2\sin^2\theta\right) abr\, dr\, d\theta =$

$\frac{ab}{4}\int_0^{2\pi} a^2\cos^2\theta + b^2\sin^2\theta\, d\theta = \frac{\pi ab\left(a^2+b^2\right)}{4}$

9. $\begin{vmatrix} \sin\phi\cos\theta & \rho\cos\phi\cos\theta & -\rho\sin\phi\sin\theta \\ \sin\phi\sin\theta & \rho\cos\phi\sin\theta & \rho\sin\phi\cos\theta \\ \cos\phi & -\rho\sin\phi & 0 \end{vmatrix} = \cos\phi \begin{vmatrix} \rho\cos\phi\cos\theta & -\rho\sin\phi\sin\theta \\ \rho\cos\phi\sin\theta & \rho\sin\phi\cos\theta \end{vmatrix} +$

$\rho\sin\phi \begin{vmatrix} \sin\phi\cos\theta & -\rho\sin\phi\sin\theta \\ \sin\phi\sin\theta & \rho\sin\phi\cos\theta \end{vmatrix} = \rho^2\cos\phi\left(\sin\phi\cos\phi\cos^2\theta + \sin\phi\cos\phi\sin^2\theta\right) +$

$\rho^2\sin\phi\left(\sin^2\phi\cos^2\theta + \sin^2\phi\sin^2\theta\right) = \rho^2\sin\phi\cos^2\phi + \rho^2\sin^3\phi =$

$\rho^2\sin\phi\left(\cos^2\phi + \sin^2\phi\right) = \rho^2\sin\phi$

11. $J(u,v,w) = \begin{vmatrix} a & 0 & 0 \\ 0 & b & 0 \\ 0 & 0 & c \end{vmatrix} = abc$. The transformation takes the $\frac{x^2}{a^2} + \frac{y^2}{b^2} + \frac{z^2}{c^2} = 1$ region in the xyz–space into the $u^2 + v^2 + w^2 = 1$ region in the uvw–space which is a unit sphere with $V = \frac{4}{3}\pi$.

$\therefore\ V = \iiint_R dx\, dy\, dz = \iiint_G abc\, du\, dv\, dw = \frac{4\pi abc}{3}$

13. $J(u,v,w) = \begin{vmatrix} 1 & 0 & 0 \\ -v/u^2 & 1/u & 0 \\ 0 & 0 & 1/3 \end{vmatrix} = \frac{1}{3u}$; $\int\int_R\int x^2y + 3xyz\ dx\ dy\ dz =$

$\int\int_G\int u^2\left(\frac{v}{u}\right) + 3u\,\frac{v}{u}\frac{w}{3}\,J(u,v,w)\ du\ dv\ dw = \frac{1}{3}\int_0^3\int_0^2\int_1^2 v + \frac{vw}{u}\,du\ dv\ dw =$

$\frac{1}{3}\int_0^3\int_0^2 v + vw\ln 2\ dv\ dw = \frac{1}{3}\int_0^3 2 + (\ln 4)w\ dw = 2 + \ln 8$

13.M MISCELLANEOUS EXERCISES

1.

a) $\int_0^4\int_{-\sqrt{4-y}}^{(y-4)/2} dx\ dy = \int_0^4 -\sqrt{4-y} + \frac{y}{2}\ dy =$

$\left[-\frac{2(4-y)^{3/2}}{3} - 2y + \frac{y^2}{4}\right]_0^4 = \frac{4}{3}$

$\int_{-2}^0\int_{2x+4}^{4-x^2} dy\ dx = \int_{-2}^0 -2x - x^2\ dx = \frac{4}{3}$

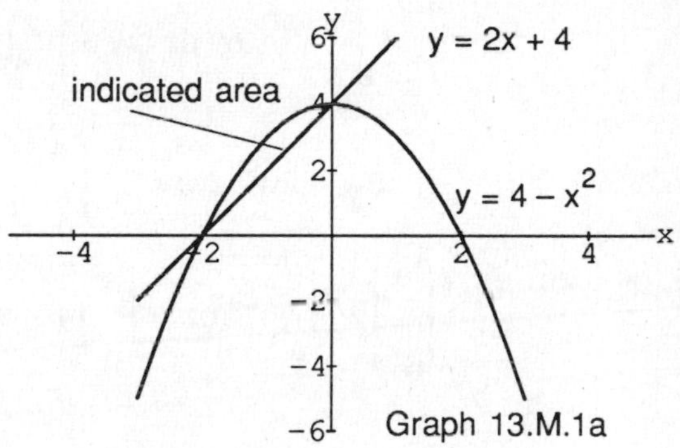

Graph 13.M.1a

b) $\int_{-1}^1\int_{u^2}^1 dv\ du = \int_{-1}^1 (1-u^2)\ du = \frac{4}{3}$

$\int_0^1\int_{-\sqrt{v}}^{\sqrt{v}} du\ dv = \int_0^1 2\sqrt{v}\ dv = \left[\frac{4}{3}v^{3/2}\right]_0^1 = \frac{4}{3}$

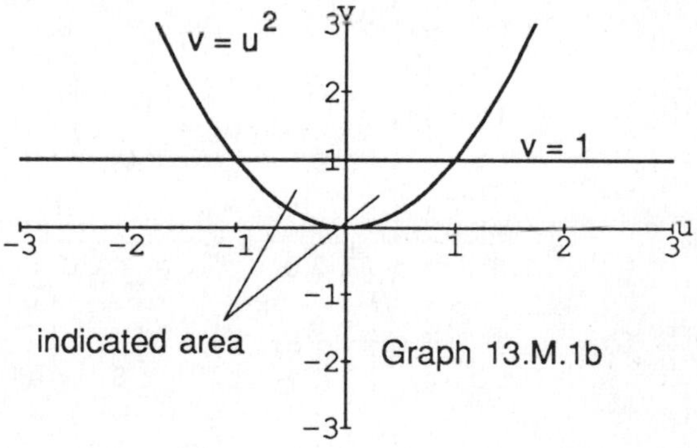

Graph 13.M.1b

c) $\int_0^{3/2}\int_{-\sqrt{9-4t^2}}^{\sqrt{9-4t^2}} t\,ds\,dt = \int_0^{3/2}\left[\,ts\,\right]_{-\sqrt{9-4t^2}}^{\sqrt{9-4t^2}}dt =$

$\int_0^{3/2} 2t\sqrt{9-4t^2}\,dt = \frac{9}{2}$

$\int_{-3}^{3}\int_0^{(9-s^2)^{1/2}/2} t\,dt\,ds = \frac{1}{2}\int_{-3}^{3}\left[t^2\right]_0^{(9-s^2)^{1/2}/2} ds =$

$\frac{1}{8}\int_{-3}^{3}\left(9-s^2\right)ds = \frac{9}{2}$

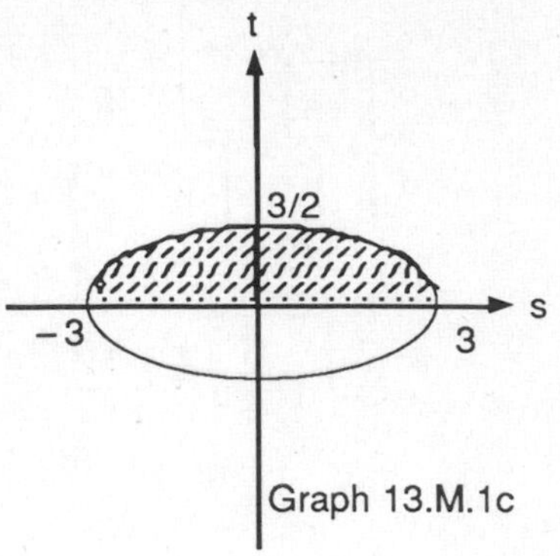

Graph 13.M.1c

d) $\int_0^2\int_0^{4-w^2} 2w\,dz\,dw = \int_0^2 [2wz]_0^{4-w^2}\,dw =$

$\int_0^2 8w - 2w^3\,dw = 8,\ \int_0^4\int_0^{\sqrt{4-z}} 2w\,dw\,dz =$

$\int_0^4 \left[w^2\right]_0^{\sqrt{4-z}} dz = \int_0^4 (4-z)\,dz = 8$

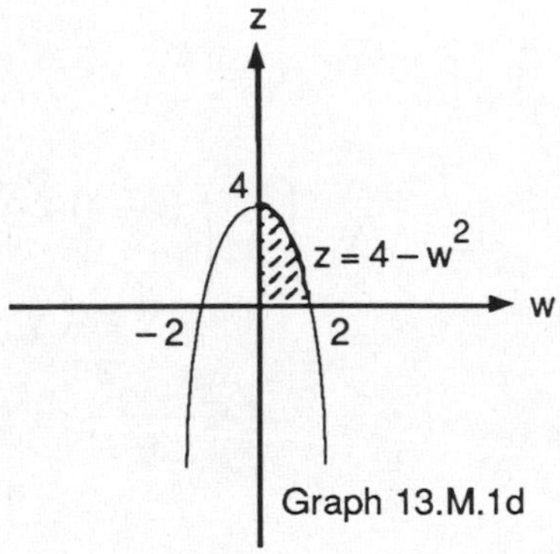

Graph 13.M.1d

3. $\int_1^{\infty}\int_0^{1/(x\sqrt{x^2-1})} 2\,dy\,dx = \lim_{b\to\infty}\int_1^b \frac{2}{x\sqrt{x^2-1}}dx = \lim_{b\to\infty} 2\left(\arccos\left(\frac{1}{b}\right) - \arccos(1)\right) = \pi$

4. $\int_0^1\int_0^y e^{-x^2}\,dx\,dy = \int_0^1\int_x^1 e^{-x^2}\,dy\,dx = \int_0^1 e^{-x^2}\,dx - \int_0^1 e^{-x^2}(x)\,dx = \frac{\sqrt{\pi}\,\text{erf}(1)}{2} + \frac{1}{2}\left[e^{-x^2}\right]_0^1 =$

$\frac{\sqrt{\pi}\,\text{erf}(1)}{2} + \frac{1-e}{2e} = \frac{1-e+e\sqrt{\pi}\,\text{erf}(1)}{2e}$

5. $\int_0^{\infty}\frac{e^{-ax}-e^{-bx}}{x}dx = \int_0^{\infty}\int_a^b e^{-xy}\,dy\,dx = \int_a^b\int_0^{\infty} e^{-xy}\,dx\,dy = \int_a^b \lim_{t\to\infty}\int_0^t e^{-xy}\,dx\,dy =$

$\int_a^b \lim_{t\to\infty}\left[-\frac{e^{-xy}}{y}\right]_0^t dy = \int_a^b \frac{1}{y}\,dy = [\ln y]_a^b = \ln\left(\frac{b}{a}\right)$

7. $$\int_0^a \int_0^b e^{\max(b^2x^2, a^2y^2)}\, dy\, dx = \int_0^a \int_0^{bx/a} e^{b^2x^2}\, dy\, dx + \int_0^b \int_0^{ay/b} e^{a^2y^2}\, dx\, dy =$$

$$\frac{1}{2ab}\int_0^a e^{b^2x^2}(2b^2x)\, dx + \frac{1}{2ab}\int_0^b e^{a^2y^2}(2a^2y)\, dy = \frac{e^{a^2b^2} - 1}{ab}$$

9. a) $$\int_0^x \int_0^u e^{m(x-t)} f(t)\, dt\, du = \int_0^x \int_t^x e^{m(x-t)} f(t)\, du\, dt = \int_0^x (x-t)\, e^{m(x-t)} f(t)\, dt$$

b) $$\int_0^x \int_0^v \int_0^u e^{m(x-t)} f(t)\, dt\, du\, dv = \int_0^x \int_t^x \int_t^v e^{m(x-t)} f(t)\, du\, dv\, dt =$$

$$\int_0^x \int_t^x (v-t)\, e^{m(x-t)} f(t)\, dv\, dt = \int_0^x \Big[\tfrac{1}{2}(v-t)^2\, e^{m(x-t)} f(t)\Big]_t^x dt =$$

$$\int_0^x \frac{(x-t)^2}{2}\, e^{m(x-t)} f(t)\, dt$$

11. a) $$V = \int_{-3}^2 \int_x^{6-x^2} x^2\, dy\, dx$$

b) $$V = \int_{-3}^2 \int_x^{6-x^2} \int_0^{x^2} dz\, dy\, dx$$

c) $$V = \int_{-3}^2 \int_x^{6-x^2} x^2\, dy\, dx = \int_{-3}^2 6x^2 - x^4 - x^3\, dx = \left[2x^3 - \frac{x^5}{5} - \frac{x^4}{4}\right]_{-3}^2 = \frac{125}{4}$$

13. $$M = \int_1^2 \int_{2/x}^2 dy\, dx = \int_1^2 2 - \frac{2}{x}\, dx = 2 - \ln 4$$

$$M_y = \int_1^2 \int_{2/x}^2 x\, dy\, dx = \int_1^2 x\left[2 - \frac{2}{x}\right] dx = 1$$

$$M_x = \int_1^2 \int_{2/x}^2 y\, dy\, dx = \int_1^2 2 - \frac{2}{x^2}\, dx = 1 \Rightarrow \overline{x} = \frac{1}{2 - \ln 4},\ \overline{y} = \frac{1}{2 - \ln 4}$$

15. $$M = \int_0^4 \int_{-2y}^{2y-y^2} dx\, dy = \int_0^4 4y - y^2\, dy = \frac{32}{3},\ M_x = \int_0^4 \int_{-2y}^{2y-y^2} y\, dx\, dy =$$

$$\int_0^4 4y^2 - y^3\, dy = \left[\frac{4y^3}{3} - \frac{y^4}{4}\right]_0^4 = \frac{64}{3},\ M_y = \int_0^4 \int_{-2y}^{2y-y^2} x\, dx\, dy =$$

$$\int_0^4 \frac{(2y-y^2)^2}{2} - 2y^2\, dy = \left[\frac{y^5}{10} - \frac{y^4}{2}\right]_0^4 = -\frac{128}{5},\ \overline{x} = \frac{M_y}{M} = -\frac{12}{5},\ \overline{y} = \frac{M_x}{M} = 2$$

17. $M = \int_{-1}^{1}\int_{-1}^{1}\left(x^2 + y^2 + \frac{1}{3}\right) dy\, dx = \int_{-1}^{1} 2x^2 + \frac{4}{3} dx = 4$

$M_x = \int_{-1}^{1}\int_{-1}^{1} y\left(x^2 + y^2 + \frac{1}{3}\right) dy\, dx = \int_{-1}^{1} 0\, dx = 0$

$M_y = \int_{-1}^{1}\int_{-1}^{1} x\left(x^2 + y^2 + \frac{1}{3}\right) dy\, dx = \int_{-1}^{1} x\left(2x^2 + \frac{4}{3}\right) dx = 0$

19. Place the ΔABC with its vertices at A(0,0), B(b,0) and C(a,h). The line through the points A and B is $y = \frac{h}{a}x$ while the line through points C and B is $y = \frac{h}{a-b}(x - b)$. M = $\int_0^h \int_{ay/h}^{(a-b)y/h + b} \delta\, dx\, dy = b\delta \int_0^h 1 - \frac{y}{h}\, dy = \frac{\delta bh}{2}$, $I_x = \int_0^h \int_{ay/h}^{(a-b)y/h + b} y^2\, \delta\, dx\, dy =$ $b\delta \int_0^h y^2 - \frac{y^3}{h}\, dy = \frac{\delta b h^3}{12}$, $R_x = \sqrt{\frac{I_x}{M}} = \frac{h}{\sqrt{6}}$

21. a) $M = 2\int_0^{\pi/2}\int_1^{1+\cos\theta} r\, dr\, d\theta = \int_0^{\pi/2} 2\cos\theta + \frac{1 + \cos 2\theta}{2}\, d\theta = \frac{8 + \pi}{4}$

$M_y = \int_{-\pi/2}^{\pi/2}\int_1^{1+\cos\theta} \cos\theta\, r^2\, dr\, d\theta = \int_{-\pi/2}^{\pi/2}\left(\cos^2\theta + \cos^3\theta + \frac{\cos^4\theta}{3}\right) d\theta = \frac{32 + 15\pi}{24} \Rightarrow$

$\overline{x} = \frac{15\pi + 32}{6\pi + 48}$, $\overline{y} = 0$ by symmetry

b)

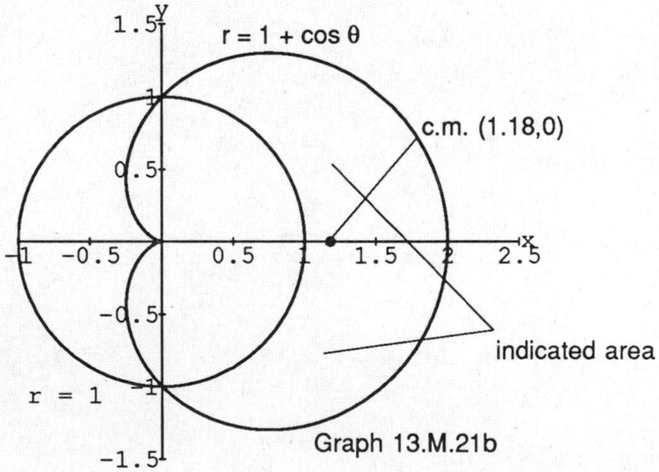

Graph 13.M.21b

23. $M = \int_0^{\pi/2}\int_1^3 r\, dr\, d\theta = 4\int_0^{\pi/2} d\theta = 2\pi$, $M_y = \int_0^{\pi/2}\int_1^3 r^2 \cos\theta\, dr\, d\theta = \frac{26}{3}\int_0^{\pi/2}\cos\theta\, d\theta = \frac{26}{3} \Rightarrow$

$\overline{x} = \frac{13}{3\pi}$, $\overline{y} = \frac{13}{3\pi}$ by symmetry

25. $A = \int_{-\theta}^{\theta} \int_{b\sec\theta}^{a} r\,dr\,d\theta = \int_{-\theta}^{\theta} \frac{a^2}{2} - \frac{b^2}{2\cos^2\theta}\,d\theta = \left[\frac{a^2\theta b^2 \sin\theta}{2\ 2\cos\theta}\right]_{-\theta}^{\theta} = a^2\theta - b^2\tan\theta =$

$a^2\cos^{-1}\left(\frac{b}{a}\right) - b\sqrt{a^2-b^2}$, $I_o = \int_{-\theta}^{\theta} \int_{b\sec\theta}^{a} r^3\,dr\,d\theta = \frac{1}{4}\int_{-\theta}^{\theta} a^4 - b^4(1+\tan^2\theta)\sec^2\theta\,d\theta =$

$\frac{1}{4}\left[a^4\theta - b^4\tan\theta - \frac{b^4\tan^3\theta}{3}\right]_{-\theta}^{\theta} = \frac{a^4\theta}{2} - \frac{b^4\tan\theta}{2} - \frac{b^4\tan^3\theta}{6} = \frac{1}{2}a^4\cos^{-1}\left(\frac{b}{a}\right) - \frac{1}{2}b^3\sqrt{a^2-b^2} -$

$\frac{1}{6}b^3(a^2-b^2)^{3/2}$

27. $\int_0^{a\sin\beta} \int_{y\cot\beta}^{\sqrt{a^2-y^2}} \ln(x^2+y^2)\,dx\,dy = \int_0^{\beta}\int_0^{a} 2\ln(r)\,r\,dr\,d\theta = \frac{1}{2}\int_0^{\beta}\left[2r^2\ln(r) - r^2\right]_0^a d\theta =$

$\frac{1}{2}\int_0^{\beta} (2a^2\ln(a) - a^2) - \lim_{b\to 0^+} 2b^2\ln(b)\,d\theta = \frac{a^2}{2}\int_0^{\beta} 2\ln(a) - 1\,d\theta = \frac{a^2\beta}{2}(2\ln(a)-1) = a^2\beta\left(\ln(a) - \frac{1}{2}\right)$

29. $Q = \int_0^{2\pi}\int_0^R kr^2(1-\sin\theta)\,dr\,d\theta = \frac{kR^3}{3}\int_0^{2\pi}(1-\sin\theta)\,d\theta = \frac{kR^3}{3}\left[\theta+\cos\theta\right]_0^{2\pi} = \frac{2\pi kR^3}{3}$

31. $V = 4\int_0^2 \int_0^{\sqrt{4-x^2}} \int_0^{4-x^2} dz\,dy\,dx = 4\int_0^2\int_0^{\sqrt{4-x^2}} 4-x^2\,dy\,dx = 4\int_0^2 \left(4-x^2\right)^{3/2} dx = 12\pi$

33. a) $\int_{-1}^{1} \int_{-\sqrt{1-x^2}}^{\sqrt{1-x^2}} \int_0^{\sqrt{4-x^2-y^2}} \sqrt{x^2+y^2}\,dz\,dy\,dx$

b) $\int_0^{2\pi}\int_0^{\pi/6}\int_0^{2} \rho^3\sin^2\phi\,d\rho\,d\phi\,d\theta + \int_0^{2\pi}\int_{\pi/6}^{\pi/2}\int_0^{\csc\phi} \rho^3\sin^2\phi\,d\rho\,d\phi\,d\theta$

35. $\int_0^1 \int_{\sqrt{1-x^2}}^{\sqrt{3-x^3}} \int_1^{\sqrt{4-x^2-y^2}} yxz^2\,dz\,dy\,dx + \int_0^{\sqrt{3}} \int_0^{\sqrt{3-x^2}} \int_1^{\sqrt{4-x^2-y^2}} yxz^2\,dz\,dy\,dx$

37. $V = \int_{-1}^{1} \int_{-\sqrt{1-x^2}}^{\sqrt{1-x^2}} \int_0^{2-x-y} dz\,dy\,dx = \int_{-1}^{1}\int_{-\sqrt{1-x^2}}^{\sqrt{1-x^2}} 2-x-y\,dy\,dx =$

$\int_{-1}^{1} 4\sqrt{1-x^2} - 2x\sqrt{1-x^2}\,dx = \left[2x\sqrt{1-x^2} + \frac{2(1-x^2)^{3/2}}{3} + 2\arcsin x\right]_{-1}^{1} = 2\pi$

39. $V = \int_0^{\pi/2}\int_1^2\int_0^{r^2\sin\theta\cos\theta} r\,dz\,dr\,d\theta = \int_0^{\pi/2}\int_1^2 r^3\sin\theta\cos\theta\,dr\,d\theta =$

$\frac{15}{4}\int_0^{\pi/2}\sin\theta\cos\theta\,d\theta = \frac{15}{8}$

41. $V = 4\int_0^1\int_0^{\sqrt{1-x^2}}\int_{2x^2+2y^2}^{3-x^2-y^2} dz\,dy\,dx = 4\int_0^{\pi/2}\int_0^1\int_{2r^2}^{3-r^2} r\,dz\,dr\,d\theta =$

$4\int_0^{\pi/2}\int_0^1\left(3r - 3r^3\right)dr\,d\theta = 3\int_0^{\pi/2} d\theta = \frac{3\pi}{2}$

43. a) The radius of the sphere is 2; the radius of the hole is 1.

b) $V = 2\int_0^{2\pi}\int_0^{\sqrt{3}}\int_1^{\sqrt{4-z^2}} r\,dr\,dz\,d\theta = \int_0^{2\pi}\int_0^{\sqrt{3}} 3 - z^2\,dz\,d\theta = 2\sqrt{3}\int_0^{2\pi} d\theta = 4\sqrt{3}\pi$

45. $V = 4\int_0^1\int_0^{\sqrt{1-x^2}}\int_{x^2+y^2}^{(x^2+y^2+1)/2} dz\,dy\,dx = 4\int_0^{\pi/2}\int_0^1\int_{r^2}^{(r^2+1)/2} r\,dz\,dr\,d\theta =$

$4\int_0^{\pi/2}\int_0^1\left(\frac{r^2}{2}+\frac{1}{2}-r^2\right)r\,dr\,d\theta = 4\int_0^{\pi/2}\frac{1}{8}\,d\theta = \frac{\pi}{4}$

47. $I_z = \int_0^{2\pi}\int_0^{\pi/3}\int_0^2 (\rho\sin\phi)^2\rho^2\sin\phi\,d\rho\,d\phi\,d\theta = \int_0^{2\pi}\int_0^{\pi/3}\int_0^2 \rho^4\sin^3\phi\,d\rho\,d\phi\,d\theta =$

$\frac{32}{5}\int_0^{2\pi}\int_0^{\pi/3}\sin\phi - \cos^2\phi\sin\phi\,d\phi\,d\theta = \frac{32}{5}\int_0^{2\pi}\left[-\cos\phi + \frac{\cos^3\phi}{3}\right]_0^{\pi/3} d\theta = \frac{4}{3}\int_0^{2\pi} d\theta = \frac{8\pi}{3}$

49. $I_z = \int_0^{2\pi}\int_0^{\pi}\int_0^{1-\cos\theta} (\rho\sin\phi)^2\rho^2\sin\phi\,d\rho\,d\phi\,d\theta = \int_0^{2\pi}\int_0^{\pi}\int_0^{1-\cos\theta}\rho^4\sin^3\phi\,d\rho\,d\phi\,d\theta =$

$\frac{1}{5}\int_0^{2\pi}\int_0^{\pi}(1-\cos\phi)^5\sin^3\phi\,d\phi\,d\theta = \int_0^{2\pi}\int_0^{\pi}(1-\cos\phi)^6(1+\cos\phi)\sin\phi\,d\phi\,d\theta =$

$\frac{1}{5}\int_0^{2\pi}\int_0^2 (u)^6(2-u)\,du\,d\theta = \frac{32}{35}\int_0^{2\pi} d\theta = \frac{64\pi}{35}$, where $u = 1 - \cos\phi$

51. If $x = u + y$ and $y = v \Rightarrow x = u + v$ and $y = v$, then $J(u,v) = \begin{vmatrix} 1 & 1 \\ 0 & 1 \end{vmatrix} = 1$.

$\therefore \int_0^{\infty}\int_0^{x} e^{-sx} f(x-y,y)\,dy\,dx = \int_0^{\infty}\int_0^{\infty} e^{-s(u+v)} f(u,v)\,du\,dv$

CHAPTER 14

INTEGRATION IN VECTOR FIELDS

SECTION 14.1 LINE INTEGRALS

1. $\mathbf{r} = t\,\mathbf{i} + (1-t)\,\mathbf{j} \Rightarrow x = t,\ y = 1 - t \Rightarrow$
$y = 1 - x \Rightarrow c$

3. $\mathbf{r} = (2\cos t)\,\mathbf{i} + (2\sin t)\,\mathbf{j} \Rightarrow x = 2\cos t,$
$y = 2\sin t \Rightarrow x^2 + y^2 = 4 \Rightarrow g$

5. $\mathbf{r} = t\,\mathbf{i} + t\,\mathbf{j} + t\,\mathbf{k} \Rightarrow x = t,\ y = t,\ z = t \Rightarrow d$

7. $\mathbf{r} = (t^2 - 1)\,\mathbf{j} + 2t\,\mathbf{k} \Rightarrow y = t^2 - 1,\ z = 2t \Rightarrow$
$y = \frac{z^2}{4} - 1 \Rightarrow f$

9. $\mathbf{r} = t\,\mathbf{i} + (1-t)\,\mathbf{j} \Rightarrow x = t,\ y = 1 - t,\ z = 0 \Rightarrow f(g(t),h(t),k(t)) = 1.\ \frac{dx}{dt} = 1,\ \frac{dy}{dt} = -1,\ \frac{dz}{dt} = 0 \Rightarrow$

$$\sqrt{\left(\frac{dx}{dt}\right)^2 + \left(\frac{dy}{dt}\right)^2 + \left(\frac{dz}{dt}\right)^2}\ dt = \sqrt{2}\ dt \Rightarrow \int_C f(x,y,z)\ ds = \int_0^1 \sqrt{2}\ dt = \sqrt{2}$$

11. $\mathbf{r} = 2t\,\mathbf{i} + t\,\mathbf{j} + (2 - 2t)\,\mathbf{k} \Rightarrow x = 2t,\ y = t,\ z = 2 - 2t \Rightarrow f(g(t),h(t),k(t)) = 2t^2 - t + 2.\ \frac{dx}{dt} = 2,\ \frac{dy}{dt} = 1,\ \frac{dz}{dt} = -2$

$$\Rightarrow \sqrt{\left(\frac{dx}{dt}\right)^2 + \left(\frac{dy}{dt}\right)^2 + \left(\frac{dz}{dt}\right)^2}\ dt = 3\ dt \Rightarrow \int_C f(x,y,z)\ ds = \int_0^1 (2t^2 - t + 2)3\ dt = \frac{13}{2}$$

13. $\mathbf{r} = \mathbf{i} + \mathbf{j} + t\,\mathbf{k} \Rightarrow x = 1,\ y = 1,\ z = t \Rightarrow f(g(t),h(t),k(t)) = 3t\sqrt{4 + t^2}.\ \frac{dx}{dt} = 0,\ \frac{dy}{dt} = 0,\ \frac{dz}{dt} = 1 \Rightarrow$

$$\sqrt{\left(\frac{dx}{dt}\right)^2 + \left(\frac{dy}{dt}\right)^2 + \left(\frac{dz}{dt}\right)^2}\ dt = 1\ dt = dt \Rightarrow \int_C f(x,y,z)\ ds = \int_{-1}^1 3t\sqrt{4 + t^2}\ dt = 0$$

15. C_1: $\mathbf{r} = t\,\mathbf{i} + t^2\,\mathbf{j} \Rightarrow x = t,\ y = t^2,\ z = 0 \Rightarrow f(g(t),h(t),k(t)) = t + \sqrt{t^2} = 2t$ since $0 \le t \le 1$. $\frac{dx}{dt} = 1,\ \frac{dy}{dt} = 2t,$

$$\frac{dz}{dt} = 0 \Rightarrow \sqrt{\left(\frac{dx}{dt}\right)^2 + \left(\frac{dy}{dt}\right)^2 + \left(\frac{dz}{dt}\right)^2}\ dt = \sqrt{1 + 4t^2}\ dt \Rightarrow \int_{C_1} f(x,y,z)\ ds = \int_0^1 2t\sqrt{1 + 4t^2}\ dt =$$

$\frac{1}{6}\left(5^{3/2}\right) - \frac{1}{6} = \frac{5}{6}\sqrt{5} - \frac{1}{6}$. C_2: $\mathbf{r} = \mathbf{i} + \mathbf{j} + t\,\mathbf{k} \Rightarrow x = 1,\ y = 1,\ z = t \Rightarrow f(g(t),h(t),k(t)) = 2 - t^2.\ \frac{dx}{dt} = 0,$

$$\frac{dy}{dt} = 0,\ \frac{dz}{dt} = 1 \Rightarrow \sqrt{\left(\frac{dx}{dt}\right)^2 + \left(\frac{dy}{dt}\right)^2 + \left(\frac{dz}{dt}\right)^2}\ dt = 1\ dt = dt \Rightarrow \int_{C_2} f(x,y,z)\ ds = \int_0^1 (2 - t^2)\ dt =$$

$$\frac{5}{3}.\ \therefore \int_C f(x,y,z)\ ds = \int_{C_1} f(x,y,z)\ ds + \int_{C_2} f(x,y,z)\ ds = \frac{5}{6}\sqrt{5} + \frac{3}{2}.$$

17. $\mathbf{r} = t\,\mathbf{i} + t\,\mathbf{j} + t\,\mathbf{k} \Rightarrow x = t,\ y = t,\ z = t \Rightarrow f(g(t),h(t),k(t)) = \frac{t+t+t}{t^2+t^2+t^2} = \frac{1}{t}.\ \frac{dx}{dt} = 1,\ \frac{dy}{dt} = 1,\ \frac{dz}{dt} = 1 \Rightarrow$

$$\sqrt{\left(\frac{dx}{dt}\right)^2 + \left(\frac{dy}{dt}\right)^2 + \left(\frac{dz}{dt}\right)^2}\,dt = \sqrt{3}\,dt \Rightarrow \int_C f(x,y,z)\,ds = \int_a^b \frac{1}{t}\left(\sqrt{3}\,dt\right) = \sqrt{3}\ln|b| - \sqrt{3}\ln|a|$$

$$= \sqrt{3}\ln\left|\frac{b}{a}\right|$$

19. $\delta(x,y,z) = 2 - z$, $\mathbf{r} = (\cos t)\,\mathbf{j} + (\sin t)\,\mathbf{k}$, $0 \le t \le \pi$, $ds = dt$, $x = 0$, $y = \cos t$, $z = \sin t$, and $M = 2\pi - 2$ are all given or found in Example 3 in the text, page 946. $I_x = \int_C (y^2 + z^2)\delta\,ds$

$$= \int_0^\pi (\cos^2 t + \sin^2 t)(2 - \sin t)\,dt = \int_0^\pi (2 - \sin t)\,dt = 2\pi - 2 \Rightarrow R_x = \sqrt{\frac{I_x}{M}} = \sqrt{\frac{2\pi - 2}{2\pi - 2}} = 1$$

21. Let δ be constant. Let $x = a\cos t$, $y = a\sin t$. Then $\frac{dx}{dt} = -a\sin t$, $\frac{dy}{dt} = a\cos t$, $0 \le t \le 2\pi$, $\frac{dz}{dt} = 0 \Rightarrow$

$$\sqrt{\left(\frac{dx}{dt}\right)^2 + \left(\frac{dy}{dt}\right)^2 + \left(\frac{dz}{dt}\right)^2}\,dt = a\,dt.\ \therefore\ I_z = \int_C (x^2 + y^2)\delta\,ds = \int_0^{2\pi} (a^2\sin^2 t + a^2\cos^2 t)a\delta\,dt =$$

$$\int_0^{2\pi} a^3\delta\,dt = 2\pi a^3\delta.\ M = \int_C \delta(x,y,z)\,ds = \int_0^{2\pi} \delta a\,dt = 2\pi\delta a.\ R_z = \sqrt{\frac{I_z}{M}} = \sqrt{\frac{2\pi a^3\delta}{2\pi a\delta}} = a.$$

23. a) $\mathbf{r} = (\cos t)\,\mathbf{i} + (\sin t)\,\mathbf{j} + t\,\mathbf{k} \Rightarrow x = \cos t,\ y = \sin t,\ z = t \Rightarrow \frac{dx}{dt} = -\sin t,\ \frac{dy}{dt} = \cos t,\ \frac{dz}{dt} = 1 \Rightarrow$

$$\sqrt{\left(\frac{dx}{dt}\right)^2 + \left(\frac{dy}{dt}\right)^2 + \left(\frac{dz}{dt}\right)^2}\,dt = \sqrt{2}\,dt.\ M = \int_C \delta(x,y,z)\,ds = \int_0^{2\pi} \delta\sqrt{2}\,dt = 2\pi\delta\sqrt{2}.$$

$$I_z = \int_C (x^2 + y^2)\delta\,ds = \int_0^{2\pi} (\cos^2 t + \sin^2 t)\delta\sqrt{2}\,dt = \int_0^{2\pi} \delta\sqrt{2}\,dt = 2\pi\delta\sqrt{2}.\ R_z = \sqrt{\frac{I_z}{M}} =$$

$$\sqrt{\frac{2\pi\delta\sqrt{2}}{2\pi\delta\sqrt{2}}} = 1$$

b) $M = \int_C \delta(x,y,z)\,ds = \int_0^{4\pi} \delta\sqrt{2}\,dt = 4\pi\delta\sqrt{2}$. $I_z = \int_C (x^2 + y^2)\delta\,ds = \int_0^{4\pi} \delta\sqrt{2}\,dt = 4\pi\delta\sqrt{2}$.

$$R_z = \sqrt{\frac{I_z}{M}} = \sqrt{\frac{4\pi\delta\sqrt{2}}{4\pi\delta\sqrt{2}}} = 1$$

25. a) $\mathbf{r} = \sqrt{2}\,t\,\mathbf{i} + \sqrt{2}\,t\,\mathbf{j} + (4 - t^2)\,\mathbf{k},\ 0 \le t \le 1 \Rightarrow x = \sqrt{2}\,t,\ y = \sqrt{2}\,t,\ z = 4 - t^2 \Rightarrow \frac{dx}{dt} = \sqrt{2},\ \frac{dy}{dt} = \sqrt{2},\ \frac{dz}{dt} = -2t$

$\Rightarrow \sqrt{\left(\frac{dx}{dt}\right)^2 + \left(\frac{dy}{dt}\right)^2 + \left(\frac{dz}{dt}\right)^2}\,dt = \sqrt{4 + 4t^2}\,dt \Rightarrow M = \int_C \delta(x,y,z)\,ds = \int_0^1 3t\sqrt{4 + 4t^2}\,dt =$

$4\sqrt{2} - 2$

b) $M = \int_C \delta(x,y,z)\,ds = \int_0^1 \sqrt{4 + 4t^2}\,dt = \sqrt{2} + \ln(1 + \sqrt{2})$

27. $\mathbf{r} = t\,\mathbf{i} + \frac{2\sqrt{2}}{3}t^{3/2}\,\mathbf{j} + \frac{t^2}{2}\,\mathbf{k},\ 0 \le t \le 2 \Rightarrow x = t,\ y = \frac{2\sqrt{2}}{3}t^{3/2},\ z = \frac{t^2}{2} \Rightarrow \frac{dx}{dt} = 1,\ \frac{dy}{dt} = \sqrt{2}\,t^{1/2},\ \frac{dz}{dt} = t \Rightarrow$

$\sqrt{\left(\frac{dx}{dt}\right)^2 + \left(\frac{dy}{dt}\right)^2 + \left(\frac{dz}{dt}\right)^2}\,dt = \sqrt{(t+1)^2}\,dt = |t + 1|\,dt = (t + 1)dt$ on the domain given. Then $M_{yz} =$

$\int_C x\delta\,ds = \int_0^2 t\left(\frac{1}{t+1}\right)(t + 1)\,dt = \int_0^2 t\,dt = 2.\ M = \int_C \delta\,ds = \int_0^2 \frac{1}{t+1}(t + 1)\,dt = \int_0^2 dt = 2$

$M_{xz} = \int_C y\delta\,ds = \int_0^2 \frac{2\sqrt{2}}{3}t^{3/2}\left(\frac{1}{t+1}\right)(t + 1)\,dt = \int_0^2 \frac{2\sqrt{2}}{3}t^{3/2}\,dt = \frac{32}{15}.\ M_{xy} = \int_C z\delta\,ds =$

$\int_0^2 \frac{t^2}{2}\left(\frac{1}{t+1}\right)(t + 1)\,dt = \int_0^2 \frac{t^2}{2}\,dt = \frac{4}{3}.\ \therefore\ \bar{x} = M_{yz}/M = \frac{2}{2} = 1,\ \bar{y} = M_{xz}/M = \frac{32/15}{2} = \frac{16}{15},\ \bar{z} = M_{xy}/M =$

$\frac{4/3}{2} = \frac{2}{3}.\ I_x = \int_C (y^2 + z^2)\delta\,ds = \int_0^2 \left(\frac{8}{9}t^3 + \frac{t^4}{4}\right)dt = \frac{232}{45}.\ I_y = \int_C (x^2 + z^2)\delta\,ds =$

$\int_0^2 \left(t^2 + \frac{t^4}{4}\right)dt = \frac{64}{15}\ \ I_z = \int_C (y^2 + x^2)\delta\,ds = \int_0^2 \left(t^2 + \frac{8}{9}t^3\right)dt = \frac{56}{9}.\ R_x = \sqrt{I_x/M}$

$= \sqrt{\frac{232/45}{2}} = \frac{2}{3}\sqrt{\frac{29}{5}}.\ R_y = \sqrt{I_y/M} = \sqrt{\frac{64/15}{2}} = 4\sqrt{\frac{2}{15}}.\ R_z = \sqrt{I_z/M} = \sqrt{\frac{56/9}{2}} = \frac{2}{3}\sqrt{7}$

SECTION 14.2 VECTOR FIELDS, WORK, CIRCULATION, AND FLUX

1. $|\mathbf{F}|$ is inversely proportional to the square of the distance from (x,y) to the origin $\Rightarrow \sqrt{(M(x,y))^2 + (N(x,y))^2} = \frac{k}{x^2+y^2}$, $k > 0$. **F** points toward the origin $\Rightarrow$ F is in the direction of $\mathbf{n} = \frac{-x}{\sqrt{x^2+y^2}}\mathbf{i} - \frac{y}{\sqrt{x^2+y^2}}\mathbf{j} \Rightarrow$ $\mathbf{F} = a\mathbf{n}$, $a > 0$. Then $M(x,y) = \frac{-ax}{\sqrt{x^2+y^2}}$, $N(x,y) = \frac{-ay}{\sqrt{x^2+y^2}} \Rightarrow \sqrt{(M(x,y))^2 + (N(x,y))^2} = a \Rightarrow$ $a = \frac{k}{x^2+y^2} \Rightarrow \mathbf{F} = \frac{-kx}{(x^2+y^2)^{3/2}}\mathbf{i} - \frac{ky}{(x^2+y^2)^{3/2}}\mathbf{j}$, $k > 0$.

3. a) **F** vertical $\Rightarrow M(x,y) = 0 \Rightarrow \frac{x^2+2y^2-4}{4} = 0 \Rightarrow$ the points are on the ellipse $x^2 + 2y^2 = 4$

   b) **F** horizontal $\Rightarrow N(x,y) = 0 \Rightarrow \frac{y-x^2}{4} = 0 \Rightarrow$ the points are on the parabola $y = x^2$

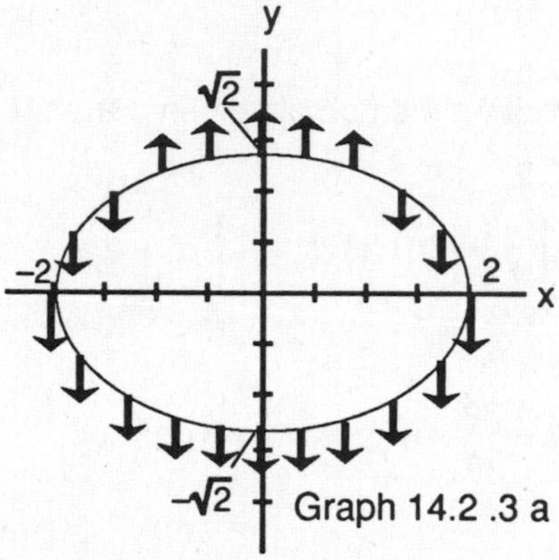

Graph 14.2.3 a

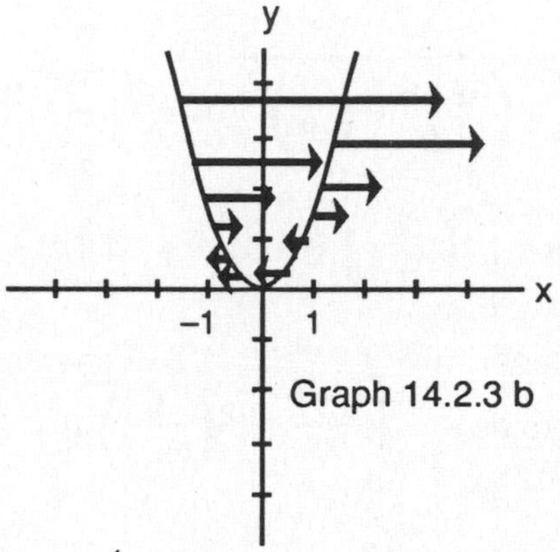

Graph 14.2.3 b

5. a) $\mathbf{F} = 3t\,\mathbf{i} + 2t\,\mathbf{j} + 4t\,\mathbf{k}$, $\frac{d\mathbf{r}}{dt} = \mathbf{i} + \mathbf{j} + \mathbf{k} \Rightarrow \mathbf{F}\cdot\frac{d\mathbf{r}}{dt} = 9t \Rightarrow W = \int_0^1 9t\,dt = \frac{9}{2}$

   b) $\mathbf{F} = 3t^2\,\mathbf{i} + 2t\,\mathbf{j} + 4t^4\,\mathbf{k}$, $\frac{d\mathbf{r}}{dt} = \mathbf{i} + 2t\,\mathbf{j} + 4t^3\,\mathbf{k} \Rightarrow \mathbf{F}\cdot\frac{d\mathbf{r}}{dt} = 7t^2 + 16t^7 \Rightarrow W = \int_0^1 (7t^2 + 16t^7)\,dt = \frac{13}{3}$

   c) $\mathbf{F}_1 = 3t\,\mathbf{i} + 2t\,\mathbf{j}$, $\frac{d\mathbf{r}_1}{dt} = \mathbf{i} + \mathbf{j} \Rightarrow \mathbf{F}_1\cdot\frac{d\mathbf{r}_1}{dt} = 5t \Rightarrow W_1 = \int_0^1 5t\,dt = \frac{5}{2}$. $\mathbf{F}_2 = 3\,\mathbf{i} + 2\,\mathbf{j} + 4t\,\mathbf{k}$, $\frac{d\mathbf{r}_2}{dt} = \mathbf{k} \Rightarrow$ $\mathbf{F}_2\cdot\frac{d\mathbf{r}_2}{dt} = 4t \Rightarrow W_2 = \int_0^1 4t\,dt = 2$. $\therefore$ $W = W_1 + W_2 = \frac{9}{2}$

7. a) $\mathbf{F} = \sqrt{t}\,\mathbf{i} - 2t\,\mathbf{j} + \sqrt{t}\,\mathbf{k}$, $\frac{d\mathbf{r}}{dt} = \mathbf{i} + \mathbf{j} + \mathbf{k} \Rightarrow \mathbf{F}\cdot\frac{d\mathbf{r}}{dt} = 2\sqrt{t} - 2t \Rightarrow W = \int_0^1 (2\sqrt{t} - 2t)\,dt = \frac{1}{3}$

   b) $\mathbf{F} = t^2\,\mathbf{i} - 2t\,\mathbf{j} + t\,\mathbf{k}$, $\frac{d\mathbf{r}}{dt} = \mathbf{i} + 2t\,\mathbf{j} + 4t^3\,\mathbf{k} \Rightarrow \mathbf{F}\cdot\frac{d\mathbf{r}}{dt} = 4t^4 - 3t^2 \Rightarrow W = \int_0^1 (4t^4 - 3t^2)\,dt = -\frac{1}{5}$

7. (Continued)

c) $\mathbf{F}_1 = -2t\,\mathbf{j} + \sqrt{t}\,\mathbf{k}, \frac{d\mathbf{r}_1}{dt} = \mathbf{i} + \mathbf{j} \Rightarrow \mathbf{F}_1 \cdot \frac{d\mathbf{r}_1}{dt} = -2t \Rightarrow W_1 = \int_0^1 -2t\,dt = -1.$ $\mathbf{F}_2 = \sqrt{t}\,\mathbf{i} - 2\,\mathbf{j} + \mathbf{k},$

$\frac{d\mathbf{r}_2}{dt} = \mathbf{k} \Rightarrow \mathbf{F}_2 \cdot \frac{d\mathbf{r}_2}{dt} = 1 \Rightarrow W_2 = \int_0^1 dt = 1.$ $\therefore$ $W = W_1 + W_2 = 0$

9. a) $\mathbf{F} = (3t^2 - 3t)\,\mathbf{i} + 3t\,\mathbf{j} + \mathbf{k}, \frac{d\mathbf{r}}{dt} = \mathbf{i} + \mathbf{j} + \mathbf{k} \Rightarrow \mathbf{F} \cdot \frac{d\mathbf{r}}{dt} = 3t^2 + 1 \Rightarrow W = \int_0^1 (3t^2 + 1)\,dt = 2$

b) $\mathbf{F} = (3t^2 - 3t)\,\mathbf{i} + 3t^4\,\mathbf{j} + \mathbf{k}, \frac{d\mathbf{r}}{dt} = \mathbf{i} + 2t\,\mathbf{j} + 4t^3\,\mathbf{k} \Rightarrow \mathbf{F} \cdot \frac{d\mathbf{r}}{dt} = 6t^5 + 4t^3 + 3t^2 - 3t \Rightarrow$

$W = \int_0^1 \left(6t^5 + 4t^3 + 3t^2 - 3t\right) dt = \frac{3}{2}$

c) $\mathbf{F}_1 = (3t^2 - 3t)\,\mathbf{i} + \mathbf{k}, \frac{d\mathbf{r}_1}{dt} = \mathbf{i} + \mathbf{j} \Rightarrow \mathbf{F}_1 \cdot \frac{d\mathbf{r}_1}{dt} = 3t^2 - 3t \Rightarrow W_1 = \int_0^1 (3t^2 - 3t)\,dt = -\frac{1}{2}$

$\mathbf{F}_2 = 3t\,\mathbf{j} + \mathbf{k}, \frac{d\mathbf{r}_2}{dt} = \mathbf{k} \Rightarrow \mathbf{F}_2 \cdot \frac{d\mathbf{r}_2}{dt} = 1 \Rightarrow W_2 = \int_0^1 dt = 1.$ $\therefore$ $W = W_1 + W_2 = \frac{1}{2}$

11. $\mathbf{F} = t^3\,\mathbf{i} + t^2\,\mathbf{j} - t^3\,\mathbf{k}, \frac{d\mathbf{r}}{dt} = \mathbf{i} + 2t\,\mathbf{j} + \mathbf{k} \Rightarrow \mathbf{F} \cdot \frac{d\mathbf{r}}{dt} = 2t^3 \Rightarrow W = \int_0^1 2t^3\,dt = \frac{1}{2}$

13. $\mathbf{F} = t\,\mathbf{i} + (\sin t)\,\mathbf{j} + (\cos t)\,\mathbf{k}, \frac{d\mathbf{r}}{dt} = (\cos t)\,\mathbf{i} - (\sin t)\,\mathbf{j} + \mathbf{k} \Rightarrow \mathbf{F} \cdot \frac{d\mathbf{r}}{dt} = t\cos t - \sin^2 t + \cos t \Rightarrow$

$W = \int_0^{2\pi} (t\cos t - \sin^2 t + \cos t)\,dt = -\pi$

15. $\mathbf{F} = -4t^3\,\mathbf{i} + 8t^2\,\mathbf{j} + 2\,\mathbf{k}, \frac{d\mathbf{r}}{dt} = \mathbf{i} + 2t\,\mathbf{j} \Rightarrow \mathbf{F} \cdot \frac{d\mathbf{r}}{dt} = 12t^3 \Rightarrow \text{Flow} = \int_0^2 12t^3\,dt = 48$

17. $\mathbf{F} = (\cos t - \sin t)\,\mathbf{i} + (\cos t)\,\mathbf{k}, \frac{d\mathbf{r}}{dt} = (-\sin t)\,\mathbf{i} + (\cos t)\,\mathbf{k} \Rightarrow \mathbf{F} \cdot \frac{d\mathbf{r}}{dt} = -\sin t\cos t + 1 \Rightarrow$

$\text{Flow} = \int_0^{\pi} (-\sin t\cos t + 1)\,dt = \pi$

19. a) $\mathbf{F}_1 = (\cos t)\,\mathbf{i} + (\sin t)\,\mathbf{j},\ \frac{d\mathbf{r}}{dt} = (-\sin t)\,\mathbf{i} + (\cos t)\,\mathbf{j} \Rightarrow \mathbf{F}_1 \cdot \frac{d\mathbf{r}}{dt} = 0 \Rightarrow \text{Circulation}_1 = 0.$

$M = \cos t,\ N = \sin t,\ dx = -\sin t\,dt,\ dy = \cos t\,dt \Rightarrow \text{Flux}_1 = \int_C M dy - N dx =$

$\int_0^{2\pi} (\cos^2 t + \sin^2 t)\,dt = \int_0^{2\pi} dt = 2\pi.$ $\mathbf{F}_2 = (-\sin t)\,\mathbf{i} + (\cos t)\,\mathbf{j},\ \frac{d\mathbf{r}}{dt} = (-\sin t)\,\mathbf{i} + (\cos t)\,\mathbf{j} \Rightarrow \mathbf{F}_2 \cdot \frac{d\mathbf{r}}{dt} =$

$1 \Rightarrow \text{Circ}_2 = \int_0^{2\pi} dt = 2\pi.$ $M = -\sin t,\ N = \cos t,\ dx = -\sin t\,dt,\ dy = \cos t\,dt \Rightarrow \text{Flux}_2 = \int_C M\,dy - N\,dx =$

$\int_0^{2\pi} (-\sin t \cos t + \sin t \cos t)\,dt\ \ 0$

b) $\mathbf{F}_1 = (\cos t)\,\mathbf{i} + (4 \sin t)\,\mathbf{j},\ \frac{d\mathbf{r}}{dt} = (-\sin t)\,\mathbf{i} + (4 \cos t)\,\mathbf{j} \Rightarrow \mathbf{F}_1 \cdot \frac{d\mathbf{r}}{dt} = 15 \sin t \cos t \Rightarrow \text{Circ}_1 =$

$\int_0^{2\pi} 15 \sin t \cos t\,dt = 0.$ $M = \cos t,\ N = 4 \sin t,\ dx = -\sin t,\ dy = 4 \cos t \Rightarrow \text{Flux}_1 =$

$\int_C M dy - N dx = \int_0^{2\pi} (4 \cos^2 t + 4 \sin^2 t)\,dt = 8\pi.$ $\mathbf{F}_2 = (-4 \sin t)\,\mathbf{i} + (\cos t)\,\mathbf{j},\ \frac{d\mathbf{r}}{dt} = (-\sin t)\,\mathbf{i} + (4 \cos t)\,\mathbf{j}$

$\Rightarrow \mathbf{F}_2 \cdot \frac{d\mathbf{r}}{dt} = 4 \Rightarrow \text{Circ}_2 = \int_0^{2\pi} 4\,dt = 8\pi.$ $M = -4 \sin t,\ N = \cos t,\ dx = -\sin t\,dt,\ dy = 4 \cos t\,dt \Rightarrow \text{Flux}_2 =$

$\int_0^{2\pi} (-16 \sin t \cos t + \sin t \cos t)\,dt = \int_0^{2\pi} (-15 \sin t \cos t)\,dt = 0$

21. $\mathbf{F}_1 = (a \cos t)\,\mathbf{i} + (a \sin t)\,\mathbf{j},\ \frac{d\mathbf{r}_1}{dt} = (-a \sin t)\,\mathbf{i} + (a \cos t)\,\mathbf{j} \Rightarrow \mathbf{F}_1 \cdot \frac{d\mathbf{r}_1}{dt} = 0 \Rightarrow \text{Circ}_1 = 0.$ $M_1 = a \cos t,\ N_1 =$

$a \sin t,\ dx = -a \sin t\,dt,\ dy = a \cos t\,dt \Rightarrow \text{Flux}_1 = \int_C M_1 dy - N_1 dx = \int_0^{\pi} (a^2 \cos^2 t + a^2 \sin^2 t)\,dt =$

$\int_0^{\pi} a^2\,dt = a^2\pi.$

$\mathbf{F}_2 = t\,\mathbf{i},\ \frac{d\mathbf{r}_2}{dt} = \mathbf{i} \Rightarrow \mathbf{F}_2 \cdot \frac{d\mathbf{r}_2}{dt} = t \Rightarrow \text{Circ}_2 = \int_{-a}^{a} t\,dt = 0.$ $M_2 = t,\ N_2 = 0,\ dx = dt,\ dy = 0 \Rightarrow \text{Flux}_2 =$

$\int_C M_2 dy - N_2 dx = \int_{-a}^{a} 0\,dt = 0.$ $\therefore\ \text{Circ} = \text{Circ}_1 + \text{Circ}_2 = 0,\ \text{Flux} = \text{Flux}_1 + \text{Flux}_2 = a^2\pi$

23. $\mathbf{F_1} = (-a \sin t)\,\mathbf{i} + (a \cos t)\,\mathbf{j}$, $\frac{d\mathbf{r_1}}{dt} = (-a \sin t)\,\mathbf{i} + (a \cos t)\,\mathbf{j} \Rightarrow \mathbf{F_1} \cdot \frac{d\mathbf{r_1}}{dt} = a^2 \sin^2 t + a^2 \cos^2 t = a^2 \Rightarrow$

$\text{Circ}_1 = \int_0^{\pi} a^2\,dt = a^2\pi.$

$\mathbf{F_2} = t\,\mathbf{j}$, $\frac{d\mathbf{r_2}}{dt} = \mathbf{i} \Rightarrow \mathbf{F_2} \cdot \frac{d\mathbf{r_2}}{dt} = 0 \Rightarrow \text{Circ}_2 = 0$. $\therefore$ $\text{Circ} = \text{Circ}_1 + \text{Circ}_2 = a^2\pi$

$M_1 = -a \sin t$, $N_1 = a \cos t$, $dx = -a \sin t$, $dy = a \cos t \Rightarrow \text{Flux}_1 = \int_C M_1 dy - N_1 dx =$

$\int_0^{\pi} (-a^2 \sin t \cos t + a^2 \sin t \cos t)\,dt = 0$. $M_2 = 0$, $N_2 = t$, $dx = dt$, $dy = 0 \Rightarrow \text{Flux}_2 =$

$\int_C M_2 dy - N_2 dx = \int_{-a}^{a} -t\,dt = 0$. $\therefore$ $\text{Flux} = \text{Flux}_1 + \text{Flux}_2 = 0$

25. a) $x^2 + y^2 = 1 \Rightarrow \mathbf{r} = (\cos t)\,\mathbf{i} + (\sin t)\,\mathbf{j}$, $0 \le t \le \pi \Rightarrow x = \cos t$, $y = \sin t \Rightarrow \mathbf{F} = (\cos t + \sin t)\,\mathbf{i} - \mathbf{j}$ and $\frac{d\mathbf{r}}{dt} = (-\sin t)\,\mathbf{i} + (\cos t)\,\mathbf{j} \Rightarrow \mathbf{F} \cdot \frac{d\mathbf{r}}{dt} = -\sin t \cos t - \sin^2 t - \cos t \Rightarrow \text{Flow} =$

$\int_0^{\pi} (-\sin t \cos t - \sin^2 t - \cos t)\,dt = -\frac{1}{2}\pi$

b) $\mathbf{r} = -t\,\mathbf{i}$, $-1 \le t \le 1 \Rightarrow x = -t$, $y = 0 \Rightarrow \mathbf{F} = -t\,\mathbf{i} - t^2\,\mathbf{j}$ and $\frac{d\mathbf{r}}{dt} = -\mathbf{i} \Rightarrow \mathbf{F} \cdot \frac{d\mathbf{r}}{dt} = t \Rightarrow \text{Flow} = \int_{-1}^{1} t\,dt =$

0

c) $\mathbf{r_1} = (1 - t)\,\mathbf{i} - t\,\mathbf{j}$, $0 \le t \le 1 \Rightarrow \mathbf{F_1} = (1 - 2t)\,\mathbf{i} - (1 - 2t + 2t^2)\,\mathbf{j}$ and $\frac{d\mathbf{r_1}}{dt} = -\mathbf{i} - \mathbf{j} \Rightarrow \mathbf{F_1} \cdot \frac{d\mathbf{r_1}}{dt} = 2t^2 \Rightarrow$

$\text{Flow}_1 = \int_0^1 2t^2\,dt = \frac{2}{3}$. $\mathbf{r_2} = -t\,\mathbf{i} + (t - 1)\,\mathbf{j}$, $0 \le t \le 1 \Rightarrow \mathbf{F_2} = -\mathbf{i} - (2t^2 - 2t + 1)\,\mathbf{j}$ and $\frac{d\mathbf{r_2}}{dt} = -\mathbf{i} +$

$\mathbf{j} \Rightarrow \mathbf{F_2} \cdot \frac{d\mathbf{r_2}}{dt} = 2t - 2t^2 \Rightarrow \text{Flow}_2 = \int_0^1 (2t - 2t^2)\,dt = \frac{1}{3}$. $\therefore$ $\text{Flow} = \text{Flow}_1 + \text{Flow}_2 = 1$

27. $\mathbf{r_1} = (\cos t)\,\mathbf{i} + (\sin t)\,\mathbf{j} + t\,\mathbf{k}$, $0 \le t \le \frac{\pi}{2} \Rightarrow \mathbf{F_1} = (2 \cos t)\,\mathbf{i} + 2t\,\mathbf{j} + (2 \sin t)\,\mathbf{k}$ and $\frac{d\mathbf{r_1}}{dt} = (-\sin t)\,\mathbf{i} + (\cos t)\,\mathbf{j} +$

$\mathbf{k} \Rightarrow \mathbf{F_1} \cdot \frac{d\mathbf{r_1}}{dt} = -2 \sin t \cos t + 2t \cos t + 2 \sin t \Rightarrow \text{Circ}_1 = \int_0^{\pi/2} (-2 \sin t \cos t + 2t \cos t + 2 \sin t)\,dt =$

$\pi - 1$. $\mathbf{r_2} = \mathbf{j} + \frac{\pi}{2}(1 - t)\,\mathbf{k}$, $0 \le t \le 1 \Rightarrow \mathbf{F_2} = \pi(1 - t)\,\mathbf{j} + 2\,\mathbf{k}$ and $\frac{d\mathbf{r_2}}{dt} = -\frac{\pi}{2}\mathbf{k} \Rightarrow \mathbf{F_2} \cdot \frac{d\mathbf{r_2}}{dt} = -\pi \Rightarrow$

27. (Continued)

$\text{Circ}_2 = \int_0^1 -\pi\, dt = -\pi$. $\mathbf{r}_3 = t\,\mathbf{i} + (1-t)\,\mathbf{j},\ 0 \le t \le 1 \Rightarrow \mathbf{F}_3 = 2t\,\mathbf{i} + 2(1-t)\,\mathbf{k}$ and $\frac{d\mathbf{r}_3}{dt} = \mathbf{i} - \mathbf{j} \Rightarrow$

$\mathbf{F}_3 \cdot \frac{d\mathbf{r}_3}{dt} = 2t \Rightarrow \text{Circ}_3 = \int_0^1 2t\, dt = 1$. $\therefore$ $\text{Circ} = \text{Circ}_1 + \text{Circ}_2 + \text{Circ}_3 = 0$

SECTION 14.3 GREEN'S THEOREM IN THE PLANE

1. Equation 15: $M = -y = -a \sin t$, $N = x = a \cos t$, $dx = -a \sin t\, dt$, $dy = a \cos t\, dt \Rightarrow \frac{\partial M}{\partial x} = 0$, $\frac{\partial M}{\partial y} = -1$,

$\frac{\partial N}{\partial x} = 1$, $\frac{\partial N}{\partial y} = 0 \Rightarrow \oint_C M dy - N dx = \int_0^{2\pi} ((-a \sin\ t)(a \cos t)\, dt - (a \cos t)(-a \sin t)\ )dt =$

0. $\iint_R \left(\frac{\partial M}{\partial x} + \frac{\partial N}{\partial y}\right) dx\, dy = \iint_R 0\, dx\, dy = 0$

Equation 16: $\oint_C M dx + N dy = \int_0^{2\pi} ((-a \sin t)(-a \sin t) + (a \cos t)(a \cos t))\, dt = 2\pi a^2$

$\iint_R \left(\frac{\partial N}{\partial x} - \frac{\partial M}{\partial y}\right) dx\, dy = \int_{-a}^{a} \int_{-\sqrt{a^2-x^2}}^{\sqrt{a^2-x^2}} 2\, dy\, dx \quad \int_{-a}^{a} 4\sqrt{a^2 - x^2}\, dx = 2a^2\pi$

3. $M = 2x = 2a \cos t$, $N = -3y = -3a \sin t$, $dx = -a \sin t$, $dy = a \cos t \Rightarrow \frac{\partial M}{\partial x} = 2$, $\frac{\partial M}{\partial y} = 0$, $\frac{\partial N}{\partial x} = 0$, $\frac{\partial N}{\partial y} = -3$

Equation 15: $\oint_C M dy - N dx = \int_0^{2\pi} (2a \cos t(a \cos t) + 3a \sin t(-a \sin t))dt =$

$\int_0^{2\pi} \left(2a^2 \cos^2 t - 3a^2 \sin^2 t\right) dt = -\pi a^2$. $\iint_R \left(\frac{\partial M}{\partial x} + \frac{\partial N}{\partial y}\right) = \iint_R -1\, dx\, dy =$

$\int_{-a}^{a} \int_{-\sqrt{a^2-x^2}}^{\sqrt{a^2-x^2}} -1\, dy\, dx = -\pi a^2$

Equation 16: $\oint_C M dx + N dy = \int_0^{2\pi} (2a \cos t(-a \sin t) + (-3a \sin t)(a \cos t))\, dt =$

3. (Continued)

$$\int_0^{2\pi} \left(-2a^2 \sin t \cos t - 3a^2 \sin t \cos t\right) dt = 0. \quad \iint_R 0 \, dx\, dy = 0$$

5. $M = x - y,\ N = y - x \Rightarrow \frac{\partial M}{\partial x} = 1,\ \frac{\partial M}{\partial y} = -1,\ \frac{\partial N}{\partial x} = -1,\ \frac{\partial N}{\partial y} = 1 \Rightarrow \text{Flux} = \iint_R 2\, dx\, dy = \int_0^1 \int_0^1 2\, dx\, dy$

$= 2.\ \text{Circ} = \iint_R (-1 - (-1))\, dx\, dy = 0$

7. $M = y^2 - x^2,\ N = x^2 + y^2 \Rightarrow \frac{\partial M}{\partial x} = -2x,\ \frac{\partial M}{\partial y} = 2y,\ \frac{\partial N}{\partial x} = 2x,\ \frac{\partial N}{\partial y} = 2y \Rightarrow \text{Flux} = \iint_R (-2x + 2y)\, dx\, dy$

$= \int_0^3 \int_0^x (-2x + 2y) dy\, dx = \int_0^3 (-2x^2 + x^2)\, dx = -9.\ \text{Circ} = \iint_R (2x - 2y)\, dx\, dy =$

$\int_0^3 \int_0^x (2x - 2y)\, dy\, dx = \int_0^3 x^2\, dx = 9$

9. $M = x + e^x \sin y,\ N = x + e^x \cos y \Rightarrow \frac{\partial M}{\partial x} = 1 + e^x \sin\ y,\ \frac{\partial M}{\partial y} = e^x \cos y,\ \frac{\partial N}{\partial x} = 1 + e^x \cos y,\ \frac{\partial N}{\partial y} = e^x \sin y$

$\Rightarrow \text{Flux} = \iint_R dx\, dy = \int_{-\pi/4}^{\pi/4} \int_0^{\sqrt{\cos \theta}} r\, dr\, d\theta = \int_{-\pi/4}^{\pi/4} \left(\frac{1}{2} \cos 2\theta\right) d\theta = \frac{1}{2}.$

$\text{Circ} = \iint_R (1 + e^x \cos y - e^x \cos y)\, dx\, dy = \iint_R dx\, dy =$

$\int_{-\pi/4}^{\pi/4} \int_0^{\sqrt{\cos \theta}} r\, dr\, d\theta = \int_{-\pi/4}^{\pi/4} \frac{1}{2} \cos \theta\, d\theta = \frac{1}{2}$

11. $M = xy,\ N = y^2 \Rightarrow \frac{\partial M}{\partial x} = y,\ \frac{\partial M}{\partial y} = x,\ \frac{\partial N}{\partial x} = 0,\ \frac{\partial N}{\partial y} = 2y \Rightarrow \text{Flux} = \iint_R (y + 2y)\,dy\,dx = \int_0^1 \int_{x^2}^{x} 3y\,dy\,dx$

$= \int_0^1 \left(\frac{3x^2}{2} - \frac{3x^4}{2}\right)dx = \frac{1}{5}.\ \text{Circ} = \iint_R -x\,dy\,dx = \int_0^1 \int_{x^2}^{x} -x\,dy\,dx = \int_0^1 (-x^2 + x^3)\,dx = -\frac{1}{12}$

13. $M = 3xy - \frac{x}{1+y^2},\ N = e^x + \tan^{-1}y \Rightarrow \frac{\partial M}{\partial x} = 3y - \frac{1}{1+y^2},\ \frac{\partial N}{\partial y} = \frac{1}{1+y^2} \Rightarrow \text{Flux} =$

$\iint_R \left(3y - \frac{1}{1+y^2} + \frac{1}{1+y^2}\right)dx\,dy = \iint_R 3y\,dx\,dy = \int_0^{2\pi} \int_0^{a(1+\cos\theta)} 3r\sin\theta\,dr\,d\theta = \int_0^{2\pi} a^3(1 + \cos\theta)^3 \sin\theta\,d\theta$

$= -4a^3 - (-4a^3) = 0$

15. $M = y^2,\ N = x^2 \Rightarrow \frac{\partial M}{\partial y} = 2y,\ \frac{\partial N}{\partial x} = 2x \Rightarrow \oint_C y^2\,dx + x^2\,dy = \iint_R (2x - 2y)dy\,dx =$

$\int_0^1 \int_0^{-x+1} (2x - 2y)\,dy\,dx = \int_0^1 (-3x^2 + 4x - 1)\,dx = 0.$

17. $M = 6y + x,\ N = y + 2x \Rightarrow \frac{\partial M}{\partial y} = 6,\ \frac{\partial N}{\partial x} = 2 \Rightarrow \oint_C (6y + x)dx + (y + 2x)dy = \iint_R (2 - 6)\,dy\,dx =$

$-4(\text{Area of the circle}) = -16\pi$

19. $M = 2xy^3,\ N = 4x^2y^2 \Rightarrow \frac{\partial M}{\partial y} = 6xy^2,\ \frac{\partial N}{\partial x} = 8xy^2 \Rightarrow \oint_C 2xy^3\,dx + 4x^2y^2\,dy = \iint_R (8xy^2 - 6xy^2)\,dx\,dy$

$\int_0^1 \int_0^{x^3} 2xy^2\,dy\,dx = \int_0^1 \frac{2}{3}x^{10}\,dx = \frac{2}{33}$

21. a) $M = f(x),\ N = g(y) \Rightarrow \frac{\partial M}{\partial y} = 0,\ \frac{\partial N}{\partial x} = 0 \Rightarrow \oint_C f(x)\,dx + g(y)\,dy = \iint_R 0\,dy\,dx = 0$

b) $M = ky,\ N = hx \Rightarrow \frac{\partial M}{\partial y} = k,\ \frac{\partial N}{\partial x} = h \Rightarrow \oint_C ky\,dx + hx\,dy = \iint_R (h-k)\,dx\,dy =$

$(h-k)$(Area of the region)

23. $M = -y^3,\ N = x^3 \Rightarrow \frac{\partial M}{\partial y} = -3y^2,\ \frac{\partial N}{\partial x} = 3x^2 \Rightarrow \oint_C -y^3\,dx + x^3\,dy = \iint_R (3x^2 + 3y^2)\,dx\,dy > 0$ on any region in the xy–plane enclosed by a closed curve C.

25. $M = \frac{\partial f}{\partial y},\ N = -\frac{\partial f}{\partial x} \Rightarrow \frac{\partial M}{\partial y} = \frac{\partial^2 f}{\partial y^2},\ \frac{\partial N}{\partial x} = -\frac{\partial^2 f}{\partial x^2} \Rightarrow \oint_C \frac{\partial f}{\partial y}\,dx - \frac{\partial f}{\partial x}\,dy = \iint_R \left(-\frac{\partial^2 f}{\partial x^2} - \frac{\partial^2 f}{\partial y^2}\right)dx\,dy = 0$ for such curves C.

27. a) $\nabla f = \left(\frac{2x}{x^2+y^2}\right)\mathbf{i} + \left(\frac{2y}{x^2+y^2}\right)\mathbf{j} \Rightarrow M = \frac{2x}{x^2+y^2},\ N = \frac{2y}{x^2+y^2})$. M, N are discontinuous at (0,0) $\Rightarrow$ we cannot apply Green's Theorem over C. Let C_h be the circle $x = h\cos\theta$, $y = h\sin\theta$, $0 < h \le a$ where C_1 is the circle $x = a\cos t$, $y = a\sin t$, $a > 0$. Then $\oint_C \mathbf{F}\cdot\mathbf{n}\,ds = \oint_{C_1} M\,dy - N\,dx + \oint_{C_h} M\,dy - N\,dx =$

$\oint_{C_1} \frac{2x}{x^2+y^2}\,dy - \frac{2y}{x^2+y^2}\,dx + \oint_{C_h} \frac{2x}{x^2+y^2}\,dy - \frac{2y}{x^2+y^2}\,dx$. In the first integral, let $x = a\cos t$, $y = a\sin t \Rightarrow dx = -a\sin t\,dt$, $dy = a\cos t\,dt$, $M = 2a\cos t$, $N = 2a\sin t$, $0 \le t \le 2\pi$. In the second integral, let $x = h\cos\theta$, $y = h\sin\theta \Rightarrow dx = -h\sin\theta\,d\theta$, $dy = h\cos\theta\,d\theta$, $M = 2h\cos\theta$, $N = 2h\sin\theta$, $0 \le \theta \le 2\pi$.

Then $\oint_C \mathbf{F}\cdot\mathbf{n}\,ds = \oint_{C_1} \frac{2x}{x^2+y^2}\,dy - \frac{2y}{x^2+y^2}\,dx + \oint_{C_h} \frac{2x}{x^2+y^2}\,dy - \frac{2y}{x^2+y^2}\,dx. = \int_0^{2\pi} 2\,dt + \int_{2\pi}^{0} 2\,d\theta = 0$ for every h.

b) If K is any simple closed curve surrounding C_h (K contains (0,0)), then $\oint_C \mathbf{F}\cdot\mathbf{n}\,ds = \oint_{C_1} M\,dy - N\,dx + \oint_{C_h} M\,dy - N\,dx$. $M\,dy - N\,dx$, in polar coordinates, is $\frac{2r\cos\theta}{r^2}\left(r\cos\theta\,d\theta + \sin\theta\,dr\right) - \frac{2r\sin\theta}{r^2}\left(-r\sin\theta\,d\theta + \cos\theta\,dr\right) = \frac{2r^2}{r^2}\,d\theta = 2\,d\theta$. 2θ increases by 4π as K is traversed once

27. b) (Continued)

counterclockwise from $\theta = 0$ to $\theta = 2\pi$. Then $\oint_C \mathbf{F} \cdot \mathbf{n}\, ds = 0$ (since $\oint_{C_h} M\, dy - N\, dx = -4\pi$)

when (0,0) is in the region. But $\oint_K \mathbf{F} \cdot \mathbf{n}\, ds = 4\pi$ when (0,0) is not in the region.

29. $$\int_{g_1(y)}^{g_2(y)} \frac{\partial N}{\partial x}\, dx\, dy = N\big(g_2(y),y\big) - N\big(g_1(y),y\big) \Rightarrow \int_c^d \int_{g_1(y)}^{g_2(y)} \left(\frac{\partial N}{\partial x}\, dx\right) dy = \int_c^d \big[N(g_2(y),y) - N(g_1(y),y)\big]\, dy$$

$$= \int_c^d N\big(g_2(y),y\big)\, dy - \int_c^d N\big(g_1(y),y\big)\, dy = \int_c^d N\big(g_2(y),y\big)\, dy + \int_d^c N\big(g_1(y),y\big)\, dy = \int_{C_2} N\, dy + \int_{C_1} N\, dy$$

$$= \oint_C N\, dy. \therefore \oint_C N\, dy = \iint_R \frac{\partial N}{\partial x}\, dx\, dy$$

31. Area of $R = \iint_R dy\, dx = \iint_R (1 + 0)\, dy\, dx = \oint_C -y\, dx = -\oint_C y\, dx$

33. $$\bar{x} = \frac{M_y}{M} = \frac{\iint_R x\, \delta(x,y)\, dA}{\iint_R \delta(x,y)\, dA}.\ \text{If } \delta(x,y) = 1, \text{ then } \bar{x} = \frac{\iint_R x\, dA}{\iint_R dA} = \frac{\iint_R x\, dA}{A} \Rightarrow \iint_R x\, dA = A\bar{x}$$

Now, $\iint_R x\, dA = \iint_R (x + 0)\, dy\, dx = \oint_C \frac{x^2}{2}\, dy = \frac{1}{2}\oint_C x^2\, dy$. Also, $\iint_R x\, dA =$

$\iint_R (0 + x)\, dy\, dx = -\oint_C xy\, dx$. And $\iint_R x\, dA = \iint_R \left(\frac{2}{3}x + \frac{1}{3}x\right) dy\, dx =$

$$\oint_C \frac{1}{3}x^2\, dy - \frac{1}{3}xy\, dx = \frac{1}{3}\oint_C x^2\, dy - xy\, dx.\ \therefore \frac{1}{2}\oint_C x^2\, dy = -\oint_C xy\, dx = \frac{1}{3}\oint_C x^2\, dy - xy\, dx = A\bar{x}$$

35. Area = $\frac{1}{2}\oint_C x\,dy - y\,dx$. M = x = a cos t, N = y = a sin t $\Rightarrow$ dx = −a sin t dt, dy = a cos t dt $\Rightarrow$ Area =

$$\frac{1}{2}\int_0^{2\pi}\left(a^2\cos^2 t + a^2\sin^2 t\right)dt = \frac{1}{2}\int_0^{2\pi} a^2\,dt = \pi a^2$$

37. Area = $\frac{1}{2}\oint_C x\,dy - y\,dx$. M = x = $\cos^3 t$, N = y = $\sin^3 t \Rightarrow dx = -3\cos^2 t\sin t\,dt$, $dy = 3\sin^2 t\cos t\,dt$

$$\Rightarrow \text{Area} = \frac{1}{2}\int_0^{2\pi}(3\sin^2 t\cos^2 t(\cos^2 t + \sin^2 t))\,dt = \frac{1}{2}\int_0^{2\pi}(3\sin^2 t\cos^2 t)\,dt = \frac{3\pi}{8}$$

SECTION 14.4 SURFACE AREA AND SURFACE INTEGRALS

1. $\mathbf{p} = \mathbf{k}$, $\nabla f = 2x\,\mathbf{i} + 2y\,\mathbf{j} - \mathbf{k} \Rightarrow |\nabla f| = \sqrt{(2x)^2 + (2y)^2 + (-1)^2} = \sqrt{4x^2 + 4y^2 + 1}$. $|\nabla f\cdot\mathbf{p}| = 1 \Rightarrow$

$$S = \iint_R \frac{|\nabla f|}{|\nabla f\cdot\mathbf{p}|}\,dA = \iint_R \sqrt{4x^2 + 4y^2 + 1}\,dx\,dy =$$

$$\iint_R \sqrt{4r^2\cos^2\theta + 4r^2\sin^2\theta + 1}\;r\,dr\,d\theta = \int_0^{2\pi}\int_0^{\sqrt{2}} \sqrt{4r^2 + 1}\;r\,dr\,d\theta = \frac{13}{3}\pi$$

3. $\mathbf{p} = \mathbf{k}$. $\nabla f = \mathbf{i} + 2\,\mathbf{j} + 2\,\mathbf{k} \Rightarrow |\nabla f| = 3$. $|\nabla f\cdot\mathbf{p}| = 2 \Rightarrow S = \iint_R \frac{|\nabla f|}{|\nabla f\cdot\mathbf{p}|}\,dA = \iint_R \frac{3}{2}\,dx\,dy =$

$$\int_{-1}^{1}\int_{y^2}^{2-y^2} \frac{3}{2}\,dx\,dy = \int_{-1}^{1}(3 - 3y^2)\,dy = 4$$

5. $\mathbf{p} = \mathbf{k}$. $\nabla f = 2x\,\mathbf{i} - 2\,\mathbf{j} - 2\,\mathbf{k} \Rightarrow |\nabla f| = \sqrt{(2x)^2 + (-2)^2 + (-2)^2} = \sqrt{4x^2 + 8}$. $|\nabla f\cdot\mathbf{p}| = 2 \Rightarrow S =$

$$\iint_R \frac{|\nabla f|}{|\nabla f\cdot\mathbf{p}|}\,dA = \iint_R \frac{\sqrt{4x^2 + 8}}{2}\,dx\,dy = \int_0^{2}\int_0^{3x}\sqrt{x^2 + 2}\,dy\,dx = \int_0^{2} 3x\sqrt{x^2 + 2}\,dx = 6\sqrt{6} - 2\sqrt{2}$$

7. $\mathbf{p} = \mathbf{k}$. $\nabla f = 2x\,\mathbf{i} + 2y\,\mathbf{j} + 2z\,\mathbf{k} \Rightarrow |\nabla f| = \sqrt{4x^2 + 4y^2 + 4z^2} = \sqrt{8} = 2\sqrt{2}$. $|\nabla f \cdot \mathbf{p}| = 2z \Rightarrow S =$

$$\iint_R \frac{|\nabla f|}{|\nabla f \cdot \mathbf{p}|}\,dA = \iint_R \frac{2\sqrt{2}}{2z}\,dA = \sqrt{2}\iint_R \frac{1}{z}\,dA =$$

$$\sqrt{2}\iint_R \frac{1}{\sqrt{2-(x^2+y^2)}}\,dA = \sqrt{2}\int_0^{2\pi}\int_0^1 \frac{r\,dr\,d\theta}{\sqrt{2-r^2}} = \sqrt{2}\int_0^{2\pi}(-1+\sqrt{2})\,d\theta = 2\pi(2-\sqrt{2})$$

9. $\mathbf{p} = \mathbf{k}$. $\nabla f = 2x\,\mathbf{i} + 2z\,\mathbf{k} \Rightarrow |\nabla f| = \sqrt{(2x)^2 + (2z)^2} = 2$. $|\nabla f \cdot \mathbf{p}| = 2z$ for the upper surface, $z \ge 0 \Rightarrow$

$$S = \iint_R \frac{|\nabla f|}{|\nabla f \cdot \mathbf{p}|}\,dA = 2\iint_R \frac{2}{2z}\,dA = 2\iint_R \frac{1}{z}\,dA = 2\iint_R \frac{1}{\sqrt{1-x^2}}\,dy\,dx =$$

$$4\int_{-1/2}^{1/2}\int_0^{1/2} \frac{1}{\sqrt{1-x^2}}\,dy\,dx = 2\int_{-1/2}^{1/2} \frac{1}{\sqrt{1-x^2}}\,dx = \frac{2\pi}{3}$$

11. $\mathbf{p} = \mathbf{j}$. $\nabla f = 2x\,\mathbf{i} + \mathbf{j} + 2z\,\mathbf{k} \Rightarrow |\nabla f| = \sqrt{4x^2 + 4z^2 + 1}$. $|\nabla f \cdot \mathbf{p}| = 1 \Rightarrow S = \iint_R \frac{|\nabla f|}{|\nabla f \cdot \mathbf{p}|}\,dA =$

$$\iint_R \sqrt{4x^2 + 4z^2 + 1}\,dx\,dz = \int_0^{2\pi}\int_0^1 \sqrt{4r^2+1}\;r\,dr\,d\theta = \int_0^{2\pi}\left(\frac{5\sqrt{5}-1}{12}\right)d\theta = \frac{\pi(5\sqrt{5}-1)}{6}$$

13. The bottom face of the cube is in the xy–plane $\Rightarrow z = 0 \Rightarrow g(x,y,0) = x + y$ and $f(x,y,z) = z = 0 \Rightarrow$

$$\nabla f = \mathbf{k} \Rightarrow |\nabla f| = 1.\ \ \mathbf{p} = \mathbf{k} \Rightarrow |\nabla f \cdot \mathbf{p}| = 1 \Rightarrow d\sigma = dx\,dy \Rightarrow \iint_{z=0} (x+y)\,dx\,dy = \int_0^a\int_0^a (x+y)\,dx\,dy =$$

a^3. Because of symmetry, you get a^3 over the face of the cube in the xz–plane and a^3 over the face of the cube in the yz–plane.

In the top of the cube, $g(x,y,z) = g(x,y,a) = x + y + a$ and $f(x,y,z) = z = a \Rightarrow \nabla f = \mathbf{k} \Rightarrow |\nabla f| = 1$. $\mathbf{p} = \mathbf{k} \Rightarrow$

$$|\nabla f \cdot \mathbf{p}| = 1 \Rightarrow d\sigma = dx\,dy \Rightarrow \iint_{z=a} (x+y+1)\,dx\,dy = \int_0^a\int_0^a (x+y+a)\,dx\,dy = 2a^3.$$ Because of

symmetry, the integral is $2a^3$ over each of the other two faces. $\therefore \iint_{\text{cube}} (x+y+z)\,d\sigma = 9a^3$.

15. On the faces in the coordinate planes, $g(x,y,z) = 0 \Rightarrow$ the integral over these faces is 0.

On the face, $x = a$, $f(x,y,z) = x = a$ and $g(x,y,z) = g(a,y,z) = ayz \Rightarrow \nabla f = \mathbf{i} \Rightarrow |\nabla f| = 1$. $\mathbf{p} = \mathbf{i} \Rightarrow |\nabla f \cdot \mathbf{p}| = 1$

$$\Rightarrow d\sigma = dy\,dz \Rightarrow \iint_{x=a} xyz\,d\sigma = \int_0^c \int_0^b ayz\,dy\,dz = \frac{ab^2c^2}{4}$$

On the face, $y = b$, $f(x,y,z) = y = b$ and $g(x,y,z) = g(x,b,z) = bxz \Rightarrow \nabla f = \mathbf{j} \Rightarrow |\nabla f| = 1$. $\mathbf{p} = \mathbf{j} \Rightarrow |\nabla f \cdot \mathbf{p}| = 1$

$$\Rightarrow d\sigma = dx\,dz \Rightarrow \iint_{y=b} xyz\,dx\,dz = \int_0^c \int_0^a bxz\,dz\,dx = \frac{a^2bc^2}{4}$$

On the face, $z = c$, $f(x,y,z) = z = c$ and $g(x,y,z) = g(x,y,c) = cxy \Rightarrow \nabla f = \mathbf{k} \Rightarrow |\nabla f| = 1$. $\mathbf{p} = \mathbf{k} \Rightarrow |\nabla f \cdot \mathbf{p}| = 1$

$$\Rightarrow d\sigma = dy\,dx \Rightarrow \iint_{z=c} xyz\,d\sigma = \int_0^b \int_0^a cxy\,dx\,dy = \frac{a^2b^2c}{4}$$

$$\therefore \iint_S g(x,y,z)\,d\sigma = \frac{abc(ab + ac + bc)}{4}$$

17. $\nabla f = 2\,\mathbf{i} + 2\,\mathbf{j} + \mathbf{k}$ and $g(x,y,z) = x + y + (2 - 2x - 2y) = 2 - x - y \Rightarrow |\nabla f| = 3$. $\mathbf{p} = \mathbf{k} \Rightarrow |\nabla f \cdot \mathbf{p}| = 1 \Rightarrow$

$$d\sigma = 3\,dy\,dx \Rightarrow \iint_S (x + y + z)\,d\sigma = 3\int_0^1 \int_0^{1-x} (2 - x - y)\,dy\,dx = 2$$

19. $\nabla G = 2x\,\mathbf{i} + 2y\,\mathbf{j} + 2z\,\mathbf{k} \Rightarrow |\nabla G| = \sqrt{4x^2 + 4y^2 + 4z^2} = 2a$. $\mathbf{n} = \dfrac{2x\,\mathbf{i} + 2y\,\mathbf{j} + 2z\,\mathbf{k}}{2\sqrt{x^2 + y^2 + z^2}} = \dfrac{x\,\mathbf{i} + y\,\mathbf{j} + z\,\mathbf{k}}{a} \Rightarrow$

$\mathbf{F} \cdot \mathbf{n} = \dfrac{z^2}{a}$. $|\nabla G \cdot \mathbf{k}| = 2z \Rightarrow d\sigma = \dfrac{2a}{2z}\,dA = \dfrac{a}{z}\,dA$. $\therefore$ Flux $= \displaystyle\iint_R \frac{z^2}{a}\left(\frac{a}{z}\right) dA = \iint_R z\,dA =$

$$\iint_R \sqrt{a^2 - (x^2 + y^2)}\,dx\,dy = \int_0^{\pi/2} \int_0^a \sqrt{a^2 - r^2}\,r\,dr\,d\theta = \frac{a^3\pi}{6}$$

21. $\mathbf{n} = \dfrac{x\,\mathbf{i} + y\,\mathbf{j} + z\,\mathbf{k}}{a}$, $d\sigma = \dfrac{a}{z}\,dA$ (See Exercise 17) and $\mathbf{F} \cdot \mathbf{n} = \dfrac{xy}{a} - \dfrac{xy}{a} + \dfrac{z}{a} = \dfrac{z}{a}$.

$$\therefore \text{Flux} = \iint_R \frac{z}{a}\left(\frac{a}{z}\right) dA = \iint_R 1\,dA = \frac{\pi a^2}{4}$$

23. $\mathbf{n} = \frac{x\mathbf{i} + y\mathbf{j} + z\mathbf{k}}{a}$, $d\sigma = \frac{a}{z}\,dA$ (See Exercise 17) and $\mathbf{F}\cdot\mathbf{n} = \frac{x^2}{a} + \frac{y^2}{a} + \frac{z^2}{a} = a \Rightarrow \text{Flux} = \iint_R a\left(\frac{a}{z}\right) dA$

$= \iint_R \frac{a^2}{z}\,dA = \iint_R \frac{a^2}{\sqrt{a^2 - (x^2 + y^2)}}\,dA = \int_0^{\pi/2}\int_0^a \frac{a^2}{\sqrt{a^2 - r^2}}\,r\,dr\,d\theta = \frac{a^3\pi}{2}$

25. $\nabla G = 2y\,\mathbf{j} + \mathbf{k} \Rightarrow |\nabla G| = \sqrt{4y^2 + 1} \Rightarrow \mathbf{n} = \frac{2y\,\mathbf{j} + \mathbf{k}}{\sqrt{4y^2 + 1}} \Rightarrow \mathbf{F}\cdot\mathbf{n} = \frac{2xy - 3z}{\sqrt{4y^2 + 1}}$. $\mathbf{p} = \mathbf{k} \Rightarrow |\nabla G \cdot \mathbf{k}| = 1 \Rightarrow$

$d\sigma = \sqrt{4y^2 + 1}\,dA \Rightarrow \text{Flux} = \iint_R \left(\frac{2xy - 3z}{\sqrt{4y^2 + 1}}\right)\sqrt{4y^2 + 1}\,dA = \iint_R (2xy - 3z)\,dA =$

$\iint_R (2xy - 3(4 - y^2))\,dA = \int_0^1\int_{-2}^2 (2xy - 12 + 3y^2)\,dy\,dx = -32$

27. $\nabla G = -e^x\,\mathbf{i} + \mathbf{j} \Rightarrow |\nabla G| = \sqrt{e^{2x} + 1}$. $\mathbf{p} = \mathbf{i} \Rightarrow |\nabla G \cdot \mathbf{i}| = e^x$. $\mathbf{n} = \frac{e^x\,\mathbf{i} - \mathbf{j}}{\sqrt{e^{2x} + 1}} \Rightarrow \mathbf{F}\cdot\mathbf{n} = \frac{-2e^x - 2y}{\sqrt{e^{2x} + 1}}$.

$d\sigma = \frac{\sqrt{e^{2x} + 1}}{e^x}\,dA \Rightarrow \text{Flux} = \iint_R \frac{-2e^x - 2y}{\sqrt{e^{2x} + 1}}\left(\frac{\sqrt{e^{2x} + 1}}{e^x}\right) dA = \iint_R -4\,dA = \int_0^1\int_1^2 -4\,dy\,dz$

$= -4$

29. On the face, z = a: $G(x,y,z) = G(x,y,a) = z \Rightarrow \nabla G = \mathbf{k} \Rightarrow |\nabla G| = 1$. $\mathbf{n} = \mathbf{k} \Rightarrow \mathbf{F}\cdot\mathbf{n} = 2xz = 2ax$ since $z = a$

$d\sigma = dA \Rightarrow \text{Flux} = \iint_R 2ax\,dx\,dy = \int_0^a\int_0^a 2ax\,dx\,dy = a^4$. On the face, z = 0: $G(x,y,z) = G(x,y,0) =$

$z \Rightarrow \nabla G = \mathbf{k} \Rightarrow |\nabla G| = 1$. $\mathbf{n} = -\mathbf{k} \Rightarrow \mathbf{F}\cdot\mathbf{n} = -2xz = 0$ since $z = 0 \Rightarrow \text{Flux} = \iint_R 0\,dx\,dy = 0$

On the face, x = a: $G(x,y,z) = G(a,y,z) = x \Rightarrow \nabla G = \mathbf{i} \Rightarrow |\nabla G| = 1$. $\mathbf{n} = \mathbf{i} \Rightarrow \mathbf{F}\cdot\mathbf{n} = 2xy = 2ay$ since $x = a$

$\text{Flux} = \int_0^a\int_0^a 2ay\,dy\,dz = a^4$ On the face, x = 0: $G(x,y,z) = G(0,y,z) = x \Rightarrow \nabla G = \mathbf{i} \Rightarrow |\nabla G| = 1$. $\mathbf{n} = -\mathbf{i}$

$\Rightarrow \mathbf{F}\cdot\mathbf{n} = -2xy = 0$ since $x = 0 \Rightarrow \text{Flux} = 0$. On the face, y = a: $G(x,y,z) = G(x,a,z) = y \Rightarrow \nabla G = \mathbf{j} \Rightarrow |\nabla G|$

29. (Continued)

$= 1$. $\mathbf{n} = \mathbf{j} \Rightarrow \mathbf{F} \cdot \mathbf{n} = 2yz = 2az$ since $y = a \Rightarrow$ Flux $= \int_0^a \int_0^a 2az\, dz\, dx = a^4$. On the face, $y = 0$: $G(x,y,z) =$

$G(z,0,z) = y \Rightarrow \nabla G = \mathbf{j} \Rightarrow |\nabla G| = 1$. $\mathbf{n} = -\mathbf{j} \Rightarrow \mathbf{F} \cdot \mathbf{n} = -2yz = 0$ since $y = 0 \Rightarrow$ Flux $= 0$.

$\therefore$ Total Flux $= 3a^4$

31. $\nabla F = 2x\,\mathbf{i} + 2y\,\mathbf{j} + 2z\,\mathbf{k} \Rightarrow |\nabla F| = \sqrt{4x^2 + 4y^2 + 4z^2} = 2a,\ a > 0$ $\mathbf{p} = \mathbf{k} \Rightarrow |\nabla F \cdot \mathbf{k}| = 2z$ since $z \ge 0 \Rightarrow d\sigma =$

$\frac{2a}{2z}\, dA = \frac{a}{z}\, dA$. $\therefore$ $M = \iint_S \delta\, d\sigma = \frac{\pi a^2}{2}\delta$. $M_{xy} = \iint_S z\delta\, d\sigma = \delta \iint_S z\left(\frac{a}{z}\right) dA =$

$a\delta \int_0^a \int_0^{\sqrt{a^2 - x^2}} dy\, dx = \frac{\pi a}{4}\delta$. $\therefore$ $\bar{z} = \dfrac{\frac{\pi a^3}{4}\delta}{\frac{\pi a^2}{2}\delta} = \frac{a}{2}$. Because of symmetry, $\bar{x} = \bar{y} = \frac{a}{2}$. $\therefore$ Centroid $= \left(\frac{a}{2}, \frac{a}{2}, \frac{a}{2}\right)$

33. Because of symmetry, $\bar{x} = \bar{y} = 0$. $M = \iint_S \delta\, d\sigma = \delta \iint_S d\sigma = \delta(\text{Area of S}) = 3\pi\sqrt{2}\,\delta$.

$\nabla F = 2x\,\mathbf{i} + 2y\,\mathbf{j} - 2z\,\mathbf{k} \Rightarrow |\nabla F| = \sqrt{4x^2 + 4y^2 + 4z^2} = 2\sqrt{x^2 + y^2 + z^2}$. $\mathbf{p} = \mathbf{k} \Rightarrow |\nabla F \cdot \mathbf{k}| = 2z \Rightarrow$

$d\sigma = \frac{2\sqrt{x^2 + y^2 + z^2}}{2z}\, dA = \frac{\sqrt{x^2 + y^2 + z^2}}{z}\, dA = \frac{\sqrt{x^2 + y^2 + (x^2 + y^2)}}{z}\, dA = \frac{\sqrt{2}\,\sqrt{x^2 + y^2}}{z}\, dA.$

$\therefore$ $M_{xy} = \delta \iint_S z\left(\frac{\sqrt{2}\,\sqrt{x^2 + y^2}}{z}\right) dA = \delta \iint_S \sqrt{2}\,\sqrt{x^2 + y^2}\, dA = \delta \int_0^{2\pi}\int_1^2 \sqrt{2}\, r^2\, dr\, d\theta =$

$\frac{14\pi\sqrt{2}}{3}\delta$. $\bar{z} = \dfrac{\frac{14\pi\sqrt{2}}{3}\delta}{3\pi\sqrt{2}\,\delta} = \frac{14}{9}$. $\therefore$ $(\bar{x}, \bar{y}, \bar{z}) = \left(0, 0, \frac{14}{9}\right)$. $I_z = \iint_S (x^2 + y^2)\,\delta\, d\sigma =$

$\iint_S (x^2 + y^2)\left(\frac{\sqrt{2}\,\sqrt{x^2 + y^2}}{z}\right)\delta\, dA = \delta\sqrt{2} \iint_S (x^2 + y^2)\, dA = \delta\sqrt{2} \int_0^{2\pi}\int_1^2 r^3\, dr\, d\theta = \frac{15\pi\sqrt{2}}{2}\delta.$

$R_z = \sqrt{I_z/M} = \frac{\sqrt{10}}{2}$

35. a) Let the diameter lie on the z–axis. Let $f(x,y,z) = x^2 + y^2 + z^2 - a^2$, $z \geq 0$ be the upper hemisphere. $\nabla f = 2x\,\mathbf{i} + 2y\,\mathbf{j} + 2z\,\mathbf{k} \Rightarrow |\nabla f| = \sqrt{4x^2 + 4y^2 + 4z^2} = 2a$, $a > 0$. $\mathbf{p} = \mathbf{k} \Rightarrow |\nabla f \cdot \mathbf{p}| = 2z$ since $z \geq 0 \Rightarrow d\sigma = \frac{2a}{2z}\,dA = \frac{a}{z}\,dA$. $\therefore I_z = \iint_S \delta(x^2 + y^2)\frac{a}{z}\,dA = a\delta \iint_S \frac{x^2 + y^2}{\sqrt{a^2 - (x^2 + y^2)}}\,dA =$

$a\delta \int_0^{2\pi}\int_0^a \frac{r^2}{\sqrt{a^2 - r^2}}\,r\,dr\,d\theta = a\delta \int_0^{2\pi} \frac{2}{3}a^3\,d\theta = \frac{4\pi}{3}a^4\delta$. $\therefore$ for the whole sphere, the moment of inertia is $\frac{8\pi}{3}a^4\delta$.

b) $I_L = I_{c.m.} + mh^2$ where m is the mass of the body and h is the distance between the parallel lines.

$I_{c.m.} = \frac{8\pi}{3}a^4\delta$ (from part a). $m = M = \iint_S \delta\,d\sigma = \delta \iint_S \frac{a}{z}\,dA = a\delta \iint_S \frac{1}{\sqrt{a^2 - (x^2 + y^2)}}\,dy\,dx$

$= a\delta \int_0^{2\pi}\int_0^a \frac{1}{\sqrt{a^2 - r^2}}\,r\,dr\,d\theta = a\delta \int_0^{2\pi} a\,d\theta = 2\pi a^2\delta$. $h = a \Rightarrow I_L = \frac{8\pi}{3}a^4\delta + 2\pi a^2\delta(a)^2 = \frac{14\pi}{3}a^4\delta$

37. $f_x(x,y) = 2x$, $f_y(x,y) = 2y \Rightarrow \sqrt{f_x{}^2 + f_y{}^2 + 1} = \sqrt{4x^2 + 4y^2 + 1} \Rightarrow \text{Area} = \iint_R \sqrt{4x^2 + 4y^2 + 1}\,dx\,dy$

$= \int_0^{2\pi}\int_0^{\sqrt{3}} \sqrt{4r^2 + 1}\,r\,dr\,d\theta = \frac{\pi}{6}\left(13\sqrt{13} - 1\right)$

39. $f_z(y,z) = -2y$, $f_y(y,z) = -2z \Rightarrow . \sqrt{f_y{}^2 + f_z{}^2 + 1} = \sqrt{4y^2 + 4z^2 + 1} \Rightarrow$

$\text{Area} = \iint_R \sqrt{4y^2 + 4z^2 + 1}\,dy\,dz = \int_0^{2\pi}\int_0^1 \sqrt{4r^2 + 1}\,r\,dr\,d\theta = \frac{\pi}{6}(5\sqrt{5} - 1)$

41. $y = \frac{z^2}{2} \Rightarrow f_x(x,z) = 0$, $f_z(x,z) = z \Rightarrow \sqrt{f_x{}^2 + f_z{}^2 + 1} = \sqrt{z^2 + 1} \Rightarrow \text{Area} = \int_0^2\int_0^1 \sqrt{z^2 + 1}\,dx\,dz =$

$\sqrt{5} + \frac{1}{2}\ln\left(\sqrt{5} + 2\right)$ (Note: On integrating the second time with respect to z, use the substitution

41. (Continued)
$z = \tan\theta$ which means the integration will go from 0 to $\tan^{-1} 2$.)

SECTION 14.5 THE DIVERGENCE THEOREM

1. $\frac{\partial}{\partial x}(y - x) = -1$, $\frac{\partial}{\partial y}(z - y) = -1$, $\frac{\partial}{\partial z}(y - x) = 0 \Rightarrow \nabla \cdot \mathbf{F} = -2 \Rightarrow \text{Flux} = \int_{-1}^{1}\int_{-1}^{1}\int_{-1}^{1} -2\, dx\, dy\, dz = -16$

3. $\frac{\partial}{\partial x}(y) = 0$, $\frac{\partial}{\partial y}(xy) = x$, $\frac{\partial}{\partial z}(-z) = -1 \Rightarrow \nabla \cdot \mathbf{F} = x - 1 \Rightarrow \text{Flux} = \iiint_{\text{solid}} (x - 1)\, dz\, dy\, dx =$

$$\int_0^{2\pi}\int_0^{2}\int_0^{r^2} (r\cos\theta - 1)\, dz\, r\, dr\, d\theta = -8\pi$$

5. $\frac{\partial}{\partial x}(x^2) = 2x$, $\frac{\partial}{\partial y}(-2xy) = -2x$, $\frac{\partial}{\partial z}(3xz) = 3x \Rightarrow \text{Flux} = \iiint_D 3x\, dx\, dy\, dz =$

$$\int_0^{\pi/2}\int_0^{\pi/2}\int_0^{2} 3\rho \sin\phi \cos\theta (\rho^2 \sin\phi)\, d\rho\, d\phi\, d\theta = 3\pi$$

7. $\frac{\partial}{\partial x}(2xz) = 2z$, $\frac{\partial}{\partial y}(-xy) = -x$, $\frac{\partial}{\partial z}(-z^2) = -2z \Rightarrow \nabla \cdot \mathbf{F} = -x \Rightarrow \text{Flux} = \iiint_D -x\, dV =$

$$\int_0^{2}\int_0^{\sqrt{16-4x^2}}\int_0^{4-y} -x\, dz\, dy\, dx = -\frac{40}{3}$$

9. Let $\rho = \sqrt{x^2+y^2+z^2}$. Then $\frac{\partial \rho}{\partial x} = \frac{x}{\rho}$, $\frac{\partial \rho}{\partial y} = \frac{y}{\rho}$, $\frac{\partial \rho}{\partial z} = \frac{z}{\rho} \Rightarrow \frac{\partial}{\partial x}(\rho x) = \frac{\partial \rho}{\partial x} x + \rho = \frac{x^2}{\rho} + \rho$, $\frac{\partial}{\partial y}(\rho y) = \frac{\partial \rho}{\partial y} y + \rho = \frac{y^2}{\rho} + \rho$, $\frac{\partial}{\partial z}(\rho z) = \frac{\partial \rho}{\partial z} z + \rho = \frac{z^2}{\rho} + \rho \Rightarrow \nabla \cdot \mathbf{F} = \frac{x^2+y^2+z^2}{\rho} + 3\rho = 4\rho$ since $\rho = \sqrt{x^2+y^2+z^2}$

$$\Rightarrow \text{Flux} = \iiint_D 4\rho \, dV = \iiint_D 4\sqrt{x^2+y^2+z^2} \, dx\, dy\, dz =$$

$$\int_0^{2\pi}\int_0^{\pi}\int_1^{\sqrt{2}} (4\rho)\rho^2 \sin\phi \, d\rho\, d\phi\, d\theta = 12\pi$$

11. $\frac{\partial}{\partial x}(5x^3 + 12xy^2) = 15x^2 + 12y^2$, $\frac{\partial}{\partial y}(y^3 + e^y \sin z) = 3y^2 + e^y \sin z$, $\frac{\partial}{\partial z}(5z^3 + e^y \cos z) = 15z^2 - e^y \sin z \Rightarrow$

$$\nabla \cdot \mathbf{F} = 15x^2 + 15y^2 + 15z^2 \Rightarrow \text{Flux} = \iiint_V (15x^2 + 15y^2 + 15z^2)\, dz\, dy\, dz =$$

$$\int_0^{2\pi}\int_0^{\pi}\int_1^{\sqrt{2}} 15\rho^2(\rho^2)\sin\phi\, d\rho\, d\phi\, d\theta = \int_0^{2\pi}\int_0^{\pi}(12\sqrt{2}-3)\sin\phi\, d\phi\, d\theta = \int_0^{2\pi}(24\sqrt{2}-6)d\theta = (48\sqrt{2}-12)\pi$$

13. $\frac{\partial}{\partial x}(x) = 1$, $\frac{\partial}{\partial y}(-2y) = -2$, $\frac{\partial}{\partial z}(z+3) = 1 \Rightarrow \nabla \cdot \mathbf{F} = 0 \Rightarrow$ Flux = 0 over the solid. In the xy–plane, $z = 0$, $\mathbf{n} = -\mathbf{k}$, and $\mathbf{F} = x\,\mathbf{i} - 2y\,\mathbf{j} + 3\,\mathbf{k} \Rightarrow \mathbf{F} \cdot \mathbf{n} = -3 \Rightarrow \text{Flux} = \iint_{z=0} -3\, d\sigma = -3(\text{Area of the square}) =$ -3. In the yz–plane, $x = 0$, $\mathbf{n} = -\mathbf{i}$, $\mathbf{F} = -2y\,\mathbf{j} + (z+3)\,\mathbf{k} \Rightarrow \mathbf{F} \cdot \mathbf{n} = 0 \Rightarrow$ Flux = 0.
In the xz–plane, $y = 0$, $\mathbf{n} = -\mathbf{j}$, $\mathbf{F} = x\,\mathbf{i} + (z+3)\,\mathbf{k} \Rightarrow \mathbf{F} \cdot \mathbf{n} = 0 \Rightarrow$ Flux = 0. $\therefore$ The total flux =
$-3 + 0 + 0 + 1 + (-3) + (\text{Flux of the top}) = 0 \Rightarrow$ Flux of the top = 5

15. a) $\frac{\partial}{\partial x}(x) = 1$, $\frac{\partial}{\partial y}(y) = 1$, $\frac{\partial}{\partial z}(z) = 1 \Rightarrow \nabla \cdot \mathbf{F} = 3 \Rightarrow \text{Flux} = \iiint_D 3\, dV = 3\iiint_D dV =$ 3(Volume of the solid)

b) If $\mathbf{F}$ is orthogonal to $\mathbf{n}$ at every point of S, then $\mathbf{F} \cdot \mathbf{n} = 0$ everywhere $\Rightarrow \text{Flux} = \iint_S \mathbf{F} \cdot \mathbf{n}\, d\sigma = 0$.

But the Flux is 3(Volume of the solid) $\neq 0$. $\therefore$ **F** is not orthogonal to **n** at every point.

17. Volume of D = $\iiint_D dV$. $\iint_S \mathbf{F}\cdot\mathbf{n}\,d\sigma = \iiint_D \nabla\cdot\mathbf{F}\,dV = \iiint_D 3\,dV$ where $\nabla\cdot\mathbf{F} = 3$

Then $\frac{1}{3}\iint_S \mathbf{F}\cdot\mathbf{n}\,d\sigma = \iiint_D dV$ = Volume of D.

19. a) Let $\mathbf{F} = \nabla f$. Then $\iint_S \frac{\partial f}{\partial x}\,d\sigma = \iint_S \nabla f\cdot\mathbf{n}\,d\sigma = \iiint_D \nabla\cdot\nabla f\,dV = \iiint_D 0\,dV = 0.$

b) Let $\mathbf{F} = f\nabla f$. $\frac{\partial f}{\partial n} = \nabla f\cdot\mathbf{n} \Rightarrow f\frac{\partial f}{\partial n} = f\nabla f\cdot\mathbf{n} \Rightarrow \iint_S f\frac{\partial f}{\partial n}\,d\sigma = \iint_S f\nabla f\cdot\mathbf{n}\,d\sigma = \iiint_D \nabla\cdot f\nabla f\,dV$

Since $\nabla\cdot f\nabla f = \frac{\partial f}{\partial x}\nabla f + f\frac{\partial^2 f}{\partial x^2} + \frac{\partial f}{\partial y}\nabla f + f\frac{\partial^2 f}{\partial y^2} + \frac{\partial f}{\partial z}\nabla f + f\frac{\partial^2 f}{\partial z^2} \Rightarrow (\nabla f)^2 + f\nabla^2 f = (\nabla f)^2 + 0 = (\nabla f)^2 = |\nabla f|^2$,

$\iiint_D \nabla\cdot f\nabla f\,dV = \iiint_D |\nabla f|^2\,dV$. $\therefore$ $\iint_S f\frac{\partial f}{\partial n}\,d\sigma = \iiint_D |\nabla f|^2\,dV$

21. Let $\mathbf{F} = f\nabla g$. $\frac{\partial g}{\partial n} = \nabla g\cdot\mathbf{n} \Rightarrow f\frac{\partial g}{\partial n} = f\nabla g\cdot\mathbf{n}$. Then $\iint_S f\frac{\partial g}{\partial n}\,d\sigma = \iint_S \mathbf{F}\cdot\mathbf{n}\,d\sigma = \iiint_D \nabla\cdot\mathbf{F}\,dV =$

$\iiint_D \nabla\cdot f\nabla g\,dV = \iiint_D \left(\frac{\partial f}{\partial x}\nabla g + f\frac{\partial^2 g}{\partial x^2} + \frac{\partial f}{\partial y}\nabla g + f\frac{\partial^2 g}{\partial y^2} + \frac{\partial f}{\partial z}\nabla g + f\frac{\partial^2 g}{\partial z^2}\right)dV$

$\iiint_D \left(\nabla f\cdot\nabla g + f\nabla^2 g\right)dV$

SECTION 14.6 STOKE'S THEOREM

1. $\text{curl}\,\mathbf{F} = \nabla\times\mathbf{F} = 2\,\mathbf{k}$, $\mathbf{n} = \mathbf{k} \Rightarrow \text{curl}\,\mathbf{F}\cdot\mathbf{n} = 2 \Rightarrow d\sigma = dx\,dy \Rightarrow \oint_C \mathbf{F}\cdot d\mathbf{R} = \iint_R 2\,dA =$

2(Area of the ellipse) = 4π

3. $\text{curl}\,\mathbf{F} = \nabla\times\mathbf{F} = -x\,\mathbf{i} - 2x\,\mathbf{j} + (z-1)\,\mathbf{k}$, $\mathbf{n} = \frac{\mathbf{i}+\mathbf{j}+\mathbf{k}}{\sqrt{3}} \Rightarrow \text{curl}\,\mathbf{F}\cdot\mathbf{n} = \frac{1}{\sqrt{3}}(-3x + z - 1) \Rightarrow d\sigma = \frac{\sqrt{3}}{1}\,dA \Rightarrow$

$\oint_C \mathbf{F}\cdot d\mathbf{R} = \iint_R \frac{1}{\sqrt{3}}(-3x + z - 1)\sqrt{3}\,dA = \int_0^1\int_0^{1-x} (-3x + z - 1)\,dy\,dx =$

3. (Continued)

$$\int_0^1\int_0^{1-x} (-3x + (1 - x - y) - 1)\, dy\, dx = \int_0^1\int_0^{1-x} (-4x - y)\, dy\, dx = -\frac{5}{6}$$

5. curl $\mathbf{F} = \nabla \times \mathbf{F} = (2y - 0)\,\mathbf{i} + (2z - 2x)\,\mathbf{j} + (2x - 2y)\,\mathbf{k} = 2y\,\mathbf{i} + (2z - 2x)\,\mathbf{j} + (2x - 2y)\,\mathbf{k}$,

$\mathbf{n} = \mathbf{k} \Rightarrow$ curl $\mathbf{F} \cdot \mathbf{n} = 2x - 2y \Rightarrow d\sigma = dx\, dy \Rightarrow \oint_C \mathbf{F} \cdot d\mathbf{R} = \int_{-1}^{1}\int_{-1}^{1} (2x - 2y)\, dx\, dy = 0$

7. $x = 3\cos t,\ y = 2\sin t \Rightarrow \mathbf{F} = (2\sin t)\,\mathbf{i} + (9\cos^2 t)\,\mathbf{j} + (9\cos^2 t + 16\sin^4 t)\sin e^{\sqrt{6\sin t\cos t\,(0)}}$ and $\mathbf{R} = (3\cos t)\,\mathbf{i} + (2\sin t)\,\mathbf{j} \Rightarrow d\mathbf{R} = (-3\sin t)\,dt\,\mathbf{i} + (2\cos t)\,dt\,\mathbf{j} \Rightarrow \mathbf{F} \cdot d\mathbf{R} = -6\sin^2 t\,dt + 18\cos^3 t\,dt \Rightarrow$

$$\iint_S \nabla \times \mathbf{F} \cdot \mathbf{n}\, d\sigma = \int_0^{2\pi} (-6\sin^2 t + 18\cos^3 t)\, dt = -6\pi$$

9. curl $\mathbf{F} = \nabla \times \mathbf{F} = -2x\,\mathbf{j} + 2\,\mathbf{k}$. Flux of $\nabla \times \mathbf{F} = \iint_S \nabla \times \mathbf{F} \cdot \mathbf{n}\, d\sigma = \oint_C \mathbf{F} \cdot d\mathbf{R}$. Let C be $x = a\cos t$, $y = a\sin t$. Then $\mathbf{R} = (a\cos t)\,\mathbf{i} + (a\sin t)\,\mathbf{j} \Rightarrow d\mathbf{R} = (-a\sin t)\,dt\,\mathbf{i} + (a\cos t)\,dt\,\mathbf{j}$. Then $\mathbf{F} \cdot d\mathbf{R} = ay\sin t\,dt + ax\cos t\,dt = a^2\sin^2 t\,dt + a^2\cos^2 t\,dt = a^2 dt$. $\therefore$ Flux of $\nabla \times \mathbf{F} = \oint_C \mathbf{F} \cdot d\mathbf{R} = \int_0^{2\pi} a^2\, dt = 2\pi a^2$

11. Let S_1 and S_2 be oriented surfaces that span C and that induce the same positive direction on C.

Then $\iint_{S_1} \nabla \times \mathbf{F} \cdot \mathbf{n}_1\, d\sigma_1 = \int_C \mathbf{F} \cdot d\mathbf{R} = \iint_{S_2} \nabla \times \mathbf{F} \cdot \mathbf{n}_2\, d\sigma_2$

13. a) $\mathbf{F} = M\,\mathbf{i} + N\,\mathbf{j} + P\,\mathbf{k} \Rightarrow$ curl $\mathbf{F} = \left(\frac{\partial P}{\partial y} - \frac{\partial N}{\partial z}\right)\mathbf{i} + \left(\frac{\partial M}{\partial z} - \frac{\partial P}{\partial x}\right)\mathbf{j} + \left(\frac{\partial N}{\partial x} - \frac{\partial M}{\partial y}\right)\mathbf{k}$.

$\nabla \cdot \nabla \times \mathbf{F} = \text{div(curl } \mathbf{F}) = \frac{\partial}{\partial x}\left(\frac{\partial P}{\partial y} - \frac{\partial N}{\partial z}\right) + \frac{\partial}{\partial y}\left(\frac{\partial M}{\partial z} - \frac{\partial P}{\partial x}\right) + \frac{\partial}{\partial z}\left(\frac{\partial N}{\partial x} - \frac{\partial M}{\partial y}\right) = \frac{\partial^2 P}{\partial x \partial y} - \frac{\partial^2 N}{\partial x \partial z} + \frac{\partial^2 M}{\partial y \partial z} - \frac{\partial^2 P}{\partial y \partial x} + \frac{\partial^2 N}{\partial z \partial x} - \frac{\partial^2 M}{\partial z \partial y} = 0$ if the partial derivatives are continuous.

b) $\iint_S \nabla \times \mathbf{F} \cdot \mathbf{n}\, d\sigma = \iiint_D \nabla \cdot \nabla \times \mathbf{F}\, dV = \iiint_D 0\, dV = 0$ if the divergence theorem applies.

15. $\mathbf{F} = \nabla f = -\frac{1}{2}(x^2 + y^2 + z^2)^{-3/2}(2x)\,\mathbf{i} - \frac{1}{2}(x^2 + y^2 + z^2)^{-3/2}(2y)\,\mathbf{j} - \frac{1}{2}(x^2 + y^2 + z^2)^{-3/2}(2z)\,\mathbf{k} =$
$-x(x^2 + y^2 + z^2)^{-3/2}\,\mathbf{i} - y(x^2 + y^2 + z^2)^{-3/2}\,\mathbf{j} - z(x^2 + y^2 + z^2)^{-3/2}\,\mathbf{k}$

a) $\mathbf{R} = (a \cos t)\,\mathbf{i} + (a \sin t)\,\mathbf{j},\ 0 \le t \le 2\pi \Rightarrow d\mathbf{R} = (-a \sin t)\,dt\,\mathbf{i} + (a \cos t)\,dt\,\mathbf{j} \Rightarrow$
$\mathbf{F} \cdot d\mathbf{R} = -x(x^2 + y^2 + z^2)^{-3/2}(-a \sin t)\,dt - y(x^2 + y^2 + z^2)^{-3/2}(a \cos t)\,dt =$
$-\frac{a \cos t}{a^3}(-a \sin t)\,dt - \frac{a \sin t}{a^3}(a \cos t)\,dt = 0 \Rightarrow \int_C \mathbf{F} \cdot d\mathbf{R} = 0$

b) $\oint_C \mathbf{F} \cdot d\mathbf{R} = \iint_S \text{curl}\,\mathbf{F} \cdot \mathbf{n}\,d\sigma = \iint_S \text{curl}(\nabla f) \cdot \mathbf{n}\,d\sigma = \iint_S \mathbf{0} \cdot \mathbf{n}\,d\sigma =$
$\iint_S 0\,d\sigma = 0$

17. $\iint_S \nabla \times \mathbf{F} \cdot \mathbf{n}\ d\sigma = \iint_{S_1} \nabla \times \mathbf{F} \cdot \mathbf{n}\ d\sigma + \iint_{S_2} \nabla \times \mathbf{F} \cdot \mathbf{n}\ d\sigma$. Since S_1 and S_2 are joined by the simple closed curve C, each of the above integrals will be equal to a circulation integral on C. But for one, the circulation will be counterclockwise and for the other, the circulation will be clockwise. Since the integrands are the same, the sum will be 0. $\therefore \iint_S \nabla \times \mathbf{F} \cdot \mathbf{n}\ d\sigma = 0$.

19. $r = \sqrt{x^2 + y^2} \Rightarrow r^4 = (x^2 + y^2)^2 \Rightarrow \nabla(r^4) = 4x(x^2 + y^2)\,\mathbf{i} + 4y(x^2 + y^2)\,\mathbf{j} \Rightarrow \mathbf{F} = 4x(x^2 + y^2)\,\mathbf{i} + 4y(x^2 + y^2)\,\mathbf{j}$.

Then $\oint_C \nabla(r^4) \cdot \mathbf{n}\ ds = \iint_R \nabla \cdot \mathbf{F}\,dA = \iint_R \left(4(x^2 + y^2) + 8x^2 + 4(x^2 + y^2) + 8y^2\right) dA =$
$\iint_R 16(x^2 + y^2)\,dA = 16 \iint_R x^2\,dA + 16 \iint_R y^2\,dA = 16\,I_y + 16\,I_x$.

SECTION 14.7 PATH INDEPENDENCE, POTENTIAL FUNCTIONS, AND CONSERVATIVE FIELDS

1. $\frac{\partial P}{\partial y} = x = \frac{\partial N}{\partial z}, \frac{\partial M}{\partial z} = y = \frac{\partial P}{\partial x}, \frac{\partial N}{\partial x} = z = \frac{\partial M}{\partial y} \Rightarrow$ Conservative

3. $\frac{\partial P}{\partial y} = -1 \neq \frac{\partial N}{\partial z} \Rightarrow$ Not Conservative

5. $\frac{\partial N}{\partial x} = 0 \neq \frac{\partial M}{\partial y} \Rightarrow$ Not Conservative

7. $\frac{\partial f}{\partial x} = 2x \Rightarrow f(x,y,z) = x^2 + g(y,z)$. $\frac{\partial f}{\partial y} = \frac{\partial g}{\partial y} = 3y \Rightarrow g(y,z) = \frac{3y^2}{2} + h(z) \Rightarrow f(x,y,z) = x^2 + \frac{3y^2}{2} + h(z)$. $\frac{\partial f}{\partial z} = h'(z)$ $= 4z. \Rightarrow h(z) = 2z^2 + C \Rightarrow f(x,y,z) = x^2 + \frac{3y^2}{2} + 2z^2 + C$

9. $\frac{\partial f}{\partial x} = e^{y+2z} \Rightarrow f(x,y,z) = x\,e^{y+2z} + g(y,z)$. $\frac{\partial f}{\partial y} = x\,e^{y+2z} + \frac{\partial g}{\partial y} = x\,e^{y+2z} \Rightarrow \frac{\partial g}{\partial y} = 0$. Then $f(x,y,z) = x\,e^{y+2z} +$ $h(z)$. $\frac{\partial f}{\partial z} = 2x\,e^{y+2z} + h'(z) = 2x\,e^{y+2z} \Rightarrow h'(z) = 0 \Rightarrow h(z) = C$. $\therefore$ $f(x,y,z) = x\,e^{y+2z} + C$

11. $\frac{\partial f}{\partial x} = \ln x + \sec^2(x+y) \Rightarrow f(x,y,z) = x \ln x - x + \tan(x+y) + g(y,z)$. $\frac{\partial f}{\partial y} = \sec^2(x+y) + \frac{\partial g}{\partial y} =$ $\sec^2(x+y) + \frac{y}{y^2+z^2} \Rightarrow f(x,y,z) = x \ln x - x + \tan(x+y) + \frac{1}{2}\ln(y^2+z^2) + h(z)$. $\frac{\partial f}{\partial z} = \frac{z}{y^2+z^2} + h'(z) = \frac{z}{y^2+z^2}$ $\Rightarrow h'(z) = 0 \Rightarrow h(z) = C$. $\therefore$ $f(x,y,z) = x \ln x - x + \tan(x+y) + \frac{1}{2}\ln(y^2+z^2) + C$.

13. Let $\mathbf{F}(x,y,z) = 2x\,\mathbf{i} + 2y\,\mathbf{j} + 2z\,\mathbf{k} \Rightarrow \frac{\partial P}{\partial y} = 0 = \frac{\partial N}{\partial z}, \frac{\partial M}{\partial z} = 0 = \frac{\partial P}{\partial x}, \frac{\partial N}{\partial x} = 0 = \frac{\partial M}{\partial y} \Rightarrow M\,dx + N\,dy + P\,dz$ is exact. $\frac{\partial f}{\partial x} = 2x \Rightarrow f(x,y,z) = x^2 + g(y,z)$. $\frac{\partial f}{\partial y} = \frac{\partial g}{\partial y} = 2y \Rightarrow g(y,z) = y^2 + h(z) \Rightarrow f(x,y,z) = x^2 + y^2 + h(z)$.
$\frac{\partial f}{\partial z} = h'(z) = 2z \Rightarrow h(z) = z^2 + C$. $\therefore$ $f(x,y,z) = x^2 + y^2 + z^2 + C \Rightarrow \int_{(0,0,0)}^{(2,3,-6)} 2x\,dx + 2y\,dy + 2z\,dz =$ $f(2,3,-6) - f(0,0,0) = 49$

15. Let $\mathbf{F}(x,y,z) = 2xy\,\mathbf{i} + (x^2 - z^2)\,\mathbf{j} - 2yz\,\mathbf{k} \Rightarrow \frac{\partial P}{\partial y} = -2z = \frac{\partial N}{\partial z}, \frac{\partial M}{\partial z} = 0 = \frac{\partial P}{\partial x}, \frac{\partial N}{\partial x} = 2x = \frac{\partial M}{\partial y} \Rightarrow M\,dx + N\,dy +$ $P\,dz$ is exact. $\frac{\partial f}{\partial x} = 2xy \Rightarrow f(x,y,z) = x^2y + g(y,z)$. $\frac{\partial f}{\partial y} = x^2 + \frac{\partial g}{\partial y} = x^2 - z^2 \Rightarrow \frac{\partial g}{\partial y} = -z^2 \Rightarrow g(y,z) = -yz^2 +$ $h(z) \Rightarrow f(x,y,z) = x^2y - yz^2 + h(z)$. $\frac{\partial f}{\partial z} = -2yz + h'(z) = -2yz \Rightarrow h'(z) = 0 \Rightarrow h(z) = C \Rightarrow f(x,y,z) =$ $x^2y - yz^2 + C \Rightarrow \int_{(0,0,0)}^{(1,2,3)} 2xy\,dx + (x^2 - z^2)\,dy - 2yz\,dz = f(1,2,3) - f(0,0,0) = -16$

17. Let $\mathbf{F}(x,y,z) = (\sin y \cos x)\,\mathbf{i} + (\cos y \sin x)\,\mathbf{j} + \mathbf{k} \Rightarrow \frac{\partial P}{\partial y} = 0 = \frac{\partial N}{\partial z}, \frac{\partial M}{\partial z} = 0 = \frac{\partial P}{\partial x}, \frac{\partial N}{\partial x} = \cos y \cos x = \frac{\partial M}{\partial y} \Rightarrow$ $M\,dx + N\,dy + P\,dz$ is exact. $\frac{\partial f}{\partial x} = \sin y \cos x \Rightarrow f(x,y,z) = \sin y \sin x + g(y,z)$. $\frac{\partial f}{\partial y} = \cos y \sin x + \frac{\partial g}{\partial y} =$ $\cos y \sin x \Rightarrow \frac{\partial g}{\partial y} = 0 \Rightarrow g(y,z) = h(z) \Rightarrow f(x,y,z) = \sin y \sin x + h(z)$. $\frac{\partial f}{\partial z} = h'(z) = 1 \Rightarrow h(z) = z + C \Rightarrow$

$$f(x,y,z) = \sin y \sin x + z + C \Rightarrow \int_{(1,0,0)}^{(0,1,1)} \sin y \cos x\,dx + \cos y \sin x\,dy + dz = f(0,1,1) - f(1,0,0) = 1$$

19. Let $\mathbf{F}(x,y,z) = (2\cos y)\,\mathbf{i} + \left(\frac{1}{y} - 2x \sin y\right)\mathbf{j} + \frac{1}{z}\mathbf{k} \Rightarrow \frac{\partial P}{\partial y} = 0 = \frac{\partial N}{\partial z}, \frac{\partial M}{\partial z} = 0 = \frac{\partial P}{\partial x}, \frac{\partial N}{\partial x} = -2\sin y = \frac{\partial M}{\partial y} \Rightarrow$ $M\,dx + N\,dy + P\,dz$ is exact. $\frac{\partial f}{\partial x} = 2\cos y \Rightarrow f(x,y,z) = 2x\cos y + g(y,z)$. $\frac{\partial f}{\partial y} = -2x \sin y + \frac{\partial g}{\partial y} = \frac{1}{y} - 2x\sin y$ $\Rightarrow \frac{\partial g}{\partial y} = \frac{1}{y} \Rightarrow g(y,z) = \ln y + h(z) \Rightarrow f(x,y,z) = 2x\cos y + \ln y + h(z)$. $\frac{\partial f}{\partial z} = h'(z) = \frac{1}{z} \Rightarrow h(z) = \ln z + C \Rightarrow$

$$f(x,y,z) = 2x\cos y + \ln y + \ln z + C \Rightarrow \int_{(0,2,1)}^{(1,\pi/2,2)} 2\cos y\,dx + \left(\frac{1}{y} - 2x\sin y\right)dy + \frac{1}{z}\,dz =$$

$$f\left(1,\frac{\pi}{2},2\right) - f(0,2,1) = \ln\frac{\pi}{2}$$

21. Let $\mathbf{F}(x,y,z) = \frac{1}{y}\mathbf{i} + \left(\frac{1}{z} - \frac{x}{y^2}\right)\mathbf{j} - \frac{y}{z^2}\mathbf{k} \Rightarrow \frac{\partial P}{\partial y} = -\frac{1}{z^2} = \frac{\partial N}{\partial z}, \frac{\partial M}{\partial z} = 0 = \frac{\partial P}{\partial x}, \frac{\partial N}{\partial x} = -\frac{1}{y^2} = \frac{\partial M}{\partial y} \Rightarrow$ $M\,dx + N\,dy + P\,dz$ is exact. $\frac{\partial f}{\partial x} = \frac{1}{y} \Rightarrow f(x,y,z) = \frac{x}{y} + g(y,z)$. $\frac{\partial f}{\partial y} = -\frac{x}{y^2} + \frac{\partial g}{\partial y} = \frac{1}{z} - \frac{x}{y^2} \Rightarrow \frac{\partial g}{\partial y} = \frac{1}{z} \Rightarrow$ $g(y,z) = \frac{y}{z} + h(z) \Rightarrow f(x,y,z) = \frac{x}{y} + \frac{y}{z} + h(z)$. $\frac{\partial f}{\partial z} = -\frac{y}{z^2} + h'(z) = -\frac{y}{z^2} \Rightarrow h'(z) = 0 \Rightarrow h(z) = C \Rightarrow$

$$f(x,y,z) = \frac{x}{y} + \frac{y}{z} + C \Rightarrow \int_{(1,1,1)}^{(2,2,2)} \frac{1}{y}\,dx + \left(\frac{1}{z} - \frac{x}{y^2}\right)dy - \frac{y}{z^2}\,dz = f(2,2,2) - f(1,1,1) = 0$$

23. Let $x - 1 = t$, $y - 1 = 2t$, $z - 1 = -2t$, $0 \le t \le 1 \Rightarrow dx = dt$, $dy = 2\,dt$, $dz = -2\,dt \Rightarrow$

$$\int_{(1,1,1)}^{(2,3,-1)} y\,dx + x\,dy + 4\,dz = \int_0^1 (2t+1)dt + (t+1)2\,dt + 4(-2\,dt) = \int_0^1 (4t - 5)\,dt = -3$$

25. $\frac{\partial P}{\partial y} = 0 = \frac{\partial N}{\partial z}, \frac{\partial M}{\partial z} = 2z = \frac{\partial P}{\partial x}, \frac{\partial N}{\partial x} = 0 = \frac{\partial M}{\partial y} \Rightarrow M\,dx + N\,dy + P\,dz$ is exact $\Rightarrow$ **F** is conservative $\Rightarrow$ path independence.

27. a) $\frac{\partial P}{\partial y} = 0 = \frac{\partial N}{\partial z}, \frac{\partial M}{\partial z} = 0 = \frac{\partial P}{\partial x}, \frac{\partial N}{\partial x} = -\frac{2x}{y^2} = \frac{\partial M}{\partial y} \Rightarrow$ **F** is conservative $\Rightarrow$ there exists an f so that $\mathbf{F} = \nabla f$.
$\frac{\partial f}{\partial x} = \frac{2x}{y} \Rightarrow f(x,y) = \frac{x^2}{y} + g(y) \Rightarrow \frac{\partial f}{\partial y} = -\frac{x^2}{y} + g'(y) = \frac{1 - x^2}{y^2} \Rightarrow g'(y) = \frac{1}{y^2} \Rightarrow g(y) = -\frac{1}{y} + C \Rightarrow f(x,y) = \frac{x^2}{y} -$ $\frac{1}{y} + C$. Let $C = 0$. Then $f(x,y) = \frac{x^2 - 1}{y}$. $\therefore \mathbf{F} = \nabla\left(\frac{x^2 - 1}{y}\right)$.

27. b) $\frac{\partial P}{\partial y} = \cos z = \frac{\partial N}{\partial z}, \frac{\partial M}{\partial z} = 0 = \frac{\partial P}{\partial x}, \frac{\partial N}{\partial x} = \frac{e^x}{y} = \frac{\partial M}{\partial y} \Rightarrow$ Conservative $\Rightarrow$ there exists an f so that $\mathbf{F} = \nabla f$. $\frac{\partial f}{\partial x} = e^x \ln y \Rightarrow f(x,y,z) = e^x \ln y + g(y,z) \Rightarrow \frac{\partial f}{\partial y} = \frac{e^x}{y} + \frac{\partial g}{\partial y} = \frac{e^x}{y} + \sin z \Rightarrow \frac{\partial g}{\partial y} = \sin z \Rightarrow g(y,z) = y \sin z + h(z) \Rightarrow f(x,y,z) = e^x \ln y + y \sin z + h(z) \Rightarrow \frac{\partial f}{\partial z} = y \cos z + h'(z) = y \cos z \Rightarrow h'(z) = 0 \Rightarrow h(z) = C$. Then $f(x,y,z) = e^x \ln y + y \sin z + C$. Let $C = 0$. Then $f(x,y,z) = e^x \ln y + y \sin z$. $\therefore \mathbf{F} = \nabla(e^x \ln y + y \sin z)$.

29. $\mathbf{F} = (x^2 + y)\,\mathbf{i} + (y^2 + x)\,\mathbf{j} + z e^z\,\mathbf{k} \Rightarrow \frac{\partial P}{\partial y} = 0 = \frac{\partial N}{\partial z}, \frac{\partial M}{\partial z} = 0 = \frac{\partial P}{\partial x}, \frac{\partial N}{\partial x} = 1 = \frac{\partial M}{\partial y} \Rightarrow \mathbf{F}$ is conservative. $\frac{\partial f}{\partial x} = x^2 + y \Rightarrow f(x,y,z) = \frac{x^3}{3} + xy + g(y,z) \Rightarrow \frac{\partial f}{\partial y} = x + \frac{\partial g}{\partial y} = y^2 + x \Rightarrow \frac{\partial g}{\partial y} = y^2 \Rightarrow g(y,z) = \frac{y^3}{3} + h(z)$. Then $f(x,y,z) = \frac{x^3}{3} + xy + \frac{y^3}{3} + h(z) \Rightarrow \frac{\partial f}{\partial z} = h'(z) = z e^z \Rightarrow h(z) = z e^z - e^z + C$ (let $C = 0$). $\therefore f(x,y,z) = \frac{x^3}{3} + xy + \frac{y^3}{3} + z e^z - e^z$ is a potential function for $\mathbf{F}$. Thus work $= \int_{t=a}^{t=b} \mathbf{F} \cdot d\mathbf{R} = \int_{t=a}^{t=b} M\,dx + N\,dy + P\,dz = f(1,0,1) - f(1,0,0) = 1$ regardless of the path taken.

31. a) If the differential is exact, then $\frac{\partial P}{\partial y} = \frac{\partial N}{\partial z} \Rightarrow 2ay = cy$ for all $y \Rightarrow 2a = c$. $\frac{\partial M}{\partial z} = \frac{\partial P}{\partial x} \Rightarrow 2cx = 2cx$ for all x. $\frac{\partial N}{\partial x} = \frac{\partial M}{\partial y} \Rightarrow by = 2ay$ for all $y \Rightarrow b = 2a$. $\therefore a = a, b = 2a, c = 2a$.

b) $\mathbf{F} = \nabla f \Rightarrow \mathbf{F}$ is conservative $\Rightarrow \frac{\partial P}{\partial y} = \frac{\partial N}{\partial z} \Rightarrow 2y = cy$ for all $y \Rightarrow c = 2$. $\frac{\partial M}{\partial z} = \frac{\partial P}{\partial x} \Rightarrow 2cx = 2cx$ for all x. $\frac{\partial N}{\partial x} = \frac{\partial M}{\partial y} \Rightarrow by = 2y$ for all $y \Rightarrow b = 2$. $\therefore b = 2, c = 2$.

33. $\frac{\partial P}{\partial y} = 0, \frac{\partial N}{\partial z} = 0, \frac{\partial M}{\partial z} = 0, \frac{\partial P}{\partial x} = 0, \frac{\partial N}{\partial x} = \frac{y^2 - x^2}{(x^2 + y^2)^2}, \frac{\partial M}{\partial y} = \frac{y^2 - x^2}{(x^2 + y^2)^2} \Rightarrow \text{curl}\,\mathbf{F} = \left(\frac{y^2 - x^2}{(x^2 + y^2)^2} - \frac{y^2 - x^2}{(x^2 + y^2)^2}\right)\mathbf{k} = \mathbf{0}$. $x^2 + y^2 = 1 \Rightarrow \mathbf{r} = (a \cos t)\,\mathbf{i} + (a \sin t)\,\mathbf{j} \Rightarrow d\mathbf{r} = (-a \sin t)\,\mathbf{i} + (a \cos t)\,\mathbf{j} \Rightarrow \mathbf{F} = \frac{-a \sin t}{a^2}\,\mathbf{i} + \frac{a \cos t}{a^2}\,\mathbf{j} + z\,\mathbf{k} \Rightarrow \mathbf{F} \cdot d\mathbf{r} = \frac{a^2 \sin^2 t}{a^2} + \frac{a^2 \cos^2 t}{a^2} = 1 \Rightarrow \int_C \mathbf{F} \cdot d\mathbf{r} = \int_0^{2\pi} 1\,dt = 2\pi$

SECTION 14.M MISCELLANEOUS EXERCISES

1. Path 1: $\mathbf{r} = t\mathbf{i} + t\mathbf{j} + t\mathbf{k} \Rightarrow x = t,\ y = t,\ z = t,\ 0 \le t \le 1 \Rightarrow f(g(t),h(t),h(t)) = 3 - 3t^2$ and $\frac{dx}{dt} = 1,\ \frac{dy}{dt} = 1,$

$\frac{dz}{dt} = 1 \Rightarrow \sqrt{\left(\frac{dx}{dt}\right)^2 + \left(\frac{dy}{dt}\right)^2 + \left(\frac{dz}{dt}\right)^2}\ dt = \sqrt{3}\ dt \Rightarrow \int_C f(x,y,z)\ ds = \int_0^1 \sqrt{3}\left(3 - 3t^2\right) dt = 2\sqrt{3}$

Path 2: $\mathbf{r}_1 = t\mathbf{i} + t\mathbf{j},\ 0 \le t \le 1 \Rightarrow x = t,\ y = t,\ z = 0 \Rightarrow f(g(t),h(t),h(t)) = 2t - 3t^2 + 3$ and $\frac{dx}{dt} = 1,\ \frac{dy}{dt} = 1,$

$\frac{dz}{dt} = 0 \Rightarrow \sqrt{\left(\frac{dx}{dt}\right)^2 + \left(\frac{dy}{dt}\right)^2 + \left(\frac{dz}{dt}\right)^2}\ dt = \sqrt{2}\ dt \Rightarrow \int_{C_1} f(x,y,z)\ ds = \int_0^1 \sqrt{2}\left(2t - 3t^2 + 3\right) dt =$

$3\sqrt{2}$. $\mathbf{r}_2 = \mathbf{i} + \mathbf{j} + t\mathbf{k} \Rightarrow x = 1,\ y = 1,\ z = t \Rightarrow f(g(t),h(t),h(t)) = 2 - 2t$ and $\frac{dx}{dt} = 0,\ \frac{dy}{dt} = 0,\ \frac{dz}{dt} = 1 \Rightarrow$

$\sqrt{\left(\frac{dx}{dt}\right)^2 + \left(\frac{dy}{dt}\right)^2 + \left(\frac{dz}{dt}\right)^2}\ dt = dt \Rightarrow \int_{C_2} f(x,y,z)\ ds = \int_0^1 (2 - 2t)\ dt = 1.\ \therefore\ \int_C f(x,y,z)\ ds =$

$\int_{C_1} f(x,y,z)\ ds + \int_{C_2} f(x,y,z)\ ds = 3\sqrt{2} + 1$

3. $\mathbf{r} = (a\cos t)\mathbf{j} + (a\sin t)\mathbf{k} \Rightarrow x = 0,\ y = a\cos t,\ z = a\sin t \Rightarrow f(g(t),h(t),h(t)) = \sqrt{a^2\sin^2 t} = a\,|\sin t|$ and

$\frac{dx}{dt} = 0,\ \frac{dy}{dt} = -a\sin t,\ \frac{dz}{dt} = a\cos t \Rightarrow \sqrt{\left(\frac{dx}{dt}\right)^2 + \left(\frac{dy}{dt}\right)^2 + \left(\frac{dz}{dt}\right)^2}\ dt = a\ dt \Rightarrow \int_C f(x,y,z)\ ds =$

$\int_0^{2\pi} a^2\,|\sin t|\ dt = \int_0^{\pi} a^2 \sin t\ dt + \int_{\pi}^{2\pi} -a^2\sin t\ dt = 4a^2$

5. Let $x = t \Rightarrow \mathbf{r} = t\mathbf{i} + 2\sqrt{t}\,\mathbf{j},\ 0 \le t \le 4 \Rightarrow f(g(t),h(t),k(t)) = 2\sqrt{t},\ \frac{dx}{dt} = dt,\ \frac{dy}{dt} = t^{-1/2},\ \frac{dz}{dt} = 0 \Rightarrow$

$\sqrt{\left(\frac{dx}{dt}\right)^2 + \left(\frac{dy}{dt}\right)^2 + \left(\frac{dz}{dt}\right)^2} = \sqrt{1 + t^{-1}}\ dt \Rightarrow \int_C f(x,y,z)\ ds = \int_0^4 2\sqrt{t}\sqrt{1 + t^{-1}}\ dt = \int_0^4 2\sqrt{t + 1}\ dt =$

$\frac{4}{3}\left(\sqrt{125} - 1\right) \approx 13.57.$

7. $\mathbf{r} = (e^t\cos t)\mathbf{i} + (e^t\sin t)\mathbf{j} + e^t\mathbf{k},\ 0 \le t \le \ln 2 \Rightarrow x = e^t\cos t,\ y = e^t\sin t,\ z = e^t \Rightarrow \frac{dx}{dt} =$

$(e^t\cos t - e^t\sin t),\ \frac{dy}{dt} = (e^t\sin t + e^t\cos t),\ \frac{dz}{dt} = e^t \Rightarrow \sqrt{\left(\frac{dx}{dt}\right)^2 + \left(\frac{dy}{dt}\right)^2 + \left(\frac{dz}{dt}\right)^2}\ dt =$

$\sqrt{(e^t\cos t - e^t\sin t)^2 + (e^t\sin t + e^t\cos t)^2 + \left(e^t\right)^2}\ \ dt = \sqrt{3e^{2t}}\ dt = \sqrt{3}\,e^t\ dt.\ \ \delta\ ds = \sqrt{3}\,e^t\ dt$

7. (Continued)

Then $M = \int_C \delta\, ds = \int_0^{\ln 2} \sqrt{3}\, e^t\, dt = \sqrt{3}$. $M_{xy} = \int_C z\delta\, ds = \int_0^{\ln 2} \sqrt{3}\, e^t\, (e^t)\, dt = \int_0^{\ln 2} \sqrt{3}\, e^{2t}\, dt$

$= \frac{3\sqrt{3}}{2} \Rightarrow \bar{z} = M_{xy}/M = \frac{3\sqrt{3}/2}{\sqrt{3}} = \frac{3}{2}$. $I_z = \int_C (x^2 + y^2)\delta\, ds = \int_0^{\ln 2} (e^{2t}\cos^2 t + e^{2t}\sin^2 t)\sqrt{3}\, e^t\, dt =$

$\int_0^{\ln 2} \sqrt{3}\, e^{3t}\, dt = \frac{7\sqrt{3}}{3} \Rightarrow R_z = \sqrt{I_z/M} = \sqrt{\frac{7\sqrt{3}/3}{\sqrt{3}}} = \sqrt{\frac{7}{3}}$

9. a) $x^2 + y^2 = 1 \Rightarrow \mathbf{r} = (\cos t)\,\mathbf{i} + (\sin t)\,\mathbf{j},\ 0 \le t \le \pi \Rightarrow x = \cos t,\ y = \sin t \Rightarrow \mathbf{F} = (\cos t + \sin t)\,\mathbf{i} - \mathbf{j}$ and $\frac{d\mathbf{r}}{dt} = (-\sin t)\,\mathbf{i} + (\cos t)\,\mathbf{j} \Rightarrow \mathbf{F} \cdot \frac{d\mathbf{r}}{dt} = -\sin t\cos t - \sin^2 t - \cos t \Rightarrow$ Flow $=$

$\int_0^{\pi} (-\sin t\cos t - \sin^2 t - \cos t)\, dt = -\frac{1}{2}\pi$

b) $\mathbf{r} = -t\,\mathbf{i},\ -1 \le t \le 1 \Rightarrow x = -t,\ y = 0 \Rightarrow \mathbf{F} = -t\,\mathbf{i} - t^2\,\mathbf{j}$ and $\frac{d\mathbf{r}}{dt} = -\mathbf{i} \Rightarrow \mathbf{F} \cdot \frac{d\mathbf{r}}{dt} = t \Rightarrow$ Flow $= \int_{-1}^{1} t\, dt =$

0

c) $\mathbf{r}_1 = (1 - t)\,\mathbf{i} - t\,\mathbf{j},\ 0 \le t \le 1 \Rightarrow \mathbf{F}_1 = (1 - 2t)\,\mathbf{i} - (1 - 2t - 2t^2)\,\mathbf{j}$ and $\frac{d\mathbf{r}_1}{dt} = -\mathbf{i} - \mathbf{j} \Rightarrow \mathbf{F}_1 \cdot \frac{d\mathbf{r}_1}{dt} = 2t^2 \Rightarrow$

$\text{Flow}_1 = \int_0^1 2t^2\, dt = \frac{2}{3}$. $\mathbf{r}_2 = -t\,\mathbf{i} + (t - 1)\,\mathbf{j},\ 0 \le t \le 1 \Rightarrow \mathbf{F}_2 = -\mathbf{i} - (2t^2 - 2t + 1)\,\mathbf{j}$ and $\frac{d\mathbf{r}_2}{dt} = -\mathbf{i} +$

$\mathbf{j} \Rightarrow \mathbf{F}_2 \cdot \frac{d\mathbf{r}_2}{dt} = -2t^2 + 2t \Rightarrow \text{Flow}_2 = \int_0^1 (-2t^2 + 2t)\, dt = \frac{1}{3}$. $\therefore$ Flow $= \text{Flow}_1 + \text{Flow}_2 = 1$

11.

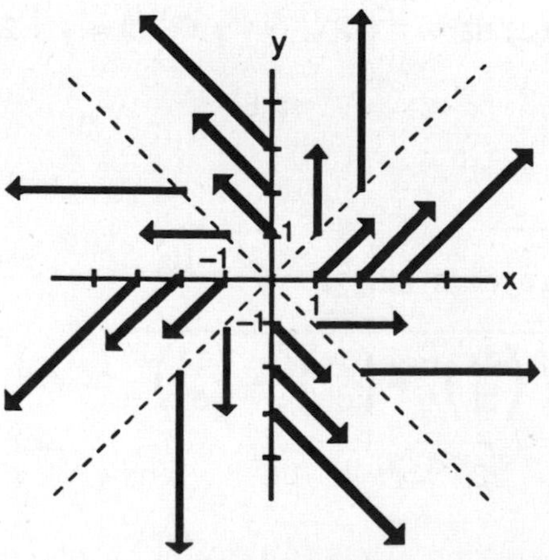

Graph 14.M.11

13. $M = 2xy + x$, $N = xy - y \Rightarrow \frac{\partial M}{\partial x} = 2y + 1$, $\frac{\partial M}{\partial y} = 2x$, $\frac{\partial N}{\partial x} = y$, $\frac{\partial N}{\partial y} = x - 1 \Rightarrow$ Flux =

$$\iint_R (2y + 1 + x - 1)\, dy\, dx = \int_0^1 \int_0^1 (2y + x)\, dy\, dx = \frac{3}{2}. \text{ Circ} = \iint_R (y - 2x)\, dy\, dx =$$

$$\int_0^1 \int_0^1 (y - 2x)\, dy\, dx = -\frac{1}{2}$$

15. Let $M = \frac{\cos y}{x}$, $N = \ln x \sin y \Rightarrow \frac{\partial M}{\partial y} = -\frac{\sin y}{x}$, $\frac{\partial N}{\partial x} = \frac{\sin y}{x}$. Then $\oint_C \ln x \sin y\, dy - \frac{\cos y}{x}\, dx =$

$$\iint_R \left(\frac{\sin y}{x} - \left(-\frac{\sin y}{x}\right)\right) dx\, dy = 0$$

17. Let $M = 8x \sin y$, $N = -8y \cos x \Rightarrow \frac{\partial M}{\partial y} = 8x \cos y$, $\frac{\partial N}{\partial x} = 8y \sin x \Rightarrow \int_C 8x \sin y\, dx - 8y \cos x\, dy =$

$$\iint_R (8y \sin x - 8x \cos y)\, dy\, dx = \int_0^{\pi/2} \int_0^{\pi/2} (8y \sin x - 8x \cos y)\, dy\, dx = 0$$

19. a) $M = \frac{2 \cos t - \sin t}{2}$, $N = \sin t \Rightarrow x = \frac{2 \cos t - \sin t}{2} \Rightarrow dx = \frac{-2 \sin t - \cos t}{2} dt$, $y = \sin t \Rightarrow dy = \cos t\, dt$

$$\text{Area} = \frac{1}{2} \oint_C x\, dy - y\, dx = \frac{1}{2} \int_0^{2\pi} \left(\left(\frac{2 \cos t - \sin t}{2}\right) \cos t - \sin t \left(\frac{-2 \sin t - \cos t}{2}\right)\right) dt = \frac{1}{2} \int_0^{2\pi} dt = \pi$$

b) $x = \frac{2 \cos t - \sin t}{2} = \frac{2\sqrt{1 - \sin^2 t} - \sin t}{2} = \frac{2\sqrt{1 - y^2} - y}{2} \Rightarrow 2x = 2\sqrt{1 - y^2} - y \Rightarrow 2x + y = 2\sqrt{1 - y^2} \Rightarrow$
$4x^2 + 4xy + y^2 = 2(1 - y^2) \Rightarrow 4x^2 + 4xy + 3y^2 = 2 \Rightarrow B^2 - 4AC = -32 \Rightarrow$ Ellipse

21. Let $z = 1 - x - y \Rightarrow f_x(x,y) = -1$, $f_y(x,y) = -1 \Rightarrow \sqrt{f_x{}^2 + f_y{}^2 + 1} = \sqrt{3} \Rightarrow \text{Area} = \iint_R \sqrt{3}\, dx\, dy =$

$\sqrt{3}$(Area of the circular region in the xy–plane) $= \pi\sqrt{3}$

23. $\nabla f = 2x\,\mathbf{i} + 2y\,\mathbf{j} + 2z\,\mathbf{k}$, $\mathbf{p} = \mathbf{k} \Rightarrow |\nabla f| = \sqrt{4x^2 + 4y^2 + 4z^2} = 2\sqrt{x^2 + y^2 + z^2} = 2$ and $|\nabla f \cdot \mathbf{p}| = |2z| = 2z$ since $z \ge 0 \Rightarrow S = \iint_R \frac{2}{2z}\,dA = \iint_R \frac{1}{z}\,dA = \iint_R \frac{1}{\sqrt{1 - x^2 - y^2}}\,dx\,dy =$

$$\int_0^{2\pi}\int_0^{1/2} \frac{1}{\sqrt{1 - r^2}}\,r\,dr\,d\theta = \int_0^{2\pi}\left(1 - \frac{\sqrt{3}}{2}\right)d\theta = 2\pi - \pi\sqrt{3}$$

25. $\frac{x}{a} + \frac{y}{b} + \frac{z}{c} = 1 \Rightarrow$ x–intercept = a, y–intercept = b, z–intercept = c. $F = \frac{x}{a} + \frac{y}{b} + \frac{z}{c} \Rightarrow \nabla F = \frac{1}{a}\mathbf{i} + \frac{1}{b}\mathbf{j} + \frac{1}{c}\mathbf{k} \Rightarrow$

$|\nabla F| = \sqrt{\frac{1}{a^2} + \frac{1}{b^2} + \frac{1}{c^2}}$. $\mathbf{p} = \mathbf{k} \Rightarrow |\nabla F \cdot \mathbf{k}| = \frac{1}{c}$ if since $c > 0$. Area $= \iint_R \frac{\sqrt{\frac{1}{a^2} + \frac{1}{b^2} + \frac{1}{c^2}}}{1/c}\,dA =$

$$c\sqrt{\frac{1}{a^2} + \frac{1}{b^2} + \frac{1}{c^2}} \iint_R dA = \frac{1}{2}abc\sqrt{\frac{1}{a^2} + \frac{1}{b^2} + \frac{1}{c^2}}$$

27. a) $\nabla f = 2y\,\mathbf{j} - \mathbf{k}$, $\mathbf{p} = \mathbf{k} \Rightarrow |\nabla f| = \sqrt{4y^2 + 1}$ and $|\nabla f \cdot \mathbf{p}| = 1 \Rightarrow d\sigma = \sqrt{4y^2 + 1}\,dx\,dy \Rightarrow$

$$\iint_S g(x,y,z)\,d\sigma = \iint_S \frac{z}{\sqrt{4y^2 + 1}}\sqrt{4y^2 + 1}\,dx\,dy = \iint_S y(y^2 - 1)\,d\sigma =$$

$$\int_{-1}^{1}\int_0^3 (y^3 - y)\,dx\,dy = 0$$

b) $$\iint_S g(x,y,z)\,d\sigma = \iint_S \frac{z}{\sqrt{4y^2 + 1}}\sqrt{4y^2 + 1}\,dx\,dy = \int_{-1}^{1}\int_0^3 (y^2 - 1)\,dx\,dy = -4$$

29. On the face, $z = 1$: $G(x,y,z) = G(x,y,1) = z \Rightarrow \nabla G = \mathbf{k} \Rightarrow |\nabla G| = 1$. $\mathbf{n} = \mathbf{k} \Rightarrow \mathbf{F} \cdot \mathbf{n} = 2xz = 2x$ since $z = 1$

$d\sigma = dA \Rightarrow \text{Flux} = \iint_R 2x\,dx\,dy = \int_0^1\int_0^1 2x\,dx\,dy = 1$. On the face, $z = 0$: $G(x,y,z) = G(x,y,0) =$

$z \Rightarrow \nabla G = \mathbf{k} \Rightarrow |\nabla G| = 1$. $\mathbf{n} = -\mathbf{k} \Rightarrow \mathbf{F} \cdot \mathbf{n} = -2xz = 0$ since $z = 0 \Rightarrow \text{Flux} = \iint_R 0\,dx\,dy = 0$

On the face, $x = 1$: $G(x,y,z) = G(1,y,z) = x \Rightarrow \nabla G = \mathbf{i} \Rightarrow |\nabla G| = 1$. $\mathbf{n} = \mathbf{i} \Rightarrow \mathbf{F} \cdot \mathbf{n} = 2xy = 2y$ since $x = 1$

29. (Continued)

Flux = $\int_0^1 \int_0^1 2y\, dy\, dz = 1$ On the face, x = 0: $G(x,y,z) = G(0,y,z) = x \Rightarrow \nabla G = \mathbf{i} \Rightarrow |\nabla G| = 1$. $\mathbf{n} = -\mathbf{i}$

$\Rightarrow \mathbf{F} \cdot \mathbf{n} = -2xy = 0$ since $x = 0 \Rightarrow$ Flux = 0. On the face, y = 1: $G(x,y,z) = G(x,1,z) = y \Rightarrow \nabla G = \mathbf{j} \Rightarrow |\nabla G|$

$= 1$. $\mathbf{n} = \mathbf{j} \Rightarrow \mathbf{F} \cdot \mathbf{n} = 2yz = 2z$ since $y = 1 \Rightarrow$ Flux = $\int_0^1 \int_0^1 2z\, dz\, dx = 1$. On the face, y = 0: $G(x,y,z) =$

$G(z,0,z) = y \Rightarrow \nabla G = \mathbf{j} \Rightarrow |\nabla G| = 1$. $\mathbf{n} = -\mathbf{j} \Rightarrow \mathbf{F} \cdot \mathbf{n} = -2yz = 0$ since $y = 0 \Rightarrow$ Flux = 0.

$\therefore$ Total Flux = 3

31. Because of symmetry $\bar{x} = \bar{y} = 0$. Let $F(x,y,z) = x^2 + y^2 + z^2 = 25 \Rightarrow \nabla F = 2x\,\mathbf{i} + 2y\,\mathbf{j} + 2z\,\mathbf{k} \Rightarrow$

$|\nabla F| = \sqrt{4x^2 + 4y^2 + 4z^2} = 10$, $\mathbf{p} = \mathbf{k} \Rightarrow |\nabla F \cdot \mathbf{p}| = 2z$ since $z \geq 0 \Rightarrow M = \iint_R \delta(x,y,z)\, d\sigma =$

$\iint_R z\left(\frac{10}{2z}\right) dA = \iint_R 5\, dA = 5(\text{Area of the circular region}) = 80\pi$. $M_{xy} = \iint_R z\delta\, d\sigma =$

$\iint_R z\, dA = \iint_R 5\sqrt{25 - x^2 - y^2}\, dx\, dy = \int_0^{2\pi} \int_0^4 5\sqrt{25 - r^2}\, r\, dr\, d\theta = \int_0^{2\pi} \frac{490}{3}\, d\theta = \frac{980}{3}\pi$

$\therefore\ \bar{z} = \dfrac{\frac{980}{3}\pi}{80\pi} = \frac{49}{12}$. Thus $(\bar{x},\bar{y},\bar{z}) = \left(0,0,\frac{49}{12}\right)$. $I_z = \iint_R (x^2 + y^2)\delta\, d\sigma = \iint_R 5(x^2 + y^2)\, dx\, dy =$

$\int_0^{2\pi} \int_0^4 5r^3\, dr\, d\theta = \int_0^{2\pi} 320\, d\theta = 640\pi$. $R_z = \sqrt{I_z/M} = \sqrt{\frac{640\pi}{80\pi}} = 2\sqrt{2}$

33. $\frac{\partial}{\partial x}(2xy) = 2y$, $\frac{\partial}{\partial y}(2yz) = 2z$, $\frac{\partial}{\partial z}(2xz) = 2x \Rightarrow \nabla \cdot \mathbf{F} = 2x + 2y + 2z \Rightarrow$ Flux $= \iiint_D (2x + 2y + 2z)\, dV$

$\int_0^1 \int_0^1 \int_0^1 (2x + 2y + 2z)\, dx\, dy\, dz = \int_0^1 \int_0^1 (1 + 2y + 2z)\, dy\, dz = \int_0^1 (2 + 2z)\, dz = 3$

35. $\frac{\partial}{\partial x}(-2x) = -2, \frac{\partial}{\partial y}(-3y) = -3, \frac{\partial}{\partial z}(z) = 1 \Rightarrow \nabla \cdot \mathbf{F} = -4 \Rightarrow \text{Flux} = \iiint_D -4\, dV =$

$$-4\int_0^{2\pi}\int_0^1\int_{r^2}^{\sqrt{2-r^2}} dz\, r\, dr\, d\theta = -4\int_0^{2\pi}\int_0^1 (r\sqrt{2-r^2} - r^3)\, dr\, d\theta = -4\int_0^{2\pi}\left(-\frac{7}{12}+\frac{2}{3}\sqrt{2}\right) d\theta =$$

$$\frac{2}{3}\pi\left(7-8\sqrt{2}\right)$$

37. Let $f(x,y,z) = y - \ln x \Rightarrow \nabla f = -\frac{1}{x}\mathbf{i} + \mathbf{j}$. $\nabla f \cdot \mathbf{j} = 1 \Rightarrow$ **n** is in the direction of ∇f since $\mathbf{n} \cdot \mathbf{j} > 0$. $|\nabla f| =$ $\sqrt{\frac{1}{x^2}+1} = \frac{\sqrt{1+x^2}}{x}$ since $1 \le x \le e$. $\therefore \mathbf{n} = \frac{\nabla f}{|\nabla f|} = \frac{x\left(-\frac{1}{x}\mathbf{i}+\mathbf{j}\right)}{\sqrt{x^2+1}} = -\frac{1}{\sqrt{x^2+1}}\mathbf{i} + \frac{x}{\sqrt{x^2+1}}\mathbf{j} \Rightarrow \mathbf{F}\cdot\mathbf{n} =$ $\frac{2yx}{\sqrt{x^2+1}} = \frac{2x\ln x}{\sqrt{x^2+1}}$. $d\sigma = \frac{|\nabla f|}{|\nabla f \cdot \mathbf{j}|}dA$ where $\mathbf{p} = \mathbf{j} \Rightarrow |\nabla f \cdot \mathbf{j}| = 1 \Rightarrow d\sigma = \frac{\sqrt{1+x^2}}{x}dA$. $\therefore$ Flux =

$$\iint_S \mathbf{F}\cdot\mathbf{n}\, d\sigma = \iint_R \left(\frac{2x\ln x}{\sqrt{x^2+1}}\right)\left(\frac{\sqrt{1+x^2}}{x}\right)dA = \int_0^1\int_1^e 2\ln x\, dx\, dz = \int_0^1 2\, dz = 2$$

39. $\mathbf{F} = 3xz^2\,\mathbf{i} + y\,\mathbf{j} - z^3\,\mathbf{k} \Rightarrow \nabla\cdot\mathbf{F} = 3z^2 + 1 - 3z^2 = 1$. $\therefore$ Flux $= \iint_S \mathbf{F}\cdot\mathbf{n}\, d\sigma = \iiint_D \nabla\cdot\mathbf{F}\, dV =$

$$= \int_0^4\int_0^{\sqrt{16-x^2}}\int_0^{y/2} 1\, dz\, dy\, dx = \frac{8}{3}.$$

41. $\nabla f = 2\,\mathbf{i} + 6\,\mathbf{j} - 3\,\mathbf{k} \Rightarrow \nabla \times \mathbf{F} = -2y\,\mathbf{k}$. $\mathbf{n} = \frac{2\,\mathbf{i}+6\,\mathbf{j}-3\,\mathbf{k}}{\sqrt{4+36+9}} = \frac{2\,\mathbf{i}+6\,\mathbf{j}-3\,\mathbf{k}}{7} \Rightarrow \nabla\times\mathbf{F}\cdot\mathbf{n} = \frac{6}{7}y$. $\mathbf{p} = \mathbf{k} \Rightarrow$ $|\nabla f\cdot\mathbf{p}| = 3 \Rightarrow d\sigma = \frac{7}{3}dA \Rightarrow \oint_C \mathbf{F}\cdot d\mathbf{r} = \iint_R \frac{6}{7}y\, d\sigma = \iint_R \frac{6}{7}y\left(\frac{7}{3}dA\right) = \iint_R 2y\, dx\, dy =$

$$\int_0^{2\pi}\int_0^1 2r\sin\theta\; r\, dr\, d\theta = \int_0^{2\pi}\frac{2}{3}\sin\theta\, d\theta = 0$$

43. $\frac{\partial P}{\partial y} = 0 = \frac{\partial N}{\partial z}, \frac{\partial M}{\partial z} = 0 = \frac{\partial P}{\partial x}, \frac{\partial N}{\partial x} = 0 = \frac{\partial M}{\partial y} \Rightarrow$ Conservative

45. $\frac{\partial P}{\partial y} = 0 \neq ye^z = \frac{\partial N}{\partial z} \Rightarrow$ Not Conservative

47. $\frac{\partial f}{\partial x} = 2 \Rightarrow f(x,y,z) = 2x + g(y,z)$. $\frac{\partial f}{\partial y} = \frac{\partial g}{\partial y} = 2y + z \Rightarrow g(y,z) = y^2 + zy + h(z) \Rightarrow f(x,y,z) = 2x + y^2 + zy + h(z)$.
$\frac{\partial f}{\partial z} = y + h'(z) = y + 1 \Rightarrow h'(z) = 1 \Rightarrow h(z) = z + C$. $\therefore$ $f(x,y,z) = 2x + y^2 + zy + z + C$

49. $\frac{\partial P}{\partial y} = -\frac{1}{2}(x+y+z)^{-3/2} = \frac{\partial N}{\partial z}, \frac{\partial M}{\partial z} = -\frac{1}{2}(x+y+z)^{-3/2} = \frac{\partial P}{\partial x}, \frac{\partial N}{\partial x} = -\frac{1}{2}(x+y+z)^{-3/2} = \frac{\partial M}{\partial y} \Rightarrow M\,dx +$
$N\,dy + P\,dz$ is exact. $\frac{\partial f}{\partial x} = \frac{1}{\sqrt{x+y+z}} \Rightarrow f(x,y,z) = 2\sqrt{x+y+z} + g(y,z)$. $\frac{\partial f}{\partial y} = \frac{1}{\sqrt{x+y+z}} + \frac{\partial g}{\partial y} = \frac{1}{\sqrt{x+y+z}}$
$\Rightarrow \frac{\partial g}{\partial y} = 0 \Rightarrow g(y,z) = h(z) \Rightarrow f(x,y,z) = 2\sqrt{x+y+z} + h(z)$. $\frac{\partial f}{\partial z} = \frac{1}{\sqrt{x+y+z}} + h'(z) = \frac{1}{\sqrt{x+y+z}} \Rightarrow h'(z) = 0$
$\Rightarrow h(z) = C \Rightarrow f(x,y,z) = 2\sqrt{x+y+z} + C \Rightarrow$ $\int_{(-1,1,1)}^{(4,-3,0)} \frac{dx + dy + dz}{\sqrt{x+y+z}} = f(4,-3,0) - f(-1,1,1) = 0$

51. Over Path 1: $\mathbf{r} = t\,\mathbf{i} + t\,\mathbf{j} + t\,\mathbf{k} \Rightarrow x = t, y = t, z = t$ and $d\mathbf{r} = (\mathbf{i} + \mathbf{j} + \mathbf{k})\,dt \Rightarrow \mathbf{F} = 2t^2\,\mathbf{i} + \mathbf{j} + t^2\,\mathbf{k} \Rightarrow$
$\mathbf{F} \cdot d\mathbf{r} = (3t^2 + 1)\,dt \Rightarrow \text{Work} = \int_0^1 (3t^2 + 1)\,dt = 2$
Over Path 2: $\mathbf{r}_1 = t\,\mathbf{i} + t\,\mathbf{j}, 0 \le t \le 1 \Rightarrow x = t, y = t, z = 0$ and $d\mathbf{r}_1 = (\mathbf{i} + \mathbf{j})\,dt \Rightarrow \mathbf{F}_1 = 2t^2\,\mathbf{i} + \mathbf{j} + t^2\,\mathbf{k} \Rightarrow$
$\mathbf{F}_1 \cdot d\mathbf{r}_1 = (2t^2 + 1)\,dt \Rightarrow \text{Work}_1 = \int_0^1 (2t^2 + 1)\,dt = \frac{5}{3}$. $\mathbf{r}_2 = \mathbf{i} + \mathbf{j} + t\,\mathbf{k}, 0 \le t \le 1 \Rightarrow x = 1, y = 1, z = t$ and
$d\mathbf{r}_2 = \mathbf{k}\,dt \Rightarrow \mathbf{F}_2 = 2\,\mathbf{i} + \mathbf{j} + \mathbf{k} \Rightarrow \mathbf{F}_2 \cdot d\mathbf{r}_2 = dt \Rightarrow \text{Work}_2 = \int_0^1 dt = 1$. $\therefore$ $\text{Work} = \text{Work}_1 + \text{Work}_2 = \frac{8}{3}$

53. $\mathbf{F} = x\,\mathbf{i} + y\,\mathbf{j} + z\,\mathbf{k} \Rightarrow \frac{\partial P}{\partial y} = 0 = \frac{\partial N}{\partial z}, \frac{\partial M}{\partial z} = 0 = \frac{\partial P}{\partial x}, \frac{\partial N}{\partial x} = 0 = \frac{\partial M}{\partial y} \Rightarrow \mathbf{F}$ is conservative $\Rightarrow$ curl $\mathbf{F} = \mathbf{0} \Rightarrow$
circulation is 0.

55. a) $\mathbf{r} = (e^t\cos t)\,\mathbf{i} + \left(e^t\sin t\right)\mathbf{j} \Rightarrow x = e^t\cos t, y = e^t\sin t$ from $(1,0)$ to $\left(e^{2\pi},0\right) \Rightarrow 0 \le t \le 2\pi \Rightarrow$
$d\mathbf{r} = \left(e^t\cos t = e^t\sin t\right)dt\,\mathbf{i} + \left(e^t\sin t + e^t\cos t\right)dt\,\mathbf{j}$. $\mathbf{F} = \frac{x\,\mathbf{i} + y\,\mathbf{j}}{(x^2 + y^2)^{3/2}} = \frac{e^t\cos t\,\mathbf{i} + e^t\sin t\,\mathbf{j}}{\left(e^{2t}\cos^2 t + e^{2t}\sin^2 t\right)^{3/2}} =$
$\frac{\cos t}{e^{2t}}\,\mathbf{i} + \frac{\sin t}{e^{2t}}\,\mathbf{j}$. $\therefore$ $\mathbf{F} \cdot d\mathbf{r} = \left(\frac{\cos^2 t}{e^t} - \frac{\sin t\cos t}{e^t} + \frac{\sin^2 t}{e^t} + \frac{\sin t\cos t}{e^t}\right)dt = e^{-t}\,dt$. Thus the work =
$\int_0^{2\pi} e^{-t}\,dt = -e^{-2\pi} + 1$.

55. b) $\mathbf{F} = \frac{x\,\mathbf{i} + y\,\mathbf{j}}{(x^2+y^2)^{3/2}} \Rightarrow \frac{\partial f}{\partial x} = \frac{x}{(x^2+y^2)^{3/2}} \Rightarrow f(x,y,z) = -(x^2+y^2)^{-1/2} + g(y,z).$ $\frac{\partial f}{\partial y} = \frac{y}{(x^2+y^2)^{3/2}} + \frac{\partial g}{\partial y} = \frac{y}{(x^2+y^2)^{3/2}} \Rightarrow g(y,z) = C.$ $\therefore$ $f(x,y,z) = -(x^2+y^2)^{-1/2} + C$ is a potential function for $\mathbf{F} \Rightarrow$

$$\int_C \mathbf{F}\cdot d\mathbf{r} = f\left(e^{2\pi},0\right) - f(1,0) = 1 - e^{-2\pi}$$

57. $dx = (-2\sin t + 2\sin 2t)\,dt$, $dy = (2\cos t - 2\cos 2t)\,dt$. Area $= \frac{1}{2}\oint_C x\,dy - y\,dx =$

$$\frac{1}{2}\int_0^{2\pi} \big((2\cos t - \cos 2t)(2\cos t - 2\cos 2t) - (2\sin t - \sin 2t)(-2\sin t + 2\sin 2t)\big)\,dt =$$

$$\frac{1}{2}\int_0^{2\pi} (6 - 6\cos t)\,dt = 6\pi$$

59. $dx = \cos 2t\,dt$, $dy = \cos t\,dt$. Area $= \frac{1}{2}\oint_C x\,dy - y\,dx = \frac{1}{2}\int_0^{\pi}\left(\frac{1}{2}\sin 2t\cos t - \sin t\cos 2t\right)dt =$

$$\frac{1}{2}\int_0^{\pi} (-\sin t\cos^2 t + \sin t)\,dt = \frac{2}{3}$$

CHAPTER 15

DIFFERENTIAL EQUATIONS

SECTION 15.1 SEPARABLE FIRST ORDER EQUATIONS

1. First Order, Non–Linear

3. Fourth Order, Linear

5. a) $y = x^2 \Rightarrow y' = 2x,\ y'' = 2 \Rightarrow xy'' - y' = 2x - 2x = 0$

   b) $y = 1 \Rightarrow y' = 0,\ y'' = 0 \Rightarrow xy'' - y' = 0$

   c) $y = C_1x^2 + C_2 \Rightarrow y' = 2C_1x,\ y'' = 2C_1 \Rightarrow xy'' - y' = x(2C_1) - 2C_1x = 0$

7. a) $y = e^{-x} \Rightarrow y' = -e^{-x} \Rightarrow 2y' + 3y = 2(-e^{-x}) + 3e^{-x} = e^{-x}$

   b) $y = e^{-x} + e^{-3x/2} \Rightarrow y' = -e^{-x} - \frac{3}{2}e^{-3x/2} \Rightarrow 2y' + 3y = 2\left(-e^{-x} - \frac{3}{2}e^{-3x/2}\right) + 3\left(e^{-x} + e^{-3x/2}\right) = e^{-x}$

   c) $y = e^{-x} + Ce^{-3x/2} \Rightarrow y' = -e^{-x} - \frac{3}{2}Ce^{-3x/2} \Rightarrow$
   $2y' + 3y = 2\left(-e^{-x} - \frac{3}{2}Ce^{-3x/2}\right) + 3\left(e^{-x} + Ce^{-3x/2}\right) = e^{-x}$

9. $y = \frac{1}{x}\int_1^x \frac{e^t}{t}\,dt \Rightarrow y' = -\frac{1}{x^2}\int_1^x \frac{e^t}{t}\,dt + \frac{1}{x}\left(\frac{e^x}{x}\right) \Rightarrow x^2y' + xy = x^2\left(-\frac{1}{x^2}\right)\int_1^x \frac{e^t}{t}\,dt + x^2\left(\frac{e^x}{x^2}\right) + x\left(\frac{1}{x}\right)\int_1^x \frac{e^t}{t}\,dt = e^x$

11. $y'' = -32 \Rightarrow y' = -32x + C_1.\ y'(5) = 0 \Rightarrow -32(5) + C_1 = 0 \Rightarrow C_1 = 160 \Rightarrow y' = -32x + 160 \Rightarrow$
    $y = -16x^2 + 160x + C_2.\ y(5) = 400 \Rightarrow -16(5)^2 + 160(5) + C_2 = 400 \Rightarrow C_2 = 0 \Rightarrow y = 160x - 16x^2$

13. $y = 3\cos 2x - \sin 2x \Rightarrow y(0) = 3\cos 0 - \sin 0 = 3$ and $y' = -6\sin 2x - 2\cos 2x \Rightarrow y'(0) = -6\sin 0 - 2\cos 0 = -2$ and $y'' = -12\cos 2x + 4\sin 2x \Rightarrow y'' + 4y = -12\cos 2x + 4\sin 2x + 4(3\cos 2x - \sin 2x) = 0$

15. $\frac{dy}{dx} = \frac{2y^2 + 2}{x^2 - 1} \Rightarrow \frac{dy}{2(y^2 + 1)} = \frac{dx}{x^2 - 1} \Rightarrow \frac{1}{2}\int \frac{dy}{y^2 + 1} = \int \frac{dx}{x^2 - 1} \Rightarrow \frac{1}{2}\tan^{-1}y = \frac{1}{2}\ln\left|\frac{x - 1}{x + 1}\right| + C \Rightarrow$
    $\tan^{-1}y = \ln\left|\frac{x - 1}{x + 1}\right| + C'.\ y(0) = 1 \Rightarrow \tan^{-1}1 = \ln|-1| + C' \Rightarrow \frac{\pi}{4} = C'.\ \therefore \tan^{-1}y = \ln\left|\frac{x - 1}{x + 1}\right| + \frac{\pi}{4}.$

17. $\frac{dy}{dx} = \frac{e^{y+\sqrt{x}}}{\sqrt{x}\,y^3} \Rightarrow \frac{y^3}{e^y}\,dy = \frac{e^{\sqrt{x}}}{\sqrt{x}}\,dx \Rightarrow \int \frac{y^3}{e^y}\,dy = \int \frac{e^{\sqrt{x}}}{\sqrt{x}}\,dx \Rightarrow -y^3e^{-y} - 3y^2e^{-y} - 6ye^{-y} - 6e^{-y} = 2e^{\sqrt{x}} +$ C (The left hand side is done by parts.) $y\left(\ln^2 2\right) = 0 \Rightarrow -6e^0 = 2e^{\sqrt{\ln^2 2}} + C \Rightarrow C = -10.$
$\therefore -y^3e^{-y} - 3y^2e^{-y} - 6ye^{-y} - 6e^{-y} = e^{\sqrt{x}} - 10$

19. $\frac{dy}{dx} = \sqrt{x^2 - x^2y^2 + 1 - y^2} = \sqrt{(1-y^2)(x^2+1)} = \sqrt{1-y^2}\,\sqrt{x^2+1} \Rightarrow \frac{dy}{\sqrt{1-y^2}} = \sqrt{x^2+1}\,dx \Rightarrow$
$\int \frac{dy}{\sqrt{1-y^2}} = \int \sqrt{x^2+1}\;dx \Rightarrow \sin^{-1}y = \frac{1}{2}x\sqrt{x^2+1} + \frac{1}{2}\ln\left|x + \sqrt{x^2+1}\right| + C.\; y(0) = -\frac{1}{2} \Rightarrow \sin^{-1}\left(-\frac{1}{2}\right)$
$= C \Rightarrow C = -\frac{\pi}{6}.\; \therefore \sin^{-1}y = \frac{1}{2}x\sqrt{x^2+1} + \frac{1}{2}\ln\left|x + \sqrt{x^2+1}\right| - \frac{\pi}{6}$

21. $\frac{dT}{dt} = k(T - T_s) \Rightarrow \frac{dT}{T - T_s} = k\,dt \Rightarrow \int \frac{dT}{T - T_s} = \int k\,dt \Rightarrow \ln(T - T_s) = kt + C \Rightarrow T - T_s = C_1\,e^{kt}$
Given $T_s = 20°$ C and the object cools from 100°C to 40°C in 20 minutes, then $T = 100°C \Rightarrow t_0 = 0 \Rightarrow$ $100° - 20° = C_1e^{k(0)} \Rightarrow C_1 = 80° \Rightarrow T - T_s = 80\,e^{kt}.$ $t = 20$ min $\Rightarrow T = 40°C \Rightarrow 40° - 20° = 80\,e^{k(20)}$
$\Rightarrow \frac{1}{4} = e^{20k} \Rightarrow \ln\frac{1}{4} = 20\,k \Rightarrow k = \frac{\ln\frac{1}{4}}{20} \approx -0.069 \Rightarrow T - T_s = 80\,e^{-0.069\,t}.$ $T = 60°C \Rightarrow 60° - 20° =$
$80\,e^{-0.069\,t} \Rightarrow \frac{1}{2} = e^{-0.069\,t} \Rightarrow t = \frac{\ln\frac{1}{2}}{-0.069} \Rightarrow t \approx 10.05$ minutes

23. $(x^2 + y^2)\,dx + xy\,dy = 0 \Rightarrow \left(1 + \frac{y^2}{x^2}\right)dx + \frac{xy}{x^2}\,dy = 0 \Rightarrow \left(1 + \left(\frac{y}{x}\right)^2\right)dx + \frac{y}{x}\,dy = 0 \Rightarrow \frac{y}{x}\,dy =$
$-\left(1 + \left(\frac{y}{x}\right)^2\right) \Rightarrow \frac{dy}{dx} = \frac{-\left(1 + \left(\frac{y}{x}\right)^2\right)}{y/x} \Rightarrow$ Homogeneous. $v = \frac{y}{x} \Rightarrow F(v) = \frac{-(1+v^2)}{v} \Rightarrow \frac{dx}{x} + \frac{dv}{v + \frac{1+v^2}{v}} =$
$0 \Rightarrow \frac{dx}{x} + \frac{v\,dv}{1 + 2v^2} = 0 \Rightarrow \ln|x| + \frac{1}{4}\ln(1 + 2v^2) = C \Rightarrow |x|\left(1 + 2v^2\right)^{1/4} = e^C \Rightarrow x^4(1 + 2v^2) = C_1$ where
$C_1 = (e^C)^4 \Rightarrow x^4\left(1 + 2\left(\frac{y^2}{x^2}\right)\right) = C_1 \Rightarrow x^4 + 2x^2y^2 = C_1$

25. $\left(x\,e^{y/x} + y\right)dx - x\,dy = 0 \Rightarrow \left(e^{y/x} + \frac{y}{x}\right)dx - dy = 0 \Rightarrow \frac{dy}{dx} = e^{y/x} + \frac{y}{x} \Rightarrow$ Homogeneous. $v = \frac{y}{x} \Rightarrow$
$F(v) = e^v + v \Rightarrow \frac{dx}{x} + \frac{dv}{v - (e^v + v)} = 0 \Rightarrow \frac{dx}{x} + \frac{dv}{-e^v} = 0 \Rightarrow \ln|x| + e^{-v} = C \Rightarrow \ln|x| + e^{-y/x} = C$

27. $\frac{dy}{dx} = \frac{y}{x} + \cos\left(\frac{y-x}{x}\right) \Rightarrow \frac{dy}{dx} = \frac{y}{x} + \cos\left(\frac{y}{x} - 1\right) \Rightarrow$ Homogeneous. $v = \frac{y}{x} \Rightarrow F(v) = v + \cos(v-1) \Rightarrow$
$\frac{dx}{x} + \frac{dv}{v - (v + \cos(v-1))} = 0 \Rightarrow \frac{dx}{x} - \sec(v-1)\,dv = 0 \Rightarrow \ln|x| + \ln|\sec(v-1) + \tan(v-1)| = C \Rightarrow$

27. (Continued)

$$\frac{x}{\sec(v-1)+\tan(v-1)} = \pm e^C \Rightarrow \frac{x}{\sec\left(\frac{y-x}{x}\right)+\tan\left(\frac{y-x}{x}\right)} = C_1 \Rightarrow x = C_1\left(\sec\left(\frac{y-x}{x}\right)+\tan\left(\frac{y-x}{x}\right)\right)$$

$$y(2) = 2 \Rightarrow 2 = C_1\left(\sec\left(\frac{2-2}{2}\right)+\tan\left(\frac{2-2}{2}\right)\right) \Rightarrow C_1 = 2 \Rightarrow x = 2\left(\sec\left(\frac{y-x}{x}\right)+\tan\left(\frac{y-x}{x}\right)\right)$$

29. $x\,dx + 2y\,dy = 0 \Rightarrow \frac{x^2}{2} + y^2 = C.$ $2y\,dx - x\,dy = 0 \Rightarrow \frac{2\,dx}{x} - \frac{dy}{y} = 0 \Rightarrow 2\ln|x| - \ln|y| = C \Rightarrow \frac{x^2}{|y|} = e^C \Rightarrow$
$\frac{x^2}{y} = \pm e^C = C_1 \Rightarrow x^2 = C_1 y$

31. $y\,dx + x\,dy = 0 \Rightarrow \frac{dx}{x} + \frac{dy}{y} = 0 \Rightarrow \ln|x| + \ln|y| = C \Rightarrow \ln|xy| = C \Rightarrow xy = e^C = C_1$. $x\,dx - y\,dy = 0 \Rightarrow$
$\frac{x^2}{2} - \frac{y^2}{2} = C \Rightarrow x^2 - y^2 = C_2$

SECTION 15.2 EXACT DIFFERENTIAL EQUATIONS

1. $\frac{\partial M}{\partial y} = 0,\ \frac{\partial N}{\partial x} = 0 \Rightarrow$ Exact

3. $\frac{\partial M}{\partial y} = -\frac{1}{y^2},\ \frac{\partial N}{\partial x} = -\frac{1}{y^2} \Rightarrow$ Exactt

5. $\frac{\partial M}{\partial y} = e^y,\ \frac{\partial N}{\partial x} - e^y \Rightarrow$ Exact

7. $\frac{\partial M}{\partial y} = 2,\ \frac{\partial N}{\partial x} = -2 \Rightarrow$ Not Exact

9. $\frac{\partial M}{\partial y} = 1 = \frac{\partial N}{\partial x} \Rightarrow$ Exact. $\frac{\partial f}{\partial x} = x + y \Rightarrow f(x,y) = \frac{x^2}{2} + xy + k(y).$ $\frac{\partial f}{\partial y} = x + k'(y) = x + y^2 \Rightarrow k(y) = \frac{y^3}{3} + C \Rightarrow$
$f(x,y) = \frac{x^2}{2} + xy + \frac{y^3}{3} + C = C_1 \Rightarrow 3x^2 + 6xy + 2y^3 = C_2$

11. $\frac{dy}{dx} = \frac{2xy + y^2}{y - 2xy - x^2} \Rightarrow (2xy + y^2)\,dx + (x^2 + 2xy - y)dy = 0.$ $\frac{\partial M}{\partial y} = 2x + 2y = \frac{\partial N}{\partial x} \Rightarrow$ Exact. $\frac{\partial f}{\partial x} = 2xy + y^2 \Rightarrow$
$f(x,y) = x^2y + xy^2 + k(y).$ $\frac{\partial f}{\partial y} = x^2 + 2xy + k'(y) = x^2 + 2xy - y \Rightarrow k'(y) = -y \Rightarrow k(y) = -\frac{y^2}{2} + C \Rightarrow f(x,y) =$
$x^2y + xy^2 - \frac{y^2}{2} + C = C_1 \Rightarrow 2x^2y + 2xy^2 - y^2 = C_2$

13. $\frac{\partial M}{\partial y} = \frac{1}{y} + \frac{1}{x} = \frac{\partial N}{\partial x} \Rightarrow$ Exact. $\frac{\partial f}{\partial x} = e^x + \ln y + \frac{y}{x} \Rightarrow f(x,y) = e^x + x\ln y + y\ln x + k(y).$ $\frac{\partial f}{\partial y} = \frac{x}{y} + \ln x + k'(y) =$
$\frac{x}{y} + \ln x + \sin y \Rightarrow k'(y) = \sin y \Rightarrow k(y) = -\cos y + C \Rightarrow f(x,y) = e^x + x\ln y + y\ln x - \cos y + C = C_1 \Rightarrow$
$e^x + x\ln y + y\ln x - \cos y = C_2$

15. $\frac{\partial M}{\partial y} = \cos x \sin^2 y = \frac{\partial N}{\partial x} \Rightarrow$ Exact. $\frac{\partial f}{\partial x} = \cos x \int_0^y \sin^2 t\, dt \Rightarrow f(x,y) = \sin x \int_0^y \sin^2 t\, dt + k(y)$. $\frac{\partial f}{\partial y} = \sin x \sin^2 y + k'(y) = \sin x \sin^2 y \Rightarrow k'(y) = 0 \Rightarrow k(y) = C \Rightarrow f(x,y) = \sin x \int_0^y \sin^2 t\, dt + C = C_1 \Rightarrow \sin x \int_0^y \sin^2 t\, dt = C_2$

17. $\frac{\partial M}{\partial y} = \sum_{n=0}^{\infty} \frac{x^{n-1}y^{n-1}}{(n-1)!} = \frac{\partial N}{\partial x} \Rightarrow$ Exact. $\frac{\partial f}{\partial x} = \frac{1}{x} + \sum_{n=0}^{\infty} \frac{x^{n-1}y^n}{n!} = \frac{1}{x} + \frac{1}{x} + \sum_{n=1}^{\infty} \frac{x^{n-1}y^n}{n!} \Rightarrow f(x,y) = 2\ln x + \sum_{n=1}^{\infty} \frac{x^n y^n}{n(n!)} + k(y)$. $\frac{\partial f}{\partial y} = \sum_{n=1}^{\infty} \frac{x^n y^{n-1}}{n!} + k'(y) = \frac{1}{y} + \sum_{n=0}^{\infty} \frac{x^n y^{n-1}}{n!} = \frac{1}{y} + \frac{1}{y} + \sum_{n=1}^{\infty} \frac{x^n y^{n-1}}{n!} \Rightarrow k'(y) = \frac{2}{y} \Rightarrow k(y) = 2\ln y + C \Rightarrow f(x,y) = 2\ln x + \sum_{n=1}^{\infty} \frac{x^n y^n}{n(n!)} + 2\ln y + C = C_1 \Rightarrow \ln x^2y^2 + \sum_{n=1}^{\infty} \frac{x^n y^n}{n(n!)} = C_2$

19. $\rho = \frac{1}{y^2} \Rightarrow \frac{1}{y^2}\left(xy^2 + y\right) dx - \frac{1}{y^2}(x\, dy) = 0 \Rightarrow \left(x + \frac{1}{y}\right) dx - \frac{x}{y^2} dy = 0 \Rightarrow \frac{\partial M}{\partial y} = -\frac{1}{y^2} = \frac{\partial N}{\partial x} \Rightarrow$ Exact. $\frac{\partial f}{\partial x} = x + \frac{1}{y} \Rightarrow f(x,y) = \frac{x^2}{2} + \frac{x}{y} + k(y)$. $\frac{\partial f}{\partial y} = -\frac{x}{y^2} + k'(y) = -\frac{x}{y^2} \Rightarrow k'(y) = 0 \Rightarrow k(y) = C \Rightarrow f(x,y) = \frac{x^2}{2} + \frac{x}{y} + C = C_1 \Rightarrow x^2 + \frac{2x}{y} = C_2$

21. a) $\rho = \frac{1}{xy} \Rightarrow \frac{1}{xy}(y\, dx) + \frac{1}{xy}(x\, dy) = 0 \Rightarrow \frac{1}{x} dx + \frac{1}{y} dy = 0 \Rightarrow \frac{\partial M}{\partial y} = 0 = \frac{\partial N}{\partial x} \Rightarrow$ Exact. $\frac{\partial f}{\partial x} = \frac{1}{x} \Rightarrow f(x,y) = \ln|x| + k(y)$. $\frac{\partial f}{\partial y} = k'(y) = \frac{1}{y} \Rightarrow k(y) = \ln|y| + C \Rightarrow f(x,y) = \ln|x| + \ln|y| + C = C_1 \Rightarrow \ln|xy| = C_2 \Rightarrow xy = \pm e^{C_2} \Rightarrow xy = C_3$

b) $\rho = \frac{1}{(xy)^2} \Rightarrow \frac{1}{(xy)^2}(y\, dx) + \frac{1}{(xy)^2}(x\, dy) = 0 \Rightarrow \frac{1}{x^2y} dx + \frac{1}{xy^2} dy = 0 \Rightarrow \frac{\partial M}{\partial y} = -\frac{1}{x^2y^2} = \frac{\partial N}{\partial x} \Rightarrow$ Exact. $\frac{\partial f}{\partial x} = \frac{1}{x^2y} \Rightarrow f(x,y) = -\frac{1}{xy} + k(y)$. $\frac{\partial f}{\partial y} = \frac{1}{xy^2} + k'(y) = \frac{1}{xy^2} \Rightarrow k'(y) = 0 \Rightarrow k(y) = C \Rightarrow f(x,y) = -\frac{1}{xy} + C = C_1 \Rightarrow xy = C_2$

23. $\frac{\partial M}{\partial y} = 2y = \frac{\partial N}{\partial x} \Rightarrow$ Exact. $\frac{\partial f}{\partial x} = x + y^2 \Rightarrow f(x,y) = \frac{x^2}{2} + xy^2 + k(y)$. $\frac{\partial f}{\partial y} = 2xy + k'(y) = 2xy + 1 \Rightarrow k'(y) = 1 \Rightarrow k(y) = y + C \Rightarrow f(x,y) = \frac{x^2}{2} + xy^2 + y + C = C_1 \Rightarrow x^2 + 2xy^2 + 2y = C_2$. $y(0) = 2 \Rightarrow C_2 = 4 \Rightarrow x^2 + 2xy^2 + 2y = 4$

25. $\frac{\partial M}{\partial y} = -1 = \frac{\partial N}{\partial x} \Rightarrow$ Exact. $\frac{\partial f}{\partial x} = \frac{1}{x} - y \Rightarrow f(x,y) = \ln|x| - xy + k(y)$. $\frac{\partial f}{\partial y} = -x + k'(y) = \frac{1}{y} - x \Rightarrow k'(y) = \frac{1}{y} \Rightarrow k(y) = \ln|y| + C \Rightarrow f(x,y) = \ln|x| - xy + \ln|y| + C = C_1 \Rightarrow \ln|xy| = xy + C_2 \Rightarrow xy = \pm e^{xy+C_2} = \pm e^{C_2} e^{xy} \Rightarrow xy = C_3 e^{xy}$. $y(1) = 1 \Rightarrow C_3 = \frac{1}{e} \Rightarrow xy = \frac{1}{e} e^{xy} \Rightarrow xy = e^{xy-1}$

27. $\frac{\partial M}{\partial y} = \frac{1}{x}\left(\frac{\ln y}{y}\right) = \frac{\partial N}{\partial x} \Rightarrow$ Exact. $\frac{\partial f}{\partial x} = 2x + \frac{1}{x}\int_1^y \frac{\ln t}{t}\,dt \Rightarrow f(x,y) = x^2 + \ln x \int_1^y \frac{\ln t}{t}\,dt + k(y)$. $\frac{\partial f}{\partial y} = \ln x\left(\frac{\ln y}{y}\right) +$

$k'(y) = \frac{\ln x \ln y}{y} \Rightarrow k'(y) = 0 \Rightarrow k(y) = C \Rightarrow f(x,y) = x^2 + \ln x \int_1^y \frac{\ln t}{t}\,dt + C = C_1 \Rightarrow x^2 + \ln x \int_1^y \frac{\ln t}{t}\,dt = C_2$.

$y(2) = 1 \Rightarrow 2^2 + \ln 2 \int_1^1 \frac{\ln t}{t}\,dt = C_2 \Rightarrow C_2 = 4$. $\therefore\ x^2 + \ln x \int_1^y \frac{\ln t}{t}\,dt = 4$.

29. $\frac{\partial M}{\partial y} = 2y,\ \frac{\partial N}{\partial x} = ay \Rightarrow a = 2$. $\therefore\ (3x^2 + y^2)\,dx + 2xy\,dy = 0$ is exact. $\frac{\partial f}{\partial x} = 3x^2 + y^2 \Rightarrow f(x,y) = x^3 + xy^2 +$ $k(y)$. $\frac{\partial f}{\partial y} = 2xy + k'(y) = 2xy \Rightarrow k'(y) = 0 \Rightarrow k(y) = C \Rightarrow f(x,y) = x^3 + xy^2 + C = C_1 \Rightarrow x^3 + xy^2 = C_2$.

SECTION 15.3 LINEAR FIRST ORDER EQUATIONS

1. $\frac{dy}{dx} + 2y = e^{-x} \Rightarrow P(x) = 2,\ Q(x) = e^{-x} \Rightarrow \int P(x)\,dx = 2x \Rightarrow \rho(x) = e^{2x} \Rightarrow y = \frac{1}{e^{2x}}\int e^{2x}(e^{-x})\,dx \Rightarrow$

$y = \frac{1}{e^{2x}}\int e^x\,dx = \frac{1}{e^{2x}}\left(e^x + C\right) = e^{-x} + C\,e^{-2x}$

3. $\frac{dy}{dx} + \frac{3}{x}y = \frac{\sin x}{x^3} \Rightarrow P(x) = \frac{3}{x},\ Q(x) = \frac{\sin x}{x^3} \Rightarrow \int P(x)\,dx = 3\ln|x| = \ln|x|^3 \Rightarrow \rho(x) = e^{\ln|x|^3} = |x|^3 = x^3$ if $x > 0 \Rightarrow y = \frac{1}{x^3}\int x^3\left(\frac{\sin x}{x^3}\right)dx = \frac{1}{x^3}\int \sin x\,dx \Rightarrow y = \frac{1}{x^3}(-\cos x + C) = -\frac{\cos x}{x^3} + \frac{C}{x^3}$

5. $\frac{dy}{dx} + \frac{4y}{x-1} = \frac{x+1}{(x-1)^3} \Rightarrow P(x) = \frac{4}{x-1},\ Q(x) = \frac{x+1}{(x-1)^3} \Rightarrow \int P(x)\,dx = 4\ln|x-1| = \ln(x-1)^4 \Rightarrow$

$\rho(x) = e^{\ln(x-1)^4} = (x-1)^4 \Rightarrow y = \frac{1}{(x-1)^4}\int (x-1)^4\frac{x+1}{(x-1)^3}\,dx = \frac{1}{(x-1)^4}\int (x^2-1)\,dx =$

$\frac{1}{(x-1)^4}\left(\frac{x^3}{3} - x + C\right) \Rightarrow y = \frac{x^3}{3(x-1)^4} - \frac{x}{(x-1)^4} + \frac{C}{(x-1)^4}$

7. $\frac{dy}{dx} + (\cot x)\,y = \sec x \Rightarrow P(x) = \cot x,\ Q(x) = \sec x \Rightarrow \int P(x)\,dx = \ln|\sin x| \Rightarrow \rho(x) = e^{\ln|\sin x|} = \sin x$ if $\sin x > 0 \Rightarrow y = \frac{1}{\sin x}\int \sin x(\sec x)\,dx = \frac{1}{\sin x}\int \tan x\,dx = \frac{1}{\sin x}(\ln|\sec x| + C) = \csc x(\ln|\sec x| + C)$

9. $\frac{dy}{dx} + 2y = x \Rightarrow P(x) = 2,\ Q(x) = x \Rightarrow \int P(x)\,dx = 2x \Rightarrow \rho(x) = e^{2x} \Rightarrow y = \frac{1}{e^{2x}}\int e^{2x}(x\,dx) =$

9. (Continued)

$\frac{1}{e^{2x}}\left(\frac{1}{2}x\,e^{2x}-\frac{1}{4}e^{2x}+C\right)\Rightarrow y=\frac{1}{2}x-\frac{1}{4}+C\,e^{-2x}.\ y(0)=1\Rightarrow-\frac{1}{4}+C=1\Rightarrow C=\frac{5}{4}\Rightarrow$

$y=\frac{1}{2}x-\frac{1}{4}+\frac{5}{4}e^{-2x}$

11. $\frac{dy}{dx}+\frac{y}{x}=\frac{\sin x}{x}\Rightarrow P(x)=\frac{1}{x},\ Q(x)=\frac{\sin x}{x}\Rightarrow \int P(x)\,dx=\ln|x|\Rightarrow \rho(x)=e^{\ln|x|}=|x|\Rightarrow$

$y=\frac{1}{|x|}\int |x|\frac{\sin x}{x}dx=\frac{1}{x}\int x\frac{\sin x}{x}dx \text{ for } x\neq 0\Rightarrow y=\frac{1}{x}\int \sin x\,dx=\frac{1}{x}(-\cos x+C)\Rightarrow$

$y=-\frac{1}{x}\cos x+\frac{C}{x}.\ y\left(\frac{\pi}{2}\right)=1\Rightarrow C=\frac{\pi}{2}\Rightarrow y=-\frac{1}{x}\cos x+\frac{\pi}{2x}$

13. $\frac{dy}{dx}-ky=0\Rightarrow P(x)=-k,\ Q(x)=0\Rightarrow \int P(x)\,dx=-kx\Rightarrow \rho(x)=e^{-kx}\Rightarrow$

$y=\frac{1}{e^{-kx}}\int e^{-kx}(0)\,dx=Ce^{kx}.\ y(0)=y_0\Rightarrow C=y_0\Rightarrow y=y_0\,e^{kx}$

15. Steady State $=\frac{V}{R}\Rightarrow$ we want $i=\frac{1}{2}\left(\frac{V}{R}\right)\Rightarrow\frac{1}{2}\left(\frac{V}{R}\right)=\frac{V}{R}\left(1-e^{-Rt/L}\right)\Rightarrow\frac{1}{2}=1-e^{-Rt/L}\Rightarrow-\frac{1}{2}=-e^{-Rt/L}$

$\Rightarrow \ln\frac{1}{2}=-\frac{Rt}{L}\Rightarrow-\frac{L}{R}\ln\frac{1}{2}=t\Rightarrow t=\frac{L}{R}\ln 2$

17. $t=\frac{3L}{R}\Rightarrow i=\frac{V}{R}\left(1-e^{(-R/L)(3L/R)}\right)=\frac{V}{R}\left(1-e^{-3}\right)\approx 0.9502\frac{V}{R}$ or about 95% of the steady state value.

19. a) Distance Coasted $\approx\frac{v_0 m}{k}=\frac{22(5)}{1/5}=550$ ft.

b) $v=22\,e^{(-0.2/5)\,t}=22e^{-t/25}$. We want $v=1$ ft/sec $\Rightarrow 1=22\,e^{-t/25}\Rightarrow \ln(1/22)=-\frac{t}{25}\Rightarrow t=77.3$ sec

SECTION 15.4 LINEAR HOMOGENEOUS SECOND ORDER EQUATIONS

1. $r^2+2r=0\Rightarrow r_1=0,\ r_2=-2\Rightarrow y=C_1e^{0x}+C_2e^{-2x}=C_1+C_2e^{-2x}$

3. $r^2+6r+5=0\Rightarrow(r+5)(r+1)=0\Rightarrow r_1=-5,\ r_2=-1\Rightarrow y=C_1e^{-5x}+C_2{}^{-x}$

5. $r^2-4r+4=0\Rightarrow(r-2)^2=0\Rightarrow r_1=r_2=2\Rightarrow y=\left(C_1x+C_2\right)e^{2x}$

7. $r^2-10r+25=0\Rightarrow(r-5)^2=0\Rightarrow r_1=r_2=5\Rightarrow y=\left(C_1x+C_2\right)e^{5x}$

9. $r^2+r+1=0\Rightarrow r=\frac{-1\pm i\sqrt{3}}{2}\Rightarrow y=e^{-x/2}\left(C_1\cos\frac{\sqrt{3}}{2}x+C_2\sin\frac{\sqrt{3}}{2}x\right)$

11. $r^2 - 2r + 4 = 0 \Rightarrow r = 1 \pm i\sqrt{3} \Rightarrow y = e^x\left(C_1 \cos\sqrt{3}\,x + C_2 \sin\sqrt{3}\,x\right)$

13. $r^2 - 1 = 0 \Rightarrow r_1 = 1, r_2 = -1 \Rightarrow y = C_1e^x + C_2e^{-x}$. $y(0) = 1 \Rightarrow C_1 + C_2 = 1$. $y' = C_1e^x - C_2e^{-x}$. $y'(0) = -2$ $\Rightarrow -2 = C_1 - C_2$. $\therefore$ $C_1 = -\frac{1}{2}, C_2 = \frac{3}{2} \Rightarrow y = -\frac{1}{2}e^x + \frac{3}{2}e^{-x}$

15. $r^2 - 4 = 0 \Rightarrow r_1 = 2, r_2 = -2 \Rightarrow y = C_1e^{2x} + C_2e^{-2x}$. $y(0) = 0 \Rightarrow C_1 + C_2$. $y' = 2C_1e^{2x} - 2C_2e^{-2x}$. $y'(0) = 3 \Rightarrow 3 = 2C_1 - 2C_2$. $\therefore$ $C_1 = \frac{3}{4}, C_2 = -\frac{3}{4} \Rightarrow y = \frac{3}{4}e^{2x} - \frac{3}{4}e^{-2x}$

17. $r^2 + 2r + 1 = 0 \Rightarrow (r+1)^2 = 0 \Rightarrow r_1 = r_2 = -1 \Rightarrow y = \left(C_1x + C_2\right)e^{-x}$. $y(0) = 0 \Rightarrow C_2 = 0$. $y' = C_1\left(e^{-x} - x\,e^{-x}\right)$. $y'(0) = 1 \Rightarrow C_1 = 1 \Rightarrow y = x\,e^{-x}$

19. $4r^2 + 12r + 9 = 0 \Rightarrow (2r+3)^2 = 0 \Rightarrow r_1 = r_2 = -\frac{3}{2} \Rightarrow y = \left(C_1x + C_2\right)e^{-3x/2}$. $y(0) = 0 \Rightarrow C_2 = 0 \Rightarrow$ $y = C_1x\,e^{-3x/2} \Rightarrow y' = C_1\left(e^{-3x/2} - \frac{3}{2}x\,e^{-3x/2}\right)$. $y'(0) = -1 \Rightarrow C_1 = -1 \Rightarrow y = -x\,e^{-3x/2}$

21. $r^2 + 4 = 0 \Rightarrow r = \pm 2i \Rightarrow y = C_1 \cos 2x + C_2 \sin 2x$. $y(0) = 0 \Rightarrow C_1 = 0 \Rightarrow y = C_2 \sin 2x$. $y' = 2C_2 \cos 2x$. $y'(0) = 2 \Rightarrow C_2 = 1 \Rightarrow y = \sin 2x$

23. $r^2 - 2r + 3 = 0 \Rightarrow r = 1 \pm i\sqrt{2} \Rightarrow y = e^x\left(C_1 \cos\sqrt{2}\,x + C_2 \sin\sqrt{2}\,x\right)$. $y(0) = 2 \Rightarrow C_1 = 2 \Rightarrow y =$ $e^x\left(2\cos\sqrt{2}\,x + C_2 \sin\sqrt{2}\,x\right)$. $y' = e^x\left(2\cos\sqrt{2}\,x + C_2 \sin\sqrt{2}\,x\right) +$ $e^x\left(-2\sqrt{2}\sin\sqrt{2}\,x + \sqrt{2}\,C_2\cos\sqrt{2}\,x\right)$. $y'(0) = 1 \Rightarrow 1 = 2 + \sqrt{2}\,C_2 \Rightarrow C_2 = -\frac{1}{\sqrt{2}} \Rightarrow$ $y = e^x\left(2\cos\sqrt{2}\,x - \frac{1}{\sqrt{2}}\sin\sqrt{2}\,x\right)$

25. $c_1 = y\cos x - y'\sin x \Rightarrow c_1' = y'\cos x - y\sin x - y''\sin x - y'\cos x = -\sin x(y'' + y) = 0$. $c_2 = y\sin x -$ $y'\cos x \Rightarrow c_2' = y'\sin x + y\cos x + y''\cos x - y'\sin x = \cos x(y'' + y) = 0$. $\therefore$ $c_1' = c_2' = 0 \Rightarrow c_1, c_2$ are constants. $c_1 = y\cos x - y'\sin x \Rightarrow c_1 \csc x = y\cot x - y'$. $c_2 = y\sin x + y'\cos x \Rightarrow c_2 \sec x = y\tan x + y'$. $\therefore$ $c_1\csc x + c_2\sec x = y(\cot x + \tan x) \Rightarrow y = \dfrac{c_1\csc x + c_2\sec x}{\cot x + \tan x} = c_1\cos x + c_2\sin x$.

SECTION 15.5 SECOND ORDER EQUATIONS; REDUCTION OF ORDER

1. $\dfrac{d^2y}{dx^2} + \dfrac{dy}{dx} = 0 \Rightarrow r^2 + r = 0 \Rightarrow r_1 = 0,\ r_2 = -1 \Rightarrow y_h = C_1 + C_2 e^{-x} \Rightarrow u_1 = 1,\ u_2 = e^{-x} \Rightarrow D = \begin{vmatrix} 1 & e^{-x} \\ 0 & -e^{-x} \end{vmatrix}$
$= -e^{-x} \Rightarrow v'_1 = -\dfrac{u_2\,F(x)}{D} = x \Rightarrow v_1 = \displaystyle\int x\,dx = \frac{x^2}{2} + C_1.\ v'_2 = \frac{u_1\,F(x)}{D} = -x\,e^x \Rightarrow v_2 = \int x\,e^x\,dx =$
$-x\,e^x + e^x + C_2 \Rightarrow y = \left(\dfrac{x^2}{2} + C_1\right) + \left(-x\,e^x + e^x + C_2\right)e^{-x} = \dfrac{x^2}{2} - x + C_3 + C_2\,e^{-x}$

3. $\dfrac{d^2y}{dx^2} + y = 0 \Rightarrow r^2 + 1 = 0 \Rightarrow r = \pm i \Rightarrow y_h = C_1 \cos x + C_2 \sin x \Rightarrow u_1 = \cos x,\ u_2 = \sin x \Rightarrow D =$
$\begin{vmatrix} \cos x & \sin x \\ -\sin x & \cos x \end{vmatrix} = 1 \Rightarrow v'_1 = -\dfrac{u_2\,F(x)}{D} = -\sin^2 x \Rightarrow v_1 = \displaystyle\int -\sin^2 x\,dx = -\frac{1}{2}x + \frac{\sin 2x}{4} + C_1.$
$v'_2 = \dfrac{u_1\,F(x)}{D} = \cos x \sin x \Rightarrow v_2 = \displaystyle\int \sin x \cos x\,dx = \frac{\sin^2 x}{2} + C_2 \Rightarrow y = \cos x\left(-\frac{1}{2}x + \frac{\sin 2x}{4} + C_1\right) +$
$\sin x\left(\dfrac{\sin^2 x}{2} + C_2\right) = -\dfrac{1}{2}x\cos x + C_1 \cos x + C_3 \sin x$

5. $\dfrac{d^2y}{dx^2} + 2\dfrac{dy}{dx} + y = 0 \Rightarrow r^2 + 2r + 1 = 0 \Rightarrow r_1 = r_2 = -1 \Rightarrow y_h = C_1 x\,e^{-x} + C_2\,e^{-x} \Rightarrow u_1 = x\,e^{-x},\ u_2 = e^{-x}$
$\Rightarrow D = \begin{vmatrix} x\,e^{-x} & e^{-x} \\ e^{-x} - x\,e^{-x} & -e^{-x} \end{vmatrix} = -e^{-2x} \Rightarrow v'_1 = -\dfrac{u_2\,F(x)}{D} = 1 \Rightarrow v_1 = \displaystyle\int 1\,dx = x + C_1.\ v'_2 = \frac{u_1\,F(x)}{D} =$
$-x \Rightarrow v_2 = \displaystyle\int -x\,dx = -\frac{x^2}{2} + C_2 \Rightarrow y = \left(x + C_1\right)xe^{-x} + \left(-\frac{x^2}{2} + C_2\right)e^{-x} = C_1 x\,e^{-x} + C_2\,e^{-x} + \frac{1}{2}x^2\,e^{-x}$

7. $\dfrac{d^2y}{dx^2} - y = 0 \Rightarrow r^2 - 1 = 0 \Rightarrow r_1 = 1,\ r_2 = -1 \Rightarrow y_h = C_1\,e^x + C_2\,e^{-x} \Rightarrow u_1 = e^x,\ u_2 = e^{-x} \Rightarrow$
$D = \begin{vmatrix} e^x & e^{-x} \\ e^x & -e^{-x} \end{vmatrix} = -2 \Rightarrow v'_1 = -\dfrac{u_2\,F(x)}{D} = \dfrac{1}{2} \Rightarrow v_1 = \dfrac{1}{2}x + C_1.\ v'_2 = \dfrac{u_1\,F(x)}{D} = -\dfrac{1}{2}e^{2x} \Rightarrow v_2 = -\dfrac{1}{4}e^{2x} +$
$C_2 \Rightarrow y = \left(\dfrac{1}{2}x + C_1\right)e^x + \left(-\dfrac{1}{4}e^{2x} + C_2\right)e^{-x} = \dfrac{1}{2}x\,e^x + C_3\,e^x + C_2\,e^{-x}$

9. $\dfrac{d^2y}{dx^2} + 4\dfrac{dy}{dx} + 5y = 0 \Rightarrow r^2 + 4r + 5 = 0 \Rightarrow r = -2 \pm i \Rightarrow y_h = e^{-2x}\left(C_1 \cos x + C_2 \sin x\right) \Rightarrow$
$u_1 = e^{-2x}\cos x,\ u_2 = e^{-2x}\sin x \Rightarrow D = \begin{vmatrix} e^{-2x}\cos x & e^{-2x}\sin x \\ -2e^{-2x}\cos x - e^{-2x}\sin x & -2e^{-2x}\sin x + e^{-2x}\cos x \end{vmatrix} =$
$e^{-4x} \Rightarrow v'_1 = -\dfrac{u_2\,F(x)}{D} = -10\,e^{2x}\sin x \Rightarrow v_1 = \displaystyle\int -10\,e^{2x}\sin x\,dx = -2\,e^{2x}(2\sin x - \cos x) + C_1.$
$v'_2 = \dfrac{u_1\,F(x)}{D} = 10\,e^{2x}\cos x \Rightarrow v_2 = \displaystyle\int 10\,e^{2x}\cos x\,dx = 10\left(\frac{e^{2x}}{5}(2\cos x + \sin x)\right) + C_2 \Rightarrow$
$y = \left(-2\,e^{2x}(2\sin x - \cos x) + C_1\right)e^{-2x}\cos x + \left(2\,e^{2x}(2\cos x + \sin x) + C_2\right)e^{-2x}\sin x =$
$2 + C_1\,e^{-2x}\cos x + C_2\,e^{-2x}\sin x$

11. $\frac{d^2y}{dx^2} + y = 0 \Rightarrow r^2 + 1 = 0 \Rightarrow r = \pm i \Rightarrow y_h = C_1 \cos x + C_2 \sin x \Rightarrow u_1 = \cos x,\ u_2 = \sin x \Rightarrow$

$D = \begin{vmatrix} \cos x & \sin x \\ -\sin x & \cos x \end{vmatrix} = 1 \Rightarrow v'_1 = -\frac{u_2 F(x)}{D} = -\tan x \Rightarrow v_1 = \int -\tan x\, dx = -\ln(\sec x) + C_1.$

$v'_2 = \frac{u_1 F(x)}{D} = 1 \Rightarrow v_2 = \int dx = x + C_2 \Rightarrow y = \left(-\ln(\sec x) + C_1\right)\cos x + \left(x + C_2\right)\sin x = \cos x \ln(\cos x)$
$+ x \sin x + C_1 \cos x + C_2 \sin x$

13. $\frac{d^2y}{dx^2} - 3\frac{dy}{dx} - 10y = 0 \Rightarrow r^2 - 3r - 10 = 0 \Rightarrow r_1 = 5,\ r_2 = -2 \Rightarrow y_h = C_1 e^{5x} + C_2 e^{-2x}.\ y_p = C \Rightarrow$
$\frac{d^2y}{dx^2} = 0,\ \frac{dy}{dx} = 0 \Rightarrow -10C = 3 \Rightarrow C = \frac{3}{10} \Rightarrow y_p = \frac{3}{10} \Rightarrow y = \frac{3}{10} + C_1 e^{5x} + C_2 e^{-2x}$

15. $\frac{d^2y}{dx^2} - \frac{dy}{dx} = 0 \Rightarrow r^2 - r = 0 \Rightarrow r_1 = 0,\ r_2 = 1 \Rightarrow y_h = C_1 + C_2 e^x.\ y_p = B \cos x + C \sin x \Rightarrow \frac{dy}{dx} = -B \sin x +$
$C \cos x,\ \frac{d^2y}{dx^2} = -B \cos x - C \sin x \Rightarrow -B \cos x - C \sin x + B \sin x - C \cos x = \sin x \Rightarrow B = \frac{1}{2},\ C = -\frac{1}{2} \Rightarrow$
$y_p = \frac{1}{2}\cos x - \frac{1}{2}\sin x \Rightarrow y = \frac{1}{2}\cos x - \frac{1}{2}\sin x + C_1 + C_2 e^x$

17. $\frac{d^2y}{dx^2} + y = 0 \Rightarrow r^2 + 1 = 0 \Rightarrow r = \pm i \Rightarrow y_h = C_1 \cos x + C_2 \sin x.\ y_p = B \cos 3x + C \sin 3x \Rightarrow y'_p =$
$-3B \sin 3x + 3C \cos 3x \Rightarrow y''_p = -9B \cos 3x - 9C \sin 3x \Rightarrow -9B \cos 3x - 9C \sin 3x + B \cos 3x +$
$C \sin 3x = \cos 3x \Rightarrow B = -\frac{1}{8},\ C = 0 \Rightarrow y_p = -\frac{1}{8}\cos 3x \Rightarrow y = -\frac{1}{8}\cos 3x + C_1 \cos x + C_2 \sin x$

19. $\frac{d^2y}{dx^2} - \frac{dy}{dx} - 2y = 0 \Rightarrow r^2 - r - 2 = 0 \Rightarrow r_1 = 2,\ r_2 = -1 \Rightarrow y_h = C_1 e^{2x} + C_2 e^{-x}.\ y_p = B \cos x + C \sin x \Rightarrow$
$y'_p = -B \sin x + C \cos x \Rightarrow y''_p = -B \cos x - C \sin x \Rightarrow -B \cos x - C \sin x - (-B \sin x + C \cos x) -$
$2(B \cos x + C \sin x) = 20 \cos x \Rightarrow B = -6,\ C = -2 \Rightarrow y_p = -6 \cos x - 2 \sin x \Rightarrow y = -6 \cos x - 2 \sin x +$
$C_1 e^{2x} + C_2 e^{-x}$

21. $\frac{d^2y}{dx^2} - y = 0 \Rightarrow r^2 - 1 = 0 \Rightarrow r_1 = 1,\ r_2 = -1 \Rightarrow y_h = C_1 e^x + C_2 e^{-x}.\ y_p = Ax e^x + Dx^2 + Ex + F \Rightarrow$
$y'_p = A e^x + Ax e^x + 2Dx + E \Rightarrow y''_p = 2A e^x + Ax e^x + 2D \Rightarrow 2A e^x + Ax e^x + 2D -$
$\left(Ax e^x + Dx^2 + Ex + F\right) = e^x + x^2 \Rightarrow A = \frac{1}{2},\ D = -1,\ E = 0,\ F = -2 \Rightarrow y_p = \frac{1}{2} x e^x - x^2 - 2 \Rightarrow$
$y = \frac{1}{2} x e^x - x^2 - 2 + C_1 e^x + C_2 e^{-x}$

23. $\frac{d^2y}{dx^2} - \frac{dy}{dx} - 6y = 0 \Rightarrow r^2 - r - 6 = 0 \Rightarrow r_1 = 3,\ r_2 = -2 \Rightarrow y_h = C_1 e^{3x} + C_2 e^{-2x}.\ y_p = A e^{-x} + B \cos x +$
$C \sin x \Rightarrow y'_p = -A e^{-x} - B \sin x + C \cos x \Rightarrow y''_p = A e^{-x} - B \cos x - C \sin x \Rightarrow A e^{-x} - B \cos x -$
$C \sin x - \left(-A e^{-x} - B \sin x + C \cos x\right) - 6\left(A e^{-x} + B \cos x + C \sin x\right) = e^{-x} - 7 \cos x \Rightarrow A = -\frac{1}{4},$

23. (Continued)

$B = \frac{49}{50}$, $C = \frac{7}{50} \Rightarrow y_p = -\frac{1}{4}e^{-x} + \frac{49}{50}\cos x + \frac{7}{50}\sin x \Rightarrow y = -\frac{1}{4}e^{-x} + \frac{49}{50}\cos x + \frac{7}{50}\sin x + C_1 e^{3x} + C_2 e^{-2x}$

25. $\frac{d^2y}{dx^2} + 5\frac{dy}{dx} = 0 \Rightarrow r^2 + 5r = 0 \Rightarrow r_1 = 0, r_2 = -5 \Rightarrow y_h = C_1 + C_2 e^{-5x}$. $y_p = Dx^3 + Ex^2 + Fx \Rightarrow$
$y'_p = 3Dx^2 + 2Ex + F \Rightarrow y''_p = 6Dx + 2E \Rightarrow 6Dx + 2E + 5\left(3Dx^2 + 2Ex + F\right) = 15x^2 \Rightarrow D = 1, E = -\frac{3}{5}$,
$F = \frac{6}{25} \Rightarrow y_p = x^3 - \frac{3}{5}x^2 + \frac{6}{25}x \Rightarrow y = x^3 - \frac{3}{5}x^2 + \frac{6}{25}x + C_1 + C_2 e^{-5x}$

27. $\frac{d^2y}{dx^2} - 3\frac{dy}{dx} = 0 \Rightarrow r^2 - 3r = 0 \Rightarrow r_1 = 0, r_2 = 3 \Rightarrow y_h = C_1 + C_2 e^{3x}$. $y_p = Ax\,e^{3x} + Dx^2 + Ex \Rightarrow$
$y'_p = A\,e^{3x} + 3Ax\,e^{3x} + 2Dx + E \Rightarrow y''_p = 6A\,e^{3x} + 9Ax\,e^{3x} + 2D \Rightarrow 6A\,e^{3x} + 9Ax\,e^{3x} + 2D -$
$3\left(A\,e^{3x} + 3Ax\,e^{3x} + 2Dx + E\right) = e^{3x} - 12x \Rightarrow A = \frac{1}{3}, D = 2, E = \frac{4}{3} \Rightarrow y_p = \frac{1}{3}x\,e^{3x} + 2x^2 + \frac{4}{3}x \Rightarrow$
$y = \frac{1}{3}x\,e^{3x} + 2x^2 + \frac{4}{3}x + C_1 + C_2 e^{3x}$

29. $\frac{d^2y}{dx^2} - 5\frac{dy}{dx} = 0 \Rightarrow r^2 - 5r = 0 \Rightarrow r_1 = 0, r_2 = 5 \Rightarrow y_h = C_1 + C_2 e^{5x}$. $y_p = Ax^2 e^{5x} + Bx\,e^{5x} \Rightarrow$
$y'_p = (2A + 5B)x\,e^{5x} + 5Ax^2 e^{5x} + B\,e^{5x} \Rightarrow y''_p = (2A + 10B)e^{5x} + (20A + 25B)x\,e^{5x} + 25Ax^2 e^{5x} \Rightarrow$
$(2A + 10B)e^{5x} + (20A + 25B)x\,e^{5x} + 25Ax^2 e^{5x} - 5\left((2A + 5B)xe^{5x} + 5Ax^2 e^{5x} + B\,e^{5x}\right) = x\,e^{5x} \Rightarrow$
$A = \frac{1}{10}, B = -\frac{1}{25} \Rightarrow y_p = \frac{1}{10}x^2 e^{5x} - \frac{1}{25}x\,e^{5x} \Rightarrow y = \frac{1}{10}x^2 e^{5x} - \frac{1}{25}x\,e^{5x} + C_1 + C_2 e^{5x}$

31. $\frac{d^2y}{dx^2} + y = 0 \Rightarrow r^2 + 1 = 0 \Rightarrow r = \pm i \Rightarrow y_h = C_1 \cos x + C_2 \sin x$. $y_p = Ax\cos x + Bx\sin x \Rightarrow y'_p =$
$A\cos x - Ax\sin x + B\sin x + Bx\cos x \Rightarrow y''_p = -2A\sin x - Ax\cos x + 2B\cos x - Bx\sin x \Rightarrow$
$-2A\sin x - Ax\cos x + 2B\cos x - Bx\sin x + Ax\cos x + Bx\sin x = 2\cos x + \sin x$
$\Rightarrow A = -\frac{1}{2}, B = 1 \Rightarrow y_p = -\frac{1}{2}x\cos x + x\sin x \Rightarrow y = -\frac{1}{2}x\cos x + x\sin x + C_1\cos x + C_2\sin x$

33. a) $\frac{d^2y}{dx^2} - \frac{dy}{dx} = 0 \Rightarrow r^2 - r = 0 \Rightarrow r_1 = 0, r_2 = 1 \Rightarrow y_h = C_1 + C_2 e^x \Rightarrow u_1 = 1, u_2 = e^x \Rightarrow$

$$D = \begin{vmatrix} 1 & e^x \\ 0 & e^x \end{vmatrix} = e^x \Rightarrow v'_1 = -\frac{u_2 F(x)}{D} = -e^x - e^{-x} \Rightarrow v_1 = \int -e^x - e^{-x}\,dx = -e^x + e^{-x} + C_1.$$

$$v'_2 = \frac{u_1 F(x)}{D} = 1 + e^{-2x} \Rightarrow v_2 = \int \left(1 + e^{-2x}\right) dx = x - \frac{1}{2}e^{-2x} + C_2 \Rightarrow$$

$$y = \left(-e^x + e^{-x} + C_1\right) + \left(x - \frac{1}{2}e^{-2x} + C_2\right)e^x = \frac{1}{2}e^{-x} + x\,e^x + C_1 + C_3 e^x$$

b) $y_h = C_1 + C_2 e^x$. $y_p = Ax\,e^x + B\,e^{-x} \Rightarrow y'_p = A\,e^x + Ax\,e^x - B\,e^{-x} \Rightarrow y''_p = 2A\,e^x + Ax\,e^x + B\,e^{-x} \Rightarrow$
$2A\,e^x + Ax\,e^x + B\,e^{-x} - \left(A\,e^x + Ax\,e^x - B\,e^{-x}\right) = e^x + e^{-x} \Rightarrow A = 1, B = \frac{1}{2} \Rightarrow y_p = x\,e^x + \frac{1}{2}e^{-x} \Rightarrow$

33. (Continued)

$y = x e^x + \frac{1}{2} e^{-x} + C_1 + C_2 e^x$

35. a) $\frac{d^2y}{dx^2} - 4\frac{dy}{dx} - 5y = 0 \Rightarrow r^2 - 4r - 5 = 0 \Rightarrow r_1 = 5, r_2 = -1 \Rightarrow y_h = C_1 e^{5x} + C_2 e^{-x} \Rightarrow u_1 = e^{5x},$

$u_2 = e^{-x} \Rightarrow D = \begin{vmatrix} e^{5x} & e^{-x} \\ 5e^{5x} & -e^{-x} \end{vmatrix} = -6e^{4x} \Rightarrow v'_1 = -\frac{u_2 F(x)}{D} = \frac{1}{6} e^{-4x} + \frac{2}{3} e^{-5x} \Rightarrow$

$v_1 = \int \left(\frac{1}{6} e^{-4x} + \frac{2}{3} e^{-5x}\right) dx = -\frac{1}{24} e^{-4x} - \frac{2}{15} e^{-5x} + C_1.\ v'_2 = \frac{u_1 F(x)}{D} = -\frac{1}{6} e^{2x} - \frac{2}{3} e^x \Rightarrow$

$v_2 = \int \left(-\frac{1}{6} e^{2x} - \frac{2}{3} e^x\right) dx = -\frac{1}{12} e^{2x} - \frac{2}{3} e^x + C_2 \Rightarrow y = \left(-\frac{1}{24} e^{-4x} - \frac{2}{15} e^{-5x} + C_1\right) e^{5x} +$

$\left(-\frac{1}{12} e^{2x} - \frac{2}{3} e^x + C_2\right) e^{-x} = -\frac{1}{8} e^x - \frac{4}{5} + C_1 e^{5x} + C_2 e^{-x}$

b) $y_h = C_1 e^{5x} + C_2 e^{-x}.\ y_p = A e^x + B x + C \Rightarrow y'_p = A e^x + B \Rightarrow y''_p = A e^x \Rightarrow A e^x - 4(A e^x + B) -$

$5(A e^x + Bx + C) = e^x + 4 \Rightarrow A = -\frac{1}{8}, B = 0, C = -\frac{4}{5} \Rightarrow y_p = -\frac{1}{8} e^x - \frac{4}{5} \Rightarrow\ y = -\frac{1}{8} e^x - \frac{4}{5} + C_1 e^{5x} +$

$C_2 e^{-x}$

37. $\frac{d^2y}{dx^2} + y = 0 \Rightarrow r^2 + 1 = 0 \Rightarrow r = \pm i \Rightarrow y_h = C_1 \cos x + C_2 \sin x \Rightarrow u_1 = \cos x, u_2 = \sin x \Rightarrow$

$D = \begin{vmatrix} \cos x & \sin x \\ -\sin x & \cos x \end{vmatrix} = 1 \Rightarrow v'_1 = -\frac{u_2 F(x)}{D} = -\cos x \Rightarrow v_1 = \int -\cos x\, dx = -\sin x + C_1.$

$v'_2 = \frac{u_1 F(x)}{D} = \csc x - \sin x \Rightarrow v_2 = \int (\csc x - \sin x)\, dx = -\ln|\csc x + \cot x| + \cos x + C_2 \Rightarrow$

$y = (-\sin x + C_1)\cos x + (-\ln|\csc x + \cot x| + \cos x + C_2)\sin x = C_1 \cos x + C_2 \sin x -$

$\sin x(\ln|\csc x + \cot x|)$

39. $\frac{d^2y}{dx^2} - 8\frac{dy}{dx} = 0 \Rightarrow r^2 - 8r = 0 \Rightarrow r_1 = 0, r_2 = 8 \Rightarrow y_h = C_1 + C_2 e^{8x}\quad y_p = Ax e^{8x} \Rightarrow y'_p = A e^{8x} + 8Ax e^{8x}$

$\Rightarrow y''_p = 16A e^{8x} + 64Ax e^{8x} \Rightarrow 16A e^{8x} + 64Ax e^{8x} - 8(A e^{8x} + 8Ax e^{8x}) = e^{8x} \Rightarrow A = \frac{1}{8} \Rightarrow$

$y_p = \frac{1}{8} x e^{8x} \Rightarrow y = \frac{1}{8} x e^{8x} + C_1 + C_2 e^{8x}$

41. $\frac{d^2y}{dx^2} - \frac{dy}{dx} = 0 \Rightarrow r^2 - r = 0 \Rightarrow r_1 = 0, r_2 = 1 \Rightarrow y_h = C_1 + C_2 e^x.\ y_p = Dx^4 + Ex^3 + Fx^2 + Gx \Rightarrow$

$y'_p = 4Dx^3 + 3Ex^2 + 2Fx + G \Rightarrow y''_p = 12Dx^2 + 6Ex + 2F \Rightarrow 12Dx^2 + 6Ex + 2F -$

$(4Dx^3 + 3Ex^2 + 2Fx + G) = x^3 \Rightarrow D = -\frac{1}{4}, E = -1, F = -3, G = -6 \Rightarrow y_p = -\frac{1}{4}x^4 - x^3 - 3x^2 - 6x \Rightarrow$

$y = -\frac{1}{4}x^4 - x^3 - 3x^2 - 6x + C_1 + C_2 e^x$

43. $\frac{d^2y}{dx^2} + 2\frac{dy}{dx} = 0 \Rightarrow r^2 + 2r = 0 \Rightarrow r_1 = 0,\ r_2 = -2 \Rightarrow y_h = C_1 + C_2\,e^{-2x}$. $y_p = Ax^3 + Bx^2 + Cx + E\,e^x \Rightarrow$ $y'_p = 3Ax^2 + 2Bx + C + E\,e^x \Rightarrow y''_p = 6Ax + 2B + E\,e^x \Rightarrow 6Ax + 2B + E\,e^x + 2\left(3Ax^2 + 2Bx + C + E\,e^x\right)$ $= x^2 - e^x \Rightarrow A = \frac{1}{6},\ B = -\frac{1}{4},\ C = \frac{1}{4},\ E = -\frac{1}{3} \Rightarrow y_p = \frac{1}{6}x^3 - \frac{1}{4}x^2 + \frac{1}{4}x - \frac{1}{3}e^x \Rightarrow y = \frac{1}{6}x^3 - \frac{1}{4}x^2 + \frac{1}{4}x -$ $\frac{1}{3}e^x + C_1 + C_2\,e^{-2x}$

45. $\frac{d^2y}{dx^2} + y = 0 \Rightarrow r^2 + 1 = 0 \Rightarrow r = \pm i \Rightarrow y_h = C_1\cos x + C_2\sin x \Rightarrow u_1 = \cos x,\ u_2 = \sin x \Rightarrow$

$$D = \begin{vmatrix} \cos x & \sin x \\ -\sin x & \cos x \end{vmatrix} = 1 \Rightarrow v'_1 = -\frac{u_2\,F(x)}{D} = -\tan^2 x \Rightarrow v_1 = \int -\tan^2 x\,dx = -\tan x + x + C_1.$$

$v'_2 = \frac{u_1\,F(x)}{D} = \tan x \Rightarrow v_2 = \int \tan x\,dx = \ln|\sec x| + C_2 \Rightarrow y = \left(-\tan x + x + C_1\right)\cos x +$ $\left(\ln|\sec x| + C_2\right)\sin x \Rightarrow y = x\cos x + \sin x\ln(\sec x) + C_1\cos x + C_3\sin x$

47. $\frac{dy}{dx} - 3y = 0 \Rightarrow r - 3 = 0 \Rightarrow r = 3 \Rightarrow y_h = C_1\,e^{3x}$. $y_p = A\,e^x \Rightarrow y'_p = A\,e^x \Rightarrow A\,e^x - 3A\,e^x = e^x \Rightarrow A =$ $-\frac{1}{2} \Rightarrow y_p = -\frac{1}{2}e^x \Rightarrow y = C_1\,e^{3x} - \frac{1}{2}e^x$

49. $\frac{dy}{dx} - 3y = 0 \Rightarrow r - 3 = 0 \Rightarrow r = 3 \Rightarrow y_h = C_1\,e^{3x}$. $y_p = Ax\,e^{3x} \Rightarrow y'_p = A\,e^{3x} + 3Ax\,e^{3x} \Rightarrow A\,e^{3x} +$ $3Ax\,e^{3x} - 3Ax\,e^{3x} = 5\,e^{3x} \Rightarrow A = 5 \Rightarrow y_p = 5x\,e^{3x} \Rightarrow y = C_1e^{3x} + 5x\,e^{3x}$

51. $\frac{d^2y}{dx^2} + y = 0 \Rightarrow r^2 + 1 = 0 \Rightarrow r = \pm i \Rightarrow y_h = C_1\cos x + C_2\sin x \Rightarrow u_1 = \cos x,\ u_2 = \sin x \Rightarrow$

$$D = \begin{vmatrix} \cos x & \sin x \\ -\sin x & \cos x \end{vmatrix} = 1 \Rightarrow v'_1 = -\frac{u_2\,F(x)}{D} = -\sec x\tan x \Rightarrow v_1 = \int -\sec x\tan x\,dx = -\sec x +$$

C_1. $v'_2 = \frac{u_1\,F(x)}{D} = \sec x \Rightarrow v_2 = \int \sec x\,dx = \ln|\sec x + \tan x| + C_2 \Rightarrow y = C_1\cos x + C_2\sin x - 1 +$ $\sin x\ln|\sec x + \tan x|$. $y(0) = 1 \Rightarrow C_1\cos 0 + C_2\sin 0 - 1 + \sin 0\ln|\sec 0 + \tan 0| \Rightarrow C_1 = 2 \Rightarrow$ $y = 2\cos x + C_2\sin x - 1 + \sin x\ln|\sec x + \tan x| \Rightarrow y' = -2\sin x + C_2\cos x + \sec x\sin x +$ $\cos x\ln|\sec x + \tan x|$. $y'(0) = 1 \Rightarrow -2\sin 0 + C_2\cos 0 + \sec 0\sin 0 + \cos 0\ln|\sec 0 + \tan 0| \Rightarrow$ $C_2 = 1 \Rightarrow y = 2\cos x + \sin x - 1 + \sin x\ln|\sec x + \tan x|$

53. $\frac{dp}{dx} + p = 0 \Rightarrow \frac{dp}{dx} = -p \Rightarrow dp = -p\,dx \Rightarrow \frac{1}{p}dp = -dx \Rightarrow \ln|p| = -x + C \Rightarrow p = C_1e^{-x} \Rightarrow \frac{dy}{dx} = C_1\,e^{-x} \Rightarrow$ $y = -C_1e^{-x} + C_2 = C_3e^{-x} + C_2$

55. $y'' = (y')^2 \Rightarrow \frac{dp}{dx} = p^2 \Rightarrow \frac{dp}{p^2} = dx \Rightarrow -\frac{1}{p} = x + C \Rightarrow -\frac{1}{x + C} = p = \frac{dy}{dx} \Rightarrow y = -\ln(x + C) + C_1$

57. $x\frac{dp}{dx} + p = 0 \Rightarrow \frac{dp}{dx} + \frac{1}{x}p = 0 \Rightarrow P(x) = \frac{1}{x}, Q(x) = 0 \Rightarrow \int P(x)\,dx = \ln|x| \Rightarrow \rho(x) = e^{\ln|x|} = |x| \Rightarrow$
$p = \frac{1}{|x|}\int |x|\,(0)\,dx = \frac{C}{|x|} = \frac{C}{x}$ if $x > 0 \Rightarrow \frac{dy}{dx} = \frac{C}{x}$. $y' = 2$ when $x = 1 \Rightarrow 2 = C \Rightarrow \frac{dy}{dx} = \frac{2}{x} \Rightarrow y = 2\ln x + C_2$.
$y = -3$ when $x = 1 \Rightarrow -3 = 2\ln 1 + C_2 \Rightarrow C_2 = -3$. $\therefore\ y = 2\ln x - 3$.

59. $x\frac{dp}{dx} + 2p = 1 \Rightarrow \frac{dp}{dx} + \frac{2}{x}p = \frac{1}{x} \Rightarrow P(x) = \frac{2}{x}, Q(x) = \frac{1}{x} \Rightarrow \int P(x)\,dx = 2\ln|x| \Rightarrow \rho(x) = e^{2\ln|x|} = x^2 \Rightarrow$
$p = \frac{1}{x^2}\int x^2\left(\frac{1}{x}\right)dx = \frac{1}{x^2}\int x\,dx = \frac{1}{x^2}\left(\frac{x^2}{2} + C\right) = \frac{1}{2} + \frac{C}{x^2} \Rightarrow \frac{dy}{dx} = \frac{1}{2} + \frac{C}{x^2}$. $y' = 1$ when $x = 2 \Rightarrow 1 = \frac{1}{2} + \frac{C}{4}$
$\Rightarrow C = 2 \Rightarrow \frac{dy}{dx} = \frac{1}{2} + \frac{2}{x^2} \Rightarrow y = \frac{1}{2}x - \frac{2}{x} + C_1$. $y = 2$ when $x = 2 \Rightarrow 2 = 1 - 1 + C_1 \Rightarrow C_1 = 2$.
$\therefore\ y = \frac{1}{2}x - \frac{2}{x} + 2$.

61. $x\frac{dp}{dx} + p = x^2 \Rightarrow \frac{dp}{dx} + \frac{1}{x}p = x \Rightarrow P(x) = \frac{1}{x}, Q(x) = x \Rightarrow \int P(x)\,dx = \ln x$ if $x > 0 \Rightarrow \rho(x) = e^{\ln x} = x \Rightarrow$
$p = \frac{1}{x}\int x\,(x)\,dx = \frac{1}{x}\int x^2\,dx = \frac{1}{x}\left(\frac{x^3}{3} + C\right) = \frac{x^2}{3} + \frac{C}{x} \Rightarrow \frac{dy}{dx} = \frac{x^2}{3} + \frac{C}{x}$. $y' = 1$ when $x = 1 \Rightarrow C = \frac{2}{3} \Rightarrow$
$\frac{dy}{dx} = \frac{x^2}{3} + \frac{2}{3x} \Rightarrow y = \frac{x^3}{9} + \frac{2}{3}\ln x + C_1$. $y = 0$ when $x = 1 \Rightarrow C_1 = -\frac{1}{9} \Rightarrow y = \frac{x^3}{9} + \frac{2}{3}\ln x - \frac{1}{9}$

63. $m\frac{d^2s}{dt^2} = -ks$. Let $p = \frac{ds}{dt}, \frac{d^2s}{dt^2} = p\frac{dp}{ds}$. Then $mp\frac{dp}{ds} = -ks \Rightarrow \frac{1}{2}mp^2 = -\frac{1}{2}ks^2 + C \Rightarrow p = \pm\sqrt{\frac{2C - ks^2}{m}}$.
(Also, $\frac{1}{2}mp^2 + \frac{1}{2}ks^2 = C \Rightarrow C \ge 0$ since $m \ge 0, k > 0$.) $\frac{ds}{dt} = \pm\frac{\sqrt{2C - ks^2}}{\sqrt{m}} \Rightarrow \frac{ds}{\sqrt{2C - ks^2}} = \pm\frac{1}{\sqrt{m}}dt \Rightarrow$
$\frac{1}{\sqrt{k}}\sin^{-1}\left(\sqrt{\frac{k}{2C}}\,s\right) = \pm\frac{1}{\sqrt{m}}t + C_1 \Rightarrow \sin^{-1}\left(\sqrt{\frac{k}{2C}}\,s\right) = \pm\sqrt{\frac{k}{m}}\,t + C_2 \Rightarrow \sqrt{\frac{k}{2C}}\,s = \sin\left(\pm\sqrt{\frac{k}{m}}\,t + C_2\right) \Rightarrow$
$s = \sqrt{\frac{2C}{k}}\sin\left(\pm\sqrt{\frac{k}{m}}\,t + C_2\right) \Rightarrow s = A\sin\left(\pm\sqrt{\frac{k}{m}}\,t + B\right)$

65. Let $y = v\cos x \Rightarrow y' = v'\cos x - v\sin x \Rightarrow y'' = v''\cos x - 2v'\sin x - v\cos x$. Then $y'' + y = 0 \Rightarrow$
$v''\cos x - 2v'\sin x - v\cos x + v\cos x = 0 \Rightarrow v''\cos x - 2v'\sin x = 0$. Let $v' = p$, $v'' = \frac{dp}{dx} \Rightarrow \cos x\,\frac{dp}{dx} -$
$2p\sin x = 0 \Rightarrow \frac{dp}{p} = 2\tan x\,dx \Rightarrow \ln p = 2\ln(\sec x) + C \Rightarrow p = C_1(\sec^2 x) \Rightarrow \frac{dv}{dx} = C_1\sec^2 x \Rightarrow v = C_1\tan x$
$+ C_2$. $\therefore$ the general solution is $y = C_1\sin x + C_2\cos x$.

67. Let $y = ve^{-x} \Rightarrow y' = v'e^{-x} - ve^{-x} \Rightarrow y'' = v''e^{-x} - 2v'e^{-x} + ve^{-x}$. Then $y'' - y' - 2y = v''e^{-x} - 2v'e^{-x} +$
$ve^{-x} - v'e^{-x} + ve^{-x} - 2ve^{-x} = 0 \Rightarrow v''e^{-x} - 3v'e^{-x} = 0 \Rightarrow v'' - 3v' = 0$. Let $p = \frac{dv}{dx}, \frac{dp}{dx} = \frac{d^2v}{dx^2} \Rightarrow$
$\frac{dp}{dx} - 3p = 0 \Rightarrow \frac{dp}{p} = 3\,dx \Rightarrow \ln p = 3x + C \Rightarrow p = C_1e^{3x} \Rightarrow \frac{dv}{dx} = C_1e^{3x} \Rightarrow v = \frac{1}{3}C_1e^{3x} + C_2 \Rightarrow$
$v = kC_3e^{3x} + C_2$. $\therefore$ the general solution is $y = C_3e^{2x} + C_2e^{-x}$.

69. Let $y = vx \Rightarrow y' = v'x + v \Rightarrow y'' = v''x + 2v'$. $x^2y'' + xy' - y = v''x^3 + 2x^2v' + v'x^2 + xv - vx = v''x^3 + 3x^2v' = 0 \Rightarrow v'' + \frac{3}{x}v' = 0$. Let $p = v'$, $\frac{dp}{dx} = v'' \Rightarrow \frac{dp}{dx} + \frac{3}{x}p = 0 \Rightarrow \frac{dp}{p} = -\frac{3}{x}dx \Rightarrow \ln p = -3\ln x + C \Rightarrow p = C_1x^3$ $\Rightarrow \frac{dv}{dx} = C_1x^3 \Rightarrow v = \frac{C_1x^4}{4} + C_2 \Rightarrow v = C_3x^4 + C_2$. $\therefore$ the general solution is $y = C_3x^5 + C_2x$.

71. Let $y = vx \Rightarrow y' = v'x + v \Rightarrow y'' = v''x + v' + v' = v''x + 2v'$. $(1 - x^2)y'' - 2xy' + 2y = (1 - x^2)(v''x + 2v') - 2x(v'x + v) + 2xv = v''(x - x^3) + 2v'(1 - 2x^2) = 0$. Let $p = v'$, $\frac{dp}{dx} = v'' \Rightarrow (x - x^3)\frac{dp}{dx} + 2(1 - 2x^2)p = 0 \Rightarrow$ $\frac{dp}{p} = \frac{2(2x^2 - 1)}{x - x^3}dx = \left(-\frac{2}{x} + \frac{1}{1 - x} + \frac{-1}{1 + x}\right)dx \Rightarrow \ln p = \ln\left(\frac{1}{x^2(1 - x)(1 + x)}\right) + C \Rightarrow p = \frac{C_1}{x^2(1 - x^2)} \Rightarrow$ $\frac{dv}{dx} = \frac{C_1}{x^2(1 - x^2)} \Rightarrow v = C_1\left[-\frac{1}{x} + \frac{1}{2}\ln\left(\frac{1 + x}{1 - x}\right)\right] + C_2 = -\frac{C_1}{x} + \frac{C_1}{2}\ln\left(\frac{1 + x}{1 - x}\right) + C_2$. $\therefore$ the general solution is $y = -C_1 + \frac{C_1x}{2}\ln\left(\frac{1 + x}{1 - x}\right) + C_2x$.

73. $y^{(n)} = e^x$ for all $n \Rightarrow a_n(x)y^{(n)} + a_{n-1}(x)y^{(n-1)} + \cdots + a_1(x)y' + a_0(x)y = a_n(x)e^x + a_{n-1}(x)e^x + \cdots + a_1(x)e^x + a_0(x)e^x = e^x\left(a_n(x) + a_{n-1}(x) + \cdots + a_1(x) + a_0(x)\right) = e^x\sum_{k=0}^{n} a_k(x) = e^x(0) = 0$

75. $y''' - 3y'' + 2y = 0 \Rightarrow r^3 - 3r^2 + 2r = r(r^2 - 3r + 2) = r(r - 1)(r - 2) \Rightarrow r_1 = 0, r_2 = 1, r_3 = 2 \Rightarrow y_h = C_1 + C_2e^x + C_3e^{2x}$. $y_p = B\cos x + C\sin x \Rightarrow y'_p = -B\sin x + C\cos x \Rightarrow y_p'' = -B\cos x - C\sin x \Rightarrow y_p''' = B\sin x - C\cos x \Rightarrow (B\sin x - C\cos x) - 3(-B\cos x - C\sin x) + 2(B\cos x + C\sin x) = 3\sin x + \cos x \Rightarrow$ $-C + 5B = 1, B - C = 3 \Rightarrow B = -\frac{1}{2}, C = -\frac{7}{2} \Rightarrow y_p = -\frac{1}{2}\cos x - \frac{7}{2}\sin x \Rightarrow y = -\frac{1}{2}\cos x - \frac{7}{2}\sin x + C_1 + C_2e^x + C_3e^{2x}$.

77. $y^{(4)} - 8y'' + 16y = 0 \Rightarrow r^4 - 8r^2 + 16 = 0 \Rightarrow (r^2 - 4)^2 = 0 \Rightarrow r_1 = 2, r_2 = 2, r_3 = -2, r_4 = -2 \Rightarrow y_h = C_1e^{2x} + C_2xe^{2x} + C_3e^{-2x} + C_4xe^{-2x}$. $y_p = Ax + B \Rightarrow y_p' = A \Rightarrow y_p'' = y_p''' = y_p^{(4)} = 0 \Rightarrow 16(Ax + B) = 8x - 16$ $\Rightarrow 16A = 8, 16B = -16 \Rightarrow A = \frac{1}{2}, B = -1 \Rightarrow y_p = \frac{1}{2}x - 1 \Rightarrow y = \frac{1}{2}x - 1 + C_1e^{2x} + C_2xe^{2x} + C_3e^{-2x} + C_4xe^{-2x}$.

SECTION 15.6 OSCILLATION

1. $m\frac{d^2x}{dt^2} + kx = 0 \Rightarrow \frac{d^2x}{dt^2} + \frac{k}{m}x = 0$. Let $\omega = \sqrt{\frac{k}{m}}$. Then $\frac{d^2x}{dt^2} + \omega^2 x = 0 \Rightarrow r^2 + \omega^2 = 0 \Rightarrow r = \pm\, \omega i \Rightarrow$ $x = C_1 \cos \omega t + C_2 \sin \omega t$. $x(0) = x_0 \Rightarrow C_1 \cos 0 + C_2 \sin 0 = x_0 \Rightarrow C_1 = x_0 \Rightarrow x = x_0 \cos \omega t + C_2 \sin \omega t$ $\Rightarrow x' = -x_0\omega \sin \omega t + C_2\omega \cos \omega t$. $x'(0) = v_0 \Rightarrow -x_0\omega \sin 0 + C_2\omega \cos 0 = v_0 \Rightarrow C_2 = \frac{v_0}{\omega} \Rightarrow$

$x = x_0 \cos \omega t + \frac{v_0}{\omega} \sin \omega t$. $x = C \sin(\omega t + \phi)$ where $C = \sqrt{x_0^2 + \left(\frac{v_0}{\omega}\right)^2} = \frac{\sqrt{\omega^2 x_0^2 + v_0^2}}{\omega}$ and

$\phi = \tan^{-1}\left(\frac{\omega x_0}{v_0}\right) \Rightarrow x = \frac{\sqrt{\omega^2 x_0^2 + v_0^2}}{\omega} \sin\left(\omega t + \tan^{-1}\left(\frac{\omega x_0}{v_0}\right)\right)$

3. a) $L\frac{d^2i}{dt^2} + R\frac{di}{dt} + \frac{1}{C}i = \frac{dv}{dt}$. $R = 0$, $\frac{1}{LC} = \omega^2$, v = constant $\Rightarrow L\frac{d^2i}{dt^2} + \frac{1}{C}i = 0 \Rightarrow \frac{d^2i}{dt^2} + \frac{1}{LC}i = 0 \Rightarrow$ $\frac{d^2i}{dt^2} + \omega^2 i = 0 \Rightarrow r = \pm\, \omega i \Rightarrow i = C_1 \cos \omega t + C_2 \sin \omega t$

b) $L\frac{d^2i}{dt^2} + R\frac{di}{dt} + \frac{1}{C}i = \frac{dv}{dt}$. $R = 0$, $\frac{1}{LC} = \omega^2$, $v = V \sin \alpha t$, $\alpha \neq \omega \Rightarrow \frac{d^2i}{dt^2} + \omega^2 i = \frac{V\alpha}{L}\cos \alpha t \Rightarrow$ $i_h = C_1 \cos \omega t + C_2 \sin \omega t$. $i_p = A \cos \alpha t + B \sin \alpha t \Rightarrow i'_p = -A\alpha \sin \alpha t + B\alpha \cos \alpha t \Rightarrow$ $i''_p = -A\alpha^2 \cos \alpha t - B\alpha^2 \sin \alpha t \Rightarrow -A\alpha^2 \cos \alpha t - B\alpha^2 \sin \alpha t + \omega^2\left(A \cos \alpha t + B \sin \alpha t\right) =$ $\frac{V\alpha}{L}\cos \alpha t \Rightarrow A = \frac{V\alpha}{L\left(\omega^2 - \alpha^2\right)}$, $B = 0 \Rightarrow i_p = \frac{V\alpha}{L\left(\omega^2 - \alpha^2\right)}\cos \alpha t \Rightarrow i = C_1 \cos \omega t + C_2 \sin \omega t +$

$\frac{V\alpha}{L\left(\omega^2 - \alpha^2\right)}\cos \alpha t$

c) $L\frac{d^2i}{dt^2} + R\frac{di}{dt} + \frac{1}{C}i = \frac{dv}{dt}$ $R = 0$, $\frac{1}{LC} = \omega^2$, $v = V \sin \omega t$, V constant $\Rightarrow L\frac{d^2i}{dt^2} + \frac{1}{C}i = V\omega \cos \omega t \Rightarrow$ $\frac{d^2i}{dt^2} + \omega^2 i = \frac{V\omega}{L}\cos \omega t \Rightarrow i_h = C_1 \cos \omega t + C_2 \sin \omega t$. $i_p = At \cos \omega t + Bt \sin \omega t \Rightarrow i'_p = A \cos \omega t -$ $A\omega t \sin \omega t + B \sin \omega t + B\omega t \cos \omega t \Rightarrow y''_p = -2A\omega \sin \omega t + 2B\omega \cos \omega t - A\omega^2 t \cos \omega t -$ $B\omega^2 t \sin \omega t \Rightarrow -2A\omega \sin \omega t + 2B\omega \cos \omega t - A\omega^2 t \cos \omega t - B\omega^2 t \sin \omega t +$ $\omega^2\left(At \cos \omega t + Bt \sin \omega t\right) = \frac{V\omega}{L}\cos \omega t \Rightarrow A = 0$, $B = \frac{V}{2L} \Rightarrow i_p = \frac{V}{2L}t \sin \omega t \Rightarrow i = C_1 \cos \omega t +$ $C_2 \sin \omega t + \frac{V}{2L}t \sin \omega t$

d) $L\frac{d^2i}{dt^2} + R\frac{di}{dt} + \frac{1}{C}i = \frac{dv}{dt}$ $R = 50$, $L = 5$, $C = 9 \times 10^{-6}$, v constant $\Rightarrow 5\frac{d^2i}{dt^2} + 50\frac{di}{dt} + \frac{1}{9} \times 10^6\, i = 0 \Rightarrow$ $\frac{d^2i}{dt^2} + 10\frac{di}{dt} + \frac{1}{45} \times 10^6\, i = 0 \Rightarrow r^2 + 10r + \frac{1}{45} \times 10^6 = 0 \Rightarrow r = -5 \pm 5\sqrt{-\frac{7991}{9}} \approx -5 \pm 148.99\, i \Rightarrow$ $i = e^{-5t}\left(C_1 \cos(148.99)t + C_2 \sin(148.99)t\right)$

5. $\frac{d^2\theta}{dt^2} = -\frac{2k\theta}{mr^2} \Rightarrow \frac{d^2\theta}{dt^2} + \frac{2k}{mr^2}\theta = 0 \Rightarrow r^2 + \frac{2k}{mr^2} = 0 \Rightarrow r = \pm\sqrt{\frac{2k}{mr^2}}\, i \Rightarrow \theta = C_1 \cos\sqrt{\frac{2k}{mr^2}}\, t +$ $C_2 \sin\sqrt{\frac{2k}{mr^2}}\, t.$ $t = 0 \Rightarrow \theta = \theta_0 \Rightarrow C_1 = \theta_0 \Rightarrow \theta = \theta_0 \cos\sqrt{\frac{2k}{mr^2}}\, t + C_2 \sin\sqrt{\frac{2k}{mr^2}}\, t \Rightarrow$ $\frac{d\theta}{dt} = -\theta_0\sqrt{\frac{2k}{mr^2}} \sin\sqrt{\frac{2k}{mr^2}}\, t + C_2\sqrt{\frac{2k}{mr^2}} \cos\sqrt{\frac{2k}{mr^2}}\, t.$ $\frac{d\theta}{dt} = v_0$ at $t = 0 \Rightarrow C_2\sqrt{\frac{2k}{mr^2}} = v_0 \Rightarrow$ $C_2 = v_0\sqrt{\frac{mr^2}{2k}} \Rightarrow \theta = \theta_0 \cos\sqrt{\frac{2k}{mr^2}}\, t + v_0\sqrt{\frac{mr^2}{2k}} \sin\sqrt{\frac{2k}{mr^2}}\, t$

7. a) $f(t) = A \sin \alpha t,\ \alpha \neq \sqrt{\frac{k}{m}} \Rightarrow \frac{d^2x}{dt^2} + \frac{k}{m}x = \frac{k}{m}\left(A \sin \alpha t\right) \Rightarrow r^2 + \frac{k}{m} = 0 \Rightarrow r = \pm i\sqrt{\frac{k}{m}} \Rightarrow$ $x_h = C_1 \cos\sqrt{\frac{k}{m}}\, t + C_2 \sin\sqrt{\frac{k}{m}}\, t.$ $x_p = B \cos \alpha t + C \sin \alpha t \Rightarrow x'_p = -B\alpha \sin \alpha t + C\alpha \cos \alpha t \Rightarrow$ $x''_p = -B\alpha^2 \cos \alpha t - C\alpha^2 \sin \alpha t \Rightarrow -B\alpha^2 \cos \alpha t - C\alpha^2 \sin \alpha t + \frac{k}{m}\left(B \cos \alpha t + C \sin \alpha t\right) =$ $\frac{k}{m}\left(A \sin \alpha t\right) \Rightarrow B = 0,\ C = \frac{Ak}{k - m\alpha^2} \Rightarrow x_p = \frac{Ak}{k - m\alpha^2} \sin \alpha t \Rightarrow x = \frac{Ak}{k - m\alpha^2} \sin \alpha t + C_1 \cos\sqrt{\frac{k}{m}}\, t$ $+ C_2 \sin\sqrt{\frac{k}{m}}\, t.$ $x(0) = x_0 \Rightarrow C_1 = x_0 \Rightarrow x = x_0 \cos\sqrt{\frac{k}{m}}\, t + C_2 \sin\sqrt{\frac{k}{m}}\, t + \frac{Ak}{k - m\alpha^2} \sin \alpha t \Rightarrow$ $\frac{dx}{dt} = -x_0\sqrt{\frac{k}{m}} \sin\sqrt{\frac{k}{m}}\, t + C_2\sqrt{\frac{k}{m}} \cos\sqrt{\frac{k}{m}}\, t + \frac{Ak\alpha}{k - m\alpha^2} \cos \alpha t.$ $x'(0) = 0 \Rightarrow C_2\sqrt{\frac{k}{m}} +$ $\frac{Ak\alpha}{k - m\alpha^2} = 0 \Rightarrow C_2 = -\frac{Ak\alpha}{k - m\alpha^2}\sqrt{\frac{m}{k}} \Rightarrow x = x_0 \cos\sqrt{\frac{k}{m}}\, t - \frac{A\alpha\sqrt{mk}}{k - m\alpha^2} \sin\sqrt{\frac{k}{m}}\, t + \frac{Ak}{k - m\alpha^2} \sin \alpha t$

b) $f(t) = A \sin \alpha t,\ \alpha = \sqrt{\frac{k}{m}} \Rightarrow \frac{d^2x}{dt^2} + \frac{k}{m}x = \frac{k}{m}\left(A \sin \alpha t\right) \Rightarrow r^2 + \frac{k}{m} = 0 \Rightarrow r = \pm\sqrt{\frac{k}{m}}\, i \Rightarrow$ $x_h = C_1 \cos\sqrt{\frac{k}{m}}\, t + C_2 \sin\sqrt{\frac{k}{m}}\, t.$ $x_p = Bt \cos\sqrt{\frac{k}{m}}\, t + Ct \sin\sqrt{\frac{k}{m}}\, t \Rightarrow x'_p = B \cos\sqrt{\frac{k}{m}}\, t -$ $B\sqrt{\frac{k}{m}}\, t \sin\sqrt{\frac{k}{m}}\, t + C \sin\sqrt{\frac{k}{m}}\, t + C\sqrt{\frac{k}{m}}\, t \cos\sqrt{\frac{k}{m}}\, t \Rightarrow x''_p = -2B\sqrt{\frac{k}{m}} \sin\sqrt{\frac{k}{m}}\, t +$ $2C\sqrt{\frac{k}{m}} \cos\sqrt{\frac{k}{m}}\, t - B\left(\frac{k}{m}\right)t \cos\sqrt{\frac{k}{m}}\, t - C\left(\frac{k}{m}\right)t \sin\sqrt{\frac{k}{m}}\, t \Rightarrow -2B\sqrt{\frac{k}{m}} \sin\sqrt{\frac{k}{m}}\, t +$ $2C\sqrt{\frac{k}{m}} \cos\sqrt{\frac{k}{m}}\, t - B\left(\frac{k}{m}\right)t \cos\sqrt{\frac{k}{m}}\, t - C\left(\frac{k}{m}\right)t \sin\sqrt{\frac{k}{m}}\, t + \frac{k}{m}\left(Bt \cos\sqrt{\frac{k}{m}}\, t + Ct \sin\sqrt{\frac{k}{m}}\, t\right)$ $= \frac{kA}{m} \sin \alpha t \Rightarrow B = -\frac{A\alpha}{2},\ C = 0 \Rightarrow x_p = -\frac{A\alpha}{2} t \cos \alpha t \Rightarrow x = C_1 \cos\sqrt{\frac{k}{m}}\, t + C_2 \sin\sqrt{\frac{k}{m}}\, t -$ $\frac{A\alpha}{2} t \cos \alpha t.$ $x(0) = x_0 \Rightarrow C_1 = x_0 \Rightarrow x = x_0 \cos\sqrt{\frac{k}{m}}\, t + C_2 \sin\sqrt{\frac{k}{m}}\, t - \frac{A\alpha}{2} t \cos \alpha t \Rightarrow$ $\frac{dx}{dt} = -x_0\alpha \sin \alpha t + C_2\alpha \cos \alpha t - \frac{A\alpha}{2} \cos \alpha t - \frac{A\alpha^2}{2} t \sin \alpha t.$ $x'(0) = 0 \Rightarrow C_2\alpha - \frac{A\alpha}{2} = 0 \Rightarrow$ $C_2 = \frac{A}{2} \Rightarrow x = x_0 \cos \alpha t + \frac{A}{2} \sin \alpha t - \frac{A\alpha}{2} t \cos \alpha t$

SECTION 15.7 POWER SERIES SOLUTIONS

1. $y = 2e^{x-1} - 2 - (x-1) \Rightarrow y' = 2e^{x-1}$. Then $x + y = x + 2e^{x-1} - 2 - (x-1) = 2e^{x-1} - 1 = y'$. Also $y(1) = 2e^0 - 2 - (1-1) = 0$.

3. $y(x) = y(0) + y'(0)\,x + \dfrac{y''(0)\,x^2}{2!} + \cdots$. $y' = y \Rightarrow y'(0) = y(0) = 1$. $y'' = y' \Rightarrow y''(0) = y'(0) = 1$, and so on.

$$\therefore\ y(x) = 1 + x + \frac{x^2}{2!} + \frac{x^3}{3!} + \cdots = \sum_{n=0}^{\infty} \frac{x^n}{n!}$$

5. $y(x) = y(0) + y'(0)\,x + \dfrac{y''(0)\,x^2}{2!} + \cdots$. $y' = 2y \Rightarrow y'(0) = 2y(0) = 4$. $y'' = 2y' \Rightarrow y''(0) = 2y'(0) = 8$.

$y''' = 2y'' \Rightarrow y'''(0) = 2y'(0) = 16$, and so on. $\therefore\ y(x) = 2 + 4x + \dfrac{8x^2}{2!} + \dfrac{16x^3}{3!} + \cdots = \displaystyle\sum_{n=0}^{\infty} \frac{2^{n+1}x^n}{n!}$

7. $y(x) = y(0) + y'(0)\,x + \dfrac{y''(0)\,x^2}{2!} + \cdots$ $y'' = y \Rightarrow y''(0) = y(0) = 0$. $y''' = y' \Rightarrow y'''(0) = y'(0) = 1$. $y^{(4)} = y'' \Rightarrow$

$y^{(4)}(0) = y''(0) = 0$, and so on. $\therefore\ y(x) = 0 + x + 0 + \dfrac{x^3}{3!} + \cdots = \displaystyle\sum_{n=0}^{\infty} \frac{x^{2n+1}}{(2n+1)!}$

9. $y(x) = y(0) + y'(0)\,x + \dfrac{y''(0)\,x^2}{2!} + \cdots$ $y'' + y = x \Rightarrow y''(0) = -y(0) + 0 = -2$. $y''' + y' = 1 \Rightarrow y'''(0) = -y'(0) + 1 = 0$. $y^{(4)} + y'' = 0 \Rightarrow y^{(4)}(0) = -y''(0) = 2$. $y^{(5)} + y''' = 0 \Rightarrow y^{(5)}(0) = -y'''(0) = 0$. $y^{(6)} + y^{(4)} = 0 \Rightarrow y^{(6)}(0) = -y^{(4)}(0) = -2$, and so on. $\therefore\ y(x) = 2 + x - \dfrac{2x^2}{2!} + 0 + \dfrac{2x^4}{4!} + 0 - \dfrac{2x^6}{6!} + \cdots = 2 + x$

$$-2\sum_{n=1}^{\infty} \frac{(-1)^{n+1}x^{2n}}{(2n)!}$$

11. $y(x) = y(2) + y'(2)(x-2) + \dfrac{y''(2)(x-2)^2}{2!} + \cdots$ $y'' - y = -x \Rightarrow y'e(2) = y(2) - 2 = -2$. $y''' - y' = -1 \Rightarrow y'''(2) = y'(2) - 1 = -3$. $y^{(4)} - y'' = 0 \Rightarrow y^{(4)}(2) = y''(2) = -2$. $y^{(5)} - y''' = 0 \Rightarrow y^{(5)}(2) = y'''(2) = -3$. $y^{(6)} - y^{(4)} = 0 \Rightarrow y^{(6)}(2) = y^{(4)}(2) = -2$, and so on. $\therefore\ y(x) = 0 - 2(x-2) - \dfrac{2(x-2)^2}{2!} - \dfrac{3(x-2)^3}{3!} -$

$$\frac{2(x-2)^4}{4!} - \cdots = 4 - 2x - 2\sum_{n=1}^{\infty} \frac{(x-2)^{2n}}{(2n)!} - 3\sum_{n=1}^{\infty} \frac{(x-2)^{2n+1}}{(2n+1)!}$$

Note on Exercise 13: Evaluating this at $x = x_0$ does not yield a recognizable series. It would be better to evaluate it at $x = 0$. Then the answer in the text would be correct.

13. $y(x) = y(x_0) + y'(x_0) + \dfrac{y''(x_0)(x - x_0)^2}{2!} + \cdots$ $y'' + x^2y = x \Rightarrow y''(x_0) = x_0 - x_0^2a$. $y''' + 2xy + x^2y' = 1 \Rightarrow$ $y'''(x_0) = 1 - 2ax_0 - bx_0^2$. $y^{(4)} + 2y + 4xy' + x^2y'' = 0 \Rightarrow y^{(4)}(x_0) = -2a - 4bx_0 - x_0^3 + ax_0^4$ $y^{(5)} + 6y' + 6xy'' + x^2y''' = 0 \Rightarrow y^{(5)}(x_0) = -6b - 5x_0^2 + 4ax_0^3 - bx_0^4$. (This is not recognizable.)

$$\therefore\ y(x) = a + b(x - x_0) + \frac{(x_0 - ax_0^2)(x - x_0)^2}{2!} + \frac{(1 - 2ax_0 - bx_0)^2(x - x_0)^3}{3!}$$
$$+ \frac{(-2a - 4bx_0 - x_0^3 + ax_0^4)(x - x_0)^4}{4!} + \cdots$$

SECTION 15.8 SLOPE FIELDS AND PICARD'S THEOREM

1.

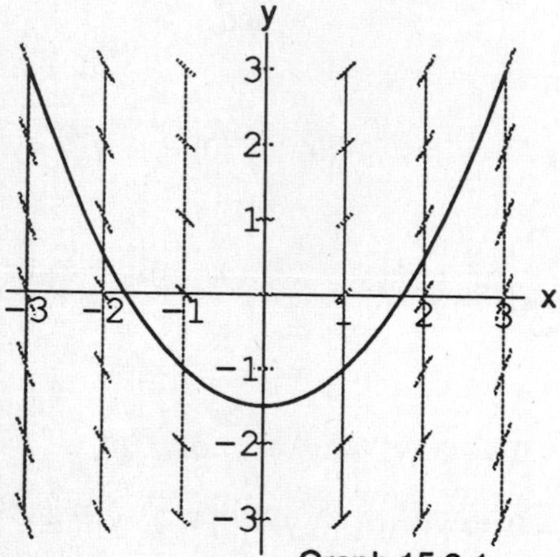

Graph 15.8.1

3.

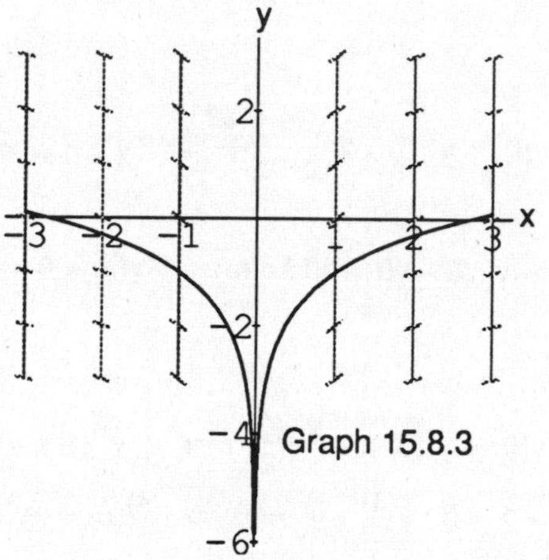

Graph 15.8.3

5.

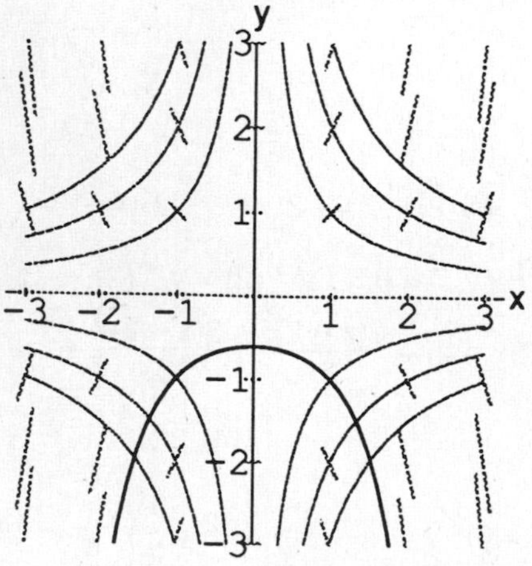

Graph 15.8.5

7.

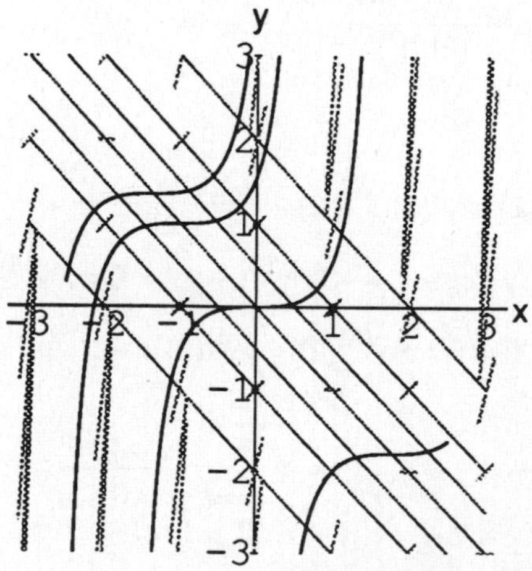

Graph 15.8.7

Exercise 7 is continued on the next page.

7. $u = x + y \Rightarrow \frac{du}{dx} = 1 + \frac{dy}{dx} \Rightarrow \frac{dy}{dx} = \frac{du}{dx} - 1.$ $\therefore \frac{du}{dx} - 1 = u^2 \Rightarrow \frac{du}{dx} = u^2 + 1 \Rightarrow \frac{du}{u^2 + 1} = dx \Rightarrow \tan^{-1}u = x + C$ $\Rightarrow \tan^{-1}(x + y) = x + C \Rightarrow x + y = \tan(x + C)$. At P(0,0), $\tan^{-1}(0 + 0) = 0 + C \Rightarrow C = 0$. $\therefore$ at (0,0), the solution is $x + y = \tan x \Rightarrow y = \tan x - x$

9.

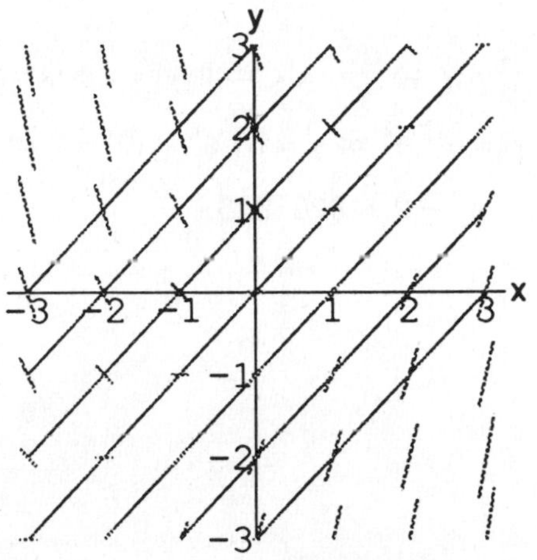

Graph 15.8.9

a) $y = x - 1$ is a solution since $y' = 1 \Rightarrow x - y = x - (x - 1) = 1 = y'$ or $y' = x - y$.

b) $y'' = 1 - y' = 1 - (x - y) = 1 - x + y$ is positive if $1 - x + y > 0$ or $y > x - 1 \Rightarrow$ concave up and negative if $y < x - 1 \Rightarrow$ concave down.

11. $y = -1 + \int_1^x (t - y(t))\,dt \Rightarrow y' = x - y,\ y(1) = -1$

13. $y = 2 - \int_0^x (1 + y(t)) \sin t\,dt \Rightarrow y' = -(1 - y)\sin x,$ $y(0) = 2$

15. $y_0 = 2,\ y_1 = 2 + \int_1^x t\,dt = 2 + \frac{1}{2}x^2 - \frac{1}{2} = \frac{1}{2}x^2 + \frac{3}{2},\ y_2 = 2 + \int_1^x \left(\frac{t^3}{2} + \frac{3t}{2}\right)dt = 2 + \left[\frac{t^4}{8} + \frac{3t^2}{4}\right]_1^x = \frac{9}{8} + \frac{x^4}{8} + \frac{3x^2}{4},\ y_3 = 2 + \int_1^x \left(\frac{9t}{8} + \frac{t^5}{8} + \frac{3t^3}{4}\right)dt = 2 + \left[\frac{9t^2}{16} + \frac{t^6}{48} + \frac{3t^4}{16}\right]_1^x = \frac{107}{48} + \frac{9x^2}{16} + \frac{x^6}{48} + \frac{3x^4}{16}$

17. $y_0 = 1,\ y_1 = 1 + \int_1^x t\,dt = 1 + \frac{x^2}{2} - \frac{1}{2} = \frac{x^2}{2} + \frac{1}{2},\ y_2 = 1 + \int_1^x \left(\frac{t^3}{2} + \frac{t}{2}\right)dt = \frac{x^4}{8} + \frac{x^2}{4} + \frac{5}{8},$ $y_3 = 1 + \int_1^x \left(\frac{t^5}{8} + \frac{t^3}{4} + \frac{5t}{8}\right)dt = \frac{x^6}{48} + \frac{x^4}{16} + \frac{5x^2}{16} + \frac{29}{48}$

19. $y_0 = 1,\ y_1 = 1 + \int_0^x (t + 1)\,dt = 1 + x + \frac{1}{2}x^2,\ y_2 = 1 + \int_0^x \left(2t + 1 + \frac{1}{2}t^2\right)dt = 1 + x + x^2 + \frac{1}{6}x^3,$ $y_3 = 1 + \int_0^x \left(2t + 1 + t^2 + \frac{1}{6}t^3\right)dt = 1 + x + x^2 + \frac{1}{3}x^3 + \frac{1}{24}x^4$

21. $y = -1 - x + (1 + x_0 + y_0)\, e^{x-x_0} \Rightarrow y\left(x_0\right) = -1 - x_0 + (1 + x_0 + y_0)\, e^0 = y_0$. $y' = -1 + (1 + x_0 + y_0)\, e^{x-x_0} \Rightarrow$ $x = y = x + \left(-1 - x + (1 + x_0 + y_0)\right) e^{x-x_0} = -1 + (1 + x_0 + y)\, e^{x-x_0} = y'$

SECTION 15.9 NUMERICAL METHODS

1.

	x_n	y_n
x_0	0	1
x_1	1/5	1.2
x_2	2/5	1.44
x_3	3/5	1.728
x_4	4/5	2.0736
x_5	1	2.48832

Exact Value: $y' = y \Rightarrow \frac{dy}{dx} = y \Rightarrow \frac{dy}{y} = dx \Rightarrow \ln|y| = x + C \Rightarrow$ $e^{\ln|y|} = e^{x+C} \Rightarrow |y| = e^C e^x \Rightarrow y = C_1\, e^x$. $y(0) = 1 \Rightarrow C_1 =$ $1 \Rightarrow y = e^x \Rightarrow y(1) = e^1 = 2.718281828\ldots$

3.

	x_n	y_n
x_0	0	1
x_1	1/5	1.22
x_2	2/5	1.4884
x_3	3/5	1.815848
x_4	4/5	2.2153346
x_5	1	2.702708163

5. $y' = x^2 + y^2 \Rightarrow x_{n+1} = x_n + h \Rightarrow y_{n+1} = y_n + h\left(x_n^{\,2} + y_n^{\,2}\right)$. $Y' = Y^2 \Rightarrow x_{n+1} = x_n + h \Rightarrow Y_{n+1} = Y_n +$ $h(Y_n^{\,2})$. $y_0 = h(x_0^{\,2} + y_0^{\,2}) = Y_0 + h(Y_0)$ since $x_0 = 0 \Rightarrow y_1 = Y_1$. $y_1 + h\left(x_1^{\,2} + y_1^{\,2}\right) > Y_1 +$ $h(Y_1^{\,2})$ since $x_1 > 0 \Rightarrow y_2 > Y_2$. And from here on up to $x = 1$, $y_{n+1} > Y_{n+1}$.

$Y' = Y^2 \Rightarrow \frac{dY}{dx} = Y^2 \Rightarrow \frac{1}{Y^2}\, dY = dx \Rightarrow -\frac{1}{Y} = x + C \Rightarrow \frac{1}{Y} = -x - C \Rightarrow Y = -\frac{1}{x + C}$. $y(0) = 1 \Rightarrow$ $-\frac{1}{C} = 1 \Rightarrow C = -1 \Rightarrow Y = -\frac{1}{x - 1} \Rightarrow y \to \infty$ as $x \to 1^-$. Since $y_{n+1} > Y_{n+1}$, $y_{n+1} \to \infty$ as $x \to 1^-$

Note: For Exercises 7–11, the Calculus Tookit was used.

7. $y = 0.571428572$

9. $y = 0.810263855$

11. a) $y = 0.841470985$

b) $y = 0.841470983$

SECTION 15.M MISCELLANEOUS EXERCISES

1. $e^{y-2}\,dx - e^{x+2y}\,dy = 0 \Rightarrow e^{-x}\,dx - \frac{e^{2y}}{e^{y-2}}\,dy = 0 \Rightarrow e^{-x}\,dx - e^{y+2}\,dy = 0 \Rightarrow -e^{-x} - e^{y+2} = C.$ $y(0) = -2$
$\Rightarrow -e^{0} - e^{-2+2} = C \Rightarrow C = -2 \Rightarrow e^{-x} + e^{y+2} = 2$

3. $\frac{dy}{dx} = \frac{x^2+y^2}{2xy} \Rightarrow \frac{dy}{dx} = \frac{1+\left(\frac{y}{x}\right)^2}{2\left(\frac{y}{x}\right)} \Rightarrow$ Homogeneous. $v = \frac{y}{x} \Rightarrow F(v) = \frac{1+v^2}{2v} \Rightarrow \frac{dx}{x} + \frac{dv}{v - \left(\frac{1+v^2}{2v}\right)} = 0 \Rightarrow$
$\frac{dx}{x} + \frac{2v\,dv}{v^2-1} = 0 \Rightarrow \ln|x| + \ln|v^2-1| = C \Rightarrow x(v^2-1) = C_1 \Rightarrow x\left(\left(\frac{y}{x}\right)^2 - 1\right) = C_1 \Rightarrow \frac{y^2}{x} - x = C_1.$ $y(5) = 0$
$\Rightarrow 0 - 5 = C_1 \Rightarrow C_1 = -5 \Rightarrow \frac{y^2}{x} - x = -5$

5. $(x^2+y)\,dx + \left(e^y + x\right)dy = 0 \Rightarrow \frac{\partial M}{\partial y} = 1 = \frac{\partial N}{\partial x} \Rightarrow$ Exact. $\frac{\partial f}{\partial x} = x^2 + y \Rightarrow f(x,y) = \frac{x^3}{3} + xy + k(y) \Rightarrow$
$\frac{\partial f}{\partial y} = x + k'(y) = e^y + x \Rightarrow k'(y) = e^y \Rightarrow k(y) = e^y + C \Rightarrow f(x,y) = \frac{x^3}{3} + xy + e^y + C \Rightarrow \frac{x^3}{3} + xy + e^y = C_1.$
$y(3) = 0 \Rightarrow 9 + 1 = C_1 \Rightarrow C_1 = 10 \Rightarrow \frac{x^3}{3} + xy + e^y = 10$

7. $(x+1)\frac{dy}{dx} + 2y = x \Rightarrow \frac{dy}{dx} + \frac{2}{x+1}y = \frac{x}{x+1} \Rightarrow P(x) = \frac{2}{x+1},\ Q(x) = \frac{x}{x+1} \Rightarrow \int P(x)\,dx = \int \frac{2}{x+1}\,dx =$
$2\ln(x+1)$. $\rho(x) = e^{2\ln(x+1)} = (x+1)^2$. Then $y = \frac{1}{(x+1)^2}\int (x+1)^2\left(\frac{x}{x+1}\right)dx = \frac{1}{(x+1)^2}\int (x^2+x)\,dx$
$= \frac{1}{(x+1)^2}\left(\frac{x^3}{3} + \frac{x^2}{2} + C\right) = \frac{1}{(x+1)^2}\left(\frac{2x^3+3x^2}{6} + C\right) = \frac{2x^3+3x^2}{6(x+1)^2} + \frac{C}{(x+1)^2}.$ $y(0) = 1 \Rightarrow C = 1$
$\therefore y = \frac{2x^3+3x^2}{6(x+1)^2} + \frac{1}{(x+1)^2} = \frac{2x^3+3x^2+6}{6(x+1)^2}$

9. $\frac{d^2y}{dx^2} - \left(\frac{dy}{dx}\right)^2 = 1.$ Let $p = \frac{dy}{dx} \Rightarrow \frac{dp}{dx} - p^2 = 1 \Rightarrow \frac{dp}{1+p^2} = dx \Rightarrow \tan^{-1} p = x + C \Rightarrow p = \tan(x+C) \Rightarrow$
$\frac{dy}{dx} = \tan(x+C) \Rightarrow y = \ln|\sec(x+C)| + C_1.$ $y'\left(\frac{\pi}{3}\right) = \sqrt{3} \Rightarrow \sqrt{3} = \tan\left(\frac{\pi}{3} + C\right) \Rightarrow \tan^{-1}\sqrt{3} = \frac{\pi}{3} + C \Rightarrow$
$\frac{\pi}{3} = \frac{\pi}{3} + C \Rightarrow C = 0 \Rightarrow y = \ln|\sec x| + C_1.$ $y\left(\frac{\pi}{3}\right) = 0 \Rightarrow 0 = \ln\left|\sec\frac{\pi}{3}\right| + C_1 \Rightarrow C_1 = -\ln 2 \Rightarrow$
$y = \ln|\sec x| - \ln 2$

11. $\frac{d^2y}{dx^2} - 4\frac{dy}{dx} + 3y = 0 \Rightarrow r^2 - 4r + 3 = 0 \Rightarrow r_1 = 3,\ r_2 = 1 \Rightarrow y = C_1 e^{3x} + C_2 e^{x} \Rightarrow y' = 3C_1 e^{3x} + C_2 e^{x}.$
$y'(0) = -2 \Rightarrow -2 = 3C_1 + C_2.$ $y(0) = 2 \Rightarrow 2 = C_1 + C_2 \Rightarrow C_1 = -2,\ C_2 = 4 \Rightarrow y = -2e^{3x} + 4e^{x}$

13. $\frac{d^2y}{dx^2} + 4\frac{dy}{dx} + 4y = 0 \Rightarrow r_1 = r_2 = -2 \Rightarrow y = \left(C_1x + C_2\right)e^{-2x}$. $y(0) = 0 \Rightarrow C_2 = 0 \Rightarrow y = C_1x\,e^{-2x} \Rightarrow$
$y' = C_1\left(e^{-2x} - 2x\,e^{-2x}\right)$. $y'(0) = 7 \Rightarrow C_1 = 7 \Rightarrow y = 7x\,e^{-2x}$

15. $\frac{d^2y}{dx^2} + 2\frac{dy}{dx} + 2y = 0 \Rightarrow r^2 + 2r + 2 = 0 \Rightarrow r = -1 \pm i \Rightarrow y = e^{-x}\left(C_1 \cos x + C_2 \sin x\right)$. $y(0) = 1 \Rightarrow$
$e^0\left(C_1 \cos 0 + C_2 \sin 0\right) = 1 \Rightarrow C_1 = 1 \Rightarrow y = e^{-x}\left(\cos x + C_2 \sin x\right) \Rightarrow y' = -e^{-x}\left(\cos x + C_2 \sin x\right) +$
$e^{-x}\left(-\sin x + C_2 \cos x\right)$. $y'(0) = 0 \Rightarrow 0 = -\left(\cos 0 + C_2 \sin 0\right) + \left(-\sin 0 + C_2 \cos 0\right) \Rightarrow C_2 = 1 \Rightarrow$
$y = e^{-x}(\cos x + \sin x)$

17. $\frac{d^2y}{dx^2} + 2\frac{dy}{dx} = 4x \Rightarrow r^2 + 2r = 0 \Rightarrow r_1 = 0, r_2 = -2 \Rightarrow y_h = C_1 + C_2\,e^{-2x}$. $y_p = Ax^2 + Bx \Rightarrow y'_p = 2Ax + B$.
$y''_p = 2A \Rightarrow 2A + 2(2Ax + B) = 4x \Rightarrow A = 1, B = -1 \Rightarrow y_p = x^2 - x \Rightarrow y = C_1 + C_2\,e^{-2x} + x^2 - x$.
$y(0) = 1 \Rightarrow C_1 + C_2 = 1$. $y' = -2C_2\,e^{-2x} + 2x - 1$. $y'(0) = -3 \Rightarrow -2C_2 - 1 = -3 \Rightarrow C_1 = 0, C_2 = 1 \Rightarrow$
$y = e^{-2x} + x^2 - x$

19. $\frac{d^2y}{dx^2} - \frac{dy}{dx} - 2y = 3\,e^{2x} \Rightarrow r^2 - r - 2 = 0 \Rightarrow r_1 = 2, r_2 = -1 \Rightarrow y_h = C_1\,e^{2x} + C_2\,e^{-x}$. $y_p = Ax\,e^{2x} \Rightarrow$
$y'_p = A\,e^{2x} + 2Ax\,e^{2x} \Rightarrow y''_p = 4A\,e^{2x} + 4Ax\,e^{2x} \Rightarrow 4A\,e^{2x} + 4Ax\,e^{2x} - \left(A\,e^{2x} + 2Ax\,e^{2x}\right) -$
$2\left(Ax\,e^{2x}\right) = 3\,e^{2x} \Rightarrow A = 1 \Rightarrow y_p = x\,e^{2x} \Rightarrow y = x\,e^{2x} + C_1\,e^{2x} + C_2\,e^{-x}$. $y(0) = -2 \Rightarrow C_1 + C_2 = -2$.
$y' = 2C_1\,e^{2x} - C_2\,e^{-x} + e^{2x} + 2x\,e^{2x}$. $y'(0) = 0 \Rightarrow 2C_1 - C_2 + 1 = 0 \Rightarrow C_1 = -1, C_2 = -1 \Rightarrow$
$y = -e^{2x} - e^{-x} + x\,e^{2x}$

21. $x^2 = Cy^3 \Rightarrow C = \frac{x^2}{y^3}$. Also $x^2 = Cy^3 \Rightarrow 2x = 3Cy^2\frac{dy}{dx} \Rightarrow 2x = 3\left(\frac{x^2}{y^3}\right)y^2\frac{dy}{dx} = \frac{3x^2}{y^2}\frac{dy}{dx} \Rightarrow \frac{dy}{dx} = \frac{2xy}{3x^2} = \frac{2y}{3x}$.
Then the differential equation for the orthogonal trajectories is $\frac{dy}{dx} = -\frac{3x}{2y} \Rightarrow 2y\,dy = -3x\,dx \Rightarrow y^2 = -\frac{3}{2}x^2 +$
C

23. $y^2 = 4C(C - x) \Rightarrow y^2 = 4C^2 - 4Cx \Rightarrow y^2 + 4Cx = 4C^2$. Also $2y\frac{dy}{dx} = -4C \Rightarrow 4y^2\left(\frac{dy}{dx}\right)^2 = 16C^2 \Rightarrow$
$y^2\left(\frac{dy}{dx}\right)^2 = 4C^2 \Rightarrow y^2\left(\frac{dy}{dx}\right)^2 = y^2 + 4Cx \Rightarrow y^2\left(\frac{dy}{dx}\right)^2 = y^2 - 2xy\frac{dy}{dx} \Rightarrow y^2\left(\frac{dy}{dx}\right)^2 + 2xy\frac{dy}{dx} - y^2 = 0 \Rightarrow$
$\left(\frac{dy}{dx}\right)^2 + 2\left(\frac{x}{y}\right)\frac{dy}{dx} - 1 = 0 \Rightarrow \frac{dy}{dx} = \frac{-\frac{2x}{y} \pm \sqrt{\frac{4x^2}{y^2} + 4}}{2} = \frac{-x \pm \sqrt{x^2 + y^2}}{y}$. Then the differential equation for
the orthogonal trajectories is $\frac{dy}{dx} = \frac{-y}{-x \pm \sqrt{x^2 + y^2}} = \frac{y}{x \mp \sqrt{x^2 + y^2}} = \left(\frac{y}{x \mp \sqrt{x^2 + y^2}}\right)\left(\frac{x \pm \sqrt{x^2 + y^2}}{x \pm \sqrt{x^2 + y^2}}\right) =$
$\frac{y\left(x \pm \sqrt{x^2 + y^2}\right)}{x^2 - (x^2 + y^2)} = \frac{x \pm \sqrt{x^2 + y^2}}{-y^2} = \frac{-x \mp \sqrt{x^2 + y^2}}{y}$, the same differential equation $\Rightarrow y^2 = 4B(B - x)$ is the
equation of the orthogonal trajectories.

25. $(x^2 - y^3)y' = 2xy \Rightarrow (x^2 - y^2)dy - 2xy\,dx = 0$. If y^n is an integrating factor, then $\frac{\partial}{\partial x}\left(y^n(x^2 - y^3)\right) = \frac{\partial}{\partial y}\left(y^n(-2xy)\right) \Rightarrow 2xy^n = -2(n+1)xy^n \Rightarrow 2 - -2(n+1) \Rightarrow n = -2$. Then $(x^2y^{-2} - y)dy - 2xy^{-1}\,dx = 0$ is exact. $\frac{\partial f}{\partial x} - -2xy^{-1} \Rightarrow f(x,y) = -x^2y^{-1} + k(y) \Rightarrow \frac{\partial f}{\partial y} = x^2y^{-2} + k'(y) = x^2y^{-2} - y \Rightarrow k'(y) = -y \Rightarrow k(y) = -\frac{y^2}{2} +$ C. $\therefore$ $f(x,y) = -x^2y^{-1} - \frac{y^2}{2} + C = 0$

27. $\frac{\partial M}{\partial y} = ax^2e^y + 2by + c$, $\frac{\partial N}{\partial x} = cy + 2 + 3x^2e^y$. If the equation is exact, then $\frac{\partial M}{\partial y} = \frac{\partial N}{\partial x} \Rightarrow ax^2e^y + 2by + c = cy + 2 + 3x^2e^y \Rightarrow a = 3,\ 2b = c,\ c = 2 \Rightarrow b = 1$. $\therefore$ $\left(3x^2e^y + y^2 + 2y\right)dx + \left(2xy + 2x + x^3e^y\right)dy = 0 \Rightarrow \frac{\partial f}{\partial x} = 3x^2e^y + y^2 + 2y \Rightarrow f(x,y) = x^3e^y + xy^2 + 2xy + k(y)$. $\frac{\partial f}{\partial y} = x^3e^y + 2xy + 2x + k'(y) = 2xy + 2x + x^3e^y \Rightarrow k'(y) = 0 \Rightarrow k(y) = C \Rightarrow f(x,y) = x^3e^y + xy^2 + 2xy + C = 0$

29. $y = uv \Rightarrow y' = u'v + v'u,\ y'' = u''v + 2u'v' + v''u$. Then $y'' + Py' + Qy = F(x) \Rightarrow u''v + 2u'v' + v''u + P(u'v + v'u) + Quv = F(x) \Rightarrow uv'' + uv'\left(P + \frac{2u'}{u}\right) + (u'' + Pu' + Qu)v = F(x) \Rightarrow uv'' + uv'\left(P + \frac{2u'}{u}\right) + (0)v = F(x) \Rightarrow v'' + v'\left(P + \frac{2u'}{u}\right) = \frac{F(x)}{u}$. For $y'e - 2y' + y = e^x$, $u(x) = e^x$ is a solution. Let $y = e^xv$. Then $y' = e^xv + v'e^x$, $y'e = v''e^x + 2e^xv' + e^xv$. The substitution yields $v'' + v'\left(-2 + \frac{2e^x}{e^x}\right) = \frac{e^x}{e^x} \Rightarrow v'' = 1$.

$w = v' \Rightarrow w' = v'' \Rightarrow w' = 1 \Rightarrow w = x + C \Rightarrow v' = x + C \Rightarrow v = \frac{x^2}{2} + Cx + C_1 \Rightarrow y = uv = e^x\left(\frac{x^2}{2} + Cx + C_1\right)$

31. If y is a solution of $(D^2 + 4)y = e^x$, then $(D - 1)(D^2 + 4)y = (D - 1)e^x = e^x - e^x = 0 \Rightarrow$ y is a solution of $(D - 1)(D^2 + 4)y = 0$. The solutions of $(D - 1)(D^2 + 4)y = 0$ are $y = C_1e^x + C_2\cos 2x + C_3\sin 2x$. $(D^2 + 4)\left(C_1e^x + C_2\cos 2x + C_3\sin 2x\right) = C_1e^x - 4C_2\cos 2x - 4C_3\sin 2x + 4C_1e^x + 4C_2\cos 2x + 4\sin 2x = 5C_1e^x = e^x \Rightarrow 5C_1 = 1 \Rightarrow C_1 = \frac{1}{5}$. $\therefore$ $y = \frac{1}{5}e^x + C_2\cos 2x + C_3\sin 2x$.

33. If $y = a_0 + a_1x + a_2x^2 + a_3x^3 + \cdots + a_nx^n + \cdots$ and $y^2 = c_0 + c_1x + c_2x^2 + \cdots + c_nx^n + \cdots$ then $c_n = \sum_{k=0}^{n} a_ka_{n-k}$ and $y' = a_1 + 2a_2x + 3a_3x^2 + \cdots + na_nx^{n-1} + \cdots$. Since $y' = x^2 + y^2$, then $a_1 = c_0$, $2a_2 = c_1$, $3a_3 = c_2 + 1$, and $na_n = c_{n-1}$ for $n \geq 4$. $y(0) = 1 \Rightarrow a_0 = 1$. $c_0 = \sum_{k=0}^{0} a_ka_{n-k} = a_0a_0 = 1 \Rightarrow a_1 = 1$. $c_1 = \sum_{k=0}^{1} a_ka_{n-k} = a_0a_1 + a_1a_0 = 2 \Rightarrow 2a_2 = 2 \Rightarrow a_2 = 1$. $c_2 = \sum_{k=0}^{2} a_ka_{n-k} = a_0a_2 + a_1a_1 + a_2a_0 = 1 + 1 + 1 = 3 \Rightarrow 3a_3 = 3 + 1 \Rightarrow a_3 = \frac{4}{3}$. $c_3 = \sum_{k=0}^{3} a_ka_{n-k} = a_0a_3 + a_1a_2 + a_2a_1 + a_3a_0 = \frac{4}{3}(1) + (1)(1) + (1)(1) + \frac{4}{3}(1) = \frac{14}{3} \Rightarrow 4a_4 = \frac{14}{3} \Rightarrow a_4 = \frac{7}{6}$. $\therefore$ $y = 1 + x + x^2 + \frac{4}{3}x^3 + \frac{7}{6}x^4 + \cdots$.

35. $\frac{dy}{dx} = x + y \Rightarrow \frac{dy}{dx} - y = x.$ $\rho(x) = e^{-x}.$ $y = \frac{1}{e^{-x}} \int e^{-x}\, x\, dx = \frac{1}{e^{-x}}\left(-x\, e^{-x} - e^{-x} + C\right) = -x - 1 + C\, e^{x}.$

$y = 0$ when $x = 0 \Rightarrow 0 = -1 + C \Rightarrow C = 1 \Rightarrow y = -x - 1 + e^{x} = -x - 1 + \left(1 + x + \frac{x^2}{2!} + \frac{x^3}{3!} + \cdots\right) = -x - 1 +$

$\sum_{n=2}^{\infty} \frac{x^n}{n!}$

37. a) $y(0) = \frac{\pi}{2}$, $y'(0) = 0 + \sin\frac{\pi}{2} = 1$, $y''(0) = 1 + (\cos\frac{\pi}{2})(1) = 1$, $y'''(0) = (-\sin\frac{\pi}{2})(1) + 1(\cos\frac{\pi}{2}) = -1$, $y^{(4)}(0) =$ $(-\cos\frac{\pi}{2})(1) - 1(\sin\frac{\pi}{2}) + (-1)\cos\frac{\pi}{2} - 1(\sin\frac{\pi}{2}) = -2.$ $\therefore\ y = \frac{\pi}{2} + x + \frac{x^2}{2!} - \frac{x^3}{3!} - \frac{2x^4}{4!} + \cdots$

b) $y(0) = -\frac{\pi}{2}$, $y'(0) = 0 + \sin\left(-\frac{\pi}{2}\right) = -1$, $y''(0) = 1 + \left(\cos\left(-\frac{\pi}{2}\right)\right)(-1) = 1$, $y'''(0) = \left(-\sin\left(-\frac{\pi}{2}\right)\right)(-1) +$ $1\left(\cos\left(-\frac{\pi}{2}\right)\right) = -1$, $y^{(4)}(0) = \left(-\cos\left(-\frac{\pi}{2}\right)\right) - 1\left(\sin\left(-\frac{\pi}{2}\right)\right) + (-1)\left(\cos\left(-\frac{\pi}{2}\right)\right) - 1\left(\sin\left(-\frac{\pi}{2}\right)\right) = 2$

$\therefore\ y = -\frac{\pi}{2} - x + \frac{x^2}{2!} - \frac{x^3}{3!} + \frac{2x^4}{4!} + \cdots$

APPENDICES

APPENDIX A.7

1. $f(x) = x,\ g(x) = x^2,\ [a,b] = [-2,0] \Rightarrow \frac{f'(c)}{g'(c)} = \frac{f(b) - f(a)}{g(b) - g(a)} \Rightarrow \frac{1}{2c} = \frac{0 - (-2)}{0 - 4} \Rightarrow \frac{1}{2c} = -\frac{1}{2} \Rightarrow c = -1$

3. $f(x) = \frac{x^3}{3} - 4x,\ g(x) = x^2,\ [a,b] = [0,3] \Rightarrow \frac{f'(c)}{g'(c)} = \frac{f(b) - f(a)}{g(b) - g(a)} \Rightarrow \frac{c^2 - 4}{2c} = \frac{-3 - 0}{9 - 0} = -\frac{1}{3} \Rightarrow 3c^2 - 4 = -2c \Rightarrow$
$3c^2 + 2c - 4 = 0 \Rightarrow c = \frac{-2 \pm \sqrt{4 - 4(-4)(3)}}{6} = \frac{-1 \pm \sqrt{13}}{3}$. Since $a \le c \le b$, $c = \frac{-1 + \sqrt{13}}{3}$.

APPENDIX A.10

1. $$\begin{vmatrix} 2 & 3 & 1 \\ 4 & 5 & 2 \\ 1 & 2 & 3 \end{vmatrix}\begin{matrix} 2 & 3 \\ 4 & 5 \\ 1 & 2 \end{matrix} = 30 + 6 + 8 - 5 - 8 - 36 = -5$$

3. $$\begin{vmatrix} 1 & 2 & 3 & 4 \\ 0 & 1 & 2 & 3 \\ 0 & 0 & 2 & 1 \\ 0 & 0 & 3 & 2 \end{vmatrix} = 1\begin{vmatrix} 1 & 2 & 3 \\ 0 & 2 & 1 \\ 0 & 3 & 2 \end{vmatrix} = 1\begin{vmatrix} 2 & 1 \\ 3 & 2 \end{vmatrix} = 1$$

5. a) $$\begin{vmatrix} 2 & -1 & 2 \\ 1 & 0 & 3 \\ 0 & 2 & 1 \end{vmatrix} = \begin{vmatrix} 2 & -5 & 2 \\ 1 & -6 & 3 \\ 0 & 0 & 1 \end{vmatrix} = 1\begin{vmatrix} 2 & -5 \\ 1 & -6 \end{vmatrix} = -7$$

b) $$\begin{vmatrix} 2 & -1 & 2 \\ 1 & 0 & 3 \\ 0 & 2 & 1 \end{vmatrix} = \begin{vmatrix} 2 & -1 & 2 \\ 1 & 0 & 3 \\ 4 & 0 & 5 \end{vmatrix} = -(-1)\begin{vmatrix} 1 & 3 \\ 4 & 5 \end{vmatrix} = -7$$

7. a) $$\begin{vmatrix} 1 & 1 & 0 & 0 \\ 0 & 0 & -2 & 1 \\ 0 & -1 & 0 & 7 \\ 3 & 0 & 2 & 1 \end{vmatrix} = \begin{vmatrix} 1 & 1 & 0 & 7 \\ 0 & 0 & -2 & 1 \\ 0 & -1 & 0 & 0 \\ 3 & 0 & 2 & 1 \end{vmatrix} = -(-1)\begin{vmatrix} 1 & 0 & 7 \\ 0 & -2 & 1 \\ 3 & 2 & 1 \end{vmatrix} = \begin{vmatrix} 1 & 0 & 7 \\ 0 & -2 & 1 \\ 0 & 2 & -20 \end{vmatrix} =$$

$$1\begin{vmatrix} -2 & 1 \\ 2 & -20 \end{vmatrix} = 38$$

7. b) $\begin{vmatrix} 1 & 1 & 0 & 0 \\ 0 & 0 & -2 & 1 \\ 0 & -1 & 0 & 7 \\ 3 & 0 & 2 & 1 \end{vmatrix} = \begin{vmatrix} 1 & 1 & 0 & 0 \\ 0 & 0 & -2 & 1 \\ 1 & 0 & 0 & 7 \\ 3 & 0 & 2 & 1 \end{vmatrix} = -1\begin{vmatrix} 0 & -2 & 1 \\ 1 & 0 & 7 \\ 3 & 2 & 1 \end{vmatrix} = -1\begin{vmatrix} 0 & -2 & 1 \\ 1 & 0 & 7 \\ 0 & 2 & -20 \end{vmatrix} =$

$-1(-1)\begin{vmatrix} -2 & 1 \\ 2 & -20 \end{vmatrix} = 38$

9. $D = \begin{vmatrix} 1 & 8 \\ 3 & -1 \end{vmatrix} = -25.\ x = \dfrac{\begin{vmatrix} 4 & 8 \\ -13 & -1 \end{vmatrix}}{-25} = \dfrac{100}{-25} = -4,\ y = \dfrac{\begin{vmatrix} 1 & 4 \\ 3 & -13 \end{vmatrix}}{-25} = \dfrac{-25}{-25} = 1$

11. $D = \begin{vmatrix} 4 & -3 \\ 3 & -2 \end{vmatrix} = 1.\ x = \dfrac{\begin{vmatrix} 6 & -3 \\ 5 & -2 \end{vmatrix}}{1} = 3,\ y = \dfrac{\begin{vmatrix} 4 & 6 \\ 3 & 5 \end{vmatrix}}{1} = 2$

13. $D = \begin{vmatrix} 2 & 1 & -1 \\ 1 & -1 & 1 \\ 2 & 2 & 1 \end{vmatrix} = \begin{vmatrix} 2 & 1 & -1 \\ 3 & 0 & 0 \\ 4 & 3 & 0 \end{vmatrix} = -1\begin{vmatrix} 3 & 0 \\ 4 & 3 \end{vmatrix} = -9.\ x = \dfrac{\begin{vmatrix} 2 & 1 & -1 \\ 7 & -1 & 1 \\ 4 & 2 & 1 \end{vmatrix}}{-9} = \dfrac{\begin{vmatrix} 2 & 1 & -1 \\ 9 & 0 & 0 \\ 6 & 3 & 0 \end{vmatrix}}{-9} =$

$\dfrac{-1\begin{vmatrix} 9 & 0 \\ 6 & 3 \end{vmatrix}}{-9} = 3,\ y = \dfrac{\begin{vmatrix} 2 & 2 & -1 \\ 1 & 7 & 1 \\ 2 & 4 & 1 \end{vmatrix}}{-9} = \dfrac{\begin{vmatrix} 2 & 2 & -1 \\ 3 & 9 & 0 \\ 4 & 6 & 0 \end{vmatrix}}{-9} = \dfrac{-1\begin{vmatrix} 3 & 9 \\ 4 & 6 \end{vmatrix}}{-9} = -2,\ z = \dfrac{\begin{vmatrix} 2 & 1 & 2 \\ 1 & -1 & 7 \\ 2 & 2 & 4 \end{vmatrix}}{-9} =$

$\dfrac{\begin{vmatrix} 3 & 0 & 9 \\ 1 & -1 & 7 \\ 4 & 0 & 18 \end{vmatrix}}{-9} = \dfrac{-1\begin{vmatrix} 3 & 9 \\ 4 & 18 \end{vmatrix}}{-9} = 2$

15. $D = \begin{vmatrix} 1 & 0 & -1 \\ 0 & 2 & -2 \\ 2 & 0 & 1 \end{vmatrix} = 2\begin{vmatrix} 1 & -1 \\ 2 & 1 \end{vmatrix} = 6.\ x = \dfrac{\begin{vmatrix} 3 & 0 & -1 \\ 2 & 2 & -2 \\ 3 & 0 & 1 \end{vmatrix}}{6} = \dfrac{2\begin{vmatrix} 3 & -1 \\ 3 & 1 \end{vmatrix}}{6} = 2,\ y = \dfrac{\begin{vmatrix} 1 & 3 & -1 \\ 0 & 2 & -2 \\ 2 & 3 & 1 \end{vmatrix}}{6} =$

15. (Continued)

$$\frac{\begin{vmatrix} 1 & 3 & -1 \\ 0 & 2 & -2 \\ 0 & -3 & 3 \end{vmatrix}}{6} = \frac{1\begin{vmatrix} 2 & -2 \\ -3 & 3 \end{vmatrix}}{6} = 0,\ z = \frac{\begin{vmatrix} 1 & 0 & 3 \\ 0 & 2 & 2 \\ 2 & 0 & 3 \end{vmatrix}}{6} = \frac{2\begin{vmatrix} 1 & 3 \\ 2 & 3 \end{vmatrix}}{6} = -1$$

17. $D = \begin{vmatrix} 2 & h \\ 1 & 3 \end{vmatrix} = 6 - h = 0 \Rightarrow h = 6.\ x: \begin{vmatrix} 8 & h \\ k & 3 \end{vmatrix} = 24 - hk = 24 - 6k = 0 \Rightarrow k = 4$

a) When $h = 6$, $k = 4$, there are infinitely many solutions.

b) When $h = 6$, $k \neq 4$, there are no solutions.

19. $au + bv + cw = 0 \Rightarrow v = \frac{-au - cw}{b}$, $au' + bv' + cw' = 0 \Rightarrow v' = \frac{-au' - cw'}{b}$, $au'' + bv'' + cw'' = 0 \Rightarrow$

$$v'' = \frac{-au'' - cw''}{b} \Rightarrow \begin{vmatrix} u & v & w \\ u' & v' & w' \\ u'' & v'' & w'' \end{vmatrix} = \begin{vmatrix} u & \frac{-au - cw}{b} & w \\ u' & \frac{-au' - cw'}{b} & w' \\ u'' & \frac{-au'' - cw''}{b} & w'' \end{vmatrix} = \frac{1}{b}\begin{vmatrix} u & -au - cw & w \\ u' & -au' - cw' & w' \\ u'' & -au'' - cw'' & w'' \end{vmatrix} =$$

$\frac{1}{b}(u(-au' - cw')w'' + u''(-au - cw)w' + w(-au'' - cw'')u' - u''(-au' - cw')w - u'(-au - cw)w'' - u(-au'' - cw'')w') = \frac{1}{b}(0) = 0$